Automated DNA Sequencing and Analysis

Automated DNA Sequencing and Analysis

Edited by
MARK D. ADAMS
Director, EST Laboratory
CHRIS FIELDS
Director, Department of Genome Informatics
J. CRAIG VENTER
President/Director
The Institute for Genomic Research
932 Clopper Road
Gaithersburg, MD 20878, USA

ACADEMIC PRESS
Harcourt Brace & Company, Publishers
London · San Diego · New York
Boston · Sydney · Toyko · Toronto

Academic Press Limited
24–28 Oval Road
London NW1 7DX

United States Edition published by
Academic Press inc.
San Diego, CA 92101

This book is printed on acid-free paper

A catalogue record for this book is available from the British Library

ISBN 0–12–717010–3

Typeset by J&L Composition Ltd, Filey, North Yorkshire
Printed and bound in Great Britain at The Bath Press, Avon

Preface

This book reviews and assesses the state of the art of automated DNA sequence analysis, from the construction of clone libraries to the development of laboratory and community databases. The book is targeted at both small laboratories that are interested in taking maximum advantage of automated sequencing resources and at groups pursuing large-scale cDNA and genomic sequencing projects. Much of the technology for automated sequencing has been developed in large laboratories, but the technology is clearly mature enough to be commonplace. Researchers in such fields as molecular genetics, cell biology, plant physiology and genome sciences are applying automated sequencing as a research tool. While automated sequencers are now in use in hundreds of laboratories, producing by far the majority of DNA sequence data, no previous publication has presented the methodologies and strategies of automated DNA sequence analysis in a way that allows them to be compared and contrasted. By taking a broad view of the process of automated sequence analysis, the present volume bridges the gap between the protocols supplied with instruments and reaction kits and the finalized data presented in the research literature.

DNA sequence analysis is a multistep process comprising sample preparation, generation of labeled fragments by sequencing reactions, electrophoretic separation of the fragments, data acquisition, assembly into a finished sequence and, most importantly, functional interpretation. Larger projects must also consider the source of the DNA to be sequenced, including having an accurate map of a DNA fragment of suitable size for sequencing. Likewise, the larger the project, the more effort must be expended to ensure that the final sequence accurately reflects the structure in the genome. The state of the art of the application of automation to DNA sequencing and analysis now covers a tremendous range of techniques and strategies.

The first software tools for interpreting DNA sequence data were developed in the late 1970s and early 1980s; software for assembling sequence fragments into contigs began to appear in the mid-1980s. The coupling of electrophoresis and sequence data acquisition in 1986, however, was the breakthrough that allowed high-throughput DNA sequencing to be considered and hence rendered the sequencing of eukaryotic genomes feasible. Automated DNA sequencers based on fluorescence detection of electrophoretically separated fragments have been available commercially since 1987; the first sequence of a gene determined by automated sequencing methods was published in 1987 (Gocayne *et al.*, *Proc. Natl Acad. Sci. U.S.A.* **84**, 8296). This advance has resulted in a new type of laboratory oriented towards production DNA sequencing where the rate-limiting step is shifted to the analysis of the data generated. The development of automation has only just begun: commercial sequencing reaction robots that automate cycle sequencing became available in 1991 and sample preparation robots were introduced in 1992. Integrated systems that carry out the entire DNA sequence analysis process without manual intervention are currently under discussion and development by several groups.

Part I presents several strategies and applications of large-scale sequencing. For each method the set-up, execution, subsequent data analysis and confirmation may vary greatly, but each has the common feature of dealing with large amounts of sequence data and attempting to synthesize it in meaningful ways. Chen presents a brief history of automated sequencing. Chapters by Church *et al.*, Huang and Mathies, and Drmanac *et al.* discuss alternatives to 'traditional' fluorescence-based sequencing. Four chapters present advantages and disadvantages of two current methods of choice for large-scale genome sequencing: shotgun and transposon-facilitated directed sequencing. Liu and Fleischmann detail the exonuclease III approach to directed sequencing. Adams and McCombie discuss issues related to cDNA (expressed sequence tag (EST)) projects in human and model organisms.

Part II focuses on sample preparation and sequencing methodology. At the beginning of the section are nine chapters dealing with construction, manipulation, and application of genomic and cDNA libraries. Each addresses potential pitfalls to be avoided that may hamper analysis of sequence data that is generated. Library methods for both mapping and sequencing of genomic regions are presented. cDNA library construction techniques are presented from several viewpoints related to libraries of sufficient quality for random selection of clones. Three robotic approaches to automated sample preparation based on magnetic beads are presented along with a discussion of the relative efficiencies of M13 and polymerase chain reaction product purification. The section closes with six chapters discussing the details of sequencing methods that represent refinements in protocols and discussion of issues that are particularly relevant for large-scale projects: gap closure, repetitive elements and final structure confirmation.

Part III reviews informatics, the key to managing and understanding DNA sequence data. Large-scale sequencing raises issues of data acquisition, assembly, integrity and interpretation over and above those that arise in small-scale sequencing projects. In addition, management and maintenance of the sequence data itself present challenges. Tibbetts provides an overview of the challenges of nucleotide base-calling and describes the use of a simulated neural network, a standard method of analyzing noisy signals, to improve the process. The chapters of Myers, Honda *et al.* and Burks *et al.* address the sequence assembly problem. Nine chapters present a range of data analysis tasks and tools. DNA sequence analysis methods fall into two broad classes: similarity-based methods and composition-based methods. Searching the public databases for a similar sequence is the most rapid method for developing good hypotheses about the function of a newly determined sequence. Sutton and Kerlavage review several common sequence similarity algorithms, including the approximate FASTA and BLAST algorithms. Compositional analysis provides a method for identifying biologically significant features of a sequence in the absence of significant similarity to sequences of known function. Fields reviews the use of both types of analysis to identify genes in anonymous sequences. The final four chapters present recent work on the crucial task of data management. Lewis reviews the requirements that most large-scale biology laboratories face in maintaining laboratory records and data, and describes a variety of approaches to solving them. Cuticchia focuses on one approach: the development of a relational database. Cherry and Cartinhour describe ACEDB, a flexible tool for building object-oriented databases. Overbeek and Price conclude the section with a discussion of the issues that arise when linking multiple databases with different contents and structures.

These chapters represent a snapshot of the rapidly moving field of automation in DNA sequencing and analysis. As clearly demonstrated, the field is progressing on many fronts. For an investigator preparing to begin a large-scale sequencing project, the next several years are likely to bring dramatic changes in methodology and instrumentation. However, the issues addressed in this volume will remain of critical importance in the design and execution of large-scale sequencing projects.

M. D. ADAMS
C. FIELDS
J. CRAIG VENTER

Acknowledgements

We would like to thank all the authors who contributed to this volume. We would especially like to thank Alison Tinsley for her excellent editorial assistance.

Contents

C. SEQUENCING METHODS

PART I

Sequencing Instruments and Strategies

CHAPTER ONE

The Efficiency of Automated DNA Sequencing

E.Y. CHEN

Advanced Center for Genetic Technology, Applied Biosystems, Division of Perkin-Elmer Corp., 850 Lincoln Centre Drive, Foster City, CA 94404, USA

1.1 INTRODUCTION

The GenBank genetic sequence database, which contained only 15 million nucleotides in 1987, has nearly doubled its size in each of the subsequent 5 years. By 1992 it reached over 120 million, with progressively more data obtained utilizing automated DNA sequencers. The historic progress leading to current DNA sequencing technology is reviewed in Fig. 1.1.

While the three-dimensional structure of DNA was known in 1953, the major sequencing work performed prior to 1965 was on RNAs such as tRNAs which were small and easy to obtain in substantial quantities. The RNAs were usually broken into smaller pieces (a few nucleotides long) and each fragment was sequenced by chromatographic methods. The first sequence of an intact molecule, an 80-base yeast tRNA, was published in 1965 (Holley *et al.*, 1965).

DNA sequencing first became possible with the discovery of restriction enzymes and DNA polymerases around 1970. For the first time well-defined fragments could be derived from a larger molecule. Methods based on primed synthesis (Wu & Taylor, 1971) and gel electrophoresis separation led to the first sequence of a genome, the 5.4 kb of bacteriophage ϕX, in 1976 (Sanger *et al.*, 1977a).

A breakthrough in the rate of sequencing came when the dideoxy chain termination (Sanger *et al.*, 1977b) and chemical degradation (Maxam & Gilbert, 1977) techniques were introduced in 1977. Several years later, the former method was used to sequence the 16.5 kb human mitochondria genome (Anderson *et al.*, 1981) and the latter was employed to achieve the analysis of the 40 kb bacteriophage T7 (Dunn & Studier, 1983). These methods provide the theoretical and practical backgrounds for modern sequencing technology. Since the dideoxy method has been used in most large-scale projects and was adopted in present-day automated fluorescent sequencing instruments, the following discussion will focus on its applications.

1.2 EFFICIENCY AND COST

Along with the major developments, Fig. 1.1 also lists estimated sequencing efficiencies, assessed as the number of nucleotides of finished sequence achieved by a skilled, dedicated person in a year. When the dideoxy method was introduced the rate was only 1.5 kb per person-year, due to the difficulties of obtaining single-stranded templates and useful primers.

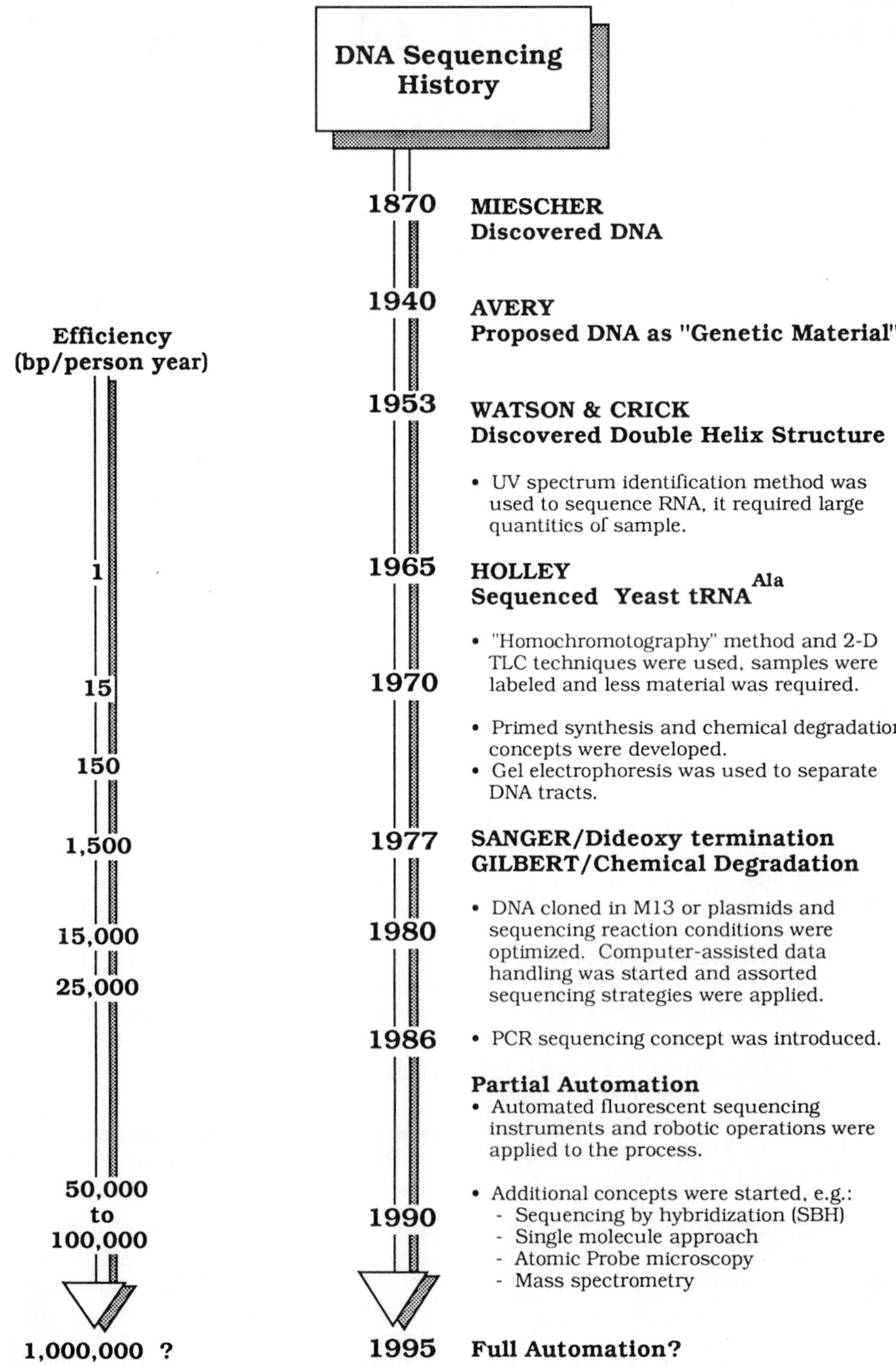

Figure 1.1 A historical view of DNA sequencing efficiency.

In a few years these problems were solved with the development of M13 cloning vectors and the availability of the oligonucleotide synthesizer. The dideoxy method thereby became universally applicable to any DNA fragment. Consequently, the rate increased by an order of magnitude to 15 kb per person-year by the early 1980s.

As a part-time effort in 1983, we set out to sequence the 66.5 kb human growth hormone locus (Chen *et al.*, 1989), using a combination of random shotgun, primer walking and nested deletion assembly strategies after subcloning DNA fragments from two cosmids into 17 plasmids. Employing optimized sequencing conditions and computer-assisted data handling, we were able to complete 98% of the project within about 2.5 person-years and the overall project in 3 person-years. (Accumulating the final 2% of the data took extra time because of the instability of some M13 clones (Chen *et al.*, 1989).) It was therefore estimated that the sequencing rate was about 20-25 kb per person-year. With an estimated cost of approximately $75 000 a year to support one person (including $25 000 salary, $15 000 overhead, $20 000 equipment and $15 000 supplies), the cost per base pair would be approximately $3 to $4. However, because manual sequencing tends to incur a high turnover rate of skilled personnel, the actual sequencing cost is even higher if the training expense is included.

Beginning in 1985, several automated sequencers were developed (Smith *et al.*, 1986; Connell *et al.*, 1987; Ansorge *et al.*, 1987; Brumbaugh *et al.*, 1988; Kambara *et al.*, 1988; Freeman *et al.*, 1990) to automate gel electrophoresis, raw data acquisition, and base-calling. Gradually computer-operated robotic workstations and more sophisticated software have also been applied to handle the sequencing reactions (Wilson *et al.*, 1988; D'Cunha *et al.*, 1990; Cathcart, 1990), prepare template samples (Mardis & Roe, 1989; Smith *et al.*, 1990) and assemble data (Staden, 1987; Chen *et al.*, 1982). It took a few years for these instruments to establish their reliable performances. By the early 1990s the resultant enhanced data throughput has increased the sequencing rate to around 100 kb per person-year. The productivity and consistency are obviously very sensitive to further technical innovation. Particularly notable has been the impact of the polymerase chain reaction (PCR) amplification technique, since it can be used to prepare sequencing samples at a much faster rate (Gyllensten, 1989 and Chen *et al.*, 1991b). Furthermore, the combination of PCR and dideoxy method has allowed the development of 'cycle sequencing' (Carothers *et al.*, 1989; Lee, 1991), a process that linearly amplifies the detectable signals, tremendously increasing the sensitivity of sequencing.

In general, it is not easy to assess the precise cost in automated sequencing. This is in part because hidden costs (Pohl & Sulston, 1992) are often overlooked by laboratory scientists, and it may also be difficult to distinguish the time spent in production sequencing from development efforts, especially when projects or methods are new. We can, however, get an estimate from the minimum cost to maintain an operating unit including one person, one sequencing instrument, and a shared robot. The cost would at present be about $150 000 a year. (This includes $50 000 salary plus overhead, $15 000 general supplies, $45 000 for sequencing kits, and $40 000 for machine depreciation over 5 years.) At a current yield of 100 kb per person-year, the net cost would then be about $1.50 per base. Even with a productivity of 150 kb per person-year (Sulston *et al.*, 1992), the minimum cost is still $1 per base; and one would need to increase efficiency to at least 300 kb per year to bring the cost down to $0.50 per base.

Of course, sequencing efficiency and cost are also directly related to the nature of the DNA analyzed. Practically, DNAs that contain more repetitive sequence or homopolymer sequence elements, or higher contents of GC require more work. Within a single project, certain regions will also tend to be more troublesome that the others. To solve specific problems, several modifications of the dideoxy method have been developed in the past decade (for reviews see: Chen *et al.*, 1991b; Hunkapiller *et al.*, 1991). For example, the use of nucleotide analogs such as inosine or deaza compounds can help eliminate 'gel-compression' problems (Jensen *et al.*, 1991); use of different cloning vectors provides an alternative when a particular M13 clone is unstable (Chen & Seeburg, 1985) and the addition of single-strand binding proteins to sequencing reactions improves the quality of the data produced from DNA templates enriched in looping structures (Chen *et al.*, 1991b). These variations are also applicable to today's automated sequencing protocols.

1.3 METHODS IN AUTOMATED SEQUENCING

On the basis of the number of fluorescent dyes used, the commercially available automated sequencers can be divided into two types, as listed in Table 1.1. The first type uses single-label, four-lane separation, and has 10- or 12-channel capacity (Brumbaugh *et al.*, 1988; Freeman *et al.*, 1990); the second type employs four-label, single-lane separation, and currently has 36-channel capacity per run (Connell *et al.*, 1987; Lee *et al.*, 1992; Hawkins *et al.*, 1992). The latter type is often the instrument of choice for large-scale or large-sample-number projects because it offers higher

Table 1.1 Automated sequencers.

A. Single color, four-lane loading (e.g. Pharmacia ALF, LI-COR 4000, Millipore BaseStation)
B. Four color, single-lane loading (ABI 373A Sequencer)

Dye chemistry	*Features*	*Comments*
Dye-primers	Four-tube reaction, single lane loading Used in cycle sequencing dC^7G replaces dG	Needs only about 0.2 μg sample Changing primer is awkward High reading accuracy (450 bp at 99% accuracy)
Dye-terminator (Taq polymerase)	Single tube reaction and loading Used in cycle sequencing dI replaces dG	Needs only about 0.2 μg sample Can use any primer Lower reading accuracy (350 bp at 90–99% accuracy) Spin column step is tedious
Dye-terminator (T7 polymerase or Sequenase)	Single tube reaction and loading 37°C reaction dNTPαS replaces dNTP Mn^{2+} substitution	Needs as much as about 2 μg sample Can use any primer High reading accuracy (450 bp at 99% accuracy)

throughput. The results from this instrument also tend to be more consistent among different samples within the same run, because single-lane separation is less sensitive to electrophoretic artifacts arising from gel distortion. The following discussion is based on the use of the ABI model 373A Automated Sequencer.

1.3.1 Dye-primer

There are alternative sample preparations for this automated sequencer, depending on the positions at which the fluorescent dyes are attached. In sequencing with dye-primers, four separate reactions are carried out for each sample, and each reaction contains a different dye-labeled primer. This set of four reactions is mixed and loaded in a single channel prior to gel electrophoresis. While the system can use either Taq polymerase or T7 DNA polymerase (sequenase), Taq polymerase is frequently used under cycle sequencing conditions, in which the signal is linearly amplified and hence only 0.1–0.2 μg of template is needed. This feature is especially important when robotic operation is used to prepare hundreds of templates in smaller quantities. The standard sequencing kit has dGTP replaced with deaza-dGTP, a substitution designed to reduce the gel compression problem caused by internal looping in DNA chains. For unknown reasons, the use of the deaza compound tends to reduce the readable length of sequencing tracts; peak broadening is observed for larger molecules (or fuzzy bands in radioactive sequencing) (K. Hennessy, personal communication; E.Y. Chen, unpublished result). Nevertheless, the average tract read with dye-primer sequencing is 450–500 bp at 98–99% accuracy. This system is, however, suitable only for sequencing projects that use pre-established universal primers since changing primers is cumbersome, requiring four separate labelings for each primer.

In laboratories analyzing large numbers of samples, a robotic system such as the ABI Catalyst 800 (Cathcart, 1990) or the Beckman Biomek 1000 (Wilson *et al.*, 1988; Civitello *et al.*, 1992) can be used to increase throughput and reliability. The catalyst robot, for example, is capable of performing a set of 24 sequencing reactions within 4 h using a 96-well thermal cycler plate equipped with an evaporation control device (Cathcart, 1990). Recent developments with this robot include the following.

(1) *Direct-load protocol* (C. Heiner, personal communication – the protocol is available from ABI). The sequencing reactions are carried out on the catalyst in 40% of the usual liquid volumes. After a brief evaporation cycle at the end of reactions, samples can be loaded directly on the gel, thus bypassing tedious ethanol precipitation.
(2) *96-Sample protocol* (K. Hennessy, personal communication – further information is available from ABI). This new design, which required a minor change of both hardware and software, enables the catalyst to perform 96 samples or 384 reactions within 18 h. Thus a single run on a catalyst robot can feed three 373A Sequencer runs per day.
(3) *Combination of template preparation and sequencing reactions.* The Catalyst equipped with a prototypic magnetic station can use the LacZBeads hybridization scheme (Fry *et al.*, 1992) to purify a fixed amount of DNA sample. Since the same robot can also carry out sequencing reactions, this allows an uninterrupted, hands-off process from

the preparation of M13 phage supernatant to gel loading.

1.3.2 Dye-terminators

One feasible way to adapt the 373A Sequencer to any sequencing primer is to move the fluorescent dyes from the 5′-end to the 3′-end – that is, to the position of the dideoxy terminators. In sequencing with dye-terminators, each of the four dideoxy nucleotide triphosphates (ddNTPs) is labeled with a different dye. Four chain extension reactions can be carried out within the same tube, sparing considerable labor. In contrast to dye-primers, dye-terminator chemistry has the advantage that it eliminates noise arising from chain termination without the incorporation of dideoxynucleotides. This chemistry, however, has major disadvantages resulting from the attachment of bulky dye molecules to the dideoxy terminator molecules. Substrate affinity is altered and a specific set of dye-labeled terminators must be tailored to each DNA polymerase. Furthermore, the sequencing profile typically shows uneven signal intensities, reducing the accuracy of base-calling.

Two sets of dye-terminators are available. The first set contains rhodamine dyes and works well with Taq polymerase (Lee *et al.*, 1992). Convenient cycle sequencing conditions are normally employed, and the sequencing reactions can be carried out with any primer and with a wide variety of templates (single stranded DNA, double stranded DNA, or PCR-generated DNA). In addition, dGTP is replaced with dITP (deoxyinosine triphosphate) in the reaction kit to reduce gel compression. At present, a major drawback is that the average reading range is only about 350 bp, with uneven signal patterns that result in accuracy rate ranging from 90 to 99% in typical runs. The current requirement for a gel filtration (spin column) step to remove the unincorporated dyes before gel loading is awkward and time consuming.

Recently, another set of dye-terminators was optimized for T7 DNA polymerase (sequenase) (Lee *et al.*, 1992; Hawkins *et al.*, 1992) to take advantage of its greater processivity and more uniform incorporation of modified nucleotides. In practical applications of this new chemistry, the ABI 373A Sequencer needs to be equipped with a five-color (530/545/560/580/610 nm) filter wheel and modified instrumental software, so that one of the previous wavelength signal channels (610 nm) can be replaced with a new one (545 nm). In addition to the dye-terminators and sequenase, the optimized reaction mixture also contains Mn^{2+} instead of Mg^{2+} (Tabor & Richardson, 1990) and four α-S-dNTPs (α-thiodeoxynucleoside triphosphates) instead of regular dNTPs. The combination of dye-ddNTP and α-S-dNTP appears to resolve gel compression very effectively (Lee *et al.*, 1992; Hawkins *et al.*, 1992). This is presumably related to the altered structure of the synthesized DNA chains, which contain both the bulkier sulfur atoms next to the phosphodiester bonds and the negatively charged dye molecules at the 3′-end. The average reading range (450-500 bp) and accuracy (98–99%) then become comparable to the results using dye-primers.

The main disadvantage of this chemistry is that it requires higher amounts of template DNA (> 2 μg), since cycle sequencing conditions cannot be used. This also affects the choice of template since single-

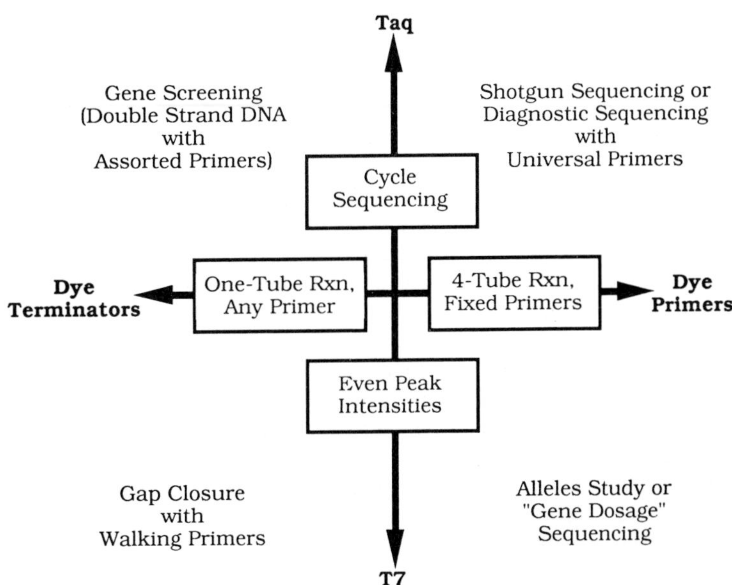

Figure 1.2 Features and application of dye-primer and dye-terminator sequencing using either Taq or T7 polymerase.

stranded M13 templates give more consistent results than double-stranded plasmids, even after the plasmids are pretreated with an additional alkaline denaturation step (Chen & Seeburg, 1985). PCR can be used to provide adequate levels of templates, but these double-stranded PCR products must again be converted to single-strand forms by one of several methods (Gyllensten, 1989; Chen *et al.*, 1991b) before sequencing. To maximize the efficiency of automated sequencing, one must start with a good knowledge of the chemistry of dye-primers and dye-terminators. Figure 1.2 outlines features that are discussed above.

Taking all of these features into consideration, the following rules-of-thumb may be useful for typical current applications.

(1) The Taq/dye-primer combination is convenient for carrying out shotgun sequencing with a large number of M13 clones, since a fixed universal primer can be used thousands of times to obtain high-quality data.
(2) The T7/dye-terminator is then ideal to close gaps between contigs since it works well with walking primers.
(3) The Taq/dye-terminator system can be used to obtain raw sequence data fast (for example, for the purpose of screening a region for genes, where 100% accuracy is not required) since it is adaptable to sequence PCR products with any primers.
(4) The T7/dye-primer system served well in studying the relative amount of different alleles in heterozygotic or complex mixtures of DNAs (e.g. 'gene dosage' comparison for a tumor tissue), since it provides peak intensities in rough proportion to the allele content.

1.4 GENOMIC DNA PROJECTS

Sequencing genomic DNA presents special problems and options, depending in part on the degree of accuracy sought. For example, searches for potential coding and promoter regions may not require 100% accuracy until such regions are defined. Thus, lower initial accuracy drives down the sequencing cost and high-accuracy sequencing can then be pursued in focused, smaller regions. (The GenBank database is estimated to contain an error frequency of 3.55% (Kristensen *et al.*, 1992); but it is generally believed that the accuracy rate is higher in the coding regions, since cDNA sequence is also available in most cases.)

In general, since the length of sequence data obtained from a single experiment is currently limited to approximately 500 bases, determination of larger genomic sequences requires a strategy to assemble those short tracts. Both ordered and random strategies have been used (for a more complete review see: Bankier *et al.*, 1987; Hunkapiller *et al.*, 1991; see also Chapters 5–9 in this volume). The ordered approach is favored in manual sequencing because it saves labor in performing gel electrophoresis by pre-sorting the template samples. Advantages of this approach include:

(1) reduction of the sequencing load;
(2) better organization of raw data from individual gel runs; and
(3) the ability to sequence highly repetitive regions.

However, the ordered approach is slow, difficult to automate, and thus far has worked for DNA fragments up to 5 or 10 kb long (Chen *et al.*, 1989, 1991a). As a result, nearly all current large-scale sequencing projects employ a random shotgun strategy, whether or not they use automated sequencers. The shotgun sequencing process is convenient, and accumulates data relatively quickly during the early phase of sequencing. Another advantage is that overdetermination of most of the sequence minimizes the errors. In practical applications, random fragments are successively sequenced until raw data equivalent to four to five times the total DNA length to be analyzed have been collected. At this stage the data can usually be assembled into a few long contigs which together cover more than 95% of the whole sequence. Small gaps between the contigs are filled using an ordered ('walking') approach (Edwards & Caskey, 1991).

This strategy, however, frequently suffers from problems both upstream of sequencing (including the need to consistently prepare clone libraries and a large number of high-quality template samples) and downstream of sequencing (sophisticated software is needed for data assembly and editing, extra effort is required to sequence difficult spots, and gap-filling can be very time-consuming). In addition, the strategy tends to fail when there are many repetitive sequences, particularly in tandem (McLean *et al.*, 1987).

1.5 FUTURE PROSPECTS

Since 1990, several new approaches have been proposed that are designed to increase sequencing rate by several orders of magnitude. They include sequencing by hybridization (SBH) to oligonucleotides (Strezoska *et al.*, 1991; Khrapko *et al.*, 1991), single-molecule sequencing (Nguyen *et al.*, 1987), atomic probe microscopy (Lindsay & Phillip, 1991), and mass spectrometry (Fenn *et al.*, 1989). Most of these approaches are at an early stage of development and have not been put to practical test. One can note, however, that the SBH technique, for example, can potentially provide a useful adjunct to conventional sequencing strategy. This is because, in principle, the oligonucleotide patterns generated from a SBH experiment could help to order random clones, greatly benefiting shotgun strategies. The cost of such a scheme, however, is prohibitive at the present time.

Other techniques have been developed that can improve the efficiency of current methods. For example, capillary electrophoresis (Swerdlow & Gesteland, 1990; Zagursky & McCormick, 1990; Smith, 1991) can expedite DNA size separation speed several-fold, and the multiplex sequencing scheme (Church & Keiffer-Higgins, 1988) can significantly reduce the number of gel runs. Despite their promising features, these methods have idiosyncratic technical problems, and have not yet been widely adopted. More recently, strings of contiguous hexamers have been demonstrated to be effective as walking primers in the presence of single-stranded binding protein (Kieleczawa *et al.*, 1992). This reduces the cost of sequencing by reusing the library of 4096 hexamers. Nevertheless, the walking primer strategy is currently labor intensive; further improvement, particularly in automation of the process, will be required before this strategy can become applicable to larger-scale projects.

Regardless of whether a revolutionary technology materializes in the long term, we must depend on existing approaches in the near future. Improvement of current automated sequencing systems most likely will include: faster robotic operation to handle clone picking, sample preparation, and sequencing reactions; longer reads of sequence from each channel on a gel; and more sophisticated computer software to assemble, edit, and compare data.

ACKNOWLEDGMENTS

I thank Dr David Schlessinger for critical reading of the manuscript, Dr Richard Cathcart for his valuable discussions, Dr Kevin Hennessy and Ms Cheryl Heiner for communicating their results prior to publication, and Ms Judy Perez-Salas for her help in preparing the manuscript.

REFERENCES

Anderson, S., Bankier, A.T., Barrell, B.G., de Bruijn, M.H.L., Coulson, A.R., Drouin, J., Eperon, I.C., Nierlich, D.P., Roe, B.A., Sanger, F., Schreier, P.H., Smith, A.J.H., Staden, R. & Young, I.G., (1981) *Nature* **290**, 457–465.

Ansorge, W., Sproat, B.S., Stegemann, J., Schwager, C. & Zenke, M. (1987) *Nucleic Acids Res.* **15**, 4596–4602.

Bankier, A.T., Weston, K.M. & Barrell, B.G. (1987) *Methods Enzymol.* **155**, 51–301.

Brumbaugh, J.A., Middendorf, L.R., Grone, D.L. & Ruth, J.L. (1988) *Proc. Natl. Acad. Sci. U.S.A.* **85**, 5610–5614.

Carothers, A.M., Urlaub, G., Mucha, J., Grunberger, D. & Chasin, L.A. (1989) *BioTechniques* **7**, 494–499.

Cathcart, R. (1990) *Nature* **347**, 310–310.

Chen, E.Y. & Seeburg, P.H. (1985) *DNA* **4**, 165–170.

Chen, E.Y., Howley, P.M., Levinson, A.D. & Seeburg, P.H. (1982) *Nature* **299**, 529–534.

Chen, E.Y., Liao, Y-C., Smith, D.H., Barrera-Saldana, H.A., Gelinas, R.E. & Seeburg, P.H. (1989) *Genomics* **4**, 479–497.

Chen, E.Y., Cheng, A., Lee, A.L., Kuang, W-J., Hiller, L-D., Green, P., Schlessinger, D., Ciccodicola, A. & D'Urson, M. (1991a) *Genomics* **10**, 792–800.

Chen, E.Y., Kuang, W-J. & Lee, A.L. (1991b) *Methods: Companion Methods Enzymol.* **3**, 3–19.

Church, G.M. and Kieffer-Higgins, S. (1988) *Science* **240**, 185–188.

Civitello, A.B., Richards, S. & Gibbs, R.A. (1992) *DNA Sequence* **3**, 17–23.

Connell, C., Fung, S., Heiner, C., Bridgham, J., Chakerian, V., Heron, E., Jones, B., Menchen, S., Mordan, W., Raff, M., Recknor, M., Smith, L., Springer, J., Woo, S. & Hunkapiller, M. (1987) *BioTechniques* **5**, 342–348.

D'Cunha, J., Berson, B.J., Brumley, Jr., R.L., Wagner, P.R. & Smith, L.M. (1990) *BioTechniques* **9**, 80–90.

Dunn, J.J. & Studier, F.W. (1983) *J. Mol. Biol.* **166**, 477–535.

Edwards, A. & Caskey, C.T. (1991) *Methods: Companion Methods Enzymol.* **3**, 41–47.

Fenn, J.B., Mann, M., Meng, C.K., Wong, S.F. & Whitehouse, C.M. (1989) *Science* **246**, 64–71.

Freeman, M., Baehler, C. & Spotts, S. (1990) *BioTechniques* **8**, 147–148.

Fry, G., Lachenmeier, E., Mayrand, E., Giusti, B., Fisher, J., Johnston-Dow, L., Cathcart, R., Finne, E. & Kilaas, L. (1992) *BioTechniques* **13**, 124–131.

Gyllensten, U.B. (1989) *BioTechniques* **7**, 700–708.

Hawkins, T., Du, Z., Halloran, N. & Wilson, R.K. (1992) *Electrophoresis* **13**, 552–559.

Holley, R.W., Apgar, J., Everett, G.A., Madison, J.T., Marquisee, M., Merrill, S.H., Penswick, J.R. & Zamir, A. (1965) *Science* **147**, 1462–1465.

Hunkapiller, T., Kaiser, R.J., Koop, B.F. & Hood, L. (1991) *Science* **254**, 59–67.

Jensen, M.A., Zagursky, R.J., Trainor, G.L., Cocuzza, A.J., Lee, A.L. & Chen, E.Y. (1991) *DNA Sequence* **1**, 233–239.

Kambara, H., Nishikawa, T., Katayama, Y. & Yamaguchi, T. (1988) *BioTechnology* **6**, 816–821.

Khrapko, K.R., Lysov, Y.P., Khorlin, A.A., Ivanov, I.B., Yershov, G.M., Vasilenko, S.K., Florentiev, V.L. & Mirzabekov, A.D. (1991) *DNA Sequence* **1**, 375–388.

Kieleczawa, J., Dunn, J.J. & Studier, F.W. (1992) *Science* **258**, 1787–1791.

Kristensen, T., Lopez, R. & Prydz, H. (1992) *DNA Sequence* **2**, 343–346.

Lee, J-S. (1991) *DNA Cell Biol.* **10**, 67–73.

Lee, L., Connell, C.R., Woo, S.L., Cheng, R.D., McArdle, B.F., Fuller, C.W., Halloran, N.D. & Wilson, R.K. (1992) *Nucleic Acids Res.* **20**, 2471–2483.

Li, C. & Tucker, P. (1993) *Nucleic Acids Res.* **21**, 1239–1244.

Lindsay, S.M. & Phillip, M. (1991) *Gen. Anal. Tech. Appl.* **8**, 8–13.

Mardis, E.R. & Roe, B.A. (1989) *BioTechniques* **7**, 840–850.

Maxam, A. & Gilbert, W. (1977) *Proc. Natl. Acad. Sci. U.S.A.* **74**, 560–564.

McLean, J.W., Tomlinson, J.E., Kuang, W-J., Eaton, D.L., Chen, E.Y., Fless, G.M., Scanu, A.M. & Lawn, R.M. (1987) *Nature* **330**, 132–137.

Nguyen, D.C., Keller, R.A., Jett, J.H. & Martin, J.C. (1987) *Anal. Chem.* **59**, 2158–2161.

Pohl, F. & Sulston, J. (1992) *Nature* **357**, 106.

Sanger, F., Air, G.M., Barrell, B.G., Brown, N.L., Coulson, A.R., Fiddes, J.C., Hutchinson III, C.A., Solcombe, P.M. & Smith, M. (1977a) *Nature* **265**, 687–695.

Sanger, F., Nicklen, S. & Coulson, A.R. (1977b) *Proc. Natl, Acad. Sci. U.S.A.* **74**, 5463–5467.

Smith, L.M. (1991) *Nature* **349**, 812–813.

Smith, L.M., Sanders, J.Z., Kaiser, R.J., Hughes, P., Dodd, C., Connell, C.R., Heiner, C., Kent, S.B.H. & Hood, L.E. (1986) *Nature* **321**, 674–679.

Smith, V., Brown, C.M., Bankier, A.T. & Barrell, B.G. (1990) *DNA Sequence* **1**, 73–78.

Sorge, J.A. & Blinderman, L.A. (1989) *Proc. Natl. Acad. Sci. U.S.A.* **86**, 9208–9212.

Staden, R. (1987) In *Nucleic Acid and Protein Sequence Analysis: A Practical Approach,* M.J. Bishop & C.J. Rawlings (eds). IRL Press, Oxford. p. 173.

Strezoska, Z., Paunesku, T., Radosavljevic, D., Labat, I., Drmanac, R. & Crkvenjakov, R. (1991) *Proc. Natl. Acad. Sci. U.S.A.* **88**, 10089–10093.

Sulston, J., Du, Z., Thomas, K., Wilson, R., Hillier, L., Staden, R., Halloran, N., Green, P., Theirry-Mieg, J., Qiu, L., Dear, S., Coulson, A., Craxton, M., Durbin, R., Berks, M., Metzstein, M., Hawkins, T., Ainscough, R. & Waterston, R. (1992) *Nature* **356**, 37–41.

Swerdlow, H. & Gesteland, R. (1990) *Nucleic Acids Res.* **18**, 1415–1419.

Tabor, S. & Richardson, C.C. (1990) *J. Biol. Chem.* **265**, 8322–8328.

Wilson, R.K., Yuesn, A.S., Clark, S.M., Spence, C., Arakelian, P. & Hood, L.E. (1988) *BioTechniques* **6**, 776–787.

Zagursky, R.J. & McCormick, R.M. (1990) *BioTechniques* **9**, 74–79.

CHAPTER TWO

Automated Multiplex Sequencing

G.M. CHURCH,[1] G. GRYAN,[1] N. LAKEY,[1] S. KIEFFER-HIGGINS,[1] L. MINTZ,[1] M. TEMPLE,[1] M. RUBENFIELD,[2] L. JAEHN,[1] H. GHAZIZADEH,[1] K. ROBISON[1] & P. RICHTERICH[2]

[1] Department of Genetics, Harvard Medical School, Howard Hughes Medical Institute, Boston, MA 02115, USA
[2] Collaborative Research Inc., Waltham, MA 01254, USA

2.1 INTRODUCTION

Automated multiplex sequencing has several advantages over other automated systems. Instead of four tags per lane (as in fluorescent gel readers), multiplexing offers 40 or more tags per lane, leading to higher efficiency in laborious tasks like DNA extractions and gels. In addition, the extra tags provide internal standards to aid automated reading with flexible gel formats. Instead of readout speeds limited by electrophoresis (1–12 h), multiplexing uses high-speed (5–15 min) reading of films or imaging plates together with highly parallel membrane probings. Such speeds would be consistent with 2600 sequence reaction sets per workstation-day. Most of the equipment required for multiplexing is not highly specialized and hence already present in molecular biology laboratories.

The extra tags can, however, require extra planning for cloning and extra steps for probings. Although current automated multiplex systems with a separate computer in each device allow efficient project management, they also require user acquaintance with at least two different computer systems (DEC-VMS and PC-DOS).

Multiplex DNA sequencing has been used on several large-scale DNA sequencing projects: *Escherichia coli* (Church & Kieffer-Higgins, 1988; Lakey *et al.*, 1993), *Salmonella* (Roth *et al.*, 1993), *Mycoplasma* (see Ohara *et al.*, 1989), *Mycobacteria* (Smith *et al.*, 1993), *Arabidopis* (Goodman *et al.*, 1992), mammalian simple sequence repeat polymorphisms (Hudson *et al.*, 1992), and human oncogenes (Cawthon *et al.*, 1990). The total for these various projects is over 9 million bases of pre-assembly data.

2.2 BASIC CONCEPTS AND VARIATIONS

Multiplexing in general consists of five steps.

Tag – natural, transposon or vector tags.
Mix – colonies or genomic fragments.
Process – growth/amplification, reactions, and gels.
Decode – DNA and/or protein probes, film scanners.
Interpret – using internal standard data.

For tagging, in addition to plasmid vectors for blunt end cloning (Church & Kieffer-Higgins, 1988) and *Bst*XI-linker cloning (G.M. Church and M. Rubenfield, unpublished), a variety of other multiplex cloning vectors have been developed. These include M13 vectors (Heller *et al.*, 1991; Chee, 1991), recombinational screening vectors (Stewart *et al.*, 1991), and multiplex transposons (Berg *et al.*, 1993; Weiss *et al.*, 1993). The processing steps are essentially the same as conventional sequencing (but about 40 times more productive), and thus multiplex sequencing has many variations and choices.

DNA – genomic, cDNA, cosmid, P1, YAC.
Fragments – restriction, shotgun, nested deletion, walking.
Amplify – plasmid, phage, PCR.
Reactions – chemical, Taq or T7 dideoxy.
Gels – gradient, direct transfer electrophoresis (DTE).
Labels – ^{32}P, ^{35}S, alkaline phosphatase, peroxidase, fluorescent, stable isotopes.

For work on nonradioactive multiplex labels see Chu *et al.* (1992), Creasey *et al.* (1991), Richterich & Church (1993), Sachleben *et al.* (1991), and Tizard *et al.* (1990)

The *Mycoplasma* group uses multiplex walking using synthetic oligomers as probes of sequence reactions performed on total genomic DNA digests. Pfeifer *et al.* (1989) have developed a ligation-mediated PCR version of such genomic sequencing. For the HMS and CRI groups, the currently favoured combination is shotgun cloning from cosmids, plasmid alkaline preparation, chemical sequencing, DTE and α-[^{32}P]dATP tailed DNA 20-mer probes. The rest of this chapter will focus on aspects of this last combination which are not covered in Church & Kieffer-Higgins (1988) or Richterich & Church (1993). Several topics below should also be relevant to DNA sequencing in general.

2.3 STARTING SYSTEM EQUIPMENT

Equipment recommended for a starting system is listed below with approximate prices, order numbers and phone numbers. Many are general-purpose molecular biology items. Additional information can be obtained from the vendors.

Approx. price	
$200	Colony mixer made from a Rocker Baxter; order No. R4193–2 (tel. 800–888–7967).
$4000	Shaker from New Brunswick; order No. M1024–1000 (tel. 201–287–1200).
$9000	Omnifuge RT for 50 ml tubes and 96-well plates from Baxter; order No. C1730–1, –20, –33.
$600	Multiplex-pipets from Hamilton; order No. 0155200 (tel. 702–786–7077).
$900	Electrotransfer device from Polytech plastics (tel. 617–666–5064).
$1350	UV crosslinker from Stratagene; order No. 400075 (tel. 800–424–5444).
$36 000	Molecular Dynamics densitometer model 300B (tel. 408–773–8343). (See Eby (1990) for others.)
$14 280	DEC VS4000/60 with 32 Mbyte RAM, 1.4 Gbyte disks, 8 Gbyte DAT tape from Compucom (tel. 404–452–1090).

Other equipment, such as electrophoresis and hybridization devices, is available from HMS machine shop (tel. 617–432–2036. Email: machshop@warren.med.harvard.edu). REPLICA, GTAC, and GA software is available from HMS Office of Technology Licensing (tel. 617–432–0920). GCG software version 7 is available from Genetics Computer Group (tel. 608–231–5200). Biotrans No. 711300 membranes are available from ICN (tel. 800–854–0530). Plex vectors can be obtained from our HMS laboratory.

2.4 SAMPLE HANDLING WITH MULTIPLEX PIPETTES AND COMBS

Multiplex pipettes and combs allow the transfer of samples from multiwell plates to other plates, gels, or membranes. Plates with 96 wells have 12 columns and 8 rows of wells spaced 9 mm apart, with a capacity of 100–1500 μl per well. Plates with 864 wells, having 3 mm spacing and 20 μl volumes, are also compatible. The multiplex pipette reduces sample order errors and promotes automation by simultaneous transfer of 12 samples in 0.2–10 μl quantities. Since the pipet has a spacing of 9 mm, higher density spacing involves interleaving. For loading gels, compatibly spaced combs must be used: sharktooth sequencing combs in 0.1, 0.2 or 0.4 mm thick material with 4.5, 3 or 2.25 mm spacings between teeth are available from Owl Scientific Plastics (tel. 617–242–9748) as well as agarose combs with 4.5 mm spacings of 1.5 mm material. The multiplex pipette typically consists of 12 syringes of 10 μl capacity with 25 mm long needles with flat tips (point style No. 3) tapered from 26 to 31 gauge. This pipette is intended to fit into thin sequencing gel wells (0.2 mm). A 27-gauge non-tapered version is easier to keep unclogged during extended use and can be used for the thinnest gels in our laboratory (0.1 mm) when entry between the gel plates is unnecessary. Versions with eight syringes load by rows rather than columns and offer less resistance and a better gripping surface. Three examples of using the pipet, all resulting in a GTACGTAC . . . 96-lane gel pattern, are given below.

(1) When 12-tip loading to 2.25 mm sequencing combs, four separate 96-well plates each for G, T, A, and C reactions are loaded with two loadings from each plate per gel (four gels).
(2) For 12-tip loading from 9 mm sequencing plates to 4.5 mm sequencing combs, plate No. 1 column order is GAGAGAGAGAGA and plate No. 2 is TCTCTCTCTCTC. Each gel receives four loadings from each plate.
(3) For 8-tip loading to 3 mm combs, a reaction row order made robotically or by use of a template overlay for manual pipetting:
Row 1 $G_1C_1A_2T_3G_4C_4A_5T_6$
Row 2 $T_1G_2C_2A_3T_4G_5C_5A_6$
Row 3 $A_1T_2G_3C_3A_4T_5G_6C_6$
and so on for 12 rows per plate.

2.5 AUTOMATIC FILM-READING SOFTWARE (REPLICA)

Automatic DNA sequence reading provided by REPLICA software applies to chemical, dideoxy, conventional or multiplex sequencing films using data from various scanners including examples of TIFF (tag image file format). A graphical user interface allows instantaneous access to gigabyte film image databases for comparisons of aligned and oriented images displayed in MOTIF UIS-X-windows under VMS. Precise use of internal standards allows reading of tightly packed lanes (3 mm) and closely spaced bands. Figure 2.1 shows the dependence of error rates on read length and laboratory protocol. The database also features single character (GTACgtac) and numerical estimates of base assignment probabilities.

The automated-reading steps are as follows.

(1) One standard and about 40 unknown films are scanned. Each film has alignment points, annotations tracking samples and probes by virtue of the hybridization pattern of the DNA ink marking present at the edges of each membrane. Typically, 88 μm × 176 μm pixels sizes are used. Lower optical resolution could be used to save disk space; however, we find that less than 15 pixels across each lane sacrifices image interpretability and is unnecessary at current modest ratios of disk cost to total project cost.
(2) The position of known bases about every 50 bp on the standard film are input manually followed by automatic completion of the standard film interpretation and a final manual checking. The internal standards are then available for determination in unknowns of: (a) lane boundaries; (b) interband distances along the electrophoretic (*Y*) axis; (c) band profile shape which can vary as a function of *Y* and is used for deconvolution band sharpening by an iterative constrained fast-Fourier-transform method (Agard *et al.*, 1981); (d) conditional and prior probabilities of certain base assignment errors which are DNA preparation and/or lane specific; (e) interlane alignment; (f) band curvature across the lane; and (g) observed resolution as a function of lane and *Y* coordinate.
(3) Unknown film readings are initiated simply by clicking on the corner positions and the automatic sequence interpretations are stored in a database. These can be checked for incorrect base calls.

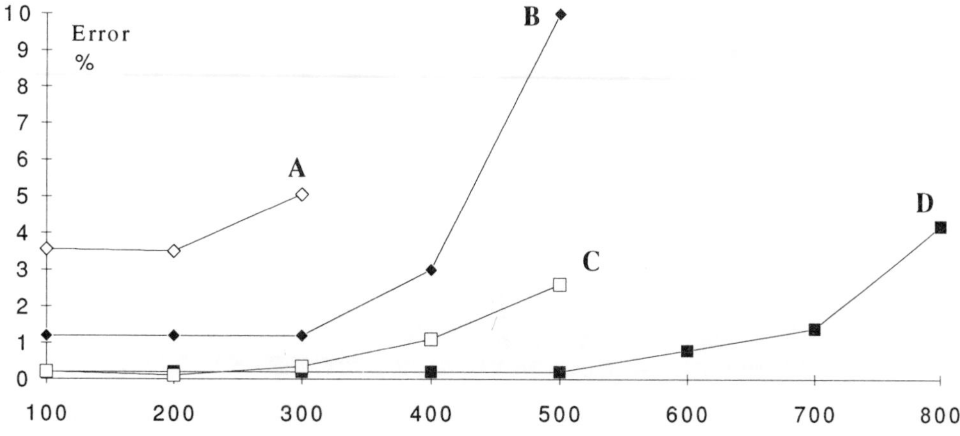

Figure 2.1 Error rate as a function of sequence read length in bases. Error rates are calculated relative to manually checked sequence (with access to both strands). They are plotted on the *Y* axis as total cumulative error (including ambiguities and compressions) with a local window of 100 bases. REPLICA automatic reads from (A) multiplex chemical sequencing on buffer gradient gels and (D) T7 dideoxy reactions on DTE gels (see original photos in Richterich & Church (1993). Lines B and C are fluorescent gel reader data (Koop *et al.*, 1993; *Sulston et al.*, 1992).

(4) If the data are part of a shotgun project they are assembled by GTAC (see below) into one or more contigs (as defined by Staden (1982)). GA (a modified version of GCG's GelAssemble) displays aligned contig subsequences and the consensus sequence. For error checking, up to four film images can be displayed horizontally (see Fig. 2.2).

2.6 AUTOMATIC SEQUENCE ASSEMBLY SOFTWARE (GTAC)

The assembly process is complicated by the presence of errors in the reading of sequencing gels, cloning artifacts, compressions, and repetitive DNA sequences. Well-described approaches to automated assembly include variants on DBAUTO and xdap (Staden, 1982; Dear & Staden 1991), or SEQAID and fa (Peltola *et al.*, 1984; Kececioglu & Meyers, 1989). For N sequences these approaches are expected to scale with N^2 (at least for low coverage) leading to long assembly times for large projects. Taking new fragment sequences in a suboptimal order, results in prematurely degrading contig ends when poor sequences become incorporated early. To address these problems GTAC assembly computation uses the following scheme:

(1) creating a hash lookup table of all 16-mers in the input;
(2) scanning this table to detect overlaps among fragments;
(3) ranking of overlapping sequences by the number of shared 16-mers;
(4) pairwise merging of sequences into multiple sequence alignments using a hybrid of the Needleman & Wunsch (1970) and Martinez (1988) methods;
(5) derivation of a local consensus; and
(6) repeating steps (4) and (5) until a single contig is derived or the remaining overlaps are weaker than a prespecified threshold.

GTAC software is aimed at large sequences: yeast and bacterial chromosomes, cosmid and P1 clones in the 50–5000 kb range. For N bases of raw data, assembly times are proportional to $N^{1.1}$. On a computer rated at 2.4 SPECmarks, assembly of 150 kb of raw data takes 40 min and 2 Mb takes 9 h. Sequence contigs 40–90 kb long are assembled from test data with four-fold minimum coverage and initial error rates up to

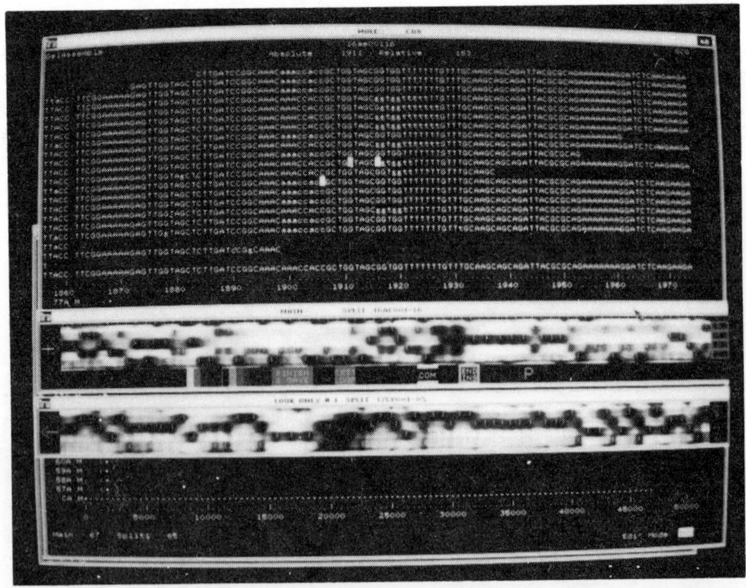

Figure 2.2 Contig sequences and aligned film images. The sequence text display is a modification of GelAssemble GCG v.7 called GA. At the top are the primary sequence reads from REPLICA which are inspected for discrepancies (shown in white capital letters) relative to the consensus on the middle line (just above the film strips). Up to four images can be displayed and aligned with the GA text. Strands orientations are indicated at the left (+> <−). At this point in the cosmid major compressions can be seen to the right of the cursor for the plus strands and to the left for the minus, with the complementary strand providing unambiguous data at these points. We find that horizontal display of full Gray scale images is more informative and compact than the vertical unaligned images or the horizontal plots in common use.

9% for nonrepetitive DNA regions and up to 3% errors for DNA regions containing dispersed repeats smaller than the sequence run lengths (e.g. human Alu repeats). Longer repeats require analysis with 'distance restraints' where the expected and observed distances between ends of clones are compared. Memory requirements scale linearly with the project size, 12+2*c* bytes (where *c* is the mean coverage) of DRAM per final base pair (at the most memory-intensive step (4) ((4) above); 32 Mbyte handles up to 500 kbp of final sequence. Currently, 16 times this amount of memory is available on compatible computers.

2.7 AUTOMATIC IDENTIFICATION OF SEQUENCE FEATURES

As an aid to biological interpretation and for quality control the program TGIF produces a quantitative display of:

(1) weak and strong blastp (Altschul *et al.*, 1991) hits to predicted reading frames longer than 50 amino acids;
(2) dicodon usage for the six reading frames (Farber *et al.*, 1992);
(3) restriction sites; and
(4) essentially exact DNA database hits.

2.8 FUTURE DIRECTIONS

One may ask, if four reactions per lane is good and 40 better, what is the upper limit? Two factors could influence limits: signal-to-noise ratio decrease during many probings, and reduced read length due to overloading. More than 55 successive reprobings have been done without significantly increased background, and signal intensity stays essentially constant during the typically used 25–30 reprobings. This indicates that much higher multiplex factors may be possible. For reduced read length due to overloading, we cannot completely rule out that such effects occur. However, the best results with 30-fold multiplexing are very similar to the ones obtained with standard (nonmultiplex) sequencing, indicating that overloading effects do not influence data quality in a significant way.

Another important aspect in multiplex sequencing is automation. Several prototypes of automated hybridization systems have been developed, but a fully integrated system that is suitable for routine high-throughput production remains to be demonstrated. Multiplexing should also be able to exploit improvements of nonmultiplex methods, such as automation of DNA preparation and sequencing reactions, higher quality and compression-free reactions, and longer reads (see Chapters 21–23, 26–30).

Finally, we note that the general notion of multiplexing is no longer limited to DNA sequencing, but now includes approaches to nucleic and protein interactions (Wang & Church, 1992), cell lineages (Walsh & Cepko, 1992), oligonucleotide synthesis (Kieffer-Higgins & Church, 1993), quantitation of relative survival of mutants in mixtures (A. Link, D. Phillips & G.M. Church, unpublished), and diagnostics and genetic linkage (Hazan *et al.*, 1992).

ACKNOWLEDGMENTS

This work was supported by DOE grant DE-FG02-87ER6065. We thank other multiplexing groups for generously sharing information.

REFERENCES

Agard, D.A., Steinberg, R.A. & Stroud, R.M. (1981) *Anal. Biochem.* **111**, 257–268.

Altschul, S.F., Gish, W., Miller, W., Myers, E.W. & Lipman, D.J. (1991) *J. Mol. Biol.* **215**, 403–410.

Berg, C.M., Wang, G., Strausbaugh, L.D. & Berg, D.E. (1993) *Methods Enzymol.* **218**, 279–306.

Cawthon, R.M., Weiss, R., Xu, G., Viskochil, D., Culver, M., Stevens, J., Robertson, M., Dunn, D., Gesteland, R., O'Connell, P. & White, R. (1990) *Cell* **62**, 193–201.

Chee, M. (1991) *Nucleic Acids Res.* **19**, 3301–3305.

Chu, T.J., Caldwell, K.D., Weiss, R.B., Gesteland, R.F. & Pitt, W.G. (1992) *Electrophoresis* **13**, 105–114.

Church, G.M. & Gilbert, W. (1984) *Proc. Natl. Acad. Sci. U.S.A.* **81**, 1991–1995.

Church, G.M. & Kieffer-Higgins, S. (1988) *Science* **240**, 185–188.

Creasey, A., D'Angio, L., Dunne, T.S., Kissinger, C., O'Keeffe, T., Perry-O'Keeffe, H., Moran, L.S., Roskey, M., Schildkraut, I., Sears, L.E. & Slatko, B. (1991) Application of a novel chemiluminescence-based DNA detection method to single-vector and multiplex DNA sequencing. *BioTechniques* **11**, 102–104.

Dear, S. & Staden, R. (1991) *Nucleic Acids Res.* **14**, 3907–3911.

Eby, M.J. (1990) *BioTechnology* **8**, 1046–1049.

Farber, R., Lapedes, A. & Sirotkin, K. (1992) *J. Mol. Biol.* **226**, 471–479.

Goodman, H.M., Hauge, B.M., Hanley, S., Hwang, I., Kochi, T., Keiffer-Higgins, S., Rubenfield, M., Church, G.M. & Gallant, P. (1992) In *Genome Sequencing Conference*. Vol. III, p. 9.

Hazan, J., Dubay, C., Pankowiak, M.P., Becuwe, N. & Weissenbach, J. (1992) *Genomics* **12**, 183–189.

Heller, C., Radley, E., Khurshid, F.A. & Beck, S. (1991) *Gene* **103**, 131–132.

Hudson, T.J., Egelstein, M., Lee, M.K., Ho, E.C., Rubenfield, M.J., Adams, C.P., Housman, D.E. & Dracopoli, N.C. (1992) *Genomics* **13**, 622–629.

Kececioglu, J. & Myers, E. (1989) A procedural interface for a fragment assembly tool. *Technical Report 89–5*. Department of Computer Science, University of Arizona, Tuscon, AZ.

Kieffer-Higgins, S. & Church, G.M. (1993) *Genome Sci. Technol.* **1**, 33. (Abstracts of Genome Sequencing and Analysis Conf. IV.)

Koop, B.F., Rowan, L., Chen, W.Q., Lee, H. & Hood, L. (1993) *BioTechniques* **14**, 442–447.

Lakey, N.D., Ghazizadeh, H., Jaehn, L., Richterich, P., Robison, K. & Church, G. (1993) *Genome Sci. Technol.* **1**, 41.

Martinez, H.M. (1988) *Nucleic Acids Res.* **16**, 1683–1691.

Needleman, S.B. & Wunsch, C.D. (1970) *J. Mol. Biol.* **48**, 443–453.

Ohara, O., Dorit, R.L. & Gilbert, W. (1989) *Proc. Natl. Acad. Sci. U.S.A.* **86**, 6883–6887.

Peltola, H., Soderlund, H. & Ukkonen, E. (1984) *Nucleic Acids Res.* **12**, 307–321.

Pfeifer, G.P., Steigerwald, S.D., Mueller, P.R., Wold, B. & Riggs, A.D. (1989) *Science* **246**, 810–813.

Richterich, P. & Church, G.M. (1993) *Methods Enzymol.* **218**, 187–222.

Roth, J.R., Lawrence, J.G., Rubenfield, M., Keiffer-Higgins, S. & Church, G.M. (1993) *J. Bacteriol.* **175**, 3303–3316.

Sachleben, R.A., Brown, G.M., Sloop, F.V., Arlinghaus, H.F., England, M.W., Foote, R.S., Larimer, F.W., Woychik, R.P., Thonnard, N. & Jacobson, K.B. (1991) *Genetic Anal. Techn. Appl.* **8**, 167–170.

Smith, D.R., Richterich, P., Rubenfield, M., Falls, K., Smyth, A., Butler, C., Lee, H.-M., Avruch, T., Capparell, N., Drill, S., Gunderson, K., Xu, Q. & Mao, J. (1993) *Genome Sci. Technol.* **1**, 20.

Staden, R. (1982) *Nucleic Acids Res.* **10**, 4731–4751.

Stewart, G.D., Hauser, M.A., Kang, H., McCann, D.P., Osamlak, M.M. & Kurnit, D.M. (1991) *Gene* **106**, 97–101.

Sulston, J. et al. (1992) *Nature* **356**, 37–41.

Tizard, R., Cate. R.L., Ramachandran, K.L., Wysk, M., Voyta, J.C., Murphy, O.J. & Bronstein, I. (1990) *Proc. Natl. Acad. Sci. U.S.A.* **87** 4514–4518.

Walsh, C. & Cepko, C.L. (1992) *Science* **255**, 434–440.

Wang, M.X. & Church, G.M. (1992) *Nature* **360**, 606–610.

Weiss, R., Dunn, D., Wheatley, W., DiSera, L., Kimball, A., Ferguson, M., Yeh, R., Milner, B., Duval, B., Lee, R., Rote, C., Cherry, J. & Gesteland, R. (1993) *Genome Sci. Technol.* **1**, 20.

CHAPTER THREE

Application of Capillary Array Electrophoresis to DNA Sequencing

X.C. HUANG & R.A. MATHIES
Chemistry Department, University of California, Berkeley, CA 94720, USA

3.1 INTRODUCTION

The Human Genome Project has generated a great deal of interest in the development of high-speed, high-throughput DNA sequencing methods (Hunkapiller *et al.*, 1991). These methods should provide faster separations, more separation lanes, easily automated loading, as well as nonradioactive and easily computerized data readout. Current automated slab gel sequencing apparatus employing fluorescence detection (Smith *et al.*, 1986; Ansorge *et al.*, 1987; Prober *et al.*, 1987) can run about 24 lanes with separation times of the order of 10 h, producing a maximum sequencing rate of about 1.3 kb h^{-1}. Capillary electrophoresis (CE) using narrow-bore (<200 μm internal diameter) gel-filled capillaries provides rapid, high-field, high-resolution separations without heating artifacts, requires small sample loads, and has a loading format that is easily automated (Cohen *et al.*, 1990; Drossman *et al.*, 1990; Swerdlow & Gesteland, 1990; Swerdlow *et al.*, 1990). A number of groups have worked on the refinement of capillary electrophoresis for DNA sequencing, and a variety of successful separation and laser-excited fluorescence detection methods have been reported (Luckey *et al.*, 1990; Swerdlow *et al.*, 1991; Chen *et al.*, 1991, 1992). However, in all of these approaches only one capillary can be run and detected at a time, so the limiting throughput of these capillary electrophoresis DNA sequencing methods is about the same as that of automated slab gel sequencing (Smith, 1991). If multiple capillaries could be run and detected in parallel, it would be an important advance.

Capillary array electrophoresis (CAE) employing laser-excited, confocal-fluorescence scanning is a method that meets the design requirements of a high-speed, high-throughput DNA sequencer (Huang *et al.* 1992a,b; Mathies & Huang, 1992). First, high-field capillary gel separations are very fast (1–2 h) and the resolution is excellent. Also electrokinetic injection is performed by simply placing the flexible capillary into the DNA sample with an electrode – a process that can be easily automated. The principal barrier to capillary array electrophoresis has been devising a method for detecting multiple capillaries without building multiple detection systems! In most CE fluorescence detection schemes, the incident laser and the emitted fluorescence are perpendicular to each other (Luckey *et al.*, 1990; Swerdlow *et al.*, 1991; Chen *et al.*, 1991, 1992), so it is difficult to configure a system to detect multiple capillaries. However, we recently developed a laser-excited, confocal-fluorescence scanner with the sensitivity and spatial resolution needed to detect bands on miniaturized DNA sequencing gels (Quesada *et al.*, 1991). This scanner utilizes an epi-illumination format where the laser is focused on the sample by a microscope objective and the fluorescence is gathered by the same objective followed by confocal detection. This 180°, retro-optical configuration is ideal for scanning arrays of microcapillaries. It has a limiting sensitivity of about 10 attomol of fluorescently labelled DNA per band and a spatial resolution of about 10 μl. We have recently used this scanner to detect DNA sequencing separations on 25 capillaries in parallel (Huang *et al.*, 1992a, b). The ability to run and detect multiple capillaries in parallel is a major advance because these separations are about 10-fold faster than typical slab gels and because about 100 'lanes' can be loaded and run in parallel with excellent band resolution and lane definition.

This chapter describes the apparatus we have developed to perform capillary array electrophoresis as well as a variety of methods for fluorescently coding or labelling the DNA sequencing fragments. For details on the construction of the one-color and two-color laser scanners, readers are referred to previous publications describing their application to high-sensitivity detection of double-stranded DNA on agarose gels using DNA : dye intercalation complexes (Glazer *et al.*, 1990; Quesada *et al.*, 1991; Rye *et al.*, 1992). The modifications to the apparatus needed for one-color and two-color detection of capillary arrays are described in Huang *et al.* (1992a,b). Here we describe in detail the methods needed to prepare and run capillary arrays. We also describe a variety of schemes for coding the sets of DNA sequencing fragments. The most obvious coding approach is to label each set of sequencing fragments with the same dye and then to separate each set of fragments in a different capillary. However, the capillary-to-capillary variation in the DNA migration velocity of about 5% makes it difficult to align the four electropherograms (Huang *et al.*, 1992a). For this reason DNA sequencing on single capillaries has typically utilized four-dye labeling (a unique dye for each set of fragments) followed by four-color detection (Luckey *et al.*, 1990; Swerdlow *et al.*, 1991). Our goal was to devise new coding methods that utilize a simpler two-color detection system. We also wanted to employ only two dye-labeled primers that could be selected to induce the same differential mobility shift of the fragments. We describe here the binary coding, permuted binary coding, four-ratio coding and two-color, two-intensity coding methods that provide a new approach to mutliplexing DNA sequencing ladders. These examples indicate that CAE, coupled with confocal-fluorescence scanning, will facilitate the development of a wide variety of high-speed, massively parallel analytical separation methods in chemistry and biology.

3.2 EXPERIMENTAL

3.2.1 Instrumentation

Figure 3.1 presents a schematic of the laser-excited, confocal-fluorescence capillary array scanner. Excitation (488 nm, 1 mW) from an argon ion laser (Model 2020, Spectra-Physics, Mountain View, CA) is reflected by a long-pass dichroic beam splitter (480DM, Omega Optical, Brattleboro, VT), passed through a 32×, N.A. (numerical aperture) 0.4 microscope objective (LD Plan-Achromat 440850, Carl Zeiss, Germany), and brought to a 10 μm diameter focus within the 100 μm internal diameter capillaries in the array. The fluorescence is collected by the objective, passed back through the first beam splitter to a second dichroic beam splitter (545LP, Omega Optical) that separates the red (>545 nm) and green (<545 nm) channels. The beams are then focused on 400 μm diameter confocal pinholes. The emission is spectrally filtered by a 40-nm band-pass filter (560DF40, Omega Optical) centered at 560 nm (red channel) and a 20-nm band-pass filter (520DF20, Omega Optical) centered at 520 nm (green channel) plus a 488-nm rejection band filter followed by photomultiplier detection. The output is amplified and filtered (Model SR560, Stanford Research Systems, Sunnyville, CA), digitized and then stored on an IBM PS/2 computer. A computer-controlled stage (Model 4000, Design Components, Franklin, MA) is used to translate the capillary array past the optical system at 2 cm s^{-1}. The fluorescence is sampled unidirectionally at 1500 Hz per channel using A 12-bit analog-to-digital convertor (DASH-16F, Metra-Byte, Tauton, MA). The image resolution is 13 μm per pixel. An image of the migrating bands is built up by accumulating periodic 1.4-s sweeps of the exposed region of capillaries. Post-acquisition image processing was performed using the NIH program Image 1.41, and the image and electropherogram displays were prepared using the programs Canvas and Kaleidagraph.

3.2.2 Preparation of capillary gel columns

Noncrosslinked polyacrylamide gel-filled capillaries were prepared using a modification of the procedure described by Cohen *et al.* (Cohen *et al.*, 1988a; Heiger *et al.*, 1990). A 3-mm wide detection window was produced in each 100 μm internal diameter, 200 μm outer diameter fused-silica capillary (Polymicro Technologies, Phoenix, AZ) by burning off the polyimide coating with a hot wire followed by cleaning with ethanol. The window was burned about 25 cm from the inlet side of the 40 cm long capillary. The inner wall of the capillaries was washed with 0.5 M NaOH,

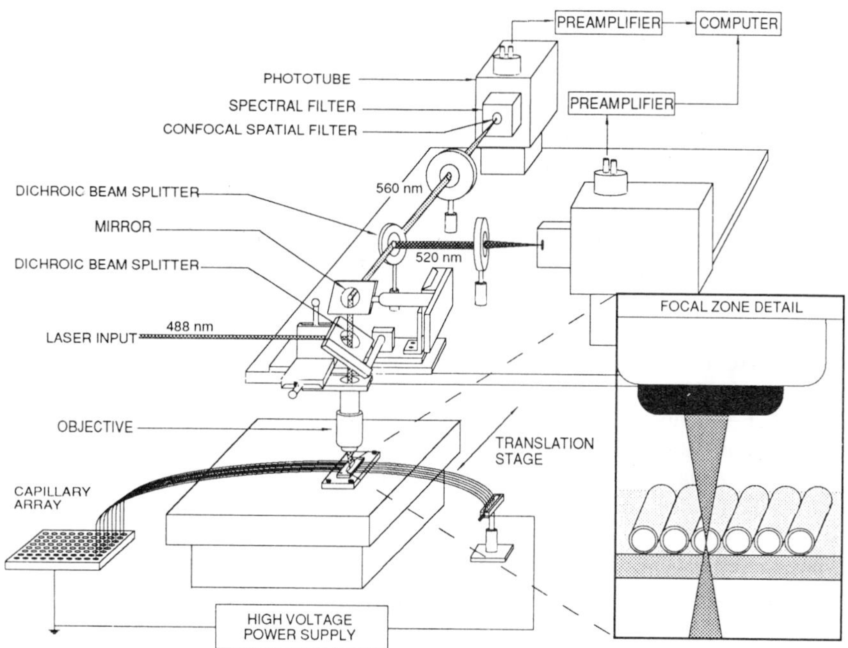

Figure 3.1 Schematic of the capillary array electrophoresis apparatus and two-color, confocal-fluorescence scanner.

01. M HCl and 0.1 M NaOH for 1 h each at room temperature, followed by rinsing with deionized water and methanol. The capillary was then treated overnight with a bifunctional reagent, γ-methacryloxypropyltrimethoxysilane (1:1 in methanol) to derivatize the walls from acrylamide binding (Cohen *et al.*, 1988a). Freshly made acrylamide gel solution (9%T, 0%C) in a 1× TBE buffer (90 mM Tris, 90 mM boric acid, 2 mM EDTA) with 7 M urea was filtered with an 0.2 μm syringe filter and degassed under vacuum for about 1 h. Five microliters of 10% (w/v) tetraethylmethylenediamine (TEMED) and 5 μl of a 10% (w/v) ammonium persulfate (APS) solution were added to the 2 ml gel solution. The solution was immediately forced into a bundle of about 10 capillaries with pressurized nitrogen and allowed to polymerize overnight in a cold room (about 4°C). Prior to use, both ends of the column were trimmed by about 1 cm and then the column was pre-electrophoresed for 30–60 min at 7 kV. The 9%T, 0%C gels were sufficiently stable that at least four consecutive DNA sequencing separations could be performed.

The capillary array was assembled in a holder mounted on the computer-controlled translation stage. To achieve uniform detection sensitivity and background, the holder was designed to keep the exposed region of the capillaries precisely in the same plane (Huang *et al.*, 1992a). Typically, the length from the inlet to the detection window was 24 cm and the applied field was about 225 V cm^{-1}.

3.2.3 Preparation of DNA sequencing samples

Chain-terminated M13mp18 DNA sequencing fragments were produced using a Sequenase 2.0 kit (United States Biochemical Corp., Cleveland, OH) and dye-labeled primers from Applied Biosystems (Foster City, CA). For runs employing permuted binary coding, four annealing solutions were prepared: tubes 1 and 2 contained 2 μl of reaction buffer, 5 μl of M13mp18 single-stranded DNA (0.05 pmol μl^{-1}), and 1.5 μl of FAM-labeled primer solution; tube 3 contained 12 μl of reaction buffer, 30 μl of M13mp18 DNA, 9.0 μl of JOE-labeled primer solution; and tube 4 contained 12 μl of reaction buffer, 30 μl of M13mp18 DNA, 3 μl of FAM-labeled primer solution, and 6 μl of JOE-labeled primer solution. The tubes were annealed by heating to 65°C for 3 min and then allowed to cool to room temperature for 30 min. When the temperature of the annealed reaction mixtures had dropped below 30°C, 1 μl of 0.1 M DTT solution, 1.5 μl of reaction buffer, and 5 μl of ddG termination mixture were added to tube 1; 1 μl of 0.1 M DTT solution, 1.5 μl of reaction buffer, and 5 μl of ddC termination mixture were added in tube 2; 6 μl of 0.1 M DTT solution, 8 μl of reaction buffer, and 30 μl of ddA termination mixture were added in tube 3; and 6 μl of 0.1 M DTT solution, 8 μl of reaction buffer, and 30 μl of ddT termination mixture were added in tube 4. The tubes were prewarmed to 37°C and Sequenase (1.5 μl of a 1:5 dilution with Sequenase dilution buffer) was added in tubes 1 and 2, and 9 μl of diluted Sequenase was added in tubes 3 and 4. The mixtures were incubated at 37°C for 5 min. The solutions in tubes 3 and 4 were divided into two parts. After addition of stop solution, the contents of tube 1 and half of tubes 3 and 4 were pooled to form the G, A, T sample. The contents of tube 2 and half of tubes 3 and 4 were pooled to form the C, A, T sample. Ethanol precipitation (100% EtOH 1×, followed by 70% EtOH 2×) was then used to remove small ions and to concentrate the sample for enhanced efficiency of sample injection.

For runs employing four-ratio coding, four annealing solutions were prepared: tube 1 contained 2 μl of reaction buffer, 5 μl of M13mp18 single-stranded DNA, and 1.5 μl of FAM-labeled primer; tube 2 contained 6 μl of reaction buffer, 15 μl of M13mp18 DNA, and 4.5 μl of JOE-labeled primer; tube 3 contained 6 μl of reaction buffer, 15 μl of M13mp18 DNA, 2.0 μl of FAM-labeled primer, and 2.5 μl of JOE-labeled primer; and tube 4 contained 6 μl of reaction buffer, 15 μl of M13mp18 DNA, 0.8 μl of FAM-labeled primer, and 3.7 μl of JOE-labeled primer. The tubes were heated to 65°C for 3 min and then allowed to cool to room temperature for 30 min. When the temperature of the annealed reaction mixtures had dropped below 30°C, 1 μl of 0.1 M DTT solution, 1.5 μl of reaction buffer, and 4 μl of ddA termination mixture were added in tube 1; 3 μl of 0.1 M DTT solution, 4 μl of reaction buffer, and 12 μl of ddT termination mixture were added in tube 2; 3 μl of 0.1 M DTT solution, 4 μl of reaction buffer, and 12 μl of ddG termination mixture were added in tube 3; and 3 μl of 0.1 M DTT solution, 4 μl of reaction buffer, and 12 μl of ddC termination mixture were added to tube 4. After prewarming the tubes to 37°C, 1.5 μl diluted Sequenase was added in tube 1, and 4.5 μl of diluted Sequenase was added in tubes 2, 3 and 4. The mixtures were incubated at 37°C for 5 min. Stop solution was then added and the samples were ethanol precipitated as described above.

For runs employing two-color, two-intensity coding, we used the Sequenase dye primer kit with a Mn^{2+} reaction buffer (United States Biochemical Corp.). Two annealing solutions were prepared: tube 1 contained 6 μl of freshly prepared Mn^{2+} reaction buffer, 10 μl of M13mp18 single-stranded DNA, and 4 μl of FAM-labeled primer; and tube 2 contained 12 μl of freshly prepared Mn^{2+} containing reaction buffer, 20 μl of M13mp18 single-stranded DNA, and 8 μl of JOE-labeled primer. The tubes were heated to 65°C for 3 min and then allowed to cool to room temperature for 30 min. When the temperature of the annealed reaction mixture had dropped below 30°C, 3 μl of ddC and 1 μl ddG termination mixture were added in tube 1; and 6 μl of ddA and 2 μl of ddT termination mixture were added in tube 2. The samples were prewarmed to 37°C, 5 μl of diluted Sequenase was added to tube 1, and 10 μl of diluted Sequenase was added to tube 2. The mixtures were incubated at 37°C for 10 min. Stop solution was added to both tubes, the contents of the tubes were pooled, and ethanol precipitation was used to prepare the sample for injection.

3.2.4 Sample injection

After the ethanol precipitations, the samples were resuspended in 4 μl 50% (v/v) formamide. The DNA sequencing sample was placed in 500 μl centrifuge tube and heated for 2 min at 90°C to denature the DNA. Electrokinetic sample injection was performed at 225 V cm^{-1}, typically for a duration of 10 s. After injection, the inlets of the capillaries were removed from the centrifuge tube and placed into a tube containing fresh 1× TBE buffer.

3.3 RESULTS AND DISCUSSION

3.3.1 Capillary array separations

Figure 3.2 presents an image obtained by confocal scanning of a 24-capillary array during electrophoretic separation of a mixture of M13mp18 DNA T-sequencing fragments (Huang *et al.*, 1992a). The horizontal dimension represents the image of the capillary array, while the vertical dimension represents the temporal passage of the fluorescent DNA fragments through the detection window. The overall data acquisition time is about 80 min after passage of the primer. The bands in all 24 lanes are well resolved and the resolution extends throughout the sequencing run with sufficient signal-to-noise ratio to detect bands more than 400 bases beyond the primer. Note that the width of each individual lane is only 100 μm and the total physical width of this array is only 4.8 mm. This result shows that it is feasible to run arrays of up to about 25 capillaries using manual assembly techniques. The physical limitations on the number of capillaries that can be run are related by the equation $N = vT/2D$, where v is the scan speed, T is the repetition period, and D is the capillary outer diameter. For example, we can run 100 capillaries (200 μm outer diameter) using a scan rate of 4 cms^{-1} and a 1-s repetition period. Automated fabrication techniques should readily permit the assembly of 100-capillary arrays. The ability to run multiple capillaries in parallel will be valuable in performing all types of high-throughput CE separations.

There are a variety of ways to label and detect the

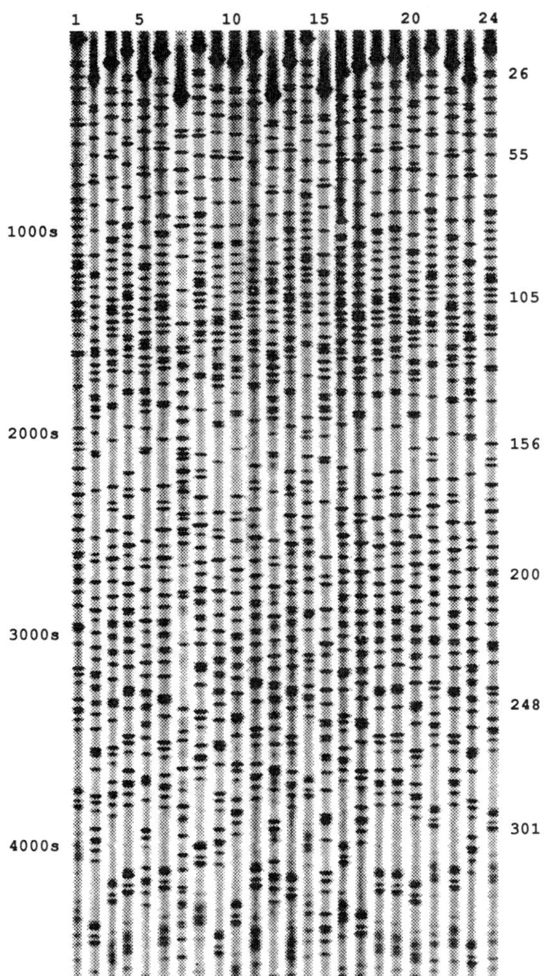

Figure 3.2 Image obtained by scanning a 24-capillary array. Fluorescent (FAM) primer-labeled M13mp18 T-sequencing fragments were separated in 100 μm internal diameter fused silica capillaries filled with 9%T, 0%C gel in 90 mM tris(borate) buffer at a field strength of 240 V cm^{-1}. The capillary length from the inlet to the detection window was 24 cm and the total length was 38 cm. Electrokinetic sample injection was performed for 10 s at 9 kV (Huang *et al.*, 1992a). Each pixel represents a 10 μm × 1 s portion of the spatial and temporal information.

four sets of DNA sequencing fragments. The simplest is to label all four sets of fragments with the same fluorescent dye. Then, the four sequencing reaction samples are introduced in four adjacent capillaries. This method is analogous to conventional slab gel sequencing with radioisotopic detection. For capillary gel electrophoresis, there are relatively large (about 5%) variations in the fragment migration time from column to column because each column is polymerized and run individually. Thus, the one-color, four-column method is not satisfactory. One way to deal with this issue is to label the four sets of DNA fragments using four different dyes followed by separation on one column and four-color detection (Smith *et al.*, 1986). The four-color, four-dye method has worked well for capillary separations (Luckey *et al.*, 1990; Swerdlow *et al.*, 1991), but correcting for the differential mobility shift of the four dyes is an added complexity – four dye-labeled primers must be synthesized, and the detection system is relatively complicated. A second approach is to label each fragment with the same dye and use different intensities to code the different bases (Ansorge *et al.*, 1990, Tabor & Richardson, 1990). This approach has been moderately successful (Chen *et al.*, 1991; Swerdlow *et al.*, 1991), but the use of modified T7 DNA polymerase (or Sequenase with Mn^{2+} containing buffer) reduces the ultimate read length (Tabor & Richardson, 1990), and it is often difficult to determine four different intensities throughout a sequencing run with satisfactory signal-to-noise ratio (Chen *et al.*, 1991; Swerdlow *et al.*, 1991; Pentoney & Konrad, 1992). Consequently, we have devised a new approach for labeling the DNA sequencing fragments that enables us to detect four sets of sequencing fragments on the same capillary using only two different dye-labeled primers and a two-color fluorescence detection system. Four variations of this method will be illustrated: binary coding; permuted binary coding; four-ratio coding; and two-color, two-intensity coding.

3.3.2 Binary coding of sequencing fragments

In the binary coding method, binary combinations of different dye-labeled primers are used to label the four sets of DNA sequencing fragments. Our approach employs a 2-bit binary coding method requiring only two different dyes. Three-bit and 4-bit binary coding methods have also been described and implemented for radioactive slab gel sequencing (Nelson *et al.*, 1992). The two-dye method is illustrated in Fig. 3.3 where a '1' denotes that the set of DNA fragments is synthesized with the corresponding dye-primer and a '0' denotes the absence of the corresponding dye-primer. In Fig. 3.3, (1,0) indicates that the T-fragments are synthesized with the FAM-labeled primer, (0,1) indicates that the G-fragments are synthesized with the JOE-labeled primer (1,1) indicates that the A-fragments are synthesized with both the FAM- and JOE-labeled primers, and the null code (0,0) indicates that no C-fragments are synthesized. Throughout this paper the first bit will indicate labeling with the green-emitting dye (FAM) and the second bit will indicate labeling with the red-emitting dye (JOE). There are two important requirements for dyes when using binary coding. First, there should be no electrophoretic mobility difference between identical DNA

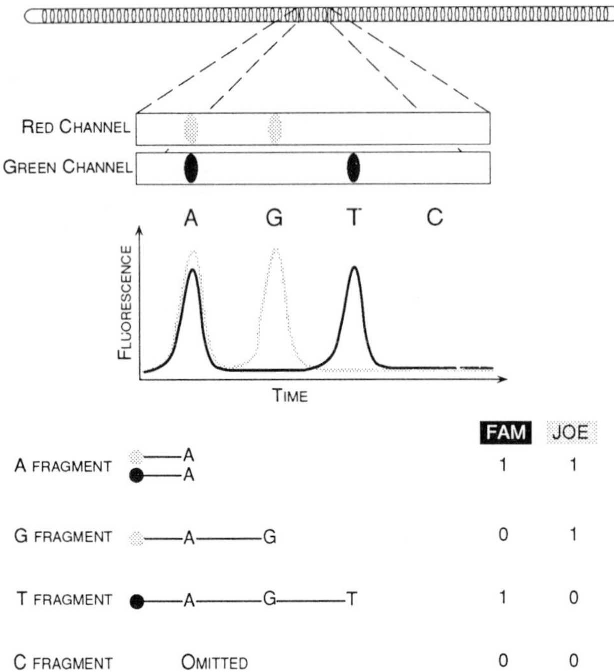

Figure 3.3 Schematic of the binary coding method for identifying four sets of DNA sequencing fragments separated on one capillary using only two dye-labeled primers and a two-color detection system.

fragments labeled with the different dyes. Second, the two dyes should have readily distinguishable fluorescence wavelength emissions (but similar excitation wavelengths). The commercially available dye-primers FAM and JOE meet the first requirement, having almost the same mobility shift (Huang *et al.*, 1992a). They do not fully meet the second requirement because FAM is also detected in the red channel. We have found that examining the ratio of outputs in the two channels can ameliorate this problem. The ratio of the signal in the two channels is independent of the amount of DNA in a band, it is insensitive to instrumental detection sensitivity fluctuations, and when there is leakage between the two detection channels, the amount of cross-talk is a constant parameter for the specified detection wavelengths. The application of this coding method to DNA sequencing is illustrated in Fig. 3.4. The bases can be easily called throughout the run by examining the image or by inspection of the plot of the green intensity divided by that in the red channel. When the band appears only in the red channel the base is a G (0,1), when the green intensity is about two times larger than the red it is a T (1,0), when the red and green intensities are nearly equal it is an A (1,1). The gaps in the sequence indicate the location of a C (0,0). Figure 3.4 also presents plots of the fluorescence intensity ratio out to 300 bases beyond the primer. Three distinct and non-overlapping ratios are observed throughout the run indicating that the individual bases can be can be accurately called. The ratio of the signals is a much more reliable parameter than the signals themselves because the ratio is insensitive to sequence-dependent termination probability. For the 209 directly observed bands in Fig. 3.4 there was only one ambiguous call. The simple binary coding method has the advantage of very robust peak identification and the chemistry is simple.

If there is residual secondary structure or anomalous migration of the sequencing fragments, the lack of direct detection of the null-coded fragment (0,0) can cause sequencing errors. This is less of an issue when resequencing known sequences to look for mutations or if the sequencing will be performed more than once, either by repeated sequencing of the same strand or by sequencing the complementary strand. In cases where the null-coded fragments cannot be tolerated, a variety of more sophisticated methods have been developed where all four sets of fragments are detected.

3.3.3 Permuted binary coding

One approach for directly detecting all four sets of fragments is to run another capillary separation in parallel with the first except that the binary coding assignment of, for example, the G- and C-fragments are exchanged or permuted. The coding of the A- and T-fragments is unchanged thereby providing constant 'marker bands' for alignment of the two

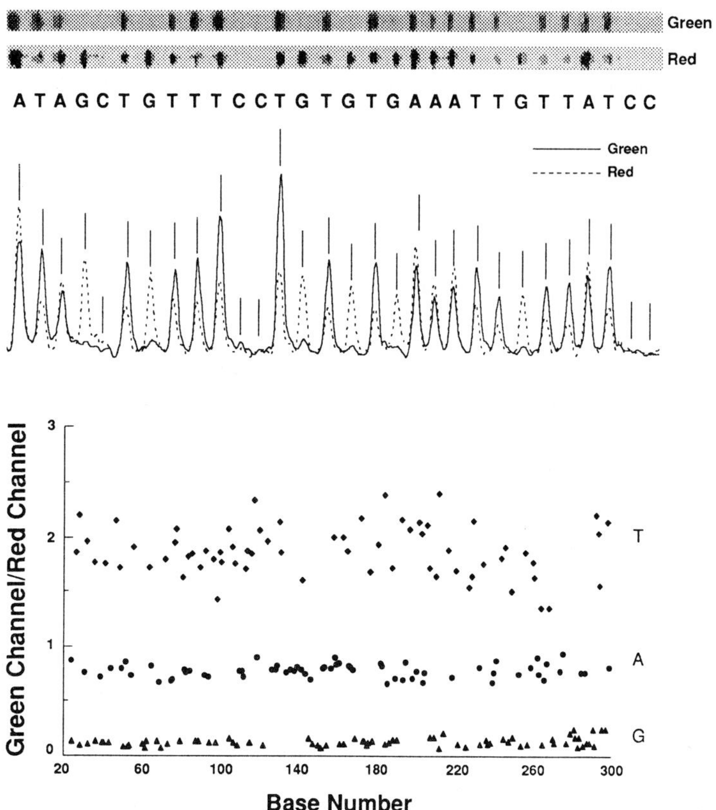

Figure 3.4 Top: Analysis of the DNA sequence from one capillary in a capillary array using binary coding. The images and one-dimensional traces represent the signal detected as a function of time from the red, 560-nm channel and from the green, 520-nm channel. G-fragments are labeled with JOE (0,1) which emits predominantly in the red channel. T-fragments are labeled with FAM (1,0) which emits in the green and red-channels at a ratio of about 2:1 for the conditions used here. A-fragments are labeled with both JOE and FAM (1,1). The molar ratio of JOE and FAM was chosen to give a green/red detection ratio of about 0.9. Gaps in the sequence indicate the location of C which is labeled with the null code (0,0). **Bottom:** Plot of the fluorescence intensity in the green channel divided by that in the red channel for each of the peaks. (◆) T-fragments labeled solely with FAM; (▲) G-fragments labeled solely with JOE; (●) A-fragments labeled with both FAM and JOE.

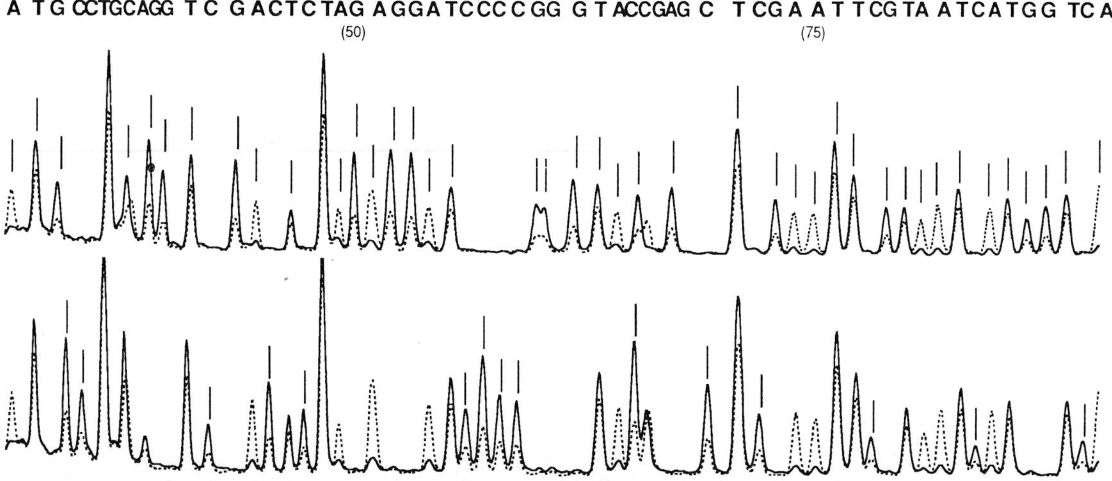

Figure 3.5 Electropherogram of an M13mp18 DNA sequencing run using the permuted binary coding method. The solid line is the fluorescence signal detected in the green channel; the dotted line is the signal detected in the red channel. In this run, the A-fragments are synthesized with the JOE-labeled primer (0,1), and T-fragments are synthesized with both the JOE- and FAM-labeled primers (1,1). G- and C-fragments are both synthesized with the FAM-labeled primer (1,0). The top set of traces presents the run where the A-, T-, and G-fragments are separated; the bottom traces present the results of the run where the A-, T-, and C-fragments are separated.

electropherograms. The two sample sets (G, A, T and C, A, T) are injected into two capillaries for simultaneous separation and detection. Since the A-fragments and T-fragments have the same coding in both separations, their electropherogram patterns are unchanged. The common peaks in each electropherogram are aligned over a region of about 20 bases to give the precise order of the peaks of interest. In the run presented in Fig. 3.5, the coding was A (0,1), T (1,1), G (1,0), and C (1,0). Even though the DNA fragments in the two runs presented have quite different absolute migration times, the relative order of the permuted fragments can be determined without difficulty. Over the 174-base separation presented here (excluding three compression regions) there was only one ambiguous call. Since the signal-to-noise ratio for this coding method is excellent, the ratios of the signals in the red and green channels provide an unambiguous indication of the sequence. The disadvantage of this approach is that two capillary runs are required, and the two runs must be compared before the full sequence can be assigned.

3.3.4 Four-ratio coding

Labeling the four sets of sequencing fragments with four different mole fraction combinations of two dyes is another useful coding method. In four-ratio coding, a unique molar ratio of FAM and JOE replaces the (0,0) code, and all the fragments are separated on one capillary. Figure 3.6 presents a portion of the data from a four-ratio sequencing run and Fig. 3.7 presents a plot of the green/red signal ratio. The ratio of signals in the green and red channel is determined by the spectroscopic filters and by the mole fraction ratio of the two dye-labeled primers. In the example presented here, the A-fragments are synthesized using only the FAM-labeled primer (1,0) and the T fragments are synthesized using only the JOE-labeled primer (0,1). The molar ratios of the FAM and JOE-labeled primers in the two mixed-dye-primer synthesized fragments (G and C) are chosen to define two new ratios that are also distinguishable from FAM alone (maximum ratio) and JOE alone (minimum ratio). The solid line in Fig. 3.7 indicates the average of the ratio for each of the indicated sets of fragments, and the shaded areas indicate the region encompassed by two standard deviations (96% probability). The four distributions are distinguished clearly with only three ambiguous calls caused by weak signal levels out of 260 bases. This excludes four compression regions that cannot be deconvoluted using our present software. The advantage of four-ratio coding is that the sequence is determined in only one run. The disadvantages are that more information is being derived from the green/ red ratio so the signal-to-noise ratio must be high, and more attention must be paid to adjusting the ratio of primers used in the synthesis of the sequencing fragments so that the ratios are distinctive.

3.3.5 Two-color, two-intensity coding

Tabor & Richardson (1990) discovered that if one uses a modified T7 DNA polymerase, or Sequenase in Mn^{2+} containing buffer, the probability of termination is

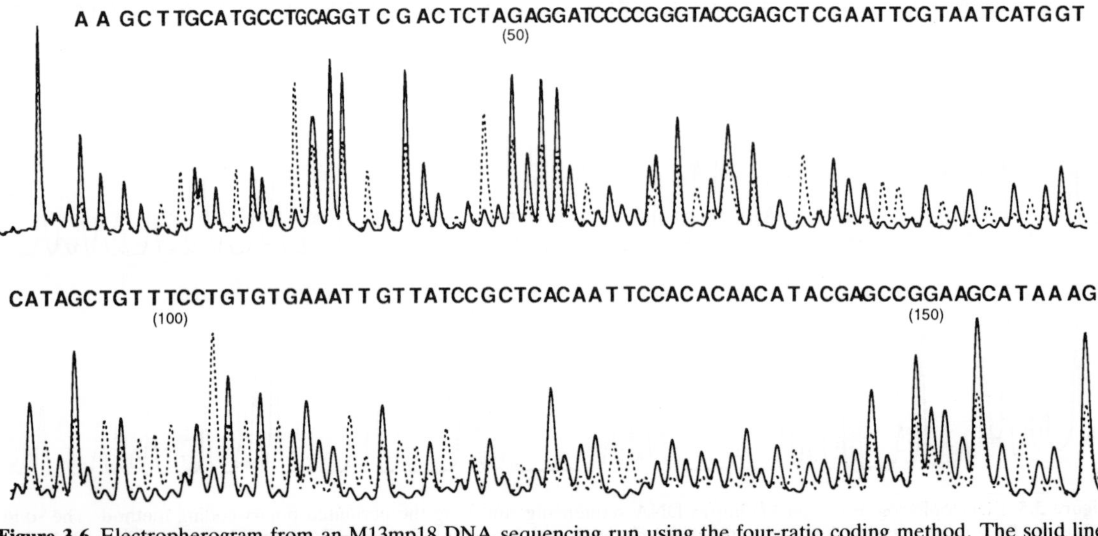

Figure 3.6 Electropherogram from an M13mp18 DNA sequencing run using the four-ratio coding method. The solid line is the fluorescence signal detected in the green channel; the dotted line is the signal detected in the red channel. A-fragments are synthesized with just the FAM-labeled primer (1,0); T-fragments are synthesized with just the JOE-labeled primer; G-fragments are synthesized with the FAM- and JOE-labeled primers in a molar ratio of 1:1.25, respectively; and C-fragments are synthesized with the FAM- and JOE-labeled primers in a molar ratio of 1:5.4, respectively.

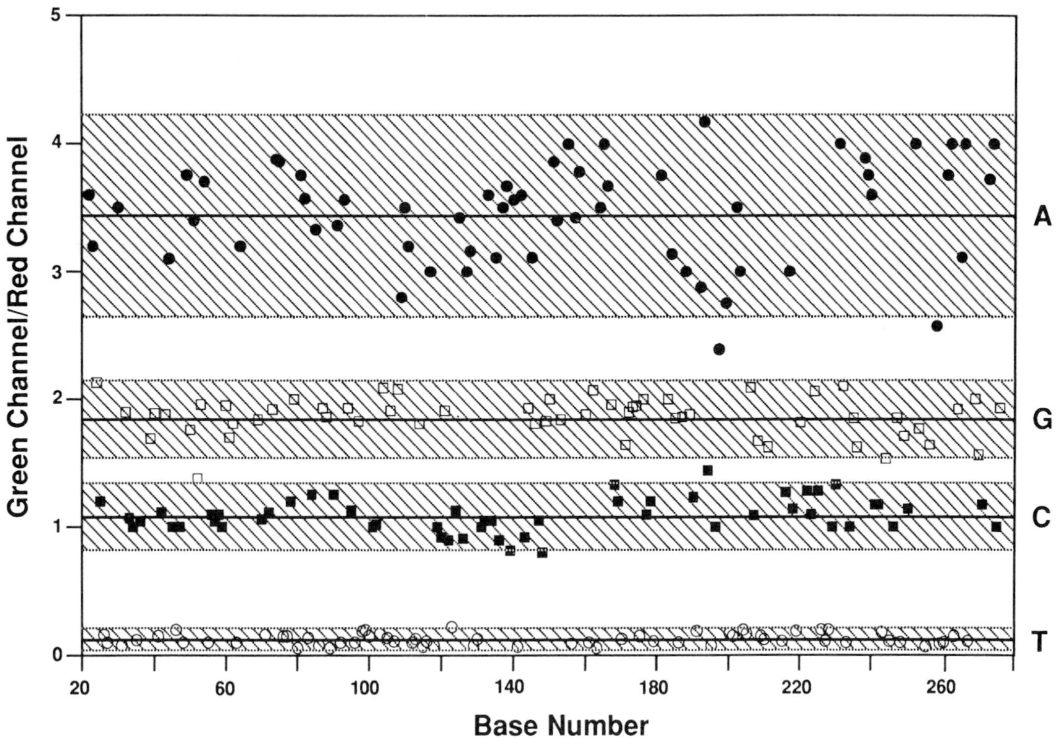

Figure 3.7 Plot of the fluorescence intensity in the green channel divided by that in the red channel for each of the peaks in the four-ratio coding run. (●) A-fragments; (□) G-fragments; (■) C-fragments; (○) T-fragments. The straight lines across data points represent the mean values of the ratio for each set of fragments; the shaded areas indicate the variance (±2σ, or 96%).

remarkably constant throughout the sequence so that the amount of a given set of DNA sequencing fragments can also be used as a tag or label. This is called 'intensity coding'. The methods we have presented thus far for coding rely on the green/red signal ratio. However, if Sequenase with a Mn^{2+} containing buffer is used, then the intensity of the signals provides an additional dimension of information for identifying the fragments (Ansorge *et al.*, 1990; Chen *et al.*, 1991; Swerdlow *et al.*, 1991). This suggests that by combining color-coding and intensity-coding a useful hybrid method could be developed.

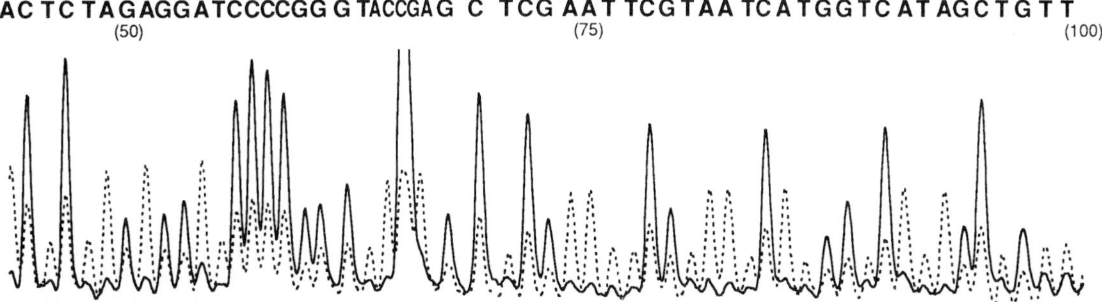

Figure 3.8 Electropherogram from an M13mp18 DNA sequencing run using the two-color, two-intensity coding method. The solid line is the fluorescence signal detected in the green channel; the dotted line is the signal detected in the red channel. A- and T-sequencing fragments are synthesized with the JOE-labeled primer; and C- and G-fragments are synthesized with the FAM-labeled primer. The sequencing fragments were produced using Sequenase with Mn^{2+} containing buffer. The ddA and ddT were added in a 3:1 molar ratio for the JOE-labeled extension; ddC and ddG were added in a 3:1 molar ratio for the FAM-labeled extension.

Figure 3.8 presents an electropherogram of a two-dye, two-intensity sequencing run. In this example, the G- and C-fragments are labeled with FAM while the A- and T-fragments are labeled with JOE. Thus, G-fragments and C-fragments can easily be distinguished from A-fragments and T-fragments by the 'color' of the peaks. The intensities of the peaks, determined by the ratio of the dideoxy terminators, then distinguish C-fragments from G-fragments. Similarly, A-fragments are distinguished from T-fragments by their peak intensities. Figure 3.9(top) presents the green/red signal ratio for this run. The fragments labeled with FAM (G and C) are clearly distinguished from those labeled with JOE (A and T). Figure 3.9(middle) presents the intensities of the FAM-coded fragments. The more intense C-fragments are clearly distinguished from the lower intensity G-fragments except in two compression regions. Similarly, Fig. 3.9(bottom) presents the intensities of the JOE-labeled fragments. The more intense A-fragments are clearly distinguished from the lower intensity T-fragments. This run produced only one ambiguous call out of 180 bases due to low signal levels. The two-color, two-intensity coding method has also been

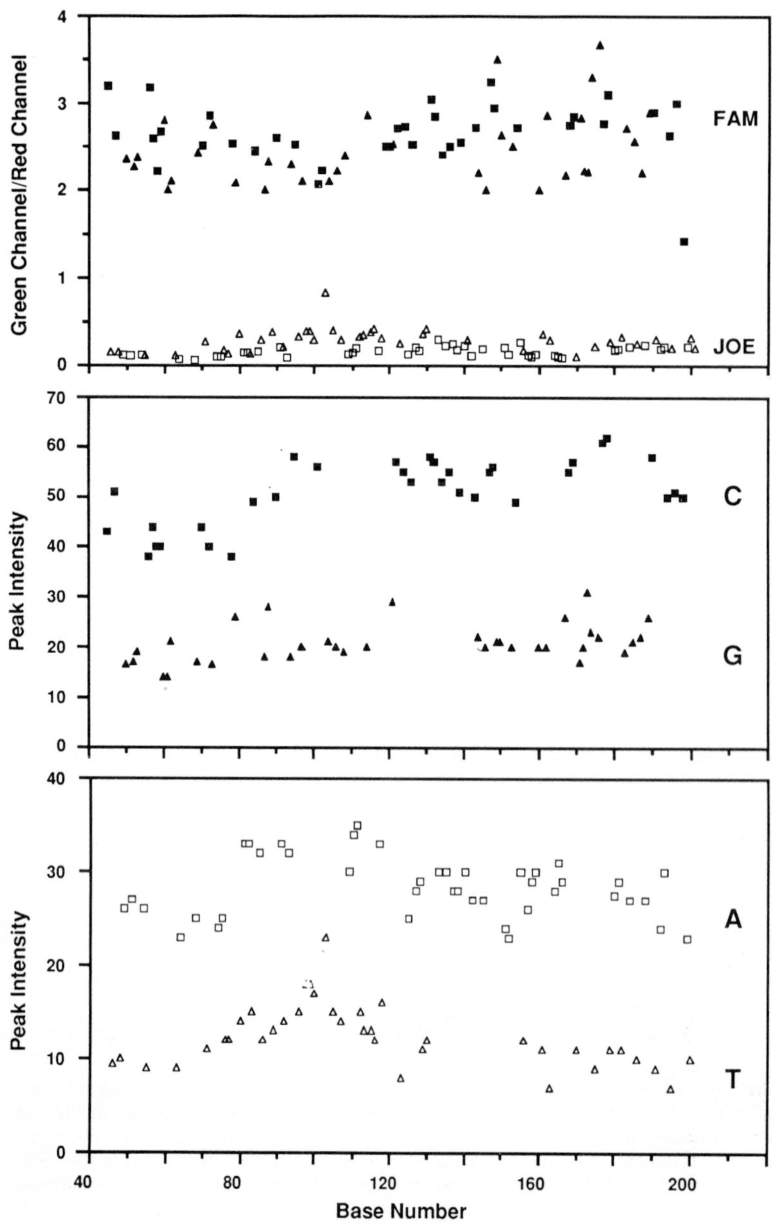

Figure 3.9 Top: Fluorescence intensity in the green channel divided by that in the red channel for the two-color, two-intensity run depicted in Fig. 3.8. **Middle:** Intensities of peaks (C and G) labeled just with FAM. **Bottom:** Intensities of peaks (A and T) labeled with just JOE. (▲) G-fragments and (■) C-fragments synthesized with the FAM-labeled primer; (△) T-fragments and (□) A-fragments synthesized with the JOE-labeled primer.

implemented by Chen *et al.* (1992). The advantage of two-color, two-intensity coding is that it is accurate, and the chemistry is relatively easy to adjust. The key disadvantage is that the length of the sequencing extension is reduced compared with the standard Sequenase procedure (Tabor & Richardson, 1990).

3.4 SUMMARY

A new technique has been developed for performing and detecting large numbers of capillary electrophoresis separations in parallel. This method, capillary array electrophoresis (CAE), is made possible through the use of a laser-excited, confocal-fluorescence scanner for the detection of the fluorescently labeled DNA fragments. With this apparatus it is possible, for the first time, to perform rapid and massively parallel electrophoretic separations using arrays of microcapillaries. It is also convenient to load these arrays because the injection ends of the capillaries can be separated, as depicted in Fig. 3.1, for rapid parallel loading of multiple samples. These traits make CAE a valuable new technique for high-speed, high-throughput DNA sequencing. CAE should also be important in the development of massively parallel, high-speed assays for sizing restriction fragments and PCR-amplified products (Cohen *et al.*, 1988b; Schwartz *et al.*, 1991; Schwartz & Ulfelder, 1992). These examples represent but a subset of the potential applications of CAE to high-speed, high-throughput separation and analysis.

The binary-coding method and its variants provide a new approach to multiplex detection of fluorescently labeled DNA fragments. For rapidly screening a known sequence for mutations, the simple binary-coding method is ideal. If a new sequence is being determined, four-ratio, or two-ratio, two-amplitude coding methods would be preferred, because all four sets of fragments are directly detected. Permuted binary coding requires two runs, but it is highly accurate. Four-ratio coding requires only one run, but the signal-to-noise ratio must be very good to determine all four ratios accurately throughout the run. Two-color, two-intensity coding involves the simplest chemistry and is accurate. The only disadvantage is that T7 DNA polymerase or Sequenase with Mn^{2+} buffer must be used which limits the length of the sequencing extension. The 'best' coding method will probably depend on the particular application.

ACKNOWLEDGMENTS

This research was supported by the Director, Office of Energy Research, Office of Health and Environmental Research of the US Department of Energy under grant DE-FG-91-ER61125. X.H. was supported by a Human Genome Distinguished Postdoctoral Fellowship sponsored by the US Department of Energy and administered by the Oak Ridge Institute for Science and Education.

REFERENCES

Ansorge, W., Sproat, B., Stegemann, J., Schwager, C. & Zenke, M. (1987) *Nucleic Acids Res.* **15**, 4593–4602.

Ansorge, W., Zimmermann, J., Schwager, C., Stegemann, J., Erfle, H. & Voss, H. (1990) *Nucleic Acids Res.* **18**, 3419–3420.

Chen, D.Y., Swerdlow, H.P., Harke, H.R., Zhang, J.Z. & Dovichi, N.J. (1991) *J. Chromatogr.* **559**, 237–246.

Chen, D., Harke, H.R. & Dovichi, N.J. (1992) *Nucleic Acids Res.* **20**, 4873–4880.

Cohen, A.S., Najarian, D., Smith, J.A. & Karger, B.L. (1988a) *J. Chromatogr.* **458**, 323–333.

Cohen, A.S., Najarian, D.R., Paulus, A., Guttman, A., Smith, J.A. & Karger, B.L. (1988b) *Proc. Natl. Acad. Sci. U.S.A.* **85**, 9660–9663.

Cohen, A.S., Najarian, D.R. & Karger, B.L. (1990) *J. Chromatogr.* **516**, 49–60.

Drossman, H., Luckey, J.A., Kostichka, A.J., D'Cunha, J. & Smith, L.M. (1990) *Anal. Chem.* **62**, 900–903.

Glazer, A.N., Peck, K. & Mathies, R.A. (1990) *Proc. Natl. Acad. Sci. U.S.A.* **87**, 3851–3855.

Heiger, D.N., Cohen, A.S. & Karger, B.L. (1990) *J. Chromatogr.* **516**, 33–48.

Huang, X.C., Quesada, M.A. & Mathies, R.A. (1992a) *Anal. Chem.* **64**, 967–972.

Huang, X.C., Quesada, M.A. & Mathies, R.A. (1992b) *Anal. Chem.* **64**, 2149–2154.

Hunkapiller, T., Kaiser, R.J., Koop, B.F. & Hood, L. (1991) *Science* **254**, 59–67.

Luckey, J.A., Drossman, H., Kostichka, A.J., Mead, D.A., D'Cunha, J., Norris, T.B. & Smith, L.M. (1990) *Nucleic Acids Res.* **18**, 4417–4421.

Mathies, R.A. & Huang, X.C. (1992) *Nature (London)* **359**, 167–169.

Nelson, M., Van Etten, J.L. & Grabherr, R. (1992) *Nucleic Acids Res.* **20**, 1345–1348.

Pentoney, S.L. & Konrad, K.D. (1992) *Proceedings of CLEO/QELS*. p. 94. Optical Society of America, Washington D.C.

Prober, J.M., Trainor, G.L., Dam, R.J., Hobbs, F.W., Robertson, C.W., Zagursky, R.J., Cocuzza, A.J., Jensen, M.A. & Baumeister, K. (1987) *Science* **238**, 336–341.

Quesada, M.A., Rye, H.S., Gingrich, J.C., Glazer, A.N. & Mathies, R.A. (1991) *BioTechniques* **10**, 616–625.

Rye, H.S., Yue, S., Quesada, M.A., Haughland, R.P., Mathies, R.A. & Glazer, A.N. (1992) *Methods Enzymol.* **217**, 414–431.

Schwartz, H.E. & Ulfelder, K.J. (1992) *Anal. Chem.* **64**, 1737–1740.

Schwartz, H.E., Ulfelder, K., Sunzeri, F.J., Busch, M.P. & Brownlee, R.G. (1991) *J. Chromatogr.* **559**, 267–283.

Smith, L.M. (1991) *Nature* **349**, 812–813.

Smith, L.M., Sanders, J.Z., Kaiser, R.J., Hughes, P., Dodd, C., Connell, C.R., Heiner, C., Kent, S.B.H. & Hood, L.E. (1986) *Nature* **321**, 674–679.

Swerdlow, H. & Gesteland, R. (1990) *Nucleic Acids Res.* **18**, 1415–1419.

Swerdlow, H., Wu, S., Harke, H. & Dovichi, N.J. (1990) *J. Chromatogr.* **516**, 61–67.

Swerdlow, H., Zhang, J.Z., Chen, D.Y., Harke, H.R., Grey, R., Wu, S., Dovichi, N.J. & Fuller, C. (1991) *Anal. Chem.* **63**, 2835–2841.

Tabor, S. & Richardson, C.C. (1990) *J. Biol. Chem.* **265**, 8322–8328.

CHAPTER FOUR

Sequencing by Hybridization

R. DRMANAC, S. DRMANAC, J. JARVIS & I. LABAT

Integral Genetics Group, Biological and Medical Research Division, Argonne National Laboratory, 9700 South Cass Avenue, Argonne, IL 60439, USA

4.1 INTRODUCTION

The DNA sequences of genes and genomes represent a basis for further advances in molecular biology, biotechnology, medical diagnostics, and the study of evolutionary processes. Unfortunately, there is a huge gap between the size of genomes and the speed of present sequencing methods based on gel separation (Maxam & Gilbert, 1977; Sanger *et al.*, 1977). Even the sequencing of bacterial genomes, which consist of 1–10 million bp (in comparison with 1–10 billion bp in vertebrates and plants), represents a challenge.

Owing to the physical and informational complexity of genomes, the way of doing biology was changed from the moment when the Human Genome Project was started (US Department of Health and Human Services, 1990). As in physics or astronomy, theoretical evaluations of the methodologies have become necessary before a final proof in practice. As a result, there are many thoughts on the strategies for complex genome analysis (US Department of Energy, 1992; Venter *et al.*, 1992). Sequencing by hybridization (SBH) is one such example. Its development would probably never be funded in the 'old' biology.

SBH is based on the fundamental chemistry of life. Because of the chemical specificity of base pairing, synthetic oligonucleotides of known formulae are able to 'read' DNA sequences when hybridized under full-match-specific conditions. Many research groups independently started development of this DNA sequencing approach (Drmanac & Crkvenjakov, 1987; Bains & Smith, 1988; Lysov *et al.*, 1988; Southern, 1988; Drmanac *et al.*, 1989; Macevicz, 1989). Nevertheless, the approach has experienced a few years of ignorance and skepticism. Now, after (i) the demonstration of accurate hybridization of short oligonucleotides (Drmanac *et al.*, 1990); (ii) successful sequencing of test DNAs (Strezoska *et al.*, 1991; Southern *et al.*, 1992) including 340 bp recently determined without error in a blind experiment (Drmanac *et al.*, 1993); (iii) various proposals for the miniaturization of hybridization arrays (Drmanac *et al.*, 1990; Drmanac & Crkvenjakov, 1990; Fodor *et al.*, 1991; Khrapko *et al.*, 1991; Eggers *et al.*, 1992); and (iv) a few informational solutions for accurate sequencing with low-redundancy and error-containing data (Bains & Smith, 1988; Drmanac *et al.*, 1989, 1991a; Pevzner, 1989; Pevzner *et al.*, 1991; Drmanac, 1992; Drmanac & Crkvenjakov, 1992), there is a strong belief that SBH is going to play an important role in the sequencing of human and other biologically or biotechnologically important genomes (Hunkapiller *et al.*, 1991; Cantor *et al.*, 1992).

In this chapter, we discuss the principles of SBH and differences from other sequencing methods, the first production line and some emerging protocols, approaches for manufacturing sequencing chips, and novel possibilities for the analysis of complex genomes opened by SBH. We will not discuss the great potential of SBH in medical diagnostics (Landegren, 1992).

4.2 SBH FEATURES

A DNA sequence can be thought of as an assembly of overlapping oligonucleotide sequences. This concept allows a DNA sequencing approach based on the compilation of constituent oligonucleotides of an appropriate size for the DNA fragments being analyzed. In this approach, there is no experimental determination of which base is at a specific position in the DNA chain; rather, continuous sequences are reconstructed from positively scored oligonucleotides.

The first sequencing techniques (Sanger *et al.*, 1965; Murray, 1970) and SBH are the only proposed methods exploiting this principle. In the old techniques, lists of very short oligonucleotides were compiled on the basis of specific fragmentations and various ionophoretic separations. In SBH, the set of constituent oligonucleotides is obtained by full-match-specific hybridization of a sufficient number of synthetic oligonucleotides. For example, only probes AAGGTT, CAAGGT, and CCAAGG from a complete set of 4096 6-mers will hybridize with the single-stranded DNA fragment 5′ AACCTTGG 3′. The continuous sequence complementary to the analyzed strand can be derived if positively scored probes are ordered with a five-base overlap and read vertically:

```
    AAGGTT
   CAAGGT
  CCAAGG
-------------
5′ CCAAGGTT 3′
```

This fingerprinting approach provides many desirable and unique features for an efficient sequencing method (Table 4.1). The major disadvantages are the difficulties in precise reading of the simple sequences (such as poly(A) stretches) (Drmanac *et al.*, 1989) and the large requirements in computation.

Two technical formats of SBH are possible: (1) based on an array of DNAs, and (2) based on an array of oligonucleotides. In format 1, oligomers, and in the format 2, DNA fragments, have to be labeled and successively hybridized. Format 1 is more appropriate for large-scale mapping and sequencing and format 2 for the sequencing of short DNA segments and diagnostic purposes. The versatility of array formats allows both immediate applications and great opportunities for future developments, which is not the case with other proposed nongel methods (Hunkapiller *et al.*, 1991; US Department of Energy, 1992). The implementation of some advantageous SBH features is discussed in following sections.

Table 4.1 SBH features.

NO

1. Experimental assignment of a base to a DNA-chain position
2. Macromolecular separation
3. Stepwise processing on DNA
4. Single molecule detection or visualization
5. Obligatory pre-specified position of the samples

YES

1. Compiling a list of constituent oligonucleotides
2. Array format (parallelism, simple to automate)
3. Equally adequate for single- and double-stranded DNA
4. Possible to miniaturize
5. Uniform and adjustable sampling of the sequences
6. Reduced effort to read similar sequences
7. Complementary to gel sequencing
8. Immediately usable

4.2.1 The first data production setup

Dot blots of M13 clones or cloned DNAs prepared by polymerase chain reaction (PCR) is an immediately usable format for extensive mapping and sequencing by hybridization of 6-mer to 8-mer probes (or rarely, 9-mers and 10-mers) (Drmanac *et al.*, 1989, 1992). High data throughput can be achieved by the parallel biological, biochemical, and mechanical treatments of a huge number of genomic or cDNA fragments using three types of dense array: an array of wells for clone storage and preparation, an array of pins for inoculation of preparative plates and spotting, and an array of DNA dots for parallel scoring of the oligonucleotide hybridizations.

Both M13 vectors and PCR allow efficient preparation of the 10^5–10^7 clones present in genomic or cDNA libraries (Drmanac *et al.*, 1992a). A protocol for managing the large number of M13 clones in 96- or 384-well plates is listed in Table 4.2. PCR includes only three major steps, technically similar to the first three steps in the M13 protocol. It eliminates vector DNA and also provides a 5- to 10-fold higher DNA concentration, but it is more expensive and has many limits on process scaling. One solution for a high PCR throughput is the BioOven (BioTherm, Fairfax, VA), which can accommodate up to six plates. A BioOven has capacity of 2000 to 10 000 PCRs per day,

Table 4.2 M13 clone managing protocol.

1. Filling the wells with a saturated host *Escherichia coli* culture diluted 1000 times
2. Inoculation of the wells by the phages from a master plate using a pin array
3. Growth under strong shaking for 16–20 h at 37°C
4. Killing the host cells by 2 h incubation at 50–60°C
5. Settling cells by keeping the plates for 2 days at room temperature
6. Evaporation of 70–80% of the liquid (1 day at room temperature) to increase phage concentration

depending whether 96-, 384-, or 864-well plates are used. SBH does not require more than 5 μl of PCR per clone. Such PCR reactions can be prepared in 20 μl wells of 864-well plates for a price of less than 10 cents each. Using BioOvens and 96-well plates, 80 000 PCRs of random cDNA clones have been prepared (unpublished results).

M13 cultures or amplified inserts can be efficiently spotted on nylon membranes by metal pins (Drmanac *et al.*, 1992). To make a dense array of samples, an *X–Y–Z* table and appropriate software are necessary. Biomek 1000 (Beckman, Fullerton, CA) is adaptable for this purpose (Drmanac *et al.*, 1992). With pins 0.3 mm in diameter, up to 144 96-well plates can be spotted by offsetting on a membrane. A filter with 7776 cDNA dots (81 plates) prepared using a Biomek1000 is shown on Fig. 4.1. In up to 25% of the PCR reactions, there is no significant amount of amplified DNA, resulting in invisible dots. Spotting is done from the PCR samples without removing the oil which prevents liquid evaporation during PCR cycling, storage and spotting. No treatment of either M13 phages or PCR products is required before spotting on membranes soaked in 0.5 M NaOH and 1.5 M NaCl. Further processing of the filters is the same as in standard Southern blot procedures. Production-size filters comprising two-, four-, or six-plate patterns and 27 648 to 82 944 dots can be prepared.

The hybridization step is simple, as in any dot-blot analysis. Conditions for efficient discrimination of mismatched targets are defined for probes as short as 6-mers (Drmanac *et al.*, 1990, 1992a). There are three basic requirements: appropriate probe concentration, adequate washing temperature and time, and a sufficient amount of spotted DNAs. A solution for very unstable hybrids and for the high influence of dangling ends (Wetmur, 1991) are groups of longer probes with a common 6–8 bases in the middle ($N_{1-2}B_{6-8}N_{1-2}$) (Drmanac *et al.*, 1989). These probes give easily detectable and very accurate hybridizations with DNA dots created by pin arrays (Fig. 4.2). All probes can be divided into four groups requiring specific conditions (Table 4.3). The required amount of DNA is 10^8 to 10^9 molecules/mm^2. By various tests, the error rate is estimated at about 3% (Strezoska *et al.*, 1991; Drmanac *et al.*, 1993), and the algorithms for sequence reconstruction can tolerate over 10% (Drmanac *et al.*, 1991a). Further tuning of the hybridization conditions and understanding of some exceptional cases is necessary. The incubation temperatures and times can vary about 20% and probe concentrations by over 100%, allowing a routine automatic processing of 20–100 hybridizations per day.

The intensities of hybridization signals vary 100-fold and can be precisely and efficiently measured by phosphor storage technology (Johnston *et al.*, 1990) if the probes are labeled either with ^{32}P or ^{33}P. Labeling with ^{33}P gives three to four times better resolution in

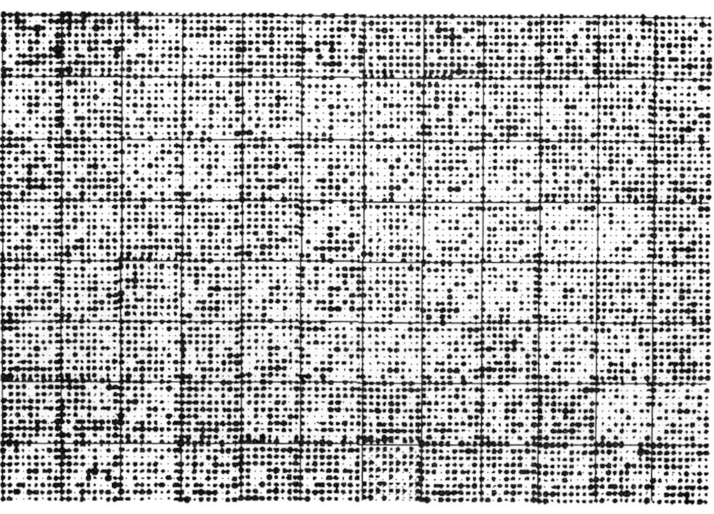

Figure 4.1 High-density dot array. PCR products of random cDNA clones from 81 96-well plates are spotted on a GeneScreen membrane (NEN, Boston, MA) by an automatic offsetting using an adapted Biomek1000. The clones are hybridized with a ^{33}P labeled 10-mer probe complementary to the amplified vector segment. The grid, consisting of points for the expected centers of dots and lines for every ninth row and column, is drawn automatically by the image-analysis program.

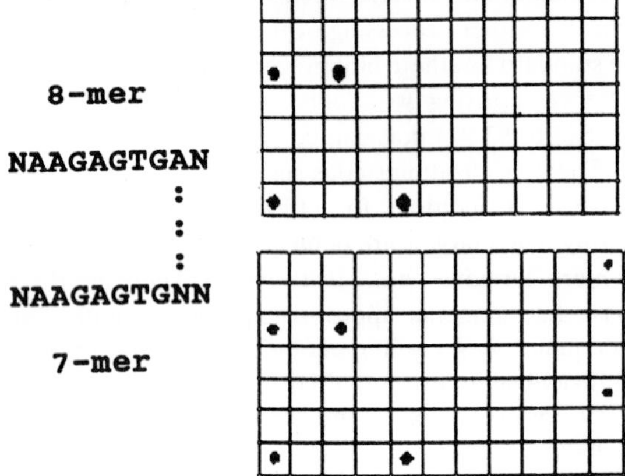

Figure 4.2 Hybridization specificity of short oligonucleotides. Patterns are shown for two probes differing only in one base hybridized with an array of 24 M13 clones (spotted in every second square). The top pattern is obtained by a group of 16 10-mers sharing the indicated 8-mer core. The bottom pattern is obtained by a group of 64 10-mers sharing the indicated 7-mer. The only difference from the previous probe group is incorporation of all four nucleotides on the position 3′A. As expected, in addition to four clones having fully matched sequences, two clones with 3′A mismatched sequences in respect to the 8-mer probe become positive with the 7-mer probe.

Table 4.3 Generalized hybridization conditions for N_2B_7N probes.

	Probe group			
	A	*B*	*C*	*D*
C+G in (B)7	0–1[a]	2–3[a]	4–5	6–7
Probe concentration (nM)	5	1	0.2	0.04
Washing (min °C^{-1})	20/2	60/8	60/15	60/20

[a] Some probes with a low duplex stability belong to group A even though they have two or three C+G, or to group B even though they have four C+G. These probes usually have alternating A and T or alternating A/T and C/G.

dense dot arrays, and it is more convenient to work with (R. Crkvenjakov, personal communication). Chemiluminescent labels are an attractive possibility, but many adaptations are necessary for the short oligonucleotides forming unstable duplexes.

There are no commercially available image-analysis programs for precisely scoring the intensities of densely arrayed dots. A DOS based program (DOTS) capable of automatically recognizing closely spaced dots on a skewed image and correcting various array imprecisions, has been developed recently by one of us (J.J.). Using filters with 144 96-well plates, a throughput of about 20 million clone-probe scores can be achieved per 8 scanning-hours on a Phosphor Imager (pixel size 176 μm; Molecular Dynamics, Sunnyvale, CA). Daily experiments can be organized differently: for example, scoring 20 probes on one million clones or 100 probes on 200 000 clones. These quantities of hybridization data are sufficient for the various large-scale mapping and sequencing applications discussed later. Presently missing components for routine use of this SBH setup are an adequate database and user-friendly software for data evaluation and processing.

4.2.2 Sequencing chips

The ability to use small quantities and a high density of samples are the most important factors for large-volume and low-cost sequencing. The array format allows miniaturization of the hybridization reactions to the physical limits of the positioning and detection of samples. A small support plate with a large number of closely spaced oligonucleotides or DNAs has been termed a 'sequencing chip' (Drmanac *et al.*, 1990). Many technical variants for producing oligonucleotide chips have been proposed: spotting separately synthesized oligonucleotides (Khrapko *et al.*, 1991), simultaneous *in situ* synthesis by a targeted physical masking at each cycle of synthesis (Southern *et al.*, 1992), a precise photolithographic removal of the photolabile blocking groups (Fodor *et al.*, 1991), and combinatorial synthesis on microbeads using a small number of columns followed by creation of fixed bead monolayers (Drmanac & Crkvenjakov, 1989; Drmanac *et al.*, 1991b). In the last approach, oligonucleotides are randomly positioned in arrays and have to be decoded by hybridization of a small set of unbound probes. The spacing of samples can be reduced from 500 μm, achievable in spotting by pin array, to 20 μm. Up to now there has been no routine synthesis of the chips. Oligonucleotide chips comprising 6-, 7-, or 8-mers can be used in SBH format 2 for kilobase sequencing, and very complex arrays (comprising up to 10^9 15-mers) for one-reaction sequencing of yeast artificial chromosomes (YACs), bacterial genomes, or even eukaryotic chromosomes. The small arrays are less suitable for complex genome sequencing due to the need to purify and label many millions of short clones.

The DNA arrays used in format 1 can also be miniaturized. Many schemes based on hybridization selection or PCR in combination with microbeads have been proposed for the preparation of chips without arraying clones into separate wells (Drmanac & Crkvenjakov, 1989, 1990). This procedure is less adequate if it is necessary to further use sequenced DNA fragments. As in the computer technology, generations of smaller and smaller chips can be expected. An approach for merging of sequencing

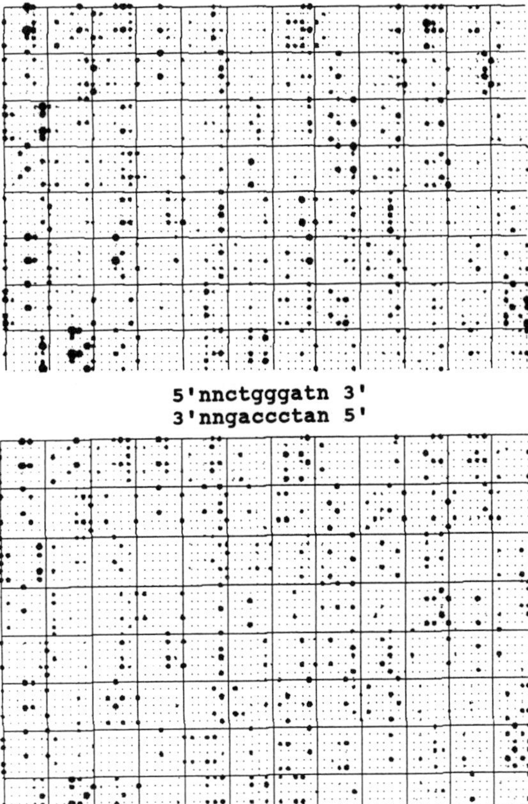

Figure 4.3 Hybridization accuracy of 7-mer probes with dot-blotted cDNAs. A filter, generated by spotting PCR products from random cDNA clones, is hybridized successively with the indicated pair of complementary probes. The filter contains pairs of dots in the strips of six rows (rows 1–4, 2–5, and 3–6) produced by spotting twice samples from 18 96-well plates.

chips and computer chips is already under development (Eggers *et al.*, 1992).

4.2.3 Catalogues and expression patterns of genes

The described capacity for collecting up to 20 million clone-probe results per day allows new measurements and inventories in molecular biology of genes and genomes. For example, it is important to know how many genes exist in a genome. Present estimates can be in error by a factor of 2 to 3. Furthermore, only about 3000 genes from 100 000 expected human genes have been cloned and studied in terms of their time and space expression patterns. The fingerprinting of random cDNA clones by 6- to 8-mer oligonucleotide probes is an efficient way to address these issues (Drmanac *et al.*, 1991c; Lennon & Lehrach, 1991).

A very small number of probes can suffice for identification of distinct genes and gene families. If 1-kb, double-stranded cDNA clones are hybridized with 7-mers two times more frequent in the transcribed than in the random sequences, 25% of the clones will be positive, on average. From a set of 40 such probes, 10 will be scored as positive per each clone. The theoretical number of distinct oligonucleotide sequence signatures (OSS) (Lennon & Lehrach, 1991) is about 10^9 (i.e. all possible combinations of 10 elements from a set of 40 elements, $40!/(10! \times 30!)$, which exceeds by 10 000 times the expected number of genes. To ensure the distinction of very similar genes (10 positively scored 7-mers per clone read only 7% of the sequence) and to allow detection of clones with low similarity, 100–200 probes have to be scored.

The intensities of hybridization signals, normalized for the variations in the amount of DNA spotted, will be used for estimating the similarity between pairs of clones. One approach is to calculate properly weighted Euclidian distances. The results have to be evaluated according to the distributions of the distances for pairs of dots representing the same DNA and for pairs of clones having no significant similarity.

We are performing a demonstration experiment of this approach on 100 000 random cDNA clones from the human brain library kindly provided by Dr Marcelo Soares, Columbia University. Hybridization images of a pair of complementary probes are shown in Fig. 4.3. Similar sizes of the pairs of dots representing the same clones spotted twice on the same membrane demonstrate a high reproducibility of the spotting and hybridization procedures. Very importantly, most of the strong dots with one probe are also positive with the other probe. These results demonstrate the specificity of hybridization, because false signals from one probe have a small chance to be obtained with the complementary probe, also. The reasoning is that sequences with strong mismatches (G/T or G/A) for one probe are the sequences with very weak (C/A or C/T) mismatches with the other. A majority of the visible discrepancies are due to weak, but detectable, signals with one of the probes or they are created by color cutoff. The calculated error rate of 3.4% per probe is similar to the error rate of 8-mers (Strezoska *et al.*, 1991; Drmanac *et al.*, 1992b). This experiment also shows that complementary (or overlapped) probes allow identification of false hybridizations and their elimination by data processing.

If the already started analysis of 100 000 clones is successful, the fingerprinting of many millions of clones combined from different tissues can be pursued. The method permits easy data integration from different laboratories when the same set of probes is used. By an exhaustive comparison of clone signatures, catalogues of genes and gene families from different tissues or species can be derived. Furthermore, if unnormalized cDNA libraries are analyzed, the expression level of every distinct gene in a tissue can be measured by counting the relative number of clones representing a gene in the library. These data are going to support and extend ongoing activity on partial sequencing of cDNA clones (Wilcox *et al.*, 1991; Adams *et al.*, 1991, 1992). A more rational selection of clones for sequencing and mapping should be possible.

4.2.4 Sequence painting

Very useful genetic data can be produced inexpensively if libraries of 1- to 2-kb long clones from chromosomal regions (YACs), chromosomes, or whole genomes are hybridized with a small set of oligonucleotides (Drmanac *et al.*, 1991c). Two types of map can be created: a 100-bp resolution physical map consisting of ordered, densely overlapped clones and a structural map (Drmanac & Crkvenjakov, 1992), e.g. compositional characteristics of the sequences and occurrence of repetitive elements, potential coding sequences, or other genome units in short consecutive segments.

The expectations are being tested by computer simulations on a set of known human genomic sequences (unpublished results). An artificial human chromosome 1 million bp long is created by connecting the longest known sequences. In this way, a reasonable representation of the human genomic sequence is provided. The distribution of 64 8-mers containing only A or T and 64 Alu-repeat-specific 7-mers is defined for this sequence with 500-bp resolution (Fig. 4.4). The resulting pictures are similar to the patterns of stained metaphase chromosomes. Genomic segments with a high frequency of W probes are like AT-rich G(iemsa) bands, and regions with many Alu repeats are like R(everse) bands. This 'sequence painting' procedure allows an informative and condensed representation of the genomic sequences. Dominant markers in the mammalian genomes are isochores of various G+C content (Bernardi, 1989) and SINE sequences. The abundant SINE repeats may be useful genome pointers and indicators of the specific types of genomic regions.

Besides AT-rich, GC-rich, and SINE-specific probes, the gene recognition probes are the most important ones. The existence of oligonucleotides predominant in the coding sequences has been documented (Claverie *et al.*, 1990; Uberbacher & Mural, 1991). Most exons can be identified with a low false positive rate by a small set of probes (E. Uberbacher, personal communication; our unpublished results). By a balanced selection, about 200 7-mers (80 coding, 32 Alu, 16 extreme A+T, 16 extreme G+C, and 56 for the other elements) can define both the order of 1.5-kb clones (using the strategy for cosmid mapping (Poustka *et al.*, 1986) and a comprehensive structural map. A library of 20 million clones representing 10 equivalents of the human genome could be analyzed in 2 years if 50% of the capacity of the described production line is steadily used.

In a real experiment, the windows from the simulations will be represented by short displacements between the neighboring clones. By comparing the results of specified sets of probes, the consecutive segments can be classified as repetitive elements, exons, HTF islands, promoters, AT-rich sequences, etc. Mammalian genes would appear as islands containing both a few segments hybridized with a significant number of probes abundant in coding sequences (potential exons) and a relatively small

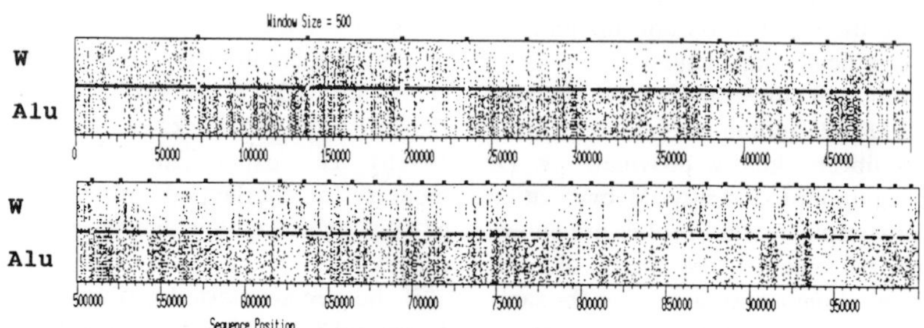

Figure 4.4 Sequence painting. Black squares and gaps in the line dividing W and Alu sections mark ends of continuous human sequences (GenBank entries HBB, GHCSA, HPRTB, FIXG, ADAG, TPA, BMYH7, ATP1A2, CRYGBC, AFP, etc.). The presence of 64 W 8-mers (A or T) and 64 7-mers specific for Alu repeats in a 500-base window is marked by a point (each probe is represented by one row of points or unmarked spaces). A high concentration of points in a window (forming a column) indicates a high similarity of that window with the consensus sequence represented by the particular probe set. Single Alu elements and various Alu clusters are apparent. Usually, the continuous genome segments belong to one type of sequence (AT-rich and Alu-poor (globin, HBB), AT-poor and Alu-rich (growth hormone, GHCSA), or AT-rich and Alu-rich (HPRT)).

number of SINEs. The knowledge provided can rationalize further genome analysis and direct cloning of the genes for genetically mapped traits. Short clones ready for sequencing will be available for any genomic region.

4.3 COMPLEMENTARITY OF SBH AND GEL METHODS: A STRATEGY FOR EFFICIENT LARGE-SCALE SEQUENCING

As pointed out in the previous two sections, data generated by 50–200 probes can be used to rationally target gel sequencing. The complementarity of the two methods can be further broadened by extending the list of probes to approximately 3000, mainly 7-mers (Drmanac, 1992). By the described SBH technology, 1 million clones (covering 10^9 bp) could be analyzed with such a number of probes within about two years in one laboratory (Drmanac *et al.*, 1992a).

The basic idea is to determine the complete and correct sequences by combining low-redundancy data from the two methods. With 3000 6-mers and 7-mers, every base is going to be read on average four times, but positively scored probes can not be correctly ordered. The missing information requires tens of thousands of additional probings (Drmanac *et al.*, 1989). On the other hand, the limiting factors for large-scale sequencing by gels are the short length of accurate reads and the need for redundant sequencing, especially in the shotgun approach. Fortunately, it is very simple to produce data with a 5–10% error rate by long gel reads of minimally overlapped clones selected on the basis of hybridization data. A solution for both problems is to combine the incorrect gel sequences and insufficient hybridization results, so that complete and accurate sequences can be assembled on the basis of parsimonious agreements with the two data sets. The combination of SBH and multiplex gel sequencing (Church & Kieffer-Higgins, 1988) is of special interest. The two methods share the filter hybridization and signal reading steps and have a matching data-production capacity.

The dominant feature of the described scheme is cyclic reinforcement. Every completed sequence allows the completion of similar sequences on the basis of hybridization data alone. For a large part of mammalian genomes (SINEs and gene families), hybridization data from 3000 probes backed by existing sequence knowledge can suffice. Also, it is possible to further reduce the gel sequencing effort by collecting hybridization data on three or four 70–90% similar genomes (Drmanac, 1992; Drmanac & Crkvenjakov, 1992). In this case, differences among the corresponding sequences mutually guide the correct assembly of constituent oligonucleotides in all genomes by parsimonious principles. Paradoxically, simultaneous sequencing of several genomes by this combinatorial scheme requires a significantly smaller effort than sequencing a single genome separately by any of the methods. Furthermore, the intrinsic errors specific to SBH as an oligonucleotide approach and to gel sequencing as a single-base approach can be eliminated.

In conclusion, SBH allows some immediately usable and many futuristic strategies for an inexpensive and (we hope) routine sequencing of complex genomes, within limits dictated primarily by the speed of readers and computers and not by biochemical reactions.

ACKNOWLEDGMENTS

We thank Nick Stavropoulos for work on the graphics of Figure 4.4, Radomir Crkvenjakov and Gregory Lennon for helpful suggestions, David E. Nadziejka for technical editing, and Diana Grygiel for technical assistance. This work was supported by the US Department of Energy, Office of Health and Environment Research, under contract No. W31–109–ENG–38.

REFERENCES

Adams, M.D., Kelley, J.M., Gocayne, J.D., Dubnick, M., Polymeropoulos, M.H., Xiao, H., Merril, C.R., Wu, A., Olde, O., Moreno, R.F., Kerlvage, A.R., McCombie, W.R. & Venter, J.C. (1991) *Science* **252**, 1651–1656.

Adams, M.D., Dubnick, M., Kerlavage, A.R., Moreno, R., Kelley, J.M., Utterback, T.R., Nagle, J.W., Fields, C. & Venter, J.C. (1992) *Nature* **355**, 632–634.

Bains, W. & Smith, G.C. (1988) *J. Theor. Biol.* **135**, 303–307.

Bernardi, G. (1989) *Annu. Rev. Genet.* **23**, 637–661.

Cantor, C.R., Mirzabekov, A. & Southern, E. (1992) *Genomics* **13**, 1378–1383.

Church, G.M. & Kieffer-Higgins, S. (1988) *Science* **240**, 185–188.

Claverie, J.-M., Sauvaget, I. & Bougueleret, L. (1990). *Methods Enzymol.* **183**, 237–252.

Drmanac, R. (1992) In *Genome Mapping and Sequencing*. R. Myers, D. Porteous & R. Roberts (eds) Cold Spring Harbor Laboratory, New York, p. 318.

Drmanac, R. & Crkvenjakov, R. (1987) *Yugoslav Patent Application 570.*

Drmanac, R. & Crkvenjakov, R. (1989) *Yugoslav Patent Application 767/89.*

Drmanac, R. & Crkvenjakov, R. (1990) *Sci. Yugoslav.* **16**, 97–107.

Drmanac, R. & Crkvenjakov, R. (1992) *Int. J. Genome Res.* **1**, 59–79.

Drmanac, R., Labat, I., Brukner, I. & Crkvenjakov, R. (1989) *Genomics* **4**, 114–128.

Drmanac, R., Strezoska, Z., Labat, I., Drmanac, S. & Crkvenjakov, R. (1990) *DNA Cell Biol.* **9**, 527–534.

Drmanac, R., Labat, I. & Crkvenjakov, R. (1991a) *J. Biomol. Struct. Dynam.* **5**, 1085–1102.

Drmanac, R., Labat, I., Strezoska, Z., Paunesku, T., Radosavljevic, D., Drmanac, S. & Crkvenjakov, R. (1991b) In *Electrophoreses, Supercomputers and the Human Genome.* C.R. Cantor & H.A. Lim (eds) World Scientific, Singapore, pp. 47–59.

Drmanac, R., Lennon, G., Drmanac, S., Labat, I., Crkvenjakov, R. & Lehrach, H. (1991c) In *Electrophoreses, Supercomputers and the Human Genome.* C.R. Cantor & H.A. Lim (eds) World Scientific, Singapore, pp. 60–74.

Drmanac, R., Drmanac, S., Labat, I., Crkvenjakov, R., Vicentic, A. & Gemmell, A. (1992) *Electrophoresis* **13**, 566–573.

Drmanac, R., Drmanac, S., Strezoska, Z., Paunesku, T., Labat, I., Zeremski, M., Snoddy, J., Funkhouser, W.R., Koop, B., Hood, L. & Crkvenjakov, R. *Science* **260**, 1649–1652.

Eggers, M., Beattie, K., Shumaker, J., Hogan, M., Hollis, M., Murphy, A., Rathman, D. & Ehrlich, D. (1992). In *Genome Mapping and Sequencing.* R. Myers, D. Porteous & R. Roberts (eds), Cold Spring Harbor Laboratory, Cold Spring Harbor, NY, p. 111.

Fodor, S.P.A., Read, J.L., Pirrung, M.C., Stryer, L., Lu, A.T. & Solas, D. (1991). *Science* **251**, 767–773.

Hunkapiller, T., Kaiser, R.J., Koop, B.F. & Hood, L. (1991) *Science* **254**, 59–67.

Johnston, R.F., Pickett, S.C. & Barker, D.L. (1990) *Electrophoresis* **11**, 335–360.

Khrapko, K.R., Lysov, Y.P., Khorlin, A.A. Ivanov, I.B., Yershov, G.M., Vasilenko, S.K., Florentiev, V.L. & Mirzabekov, A.D. (1991) *J. DNA Sequencing Mapping* **1**, 375–388.

Landegren, U. (1992) *GATA (Genetic Analysis Techniques and Applications)* **9**, 3–8.

Lennon, G.G. & Lehrach, H. (1991) *Trends Genet.* **7**, 314–317.

Lysov, Y.P., Florentiev, V.L., Khorlyn, A.A., Kraphko, K.R., Shick, V.V. & Mirzabekov, A.D. (1988) *Dokl. Acad. Nauk SSSR* **303**, 1508–1511.

Macevicz, S.C. (1989) *International Patent Application PCUS89 04741.*

Maxam, A.M. & Gilbert, W. (1977) *Proc. Natl. Acad. Sci. U.S.A.* **74**, 560–564.

Murray, K. (1970) *Biochem. J.* **118**, 831–841.

Pevzner, P.A. (1989) *J. Biomol. Struct. Dynam.* **7**, 63–73.

Pevzner, P.A., Lysov, Y.P., Khrapko, K.R., Belyavsky, A.V., Florentiev, V.L. & Mirzabekov, A.D. (1991) *J. Biomol. Struct. Dynam.* **9**, 399–410.

Poustka, A., Pohl, T., Barlow, D.P., Zehetner, A. Craig, Michiels, F., Ehrich, E., Frischauf, A.-M. & Lehrach, H. (1986) *Cold Spring Harbor Symp. Quantum Biol.* **51**, 131–139.

Sanger, F., Brownlee, G.G. & Barrell, B.G. (1965) *J. Mol. Biol.* **13**, 373–398.

Sanger, F., Nicklen, S. & Coulson, A.R. (1977) *Proc. Natl. Acad. Sci. U.S.A.* **74**, 5463–5467.

Southern, E.M. (1988) *UK Patent Application GB8810400.*

Southern, E.M., Maskos, U. & Elder, J.K. (1992) *Genomics* **13**, 1008–1017.

Strezoska, T., Paunesku, T., Radosavljevic, D., Labat, I., Drmanac, R. & Crkvenjakov, R. (1991) *Proc. Natl. Acad. Sci. U.S.A.* **8**, 10089–10093.

Uberbacher, E.C. & Mural, R.J. (1991) *Proc. Natl. Acad. Sci. U.S.A.* **88**, 11261–11265.

US Department of Energy (1992) *Human Genome 1991–1992 Program Report.* Office of Energy Research and Office of Health and Environmental Research, Washington, DC, June 1992.

US Department of Health and Human Services (1990) *Understanding Our Genetic Inheritance.* US Department of Energy Report, Washington, DC, April 1990.

Venter, J.C., Adams, M.D., Martin-Gallardo, A., McCombie, W.R. & Fields, C. (1992) *Trends Biotechnol.* **10**, 8–11.

Wetmur, J.G. (1991) *Crit. Rev. Biochem. Mol. Biol.* **26** (3/4), 227–259.

Wilcox, A.S., Khan, A.S., Hopkins, J.A. & Sikela, J.M. (1991) *Nucleic Acids Res.* **13**, 1837–1843.

CHAPTER FIVE

Shotgun Sequencing

A. MARTIN-GALLARDO,[1] J. LAMERDIN & A. CARRANO

Human Genome Center, Biology and Biotechnology Research Program, Lawrence Livermore National Laboratory, Livermore, CA, USA

5.1 INTRODUCTION

Automated sequencing technology has greatly facilitated the accomplishment of large-scale genomic sequencing. However, the amount of DNA sequence that can be read from one template in a single reaction does not usually exceed 500 bp. Thus, automated sequencing of cosmids or lambda phage clones requires the generation and subsequent subcloning of DNA fragments covering the genomic region. In order to achieve this step, several strategies have been used.

A classical, directional approach involves the construction of an accurate restriction map of the clone. Following insertion of the appropriate restriction fragments into a sequencing vector, the DNA sequence of the end(s) of the individual fragments is determined. Any remaining fragment sequence can be obtained either by primer walking, using sequence-specific oligonucleotides, or by an exonuclease III unidirectional deletion approach (McCombie *et al.*, 1991). Both the generation of restriction maps and the sequencing process require a large amount of time and manual manipulation of samples. Therefore, this strategy is not efficient with regard to either time or cost.

DNA sequencing by a novel primer walking approach, which uses strings of contiguous hexamers instead of custom-made primers, has recently been described (Kieleczawa *et al.*, 1992). Strings of three or more adjacent hexamers can specifically prime DNA sequencing reactions when the template DNA is saturated with a single-stranded binding protein. The efficiency of this method has been demonstrated in either single-stranded or denatured double-stranded templates 6.4 to 40 kb long. Since a standard hexamer preparation provides enough material to prime thousands of sequencing reactions, a library of the 4096 possible hexamers would allow rapid and economical sequencing by primer walking on templates up to at least cosmid size. While this strategy can notably increase the throughput of large-scale DNA sequencing, automation is still necessary for this method to become more cost-effective than current methods.

Recently, a transposon-facilitated sequencing method has been applied for high-throughput DNA sequencing without subcloning or primer walking (Strausbaugh *et al.*, 1990). This method introduces 'universal' mobile primer binding sites. A mapping strategy locates the sites of transposon insertion prior to sequencing.

[1] Present address: Centro Nacional de Biotecnologia, Madrid, Spain.

However, transposon insertion does not always occur randomly and, as a result, a significant number of gaps in the sequence are created (C. Martin, personal communication). Alternatively, some transposon-derived clones may be unstable and, therefore, this methodology will not be useful for large scale sequencing of some genomic contigs.

The shotgun approach, which consists of generating a random distribution of fragments by DNA shearing, allows rapid and inexpensive determination of practically every cosmid sequence. Because of its efficiency and the possibility of automating most of the steps, this strategy has been the method of choice in the majority of the large-scale sequencing projects performed to date.

5.2. DESCRIPTION OF THE METHOD

5.2.1 Overview

The shotgun sequencing method consists of several independent procedures which are described in detail below and summarized in Fig. 5.1. Briefly, the DNA is sonicated to obtain fragments of the desired size. The sonicated material is end-repaired using T4 DNA polymerase and Klenow enzyme and then, fractionated by agarose gel electrophoresis. The size-selected fragments (typically in the size range 1.5–3 kb) are purified and inserted by blunt-end ligation into a sequencing vector, such as M13. Following ligation, the DNA is used to transform an appropriate *Escherichia coli* host. Templates are sequenced using linear amplification procedures and analyzed on automated DNA sequencers. The raw data are edited to remove vector and tail sequences of low quality and are computer assembled using current software.

5.2.2 Random shearing

Sonication is the preferred method used to randomly shear DNA for shotgun cloning. A diagram of the sonication procedure is depicted in Fig. 5.2. Prior to sonication of DNA samples, the sonicator is calibrated. This step can be accomplished by monitoring the extent of fragmentation of salmon sperm DNA at timed intervals. The DNA (approximately 10 μg in a volume of 500 μl) is sonicated at 4°C using four pulses of 15 s each, with intervals of 1 min between pulses. Special care should be taken to prevent the sonicator probe from touching the walls or the bottom

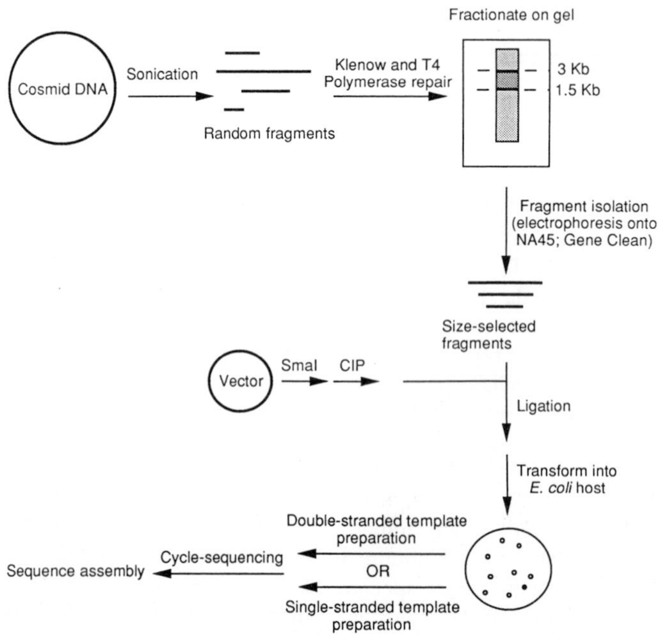

Figure 5.1 Flow diagram of the shotgun method used for automated DNA sequencing. Intact cosmid DNA (or self-ligated genomic insert) is sheared by sonication at 4°C to give an average fragment size of 2–4 kb. Sonicated DNA is end-repaired twice with a mixture of T4 DNA polymerase and Klenow enzyme and size-fractionated on an agarose gel. The DNA fragments in the desired size range (1.5–3 kb) are purified either by electrophoresis onto NA45 paper or by Gene Clean, and ligated into a sequencing vector (M13, Bluescript), which has been digested with SmaI and dephosphorylated with calf intestinal phosphatase (CIP). The ligation mixture is used to transform competent *Escherichia coli* cells. White transformants are grown, and either double-stranded or single-stranded DNA is isolated and sequenced using a cycle-sequencing protocol and ABI 373A DNA sequencers. The sequences are computer assembled using current software.

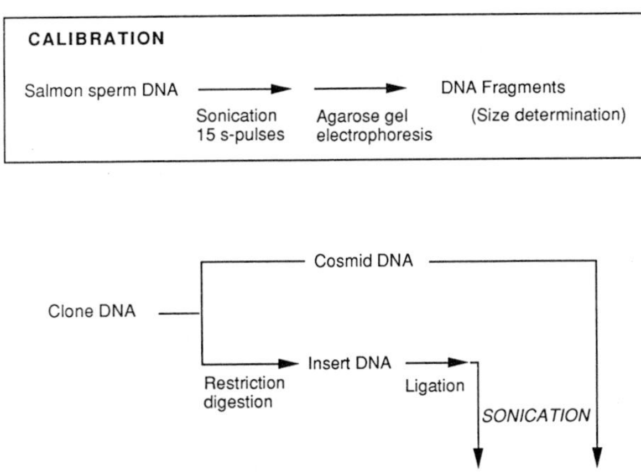

Figure 5.2 Schematic diagram of the sonication procedure. Prior to sonication of DNA samples, the sonicator is calibrated with salmon sperm DNA, by using four pulses of 15 s each, with intervals of 1 min between pulses, during which the DNA is kept at 4°C. Aliquots are removed after each pulse and analyzed by means of agarose gel electrophoresis to determine the size of the fragments. Intact cosmid DNA is sonicated without previous isolation of the genomic insert. Inserts of similar size or smaller than the vector are excised by restriction digestion and isolated before sonication. The purified insert DNA is self-ligated to reform concatemers or circular molecules. The ligation mixture is diluted and then sonicated to obtain fragments in the desired size range.

of the tube during sonication. Aliquots (10 µl) are removed after each pulse and loaded on a 0.8% agarose gel. Following electrophoresis, the extent of fragmentation is assessed by visualizing the migration of the bulk of the DNA under ultraviolet light. If the bulk of the DNA has migrated beyond the desired fragment size after the first pulse (15 s), the calibration should be repeated using shorter time pulses. The sonicator probe is cleaned in between DNA samples by boiling the probe in 5% sodium dodecyl sulfate (SDS), followed by several washes in water. Alternatively, some sonicators have removable probes that can be autoclaved.

For cosmids, in which the genomic insert is at least five times larger than the vector, isolation of insert DNA is not a necessary step prior to sonication. In addition, the cosmid vector sequences can provide an internal control for examining sequence accuracy. However, if the DNA insert is of similar or smaller size than the vector, such as bacteriophage lambda or yeast artificial chromosomes (YAC) rescued-plasmids, the insert DNA should be isolated prior to sonication (Martin-Gallardo *et al.*, in press). Although the rate of shearing does not depend on either the concentration or size of the DNA, linear DNA molecules are more rapidly sheared at the ends (Fig. 5.2). Therefore, linear DNA should be self-ligated to reform concatemers or circular molecules before sonication. Enough DNA to yield 2–5 µg of insert DNA is digested with the appropriate enzymes to excise the insert, which is then isolated by standard procedures (Sambrook *et al.*, 1989). Following precipitation, the DNA is resuspended in 10 µl of water. An equal volume of 2× ligation buffer and 1 unit of T4 DNA ligase are added to the sample, which is then incubated at 16°C overnight.

The DNA (ligation mixture or intact cosmid) is diluted with TE to a final volume of 350–500 µl, and sonicated by using short pulses with 1-min intervals on ice, as described in the calibration procedure. Since the kinetics of shearing are affected by temperature, the DNA should be kept at 4°C during sonication. Furthermore, this maximizes the probability of random shearing that, otherwise, appears to occur more frequently at AT-rich regions. An aliquot of the sonicated DNA is analyzed by agarose gel electrophoresis to verify the extent of fragmentation. After DNA repair, the size of the fragments will decrease due to the removal of single-stranded DNA tails by the exonuclease activity of the Klenow fragment and T4 DNA polymerase. Therefore, the bulk of the sonicated DNA should have a size approximately 0.5 kb larger than ultimately desired. In shotgun sequencing, the DNA is typically sonicated to give an average size of 2–4 kb.

5.2.3 DNA repair and size selection

Sonicated DNA is precipitated using sodium acetate (3 M; 1/10 vol.) and 2 vol. of ethanol, and resuspended in 20 µl of water. The DNA is then incubated at 37°C for 5–10 min to ensure that it is completely in solution.

Following the addition of 10 μl of KGB buffer (2× final concentration: 200 mM potassium glutamate, 20 mM magnesium acetate, 0.1 mg ml^{-1} BSA, 1 mM 2-mercaptoethanol, 50 mM Tris-acetate, pH 7.6) and 5 μl of 10 mM dNTPs, the DNA is end-repaired using T4 DNA polymerase and Klenow enzyme (5 units each) in a final volume of 50 μl. This step is performed twice to maximize the yield of blunt-end molecules. The DNA is extracted with phenol/chloroform and precipitated with ethanol. After resuspension in 20 μl of TE, the DNA is ready for electrophoretic separation to allow the appropriate size fragments to be selected. Usually, 1.5–3 kb fragments are isolated for subsequent cloning. We have observed that the number of clones containing double (co-ligated) inserts notably increases with fragments less than 1 kb in size. This confounds the interpretation of the results during sequence assembly since a co-ligated clone will reside in two contigs which are not adjacent in genomic DNA (Martin-Gallardo *et al.*, 1992; McCombie *et al.*, 1992). In addition, the size of the selected fragments must be larger than the sequence read length that is usually obtained from a template (400–500 bp).

The size-selected fragments can be purified using a variety of procedures. We currently use two methods: electrophoresis onto DEAE cellulose in membrane form (Schleicher and Schuell NA45), and binding to silica glass matrix (BIO 101 Gene Clean). When using NA45 membranes, the DNA is loaded on a 0.8% agarose gel made with nuclease-free, ultrapure agar and tris(borate) buffer (1× TBE). Following electrophoresis, the gel is visualized under long-wave ultraviolet light. A strip of NA45 is placed in an incision just ahead of the smallest desired fragments (1.5 kb). Another strip is placed just behind the upper-size selected fragments (3 kb) to trap larger molecules. Electrophoresis is then continued until all of the fragments between 1.5 and 3 kb have bound to the membrane. The forward strip is then removed and thoroughly washed in NET buffer (0.15 M NaCl, 0.1 mM EDTA, 20 mM Tris, pH 8) to eliminate residual agarose. The DNA is eluted by incubating the strip in 150–250 μl of high salt (2.5 M NaCl) NET buffer at 65°C for 30–45 min, and then precipitated with 2.5 vol. of ethanol. A second precipitation from 0.3 M sodium acetate is performed to remove any NaCl residue. The pelleted DNA is resuspended in water, and an aliquot is analyzed by agarose gel electrophoresis to determine the DNA concentration.

When the fragments are purified using Gene Clean silica, the sonicated, end-repaired DNA is loaded on a 0.8% agarose gel made from low-melting-point agar and Tris-acetate buffer (1× TAE). After electrophoresis, the bands in the desired size range are excised from the gel and melted by heating at 65°C for 10 min. The melted agarose is mixed with 3 vol. of 6 M NaI and 10 μl of Gene Clean are added. The DNA solution is incubated on ice for 10 min with vortexing every 2–3 min to allow DNA binding to the silica, which is then pelleted by 5 s centrifugation. After three washes in NEW buffer (NaCl, Tris, EDTA, ethanol, water; Gene Clean kit), the silica is resuspended in 60 μl of water and incubated at 55°C for 10 min to elute the DNA.

The yield of DNA recovered by using NA45 membranes is usually higher than that obtained with Gene Clean silica. However, Gene Clean preferentially binds larger DNA fragments. Thus, slow-migrating fragments of 500 bp or less that may be excised from the gel with the selected fragments would be unlikely to be recovered and cloned.

5.2.4 Cloning of shotgun fragments

Purified DNA fragments are ligated into M13 RF DNA, which has been digested with SmaI to yield blunt-ends and dephosphorylated. M13 derived plasmids (pUC and pGEM-Z) and phagemids (Bluescript) are also used as sequencing vectors. Ligation reactions are performed at a final DNA concentration of 20 ng μl^{-1}, using a ratio of 3:1 of insert to vector ends. Following the addition of 1 unit of T4 DNA ligase, the sample is incubated at 20°C for 2 h. The ligation mixture is used to transform competent *Escherichia coli* cells (DH5α F′, BRL; XL1 Blue, Stratagene) according to the supplier's instructions and the cells are plated on LB medium containing X-Gal and IPTG. When using phagemid or plasmid vectors, antibiotics are included in the plating media. White transformants are grown under the appropriate conditions to isolate either single-stranded or double-stranded DNA templates, as described (McCombie *et al.*, 1991; Fry *et al.*, 1992; Martin-Gallardo *et al.*, 1992).

5.3 APPLICATION TO LARGE-SCALE SEQUENCING

A significant number of cosmids have been sequenced by using a shotgun approach and either automated or multiplex DNA sequencing technology. Automated sequencing and the shotgun method described here were used to determine the sequence of three cosmids from human chromosome 19 (Martin-Gallardo *et al.*, 1992). Single-stranded DNA templates were sequenced using linear amplification procedures and analyzed on Applied Biosystems 373A DNA sequencers. The average sequence track length, after removal of vector and tailing sequence of low quality, was about 400 bp. The number of contigs assembled for each cosmid was more dependent upon the number of templates

Table 5.1 Summary of progress for the assembly of shotgun clones from a chromosome 19 cosmid.

No. of templates sequenced		*Sequence length (kb)*		*Contig size distribution*	
Unassembled	*Assembled*	*Unassembled*	*Assembled*	*No. of contigs*	*Length of assembled contigs (kb)*
345	267	129	29.8	35	<2.5
632	625	221	35.2	6	13.6, 10.4, 8.0, 1.7, 0.9, 0.6
850	850	331	36.7	4	14.1, 12.9, 8.0, 1.7

sequenced than on the cosmid sequence itself. The data obtained for one cosmid are shown in Table 5.1. Sequencing and assembly of 345 templates resulted in a large number of small contigs. Following further sequencing, the assembly of 632 fragments resulted in six contigs of sizes 13.6, 10.4, 8, 1.7, 0.9 and 0.6 kb. For 850 templates sequenced, four contigs of 14.1, 12.9, 8, and 1.7 kb were obtained. When additional templates (up to 1000) were sequenced, the overall redundancy was notably increased, but neither the number of gaps nor the contig lengths changed significantly (data not shown). Similar results were obtained for the other two chromosome 19 cosmids. Other groups have reported that a complete cosmid sequence can be obtained by shotgun methods exclusively (Koop *et al.*, 1991). Thus, this strategy remains an efficient approach to determining the sequence of large genomic regions.

REFERENCES

Fry, G., Lachenmeier, E., Mayrand, E., Giusti, B., Fisher, J., Johnston-Dow, L., Cathcart, R. & Finne, E. (1992) *BioTechniques* **13**, 124–131.

Kieleczawa, J., Dunn, J.J. & Studier, F.W. (1992) *Science* **258**, 1787–1791.

Koop, B.F. *et al.* (1991) In *Genome Sequencing and Analysis Conference III.*

Martin-Gallardo, A., McCombie, W.R., Gocayne, J.D., Fitzgerald, M.G., Wallace, S., Lee, B.M.B., Lamerdin, J., Trapp, S., Kelley, J.M., Liu, L-I., Dubnick, M., Johnston-Dow, L.A., Kerlavage, A.R., de Jong, P., Carrano, A., Fields, C. & Venter, J.C. (1992) *Nature Genet.* **1**, 34–39.

Martin-Gallardo, A., Marchuk, D.A., Gocayne, J., McCombie, W.R., Kerlavage, A.R., Venter, J.C., Collins, F.S. & Wallace, M.R. (1992) *DNA Sequence* **3**, 237–243.

McCombie, W.R., Kirkness, E., Fleming, J.T., Kerlavage, A.R., Iovannisci, D.M. & Martin-Gallardo, A. (1991) *Methods: Companion Methods Enzymol.* **3**, 33–40.

McCombie, W.R., Martin-Gallardo, A., Gocayne, J.D., Fitzgerald, M., Dubnick, M., Kelley, J.M., Castilla, L., Liu, L-I., Wallace, S., Trapp, S., Tagle, D., Whaley, W.L., Cheng, S., Gusella, J., Frischauf, A-M., Poutska, A., Lehrach, H., Collins, F.S., Kerlavage, A.R., Fields, C. & Venter, J.C. (1992) *Nature Genet.* **1**, 348–353.

Sambrook, J., Fritsch, E.F. & Maniatis, T. (1989) *Molecular Cloning: A Laboratory Manual*, 2nd edn. Cold Spring Harbor Laboratory Press, Cold Spring Harbor, NY.

Strausbaugh, L.D., Bourke, M.T., Sommer, M.T., Coon, M.E. & Berg, C.M. (1990) *Proc. Natl. Acad. Sci. U.S.A.* **87**, 6213–6217.

CHAPTER SIX

Shotgun Cloning as the Strategy of Choice to Generate Templates for High-throughput Dideoxynucleotide Sequencing

A. BODENTEICH, S. CHISSOE, Y.-F. WANG & B.A. ROE

Department of Chemistry and Biochemistry, University of Oklahoma, 620 Parrington Oval, Norman, OK 73019, USA

6.1 INTRODUCTION

Random shotgun cloning methods have been available for many years, but they have not become widely accepted because the rate-limiting steps in a DNA sequencing project have been sequence data collection and proofreading rather than generating subclones. With the previous, more systematic methods for template generation, fewer sequencing templates produced the data needed to yield the final, unique, unambiguous sequence when radiolabeled dideoxynucleotide sequencing methods were employed. However, with the advent of automated sequence data-collection instruments, data collection no longer is rate limiting and more efficient shotgun clone generation methods need reinvestigation.

Since the original reports describing the dideoxynucleotide method for DNA sequencing (Sanger *et al.*, 1977), many alternative methods for generating the necessary templates for radiolabeled DNA sequencing have been introduced (Sanger *et al.*, 1980; Anderson, 1981; Deininger, 1983; Bankier *et al.*, 1987; Bankier & Barrell, 1989; Schriefer *et al.*, 1990a; S. Surzyski, personal communication). These methods can be grouped into three categories: directed cloning (Sanger *et al.*, 1980; Bankier *et al.*, 1987), systematic cloning (Bankier & Barrell, 1989), and random cloning (Anderson, 1981; Deininger, 1983; Bankier *et al.*, 1987; Phadnis *et al.*, 1989; Schriefer *et al.*, 1990a; Strathmann *et al.*, 1991; Fitzgerald *et al.*, 1992; S. Surzyski, personal communication). In the directed cloning approach, fragments of the DNA of interest are digested with a series of restriction endonucleases and cloned directly into M13, pUC or other vectors with the appropriate unique cloning sites and selection methods. In the systematic cloning approach, the chimeric plasmid is linearized and a portion of the insert is digested with exonuclease III (Exo III). Subsequently, the extraneous single-stranded regions are removed by single-strand-specific nuclease digestion, and the shortened insert region is religated to yield a construct with a previously unsequenced region much closer to the 'universal' priming site. In the random cloning approach, insert fragments are generated by random cleavage via sonication (Deininger, 1983), French press (Schriefer *et al.*, 1990a), nebulizer (S. Surzyski, personal communication) or partial restriction endonuclease digestion (Bankier *et al.*, 1987; Fitzgerald *et al.*, 1992), and then shotgun cloned into the appropriate DNA sequencing vectors. One alternative to the physical shearing or enzymatic approaches is to randomly insert transposons (Phadnis *et al.*, 1989; Strathmann *et al.*, 1991) into the chimeric vector and sequence with transposon-specific primers. Through one or a combination of these approaches, a series of single-stranded or double-stranded templates is obtained and sequenced with oligonucleotide primers corresponding either to vector sequences adjacent to the insertion site or to previously determined insert sequences with custom synthetic oligonucleotide primers. A new and novel approach to 'primer walking' using strings of contiguous hexamers recently has been reported (Kieleczawa *et al.*, 1992). Alternatively, the polymerase chain reaction (PCR) (Mullis & Faloona, 1987), coupled with automated DNA synthesis methods (Caruthers, 1985) and reproducible PCR product isolation procedures (Lander & Waterman, 1988; Edwards & Caskey, 1991), provides a relatively simple, noncloning alternative to generate the necessary sequencing templates.

With the introduction of automated, fluorescence-based DNA sequencing instruments (Smith *et al.*, 1986; Ansorge *et al.*, 1987; Brumbaugh *et al.*, 1988) for sequence data collection and improved computer programs for proofreading and contig assembly (Dear & Staden, 1991), the rate-limiting step in DNA sequencing now has moved from data-collection and proofreading stages of a DNA sequencing project to the clone generation and final contig closure phases. Thus, with the ability to generate in excess of 20 000 bases of sequence data in a 24 h period on a single automated fluorescent DNA sequencing instrument with Taq polymerase-based cycle sequencing (Craxton, 1991; Chissoe *et al.*, 1991), to proofread this data rapidly, and to assemble it into a database using the TED and XBAP programs (Dear & Staden, 1991), we have investigated the overall strategy for generating the templates needed to rapidly and accurately complete both large and small DNA sequencing projects. In this chapter we present the results of these studies and the basis for our conclusion that, at present, an optimized random shotgun subcloning method is an efficient strategy for generating the large number of templates needed to complete high-throughput automated DNA sequencing.

6.2 METHODS

6.2.1 Physical shearing of intact cosmid and/or plasmid vectors via either sonication or nebulization

With the introduction of automated methods for DNA sequence data collection and proofreading, directed DNA subcloning methods became extremely labor intensive and time consuming. We recently investigated the various methods for physical DNA shearing mentioned above, and now have implemented two approaches, sonication (Deininger, 1983) and nebulization (S. Surzyski, personal communication), which yield a random fragment distribution. A central theme of this approach is to physically shear the intact chimeric vector without prior restriction digestion to eliminate the vector sequences. This approach yields two immediate benefits. First, by eliminating the restriction digestion and gel purification steps, the shotgun protocol is shortened and variations in the yield of DNA are eliminated because the highly purified chimeric cosmid or plasmid can be used directly for physical shearing. Second, the presence of known vector sequences in the shotgun library eventually produced can serve as an internal check for accuracy of the final sequence. For example, in the case of a 50 kb chimeric cosmid, approximately 15% of the sequences generated are from the vector. This advantage of a built-in checking mechanism far outweighs the disadvantage of collecting the additional vector sequence data.

6.2.1.1 Production of random DNA fragments by sonication

The preparation of random DNA fragments by sonication was performed in a Heat Systems Ultrasonics W-375 cup horn sonicator (Farmingdale, NY) filled with ice-cold water in a cold room (Deininger, 1983). It is imperative that the temperature of the sample and the water in the cup horn sonicator be maintained as close to 0°C as possible throughout the following manipulations to prevent DNA melting which can result in an uneven fragment distribution pattern. In preparation for sonication, a tube holder was prepared by removing the upper two-thirds of a 50 ml polypropylene centrifuge tube with a conical bottom and cutting a hole in the bottom to fit the 1.5 ml snap-cap conical tube containing the DNA sample. A solution containing approximately 100 μg of plasmid or cosmid DNA in 350 ml of Tris-magnesium buffer (50 mM Tris-HCl, pH 8 containing 15 mM $MgCl_2$) was prepared and distributed equally to 10 1.5 ml conical snap-cap tubes stored in an iced-water bath. After the sonication horn was filled with ice-cold water, a snap-cap tube containing 35 μl of sample was positioned in the tube holder, centered approximately 1 mm above the sonication horn, and sonicated for 5 s with the sonicator set on 'HOLD', 'CONTINUOUS' and maximum 'OUTPUT CONTROL' (= 10). The tube then was removed from the sonication horn to an iced-water bath, the ice-cold water was replaced, and another sample was positioned in the cup horn. This second sample was sonicated for 10 s, the third for 20 s, the fourth for 30 s and the fifth for 40 s with each sample being placed in an iced-water bath after sonication. The ice-cold water in the cup horn was replaced between each sample. After a brief centrifugation to collect any condensate produced during sonication, 10 µl aliquots of each sonication time point were electrophoresed on a 1% agarose gel to determine which conditions produced fragments in the desirable size range of 1–2 kb. Alternatively, fragments in the desirable size range also could be obtained by a series of individual 10 s bursts, with replacement of the ice-cold water in the cup horn between each burst. Once the sonication conditions were optimized for a given DNA sample, the remaining five microcentrifuge tubes containing 35-µl aliquots were sonicated under the optimum conditions just determined. Subsequently, each sample was centrifuged to collect any condensate produced during sonication and stored in an iced-water bath before proceeding with the end-repair step described below.

6.2.1.2 *Production of random DNA fragments by nebulization*

The method of nebulization to produce random DNA fragments was introduced to us by Dr S. Surzyski (personal communication) at Indiana University, to whom all credit for this procedure should be given. The following protocol was developed by Dr Surzyski and modified slightly to yield optimal fragment distribution for our samples. An IPI Medical Products, Inc., nebulizer (part number 4207) was purchased from a local supplier, whose name was obtained by contacting the manufacturer (IPI Medical Products Inc., 3217 North Kilpatrick, Chicago, IL 60641, USA; tel. (312) 777–0900). As suggested by Dr Surzyski, we modified the nebulizer by removing the plastic cylinder, cutting off the outside rim of the cylinder, inverting it and placing it back into the nebulizer, as shown in Fig. 6.1. The hole in the top of the nebulizer was sealed with an IsoLab, Inc., cap (number QS-T from IsoLab, Inc., Drawer 4350, Akron, OH 44303, USA). A length of Nalgene tubing (VI grade 3/16 in. internal diameter 3/16 in. wall) was used instead of the thinner walled tubing which was supplied with the nebulizer to connect the nebulizer to a pressurized nitrogen tank. A solution containing approximately 40–50 µg of plasmid or cosmid DNA in 2.0 ml of Tris-magnesium-glycerol buffer (50 mM tris-HCl, pH 8.0 containing 15 mM $MgCl_2$ and 10 to 25% glycerol) was prepared, loaded into the nebulizer cup, and nebulized at 30 p.s.i. for 150 s in an iced-water bath. These conditions give DNA pieces between 600 to greater than 1500 bp from a cosmid of 40–50 kb original size. During the nebulization, unavoidable leaks in the nebulizer can be minimized by securely tightening the nebulizer top to the sample chamber. Once the nebulization was complete, the nitrogen source was disconnected and the entire unit was placed in the rotor bucket of a table top centrifuge (Beckman GPR tabletop centrifuge) fitted with pieces of styrofoam to cushion the plastic nebulizer and briefly centrifuged at 2500 r.p.m. to collect the sample. The nebulized sample then was distributed into four 1.5 ml microcentrifuge tubes. The DNA was ethanol precipitated, collected by centrifugation, rinsed, dried, and resuspended in 35 µl of Tris-magnesium buffer (50 mM Tris-HCl, pH 8, containing 15 mM $MgCl_2$) prior to proceeding with the end-repair step described below.

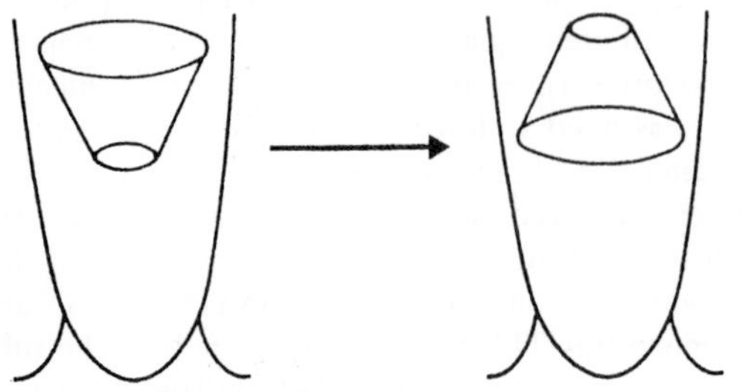

Figure 6.1 Modification of the nebulizer. The outside rim of the cylinder is cut off, inverted, and replaced in the nebulizer.

6.2.2 End repair, gel purification, phosphorylation and ligation

Because both the sonicated and nebulized fragments usually contain single-stranded ends, the samples must be made blunt-ended prior to ligation into blunt-ended vector (Deininger, 1983; Bankier *et al.*, 1987). Thus, to each tube containing 35 µl of sonicated or nebulized fragments was added 2 µl of a dNTP mixture (0.25 mM of each of the four dNTPs dissolved in TE), 3 µl of T4 DNA polymerase (NEB No. 203, 3 U μl^{-1}), and 2 µl of Klenow (NEB No. 210, 5 U μl^{-1}). After incubating for 30 min at room temperature, each sample was applied to separate wells of a 0.7% low gel temperature agarose (Schwarz/Mann Biotech) and electrophoresed for 30–60 min at 100–150 mA. Subsequently, the DNA fragments were visualized by ethidium bromide staining, excised from the gel and placed at −70°C for 15 min. The gel piece was then incubated at 65°C for 5 min, mixed with an equal volume of TE-saturated phenol, and then frozen at −70°C for 15 min. After thawing and a brief centrifugation, the aqueous phase was ether extracted and the DNA was ethanol precipitated, washed with 80% ethanol, and dried. Each sample was then dissolved in 36 µl ddH_2O and 4 µl of 10× denaturing buffer (200 mM Tris-HCl, pH 9.5, containing 1 mM EDTA and 10 mM spermidine) was added. Following an incubation at 70°C for 10 min, each sample was placed in an iced-water bath.

For the kinase treatment step, it was imperative that the solution contain sufficient rATP for the kinase reaction. Thus, both the 10 mM rATP stock solution and the 10× kinase buffer supplied by USB (500 mM Tris-HCl, pH 7.5, 100 mM $MgCl_2$, 100 mM DTT), were aliquoted separately and stored at −20°C. For the kinase reactions, 5 µl of 10× kinase buffer, 1 µl 10 mM rATP, 3 µl H_2O, and 1 µl T4 polynucleotide kinase (USB, 30 U μl^{-1}) were added to each 40 µl denatured DNA sample and then incubated at 37°C for 10–30 min. Subsequently the reactions were pooled, phenol extracted, and the DNA was recovered by ethanol precipitation, dried and dissolved in 40 µl of TE buffer (10 mM Tris-HCl, pH 7.6, 0.1 mM EDTA) yielding a typical concentration of 500–1000 ng μl^{-1}.

For the ligation step, approximately 20 ng of SmaI cut, calf intestinal alkaline phosphatase-treated M13 vector was combined with 100–1000 ng of sonicated or nebulized DNA fragments treated as described above, 1 µl of 10× T4 ligase buffer supplied by NEB (500 mM Tris-HCl, pH 7.5, 100 mM $MgCl_2$, 100 mM DTT, 10 mM ATP, and 250 µg ml^{-1} BSA) and 1 µl of T4 DNA ligase (NEB, 400 U μl^{-1}) in a final volume of 10 µl. The blunt-ended ligation mixture was then incubated at 4°C overnight in the refrigerator in a beaker of water or in an aluminium block or at 15°C for 8 h or at room temperature for 3–6 h. For gel-eluted fragments, ligation of sonicated or nebulized DNA fragment concentration was titrated over a 10–100-fold range with a fixed 20 ng of vector to obtain an optimal insert to vector ratio. In some instances, the ligation reaction efficiency could be improved by including 5% PEG (Cobianchi & Wilson, 1987).

In addition, it was useful to include several control experiments to further test the efficiency of the blunt ending process, the ligation reaction, and the quality of the vector (Bankier *et al.*, 1987). For example, we usually included parallel ligations in the absence of insert to determine the background plaques resulting from self-ligation or inefficiently phosphatased vector. It also was helpful to include a parallel ligation with a known blunt-ended insert or insert library, such as an *Alu*I digest of a standard plasmid or cosmid, to insure that the blunt-ended ligation reaction would yield sufficient insert containing plaques independent of the end-repair process. If difficulties were encountered with the end-repair process, we sometimes digested a standard plasmid or cosmid with either a 5′ or 3′ overhang producing restriction endonuclease, such as HindIII or PstI, and to divide the samples in half. Then, one-half was subjected to the repair process followed by blunt-ended ligation, while the other half was ligated directly into HindIII- or PstI-digested vector.

6.2.3 Maintenance of bacterial strains and preparation of competent bacterial cells

Bacterial strains, JM101 or XL1-Blue MRF′ (Stratagene, La Jolla, CA), were maintained as glycerol stocks stored at −70°C. These stocks were prepared by transferring 3–4 well-isolated colonies from a minimal media plate to 3 ml of 2× TY media (for XL1-Blue, 20 µg ml^{-1} tetracycline was added to this and all subsequent media) and growing at 37°C with shaking overnight. Then, 1 ml of the fresh overnight culture was transferred into 50 ml of minimal media in a 250 ml flask and grown at 37°C with shaking for 6 h. Subsequently, 12.5 ml of sterile glycerol was added to the 50 ml culture, mixed, aliquoted in 1.3 ml portions into sterile 12 ml Falcon tubes, and immediately frozen at −70°C. These bacterial stocks were viable for several years and the above growth conditions assured maintenance of the F′ episome.

For preparation of competent cells, a frozen glycerol stock was removed from −70°C, thawed at 37°C for a few minutes, directly added to 50 ml of prewarmed 2× TY media and preincubated in a 37°C water bath for 1 h without shaking. These cells then were incubated at 37°C for 2–3 h with shaking. After transferring 40 ml of this culture into a sterile 50 ml centrifuge tube, the cells were collected by centrifugation at 6000 r.p.m. for 8 min at 4°C in an RC5-B centrifuge equipped with an SS-34 rotor. The remaining 10 ml of culture was placed on ice for later use. The cell pellet was resuspended in one-half volume (20 ml) of cold 50 mM calcium chloride, incubated on ice for 20 min, and centrifuged as before. The resulting pellet was resuspended in one-tenth volume (4 ml) of cold 50 mM calcium chloride to yield the final competent cell suspension.

For transformation, the entire ligated DNA sample was added to 0.2–0.3 ml of the competent cell suspension, gently mixed, and incubated on ice for 40–60 min. This mixture was then heat shocked at 42°C for 2–5 min and 0.2 ml of noncompetent cells (from above), 25 µl of IPTG (25 mg ml^{-1} H_2O) and 25 µl of X-Gal (20 mg ml^{-1} DMF) were added. Then, 2.5 ml of lambda top agar was added, mixed by inversion, and quickly poured onto a prewarmed lambda plate. After allowing 10–20 min for the top agar to solidify, the plates were inverted and incubated overnight at 37°C. Any remaining competent cells could be stored for 1–2 days on ice at 4°C without any noticeable loss of competency.

6.2.4 Preparation of M13 vector DNA for ligation

M13 mp18 SmaI cut and dephosphorylated vector was prepared in 10 or 20 individual 1.5 ml microcentrifuge tubes with each tube containing 5 µg of M13 RF, 2 µl of 10× New England Biolabs (NEB) buffer No. 4 (500 mM potassium acetate, 200 mM Tris-acetate, 100 mM magnesium acetate, 10 mM DTT, pH 7.9 at 25°C), 3 μl of SmaI (NEB, 24 U μl^{-1}), 3 μl of calf intestine alkaline phosphatase (Boehringer Mannheim, catalog No. 1097 075, 24 U μl^{-1}), in a final reaction volume of 20 μl. After incubating at 37°C for 2–4 h, the reactions were pooled, phenol extracted, ethanol precipitated, and washed as described above. The dried SmaI cut and dephosphorylated vector was dissolved in TE to a final concentration of approximately 10 ng μl^{-1} and stored aliquoted at −70°C. Each preparation was tested for the presence of intact vector by direct transformation into competent cells. If this transformation yielded a greater than usual number of blue plaques, the linearized vector could be separated from intact vector by purification on a low gelling temperature agarose gel, as described above. Each preparation also was tested for the presence of incompletely phosphatased vector by incubation with T4 DNA ligase prior to transformation and, if necessary, was subjected to an additional CIP treatment.

6.2.5 Automated modified single-stranded DNA isolation of 48 clones

We have previously reported (Mardis & Roe, 1989; Chissoe *et al.*, 1991) the Biomek 1000 based automated isolation of 24 single-stranded M13 templates, an automated version of the procedure described originally by Eperon (1986), and now we wish to report our recent modifications which allow the simultaneous isolation of 48 single-stranded templates per Biomek 1000 robotic workstation within 3 h. In this protocol, the initial separation of phage particles from cells and cell debris in a microcentrifuge yielded lower levels of contaminating genomic DNA, and pelleting the PEG precipitated single-stranded phage DNA in flat-bottomed microtiter plate wells facilitated ssDNA resuspension. In addition, because the ethanol precipitation step also was more effective when performed in microcentrifuge tubes, the robotic workstation was programmed to perform the last transfer from the microtiter plate to microcentrifuge tubes prior to this precipitation.

Briefly, phage-infected *Escherichia coli* cultures were grown in 12 mm × 75 mm test tubes for 4–6 h, manually transferred into 1.5 ml microcentrifuge tubes, centrifuged at 13 750 r.p.m. for 15 min in a bench-top microcentrifuge, and then placed on the Biomek tablet. Subsequently, 250 μl of each culture were robotically distributed into 48 separate wells of a 96-well flat-bottom microtiter plate. This process, which was repeated two times, distributed the phage-rich supernatant from 48 individual cultures to separate wells of the microtiter plate. Then, 50 μl of 20% PEG containing 2.5 M NaCl was robotically added to each well and mixed. The microtiter plate was then covered with a sealer and incubated for 15 min at room temperature followed by a 15 min centrifugation at 2400 r.p.m. in a table-top centrifuge. The supernatant was removed by inverting the plate and gently draining on a paper towel, without dislodging the pellet. After placing the microtiter plate back onto the Biomek, 200 μl of the PEG/TE rinse solution (1:3(v/v)) was robotically added to each well. The plate then was covered with another sealer and centrifuged as before. After removing the supernatant by inverting the plate as described above and placing the microtiter plate back on the Biomek, 70 μl of the TTE solution (10 mM Tris-HCl, pH 7.6, 0.1 mM EDTA, pH 8.0, 0.5% Triton X-100) was robotically added to each well. The plate then was agitated gently and the sample from each well pair was robotically transferred to 1.5 ml microcentrifuge tubes, which were capped and placed in an 80°C water bath for 10 min. After a brief centrifugation to collect condensation, the single-stranded template DNA was precipitated by adding 250 μl of ethanol–acetate and mixed by inversion. The tubes were placed at −20°C for at least an hour, and the precipitated templates were collected by centrifugation, washed with 80% ethanol, dried, and resuspended in 20 μl of TE. The yield of single-stranded template was approximately 2–3 μg per sample.

6.2.6 *Thermus aquaticus* (Taq) DNA polymerase catalysed DNA cycle sequencing

Each base-specific fluorescent-labeled cycle sequencing reaction routinely included approximately 100 or 200 ng Biomek isolated single-stranded DNA for A and C or G and T reactions, respectively. However, the sequencing incubation temperatures and times differed slightly from those reported by others (Ansorge *et al.*, 1987; Wilson *et al.*, 1990; the protocol included with the ABI Cycle Sequencing Kit) and all reagents except template DNA were added in one pipetting step from a premix of previously aliquoted stock solutions stored at −20°C.

For the first step in preparing the reaction premix, 30 or 60 μl portions of 5× Taq reaction buffer (400 mM Tris-HCl, pH 8.9, 100 mM $(NH_4)_2SO_4$, 25 mM $MgCl_2$) were mixed with 30 or 60 μl of the respective 5× nucleotide extension/termination mix (A – 62.5 μM dATP, 250 μM dCTP, 375 μM c^7dGTP, 250 μM dTTP, 1.5 mM ddATP; C – 250 μM dATP, 62.5 μM dCTP, 375 μM c^7dGTP, 250 μM dTTP, 0.75 mM ddCTP; G – 250 μM dATP, 250 μM dCTP, 94 μM c^7dGTP, 250 μM dTTP, 0.125 mM ddGTP; T – 250 μM dATP, 250 μM dCTP, 375 μM c^7dGTP, 62.5 μM dTTP, 1.25 mM ddTTP), to yield the A, C or G, T nucleotide/buffer premixes. The deoxynucleotide and dideoxynucleotide triphosphates were purchased from Pharmacia LKB (Piscataway, NJ), while 7-deaza-2′-deoxyguanosine triphosphate was from Boehringer-Mannheim Biochemicals (Indianapolis, IN). In an attempt to minimize nucleotide triphosphate degradation due to repeated freezing and thawing, the initial 100 mM nucleotide triphosphate solutions were diluted with TE (5 mM Tris-HCl, pH 7.6, 0.1 mM EDTA, pH 8.0), aliquoted, and stored at −70°C. In addition, intermediate dilutions of 20 mM and 5 mM also were stored at −70°C. Owing to variability among commercially available nucleotide triphosphates, the exact concentrations given for all nucleotide mixes may require slight modification.

Subsequently, 30 or 60 μl of 1U μl^{-1} Perkin-Elmer Cetus AmpliTaq DNA polymerase in 1× Taq reaction buffer, 30 or 60 μl of individual fluorescent end-labeled primers (0.4 pmol μl^{-1} of Joe, Fam, Tamra, or Rox-labeled 5′-TGTAAAACGACGGCCAGT-3′ from ABI, Inc., Foster City, CA), and 60 or 120 μl of 5× nucleotide extension/termination/5× Taq reaction buffer premixes were combined to yield the 120 or 240 μl of the final A, C or G, T reaction mixes, respectively, that were sufficient for 24 template samples (protocol included with the ABI Cycle Sequencing Kit; Chissoe *et al.*, 1991).

Once the above mixes were prepared, 1 or 2 μl of single- or double-stranded DNA were pipetted into the bottom of each 0.2 ml thin-walled reaction tube (Robbins Scientific, Sunnyvale, CA), and then 4 or 8 μl of the respective reaction mixes were added to the side of each tube. Strip caps (Robbins Scientific, Sunnyvale, CA) were sealed onto the tube/retainer set and the plate was centrifuged briefly. The plate then was placed in a Perkin-Elmer/Cetus Cycler 9600 cycler the heat block of which was preheated to 95°C, and the cycling program immediately started. The cycling protocol consisted of 30 cycles of seven temperatures (30 cycles of 95°C denaturation for 4 s; 55°C annealing for 10 s; 72°C extension for 1 min; 95°C denaturation for 4 s; 72°C extension for 1 min; 95°C denaturation for 4 s; and 72°C extension for 1 min and was linked to a 4°C final soak file.

At this stage, the reactions usually were frozen and stored at −20°C for several days. Prior to pooling and precipitation, the plate was centrifuged briefly to reclaim condensation. The primer-specific reactions were pooled into 250 μl of 95% ethanol and the DNA was precipitated on ice for 10 min. The precipitated DNA was collected by centrifugation for 15 min in a microcentrifuge at 4°C. After decanting the ethanol, the pellet was rinsed with 500 μl of 80% ethanol, the tube centrifuged for 5 min, the ethanol decanted, and the DNA dried in a Speedy-Vac for 3–5 min. These pooled and dried sequencing reactions could be stored for several days at −20°C.

6.2.7 Data collection and analysis

Prior to gel loading and electrophoresis, the pooled and dried reaction products were resuspended in 3.5 μl of 1:5 50 mM EDTA/formamide (v/v) by vigorous vortexing and then heated for 2–5 min at 90°C. The BioRad Gel Loading Robot loaded the odd-numbered cycle sequencing reactions into the odd-numbered wells of a pre-electrophoresed 8 M urea/5% polyacrylamide gel which was fitted with a 36-well shark's tooth comb. The gel was electrophoresed for 5 min before the even-numbered samples were loaded robotically into the even-numbered wells. Typically, electrophoresis and data collection were for 10 h at 20 W on the ABI 373A which was fitted with a heat-distributing aluminium plate ($\frac{1}{4}$ in. thick) in contact with the outer glass gel plate in the region between the laser stop and the sample loading wells (Chissoe *et al.*, 1991). The sequence data were transferred to a SPARCStation 2 where the hard disk was mounted on the ether netted Macintosh IIcx or IIci via NFS/Share (InterCon Systems, Inc., Herndon, VA) and proofread using the TED program (Dear & Staden, 1991). Subsequently, the data were merged into contigs and edited using XBAP (Dear & Staden, 1991).

6.3 RESULTS AND DISCUSSION

Accurate and efficient DNA sequence data-collection instruments require efficient, rapid, and reproducible methods to generate the clones that will serve as the sequencing templates. Although random shotgun cloning methods have been available for many years, they have not been widely accepted, in part because manual gel reading, manual computer entry and manual proofreading are extremely time consuming and error prone. Since automated data collection now is both efficient and productive, we recently investigated two shotgun clone generation methods: sonication (Deininger, 1983) and nebulization (S. Surzyski, personal communication). Our approach was to isolate intact chimeric plasmid or cosmid using standard techniques, to generate a random M13-based library of the entire chimeric vector via either sonication or nebulization, and then to sequence a number of these single-stranded templates via Taq-based fluorescent-primer-labeled cycle sequencing after semi-automated template isolation. The sequence data were collected on an ABI 373A and transferred to a SPARCStation 2 where they were analyzed,

Table 6.1 Shotgun cosmid and plasmid sequencing statistics

Clone No.	*Clones sequenced*	*Bases sequenced*	*Redundancy coverage*	*Unique bases*	*Gaps in insert*[a]
c1	991	338 619	6.83	49 604	3
c2	935	289 788	6.09	47 623	2
c3	838	267 740	5.82	46 009	2
c4	878	270 087	6.08	44 398	2
p1	487	119 044	7.11	16 748	1
p2	185	45 622	4.90	9313	0
p3	112	34 203	4.60	7417	0
p4	99	26 793	4.29	6247	0
p5	50	17 775	3.69	4811	0
p6	41	13 765	3.15	4371	0

[a] Gaps remaining in the insert were closed by Taq terminator primer walking.

assembled into contigs and proofread with the TED and XBAP programs (Dear & Staden, 1991). Contig closure was obtained by Taq DNA polymerase catalyzed cycle sequencing with custom synthetic oligonucleotide primers and fluorescent labeled terminators.

This strategy now has been implemented to rapidly and successfully obtain the sequences of several chimeric cosmids and plasmids representing portions of the human abl and bcr genes (see Table 6.1). The number of random clones sequenced and the number of gaps remaining in the insert sequence prior to switching to closure by primer walking is indicated. It should be noted that the number of shotgun clones needed to be sequenced is directly proportional to the insert size in the original chimeric cosmid or plasmid, and inversely proportional to the length of the sequence read obtained from each of the shotgun M13 sequencing templates. It also should be noted that extreme care must be taken to remove traces of fragmented *Escherichia coli* genomic DNA which may co-purify with the cosmid or plasmid and subsequently be subcloned into sequencing vector. Although the data shown in Table 6.1 do not include any sequences of *E. coli* genomic DNA, which are found to some degree in shotgun subclone libraries, we have observed that in our best cosmid or plasmid preparations approximately 1–2% of the sequences represent *E. coli* genomic DNA contamination, and in some instances, this contamination can be as large as 10–15%. If high levels of nonchimeric cosmid or plasmid DNA are found during the initial sequence collection stages, then it is advisable to isolate another preparation of chimeric vector and to generate another shotgun library for sequencing.

A detailed examination of Table 6.1 also reveals that the entire sequence of the insert contained in plasmid clones p2 through p6 was obtained with only a few 36-lane runs on the ABI 373A with each base being sequenced at least three times to reduce sequencing artifact-induced errors. In addition, it also is interesting to observe that the redundancy obtained for each of the chimeric plasmid or cosmid vectors was proportional to their total length. By maintaining a minimum of four-fold sequence redundancy, a high degree of accuracy in the final sequence is assured. Although this level of redundancy requires the manipulation, editing, analysis, and proofreading of a large number of shotgun subclones per project, these clones easily are generated by either of the physical shearing methods and can be robotically isolated, after fragment end repair, ligation, and transformation into M13 vectors.

In our shotgun sequencing strategy, which includes automated data collection of both the insert and vector sequences, the vector sequences serve as internal controls for sequence data accuracy. Furthermore, the labor-intensive insert excision, purification, and other manipulations prior to generating a random library are not needed. These advantages far outweigh the disadvantages of the additional manipulations required to collect the already known vector sequence. The data shown in Table 6.1 also include vector sequences.

It also is clear that sequence assembly is more efficient when longer DNA sequences can be read from each template. It is our experience that even though fragments in the 600–1500 bp range are excised from agarose gels prior to shotgun subcloning, a sonication generated library usually contains smaller inserts than a nebulization-generated library. We suspect that this results from coelectrophoresis of smaller and larger fragments on nondenaturing agarose gels. However, nebulization produces a much narrower DNA fragment size range, which more readily can be controlled by altering the nebulizing conditions, i.e. a lower gas pressure produces larger fragments. Thus, shotgun libraries produced from nebulized DNA have insert sizes which are both more uniform and larger than those produced from sonicated DNA. Because the nebulizer-generated fragment library contains larger inserts, sequences to the limit of the ABI 373A software are routinely obtained. However, the data shown in Table 6.1 include sequences collected from both sonication- and nebulization-generated subcloned inserts. Thus, the average read length for each project is less than the 450–500 nucleotide ABI 373A software limitation. In addition, the clones resulting from nebulized chimeric cosmids of plasmids which contain large inserts can serve as templates for contig gap closure employing custom synthesized primers without the further manipulations as discussed below.

In the shotgun sequence approach, a stage is reached where continued shotgun sequencing becomes an inefficient strategy for final sequence completion (Edwards & Caskey, 1991). If the remaining gaps result from subcloning randomness, additional shotgun sequencing eventually will complete the entire sequence. However, this time might be better invested in the partial completion of another shotgun sequencing cosmid or plasmid project. Furthermore, if the uncloned regions result from the presence of lethal sequences or from regions high in secondary structure which consistently shear into small pieces, it could be that sequencing additional shotgun clones would not close the gaps. Thus, for each project, the efficiency of continued shotgun clone sequencing decreases dramatically once an approximately three-fold redundancy is obtained. In the strategy outlined above, a three-fold redundancy usually results in only one or two gaps in a plasmid sequencing project, while roughly 10–20 gaps remain in the larger cosmid sequencing projects. Here, additional shotgun sequencing to obtain a six- to seven-fold redundancy easily can be completed in a cosmid sequencing project, and the remaining few gaps then can be closed by an alternative approach (Edwards & Caskey, 1991). The decision to switch from sequencing shotgun clones to a more directed approach should be based on the number of gaps needing closure, the estimated length of these gaps, the overall sequence redundancy, and the density of repeated sequences, all of which affect the accuracy of the final assembled sequence. The strategies one chooses to employ for gap closure also depend on these factors and typically include custom primer synthesis followed by sequencing the appropriate shotgun subclone templates, sequencing the reverse strand of an M13 subclone after PCR amplification (Wilson *et al.*, 1990) and/or directly sequencing cosmid, plasmid, or French press (Schriefer *et al.*, 1990b) generated templates with a custom synthesized primer. Finally, as proposed earlier (Chissoe *et al.*, 1991), it may be necessary to verify the final contig assembly and/or determine the length of tandem repeated sequences by a PCR approach (Edwards & Caskey, 1991).

ACKNOWLEDGMENTS

We sincerely thank Dr Steve Surzyski at Indiana University for introducing us to his nebulizer approach to shotgun clone generation. This work was supported by grants from the NIH Human Genome Project, the Unversity of Oklahoma Graduate School, and the Oklahoma Department of Commerce Center of Excellence in Molecular Medicine Program.

REFERENCES

Anderson, S. (1981) *Nucleic Acids Res.* **9**, 3015–3027.

Ansorge, W., Sproat, B., Stegemann, J., Schwager, C. & Zenke, M. (1987) *Nucleic Acids Res.* **15**, 4593–4602.

Bankier, A.T. & Barrell, B.G. (1989) In *Nucleic Acids Sequencing: A Practical Approach*, C.J. Howe & E.S. Ward (eds). IRL Press, Oxford, pp. 37–78.

Bankier, A.T., Weston, K.M. & Barrell, B.G. (1987) *Methods Enzymol.* **155**, 51–93.

Brumbaugh, J.A., Middendorf, L.R., Grone, D.L. & Ruth, J.L. (1988). *Proc. Natl. Acad. Sci. U.S.A.* **85**, 5610–5614.

Caruthers, M.H. (1985) *Science* **230**, 281–285.

Chissoe, S.L., Wang, Y.F., Clifton, S.W., Ma, N., Sun, H.J., Lobsinger, J.S., Kenton, S.M., White, J.D. & Roe, B.A. (1991) *Methods: Companion Methods Enzymol.* **3**, 55–65.

Cobianchi, F. & Wilson, S.H. (1987) *Methods Enzymol.* **152**, 94–110.

Craxton, M. (1991). *Methods: Companion Methods Enzymol.* **3**, 20–26.

Dear, S. & Staden, R. (1991). *Nucleic Acids Res.* **19**, 3907–3911.

Deininger, P.L. (1983) *Anal. Biochem.* **129**, 216–223.

Edwards, A. & Caskey, C.T. (1991) *Methods: Companion Methods Enzymol.* **3**, 41–47.

Eperon, I.C. (1986) *Anal. Biochem.* **156**, 406–412.

Fitzgerald, M.C., Skowron, P., Van Etten, J.L., Smith, L.M. & Mead, D.A. (1992). *Nucleic Acids Res.* **20**, 3753–3762.

Kieleczawa, J., Dunn, J.J. & Studier, F.W. (1992). *Science* **258**, 1787–1791.

Lander, E.S. & Waterman, M.S. (1988) *Genomics* **2**, 231–239.

Mardis, E.R. & Roe, B.A. (1989) *BioTechniques* **7**, 736–746.

Mullis, K.B. & Faloona, F.A. (1987) *Methods Enzymol.* **155**, 335–350.

Phadnis, S.H., Huang, H.V. & Berg, D.E. (1989) *Proc. Natl. Acad.Sci. U.S.A.* **86**, 5908–5912.

Sanger, F., Nicklen, S. & Coulson, A.R. (1977) *Proc. Natl. Acad. Sci. U.S.A.* **74**, 5463–5467.

Sanger, F., Coulson, A.R., Barrell, B.G., Smith, A.J.H. & Roe, B.A. (1980) *J. Mol. Biol.* **143**, 161–178.

Schriefer, L.A., Gebauer, B.K., Qiu, L.Q.Q., Waterson,

R.H. & Wilson, R.K. (1990a) *Nucleic Acids Res.* **18**, 7455.

Schriefer, L.A., Gebauer, B.K., Qiu, L.Q.Q., Waterson, R.H. & Wilson, R.K. (1990) *Nucleic Acids Res.* **18**, 7455.

Smith, L.M., Sanders, J.Z., Kaiser, R.J., Hughes, P., Dodd, C., Connell, C.R., Heiner, C., Kent, S.B.H. & Hood, L.E. (1986) *Nature* **321**, 674–679.

Strathmann, M., Hamilton, B.A., Mayeda, C.A., Simon, M.I., Meyerowitz, E.M. & Palazzolo, M.J. (1991) *Proc. Natl. Acad. Sci. U.S.A.* **88**, 1247–1250.

Wilson, R.K., Chen, C. & Hood, L. (1990) *BioTechniques* **8**, 184–189.

CHAPTER SEVEN

Transposon-facilitated Large-scale DNA Sequencing

C.M. BERG,[1] G. WANG,[2] K. ISONO,[1] H. KASAI[2] & D.E. BERG[3]

[1] Department of Molecular and Cell Biology (U-131), The University of Connecticut, Storrs, CT 06269-2131, USA

[2] Postgraduate School for Science and Technology, and Department of Biology, Faculty of Science, Kobe University, Rokkodai, Kobe 657, Japan

[3] Departments of Molecular Microbiology and Genetics, Washington University Medical School, St Louis, MO 63130, USA

7.1 INTRODUCTION

Transposons are specialized DNA segments that can move to many sites in DNA molecules. Several bacterial transposons and their engineered derivatives are being used to provide mobile primer binding sites for sequencing DNA cloned in *Escherichia coli.* Transposon insertions permit large clones to be sequenced directly, without needing to subclone and then sort through the subclones (some of which contain vector sequences) and align contigs, or to walk with preformed primers (Berg *et al.*, 1989, 1993; Kasai *et al.*, 1992; Krishnan *et al.*, 1993). Both transposon hopping and transposon-based nested deletion strategies have been developed for accessing cloned DNAs. Nested deletions are especially appealing for analyses of large cosmid clones because deletion endpoints are inherently easier to map (by plasmid size) than insertion sites, and because endpoints in any desired region can be selected directly (for gap filling).

Although transposons can provide valuable access to all regions of cloned fragments, the automation of transposon-based strategies is in its infancy. For transposons to be widely adopted in both large- and small-scale sequencing projects, improved strategies must be developed that are efficient on libraries with low, as well as high, levels of redundancy (Kasai *et al.*, 1992; and see below), and do not depend on shotgun subcloning, such as used by others (R. Weiss and R. Gesteland, personal communication; see also Chapter 8 in this volume).

We are developing a number of transposon-based sequencing strategies for plasmids, m phage and cosmids. The properties of two very useful transposons, Tn*5* and de (also called Tn*1000*), are reviewed here because their contrasting properties make them suitable for further development as sequencing tools in a variety of m phage and plasmid/cosmid vectors. We illustrate this by describing ways in which these elements are being used in our laboratories in large-scale DNA sequencing. Additional information on these and other transposons is reviewed elsewhere (Berg *et al.*, 1989, 1993; Berg, 1989; Krishnan *et al.*, 1993).

Several considerations guided our development of transposon-based sequencing strategies.

(1) Insertions should be easy to isolate and map.
(2) Insertion sites should be random, or nearly so, in DNAs of different base compositions.
(3) Large target DNA fragments should be accessed without subcloning.
(4) Strategies should be developed for accessing

DNAs cloned in a variety of vectors, including plasmids, cosmids, and phage λ, and eventually P1, and F-based bacterial artificial chromosomes.

(5) Strategies should be flexible to accommodate both large- and small-scale sequencing operations, and to allow the initial sequencing steps to be either random or directed, but with the final (gap closure) steps always directed, for maximum efficiency.

7.2 TRANSPOSON-BASED STRATEGIES FOR SEQUENCING LARGE CLONED FRAGMENTS

Useful transposon insertions in target DNA result either from hopping from one replicon to another, to generate a simple insertion, or within the same molecule, to generate a deletion (or an inversion, which is not useful). We find that derivatives of γδ are well suited for hopping into plasmids and for generating nested deletions, while derivatives of Tn*5* are well suited for hopping into λ phage. Hopping strategies will be addressed first because they have been used most often and can be applied to pre-existing clones. Then the properties of a special cloning vector for isolating transposon-facilitated nested deletions will be described because it is likely to provide the best transposon-based system for sequencing large (e.g. cosmid) clones.

7.2.1 Hopping strategies for plasmid clones

We have shown that wild-type γδ is useful for introducing primer binding sites to plasmids in small-scale sequencing projects (Liu *et al.*, 1987; Strausbaugh *et al.*, 1990), and have constructed mini-derivatives of γδ for more efficient sequencing and for introducing marked genes into the chromosome (Berg *et al.*, 1992, 1993). γδ is the transposon of choice for plasmids because it transposes more randomly than other transposons and can be easily selected using a simple bacterial mating. This selection is based on the formation of a cointegrate as the initial product of transposition in the donor cell and resolution of the cointegrate after it has transferred to the recipient cell (Fig. 7.1). The final product is a simple insertion of γδ, bracketed by a 5 bp direct repeat of target DNA. This is the method of choice for sequencing relatively small cloned DNAs (<5 kb). It has not been used much for sequencing large DNAs, however, because mapping insertion sites in large fragments is time consuming and not readily automatable. A more user-friendly γδ-based deletion system for analysis of large DNAs is described below.

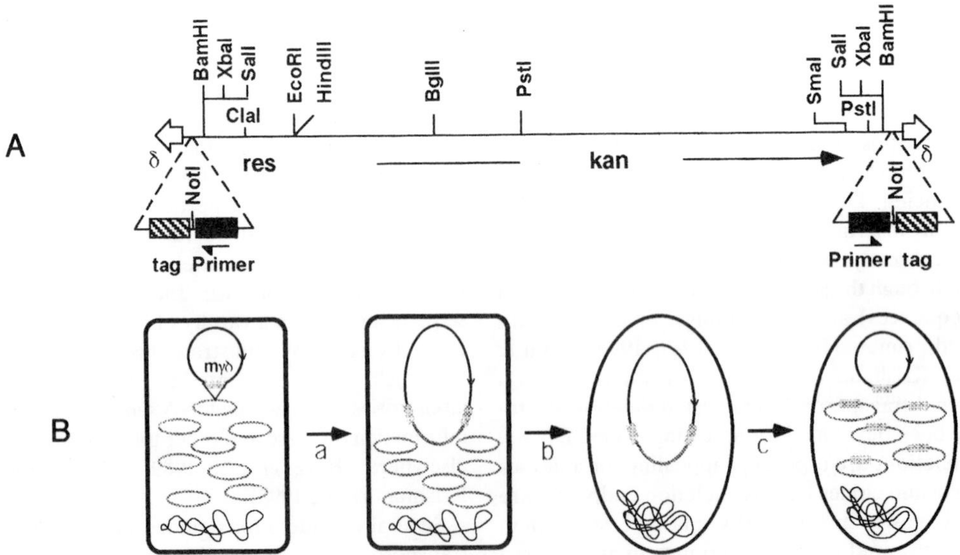

Figure 7.1 Transposition of γδ and mini-γδ. (A) The structure of mγδ-1 (1.8 kb) (Berg *et al.*, 1992) and mγδPLEX (1.9 kb) (X. Xu, G. Wang & C. M. Berg, 1993b) derivatives. Each of the four mγδPLEX derivatives has the same pair of unique primer binding sites and a different pair of tag (Church & Kieffer-Higgins, 1988) sequences, with a *Not*I site between them. (B) Selection for insertion into a nonconjugative plasmid by mobilization and conjugal transfer (from Berg *et al.*, 1993). (a) Transposition from the donor F factor (for wild-type γδ) or pOX38::mini-γδ (for mini-γδ) to a nonconjugative plasmid, such as pBR322, forming an F:: plasmid or pOX38::plasmid cointegrate. (b) Conjugation with a plasmid-free F^- recipient strain, selecting for a plasmid-borne marker (usually Amp^r). (c) Resolution of the cointegrate in the recipient to yield a plasmid containing one copy of γδ or mini-γδ, and reforming the donor molecule (F or pOX38::mini-γδ), followed by growth of the transconjugants.

7.2.2 Hopping strategies for λ clones

A strategy for efficient sequencing of overlapping λ phage clones was developed that uses insertions of Tn*5*-derived mini-transposons, called Tn*5supF*, and polymerase chain reaction (PCR) amplification of adjacent DNA segments (Phadnis *et al.*, 1989; Krishnan *et al.*, 1991, 1993; Kasai *et al.*, 1992). Tn*5* was engineered for sequencing of λ phage because it inserts quite randomly (although somewhat less so than γδ) and generates simple insertions, not cointegrates. The minimal components of Tn*5* needed for insertion into λ need have only a pair of Tn*5* inverted repeats (19 bp) and a selectable marker. Because the lengths of many λ phage clones are close to the maximum capacity of the phage heads, a small transposon is essential to obtain hops. In addition, because most λ cloning vectors kill their hosts, without lysogenizing them, the transposon-borne selectable marker must be one that can be selected during lytic growth, rather than one conferring antibiotic resistance on host cells. A selection for the suppressor tRNA gene, *supF*, has been developed (Kurnit & Seed, 1990). Tn*5supF* elements are about 300 bp long and contain only *supF*, primer binding sites and the 19 bp terminal repeats needed for transposition (Phadnis *et al.*, 1989; Kasai *et al.*, 1992). The best of these elements contains the efficient M13 forward and reverse primer binding sites and is carried in a donor plasmid with a highly active Tn*5* transposase gene (Fig. 7.2(A)) (Kasai *et al.*, 1992). Tn*5supF* is easily delivered in a single cycle of λ phage infection of donor cells, insertion does not involve a cointegrate intermediate and Tn*5supF*-containing phage are easily selected on $dnaB_{amber}$ cells (Fig. 7.2(B)). The transposition product is a simple insertion of Tn*5*, bracketed by a 9 bp direct repeat of target DNA.

7.2.3 Deletion strategy

Transposon-facilitated nested deletions show great promise for large-scale sequencing because many steps are automatable, gaps can be located and filled with ease, and repetitive DNAs can be resolved. Tn*9* has been used previously (Ahmed, 1984, 1987), but γδ is our choice for generating nested deletions since γδ transposes quite randomly and γδ-promoted deletions (that extend in either direction into cloned DNA) can be isolated, thereby accessing both target strands in a single plasmid (Wang *et al.*, 1993a,b) (Fig. 7.3). Because γδ transposes replicatively, one entire copy of γδ ends up in each of the reciprocal deletion derivatives. We ensured that both deletion derivatives would be viable by placing a plasmid replication origin

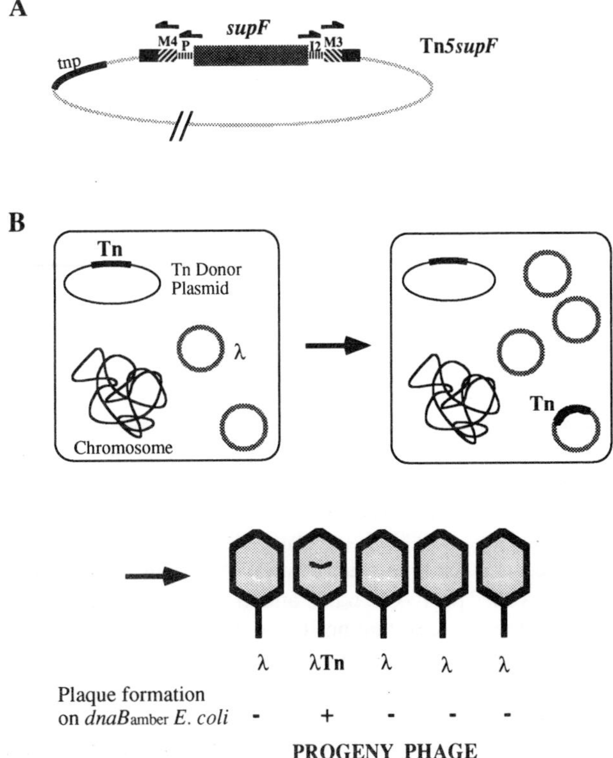

Figure 7.2 Transposition of Tn*5supF* to phage λ. (A) The Tn*5supF* donor plasmid. *supF*, suppressor tRNA gene; *tnp*, *Tn5* transposase gene; P and 12, binding sites for PCR primers used for analytical and preparative DNA amplification; M3 and M4, binding sites for PCR primers used for automated sequencing of PCR products (see Kasai *et al.*, 1992). (B) Selection for Tn*5supF* insertion into λ by plaque formation on $dnaB_{amber}$ *E. coli* strain DK21 (Kurnit & Seed, 1990; see Phadnis *et al.*, 1989; Krishnan *et al.*, 1991).

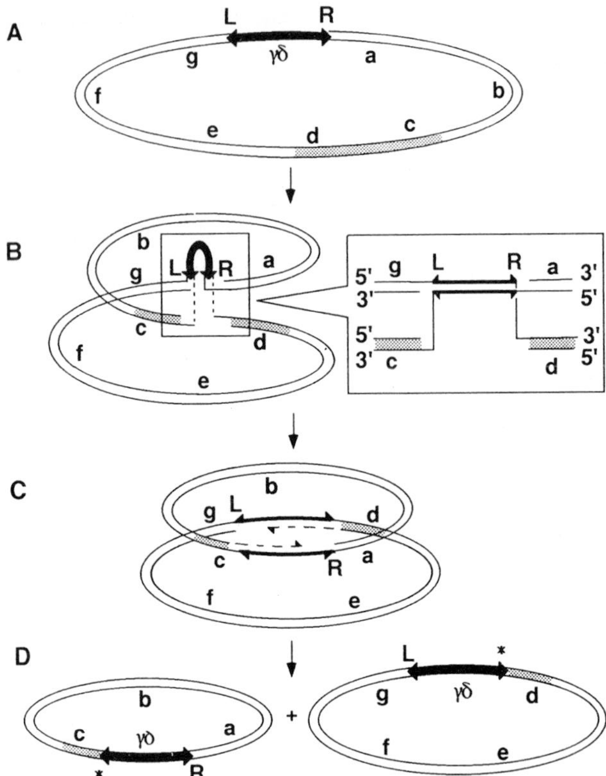

Figure 7.3 Model for transposon γδ-catalyzed deletion formation (Shapiro, 1979; Sherratt, 1989). Intramolecular transposition to a site between target positions c and d yields an inversion plasmid (not shown) or a pair of complementary deletion plasmids, each containing a copy of the γδ element. (A) Original replicon. (B) The initial stage of transposition in which transposase nicks the transposon at each 3′ end and makes staggered cuts in the target to give 5 base 5′ overhangs. Ligation of the free transposon ends with the protruding ends leaves a 5 base gap in each target strand. (C) The transposon is duplicated and the gaps filled, yielding a 5 bp duplication of the target site (one copy in each daughter molecule). (D) Each of the two complementary plasmid deletion products contains a copy of the γδ element. Normally, one daughter will contain the plasmid replication origin and the other daughter will not be recoverable. However, in the deletion factory constructs described here, both daughters can be recovered, albeit not in the same cell. It should be noted that intramolecular transposition does not involve a resolution step. Symbols: black double-headed arrow, γδ transposon; dashed lines with small half arrows, replicating γδ DNA; L and R, transposon ends that did *not* move to new sites; *, transposon ends juxtaposed to target DNA.

between the ends of γδ, as described below. Tn5 transposes by a different mechanism and cannot be easily engineered to allow recovery of both types of deletions from a single starting plasmid (see Tomcsanyi *et al.*, 1990).

The basic strategy for using transposons to isolate nested deletions efficiently involves placing a contra-selectable (conditional lethal) gene between the transposon and the cloning site (Ahmed, 1984; Tomcsanyi *et al.*, 1990; Wang *et al.*, 1994). In our γδ-based cosmid cloning vector, pDUAL (Fig. 7.4) formerly called pJANUS, *sac*B$^+$ and *strA*$^+$ were placed on each side of γδ, so that nested deletions extending into the cloned fragment in either direction could be selected easily using simple media containing sucrose or streptomycin (Wang *et al.*, 1993a,b). The selectable *kan* and *tet* genes (encoding kanamycin and tetracycline resistance, respectively) were placed between *sacB* and *strA* and the cloning site to eliminate deletions that extend beyond the ends of the cloned fragment. This selection *against* one contraselectable marker and *for* one selectable marker yields colonies containing plasmids with nested deletions that extend in the selected direction for varying distances into, but not beyond, the cloned DNA. After transposition, one transposon end always abuts a deletion endpoint and can serve as a 'universal' primer binding site for sequencing adjacent cloned DNA. Deletion endpoints can be mapped directly by plasmid size. In addition, endpoints in any region can be selected by size fractionation of pools of deletion plasmid DNAs. All regions can be sequenced using a pair of 'universal' primer binding sites engineered into γδ. Outward-facing SP6 and T7 promoters just within the γδ

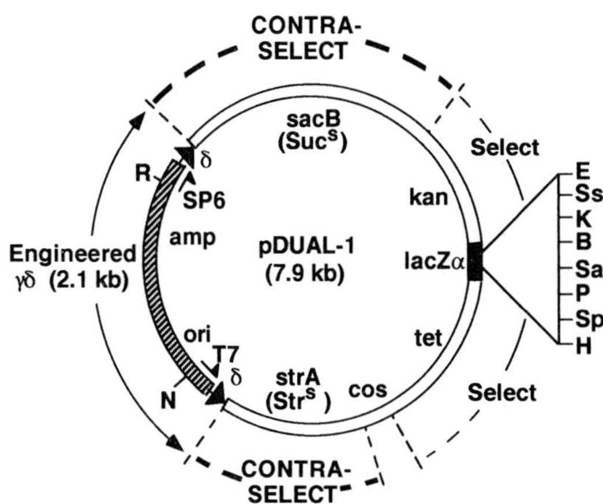

Figure 7.4 pDUAL deletion factory cosmid cloning vector (7.9 kb) (Wang *et al.*, 1993b). The cloning site shows only the restriction sites that are unique (there is a SmaI site in kan, and a second XbaI site in *lacZ*α). Sequences of the primer binding sites (bold) and γδ ends (underlined). The SP6 primer sequence (bold) and γδ end (underlined) is:

ATTTAGGTGACACTATAGAGATCCTTAACGTACGT TTTCGTTCCATTGGCCCTCAAACCCC-3′

The T7 primer sequence (bold) and gd end (underlined) is:

TAATACGACTCACTATAGGGAACTCGATCCTTAAC GTACGTTTTCGTTCCATTGGCCCTCAAACCCC-3′

The RsrII site abuts the SP6 primer sequence and the NotI site abuts the T7 primer sequence. B, BamHI; E, EcoRI; H, HindIII; K, KpnI; N, NotI; P, PstI; R, RsrII; Sa, SalI; Sp, SphI; Ss, SstI.

ends serve as sites for universal primer binding and riboprobe synthesis after γδ-mediated deletion formation.

7.2.4 Randomness of insertion

In our experience, γδ moves more randomly than other bacterial transposons. Although the 5-bp target site duplications it generates are usually A+T-rich, γδ is also useful for targets of heterogeneous or high G+C-rich base composition (Strausbaugh *et al.*, 1990; Berg *et al.*, 1992), apparently because it inserts preferentially into 5-bp long A+T-rich 'valleys' that are found throughout any DNA (Zhang, J., Strausbaugh, L.D. & Berg, C.M., in prep.). Hotspots and coldspots for γδ insertion are uncommon.

Tn*5* can also insert into dozens of sites per kilobase, although occasional sites of preferential insertion (hotspots) are found (Berg, 1989), as well as some cold regions (see below).

7.3 OUTLINE OF PROTOCOLS FOR LARGE-SCALE TRANSPOSON-BASED SEQUENCING

Transposons have been used in few large-scale sequencing projects to date. We have used Tn*5supF* to sequence λ phage clones DNA and we are beginning to use pDUAL to sequence plasmid and cosmid clones, as described below.

7.3.1 Analysis of Tn*5supF* insertions in overlapping λ clones

Tn*5supF* insertions into λ are obtained by growing λ on a Tn*5supF* donor strain and selecting for Tn*5supF*-containing phage by plaque formation on a $dnaB_{\mathrm{amber}}$ *E. coli* strain (Phadnis *et al.*, 1989) (this selection works because λ requires the host DnaB protein for its replication). The best Tn*5supF* element contains binding sites for universal M13 forward and reverse sequencing primers and is carried in a donor plasmid with a hyperactive allele of the Tn*5* transposase gene (Kasai *et al.*, 1992). It is being used to sequence major segments of the *E. coli* genome (Kasai *et al.*, 1992), taking advantage of a highly redundant library of overlapping λ phage clones (Kohara *et al.*, 1987).

In this large-scale sequencing operation, insertions were mapped and sequencing templates were prepared by PCR from aliquots of single phage plaques using Tn*5*-specific and vector-specific primers. In this approach, phage that failed to yield PCR products were not analyzed further; they contained insertions in the vector (about half of the total), or insertions that were distant from the cloning site. Preliminary tests established conditions for reproducible amplification of segments as large as 5 kb between primer binding sites in the vector and in Tn*5supF* (Krishnan *et al.*, 1991). The main impediment to routine use of Tn*5supF* insertions for complete sequencing of segments cloned in λ vectors (often 15–20 kb long) stems from an inability, at present, to amplify more than 5–6 kb reproducibly by direct PCR. When a library of overlapping phage clones is available, it is thus most efficient to focus only on easily mapped insertions within a few kilobases of the ends of any single cloned segment and obtain full coverage by exploiting the extensive overlaps among clones in the array (Kasai *et al.*, 1992). However, DNA near insertions at internal sites, far from the vector arms, can be amplified by 'crossover PCR' between Tn*5supF* insertions (Krishnan *et al.*, 1991) when necessary.

Crossover PCR products are obtained by mixing aliquots of two phages and using Tn*5supF*-specific primers, alone, to amplify DNA between the insertion sites (currently up to 3 kb). This allows mapping of some centrally located insertions that are beyond the reach of direct PCR from the vector sequences. Most Tn*5supF* insertions recovered by selection for phage growth on *dna*B_{amber} *E. coli* are in the orientation that allows *supF* to be optimally transcribed from λ phage promoters (Kersulyte *et al.*, 1992). In consequence, most pairs of insertions yield crossover PCR products that have unique left and right ends and thus can be used directly for sequencing.

The direct PCR-crossover PCR strategy has been implemented in the sequencing of a >80 kb segment of the *E. coli* genome: Tn*5supF*-containing plaques were quick-screened in an analytical PCR step, and insertions within a few kilobases of each end of the cloned segment were identified by PCR amplification using transposon- and vector-specific primers. Then segments between the λ phage ends and transposon inserts, spaced approximately every 500 bp when possible, were amplified by 'preparative PCR' using a biotinylated vector-specific primer and a nonbiotinylated Tn*5supF*-specific primer. The amplified DNA was coupled to streptavidin-coated paramagnetic beads, purified magnetically, denatured and the attached single DNA strands were used as templates for fluorescence-based automated sequencing (Kasai *et al.*, 1992).

When intervals between the two nearest insertions were in the 1 kb range, crossover PCR and sequencing about 500 bases from each end of the crossover PCR product gave one strand coverage of the entire 1 kb segment.

PCR mapping showed that the distribution of insertions within 5 kb of the ends of cloned segments was sometimes nonrandom, with some rather prominent 'cold' regions that contained few, if any, insertions (Fig. 7.5). However, insertions in the same regions in overlapping phage clones were quite random, which indicated that these cold regions may be due to long-range context effects on the expression of *supF* in the transposon when inserted at some sites (Kersulyte *et al.*, 1992).

The length and nucleotide sequence of PCR-amplified fragments were found to strongly affect product recovery when a primer with just one biotin was used. In many cases rather large amounts (up to 80%) of amplified fragments remained in the supernatant fraction after the Dynabead–DNA complexes were collected magnetically. Therefore, several further improvements were made.

(1) The coupling of biotin-containing DNA fragments and Dynabeads was reinforced by using di- and tri-biotinylated primers in preparative PCR. This gave much better sequence readings.
(2) DNA polymerases other than Klenow fragments and Sequenase were examined. We found that Bst polymerase from BioRad (Ye & Hong, 1987) gave the best results, in particular greater uniformity of heights of individual peaks. This increased the accuracy in base assignment by the computer program.
(3) The intensity of fluorescence of the sequencing reaction products was increased by using sequencing primers containing two or three FITC groups.

These changes resulted in excellent sequence readings.

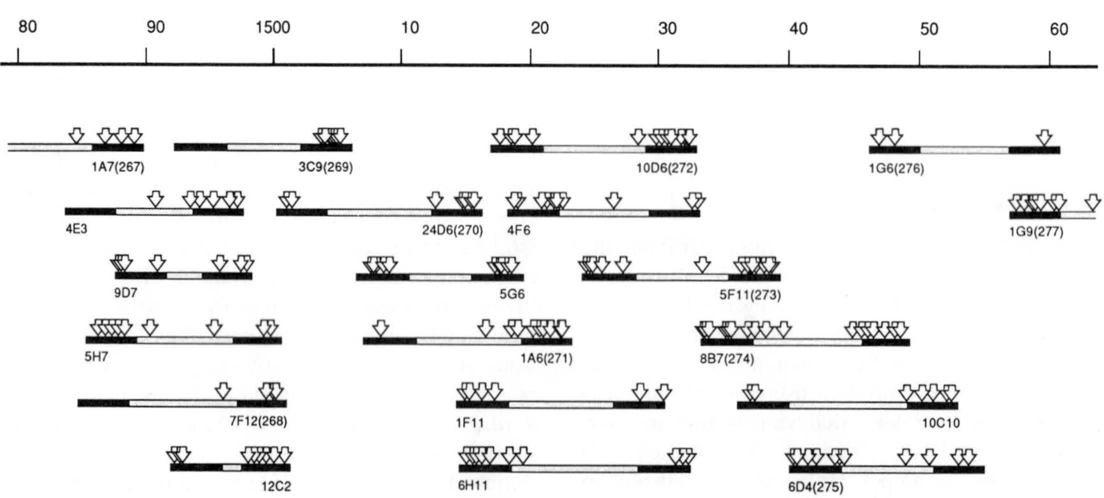

Figure 7.5 Distribution of Tn*5supF* insertions in 20 ordered λ clones from the *terC* region (min 31–33) of the *E. coli* chromosome. The top line indicates the kb coordinates used by Kohara *et al.* (1987). Horizontal bars indicate the location and extent of the cloned DNA (the λarms are not shown) and the arrows indicate the positions of the insertions, as estimated by analytical PCR. Clone names are given underneath the end attached to the right vector arm. Note that insertions in the central region of each clone were generally not mapped (modified from Kasai *et al.* (1992)).

In these experiments we found that the efficiency of recovery of preparative PCR products of more than 2 kb on Dynabeads was inversely related to their size. This potential difficulty was overcome using crossover PCR products as sequencing templates. Typically, the crossover PCR products used were 1–2 kb long and could be recovered with ease, even when used to access DNA segments 4 or 5 kb from the end of the cloned segment.

To maximize the efficiency of data collection, DNA segments were sequenced with only two- to three-fold redundancy. In most cases, this provided data of 400–500 bases per run. The accuracy of automatic base assignment was between 95 and 99%, which would be acceptable for first-pass analysis, provided that the number of bases in the sequenced region is determined very accurately. To achieve this, each sequence run contained a control reaction in which M13 phage DNA was used as template. In addition, the computer program for assigning bases was modified to cross-reference the data between control and sample runs. In this way, the number of bases within the first 400 bases of each sample were counted with an accuracy of at least 99.5%.

7.3.2 γδ-Mediated nested deletion formation in pDUAL

Several protocols were used in initial tests of pDUAL clones (Wang *et al.*, 1993b). The following are recommended as being user-friendly and potentially automatable in critical steps.

7.3.2.1 Directed sequencing strategy

(1) Transform the pDUAL clone Cam^r into a *recA*, *strA* (*rpsL*) strain that contains the transposase helper plasmid pXRD4043 (Tsai *et al.*, 1987), selecting for Kan^r Tet^r Cam^r and screening for Str^s Suc^s colonies (or transform the 5.4 kb pXRD4043 plasmid into a *recA strA* (*rpsL*) strain that contains the pDUAL clone).
(2) Spread cells from overnight cultures grown in (a) L broth (LB) plus Tet and Cam (no Suc) and (b) LB plus Kan and Cam (no Str) onto (a) L agar (LA) plus Tet and Suc, and (b) LA plus Kan and Str, to select clockwise and counterclockwise deletions, respectively. Incubate at 37°C.
(3) Scrape each selection plate, isolate supercoiled plasmid DNA and fractionate in low-melting-point agarose.
(4) Slice gels into 1–2 kb size fractions, transform a *recA strA* strain (without the transposase plasmid, pXRD4043) and plate samples from (a) and (b), above, on (a) LA plus Tet and Suc for clockwise deletions, or on (b) LA plus Kan and Str for counterclockwise deletions.
(5) Pick several individual transformants from each fraction, and isolate plasmid DNA.
(6) Sequence target DNA using the SP6 promoter primer for clockwise deletions and the T7 promoter primer for counterclockwise deletions.

Notes:

(a) Cam, which selects for the transposase helper plasmid PXRD4043, should be included in the medium in steps (1) and (2), but not in step (4).
(b) Care must be taken to avoid nicked DNA and to avoid overloading gel in step (3), otherwise a significant fraction of the plasmids recovered may be smaller than expected.

7.3.2.2 Shotgun sequencing strategy

(1) As in Section 7.3.2.1.
(2) As in Section 7.3.2.1.
(3) Pick individual colonies from selection plates and isolate plasmid DNA (usually, more than 95% of the Tet^r colonies are Kan^s, or vice versa, and have plasmids with deletion endpoints in the target DNA, making further screening unnecessary).
(4) As step (6) in Section 7.3.2.1.

Once sufficient coverage is achieved, the ends of contigs should be mapped by size of the corresponding plasmids and the gaps filled by size-fractionating large pools of Suc^r Tet^r or Str^r Kan^r colonies, as described above.

In preliminary experiments in which single colonies from the selection plates (step (2) in Section 7.3.2.1) were picked at random and the sizes of deletion plasmids determined (Fig. 7.6), some clones yielded predominantly small plasmids (large deletions) in one direction (e.g. Fig. 7.6(B)), probably because of preferential replication and segregation of small plasmids. This problem was alleviated by eliminating Amp and reducing the level of Tet (to 15 μg ml^{-1}) in the selective medium, or by selecting for deletions then isolating plasmid DNA from many pooled colonies and transforming a *recA strA* strain, thereby avoiding problems associated with plasmid segregation in cells that contain many plasmids.

7.4 CONCLUSION

We describe here several strategies for the direct, efficient use of transposon insertions to access cloned DNA in large-scale sequencing projects. The Tn*5supF*-PCR hopping method for sequencing λ phage clones used 20 overlapping ordered phage clones to sequence >80 kb of the *E. coli terC* region

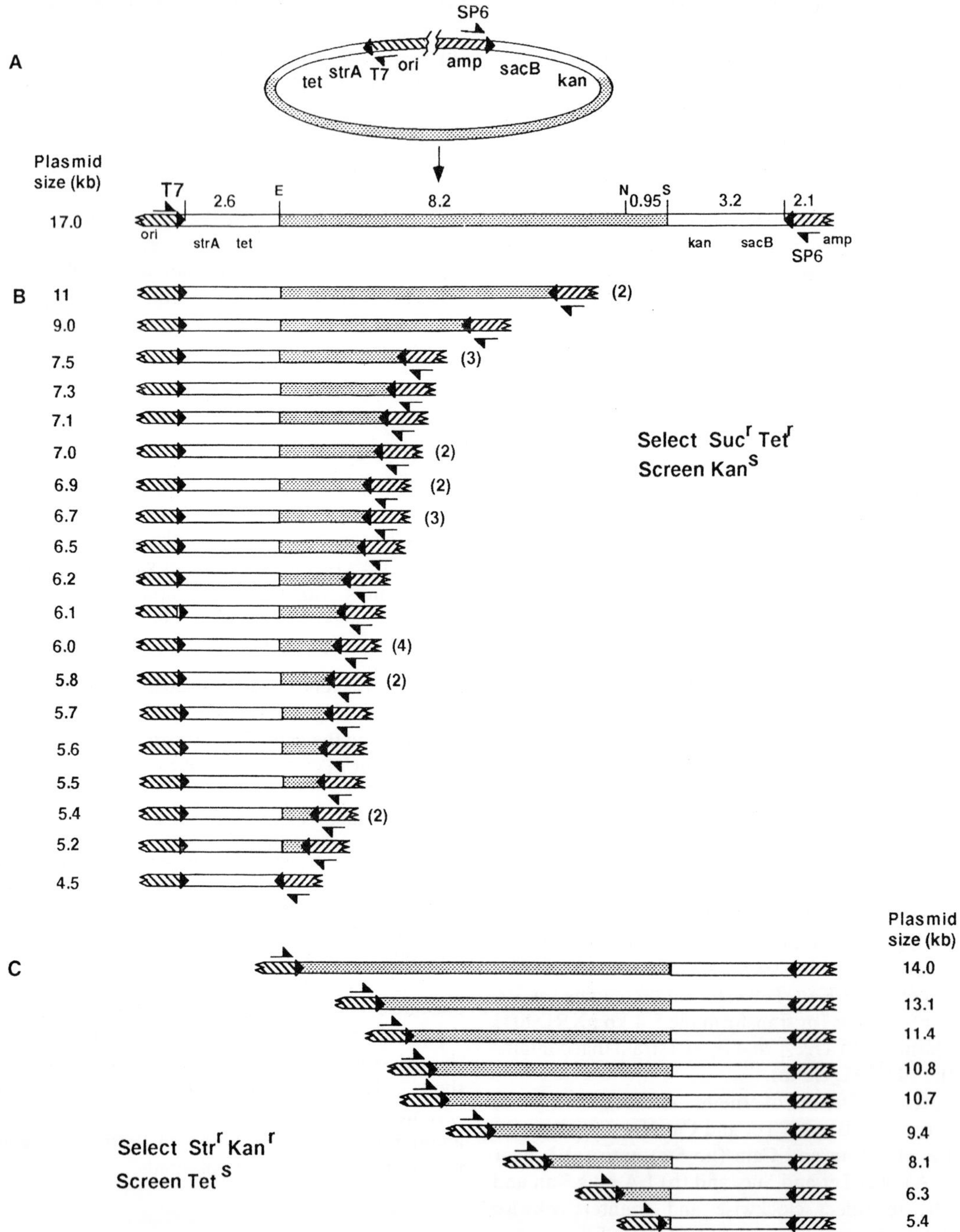

Figure 7.6 Deletions isolated in pDUAL clone pIF339 (G. Wang, M. Lalande, E. Kirkness & C. M. Berg, unpublished data). A 9.2 kb EcoRI-SalI GABRB3 gene fragment from human chromosome 15q11q13 was cloned into the EcoRI-SalI site in pDUAL. (A) Parent plasmid. (B) Clockwise (Suc^r Tet^r) deletions isolated from 22 independent parental colonies. (C) Counterclockwise (Str^r Kan^r) deletions isolated from nine independent parental colonies.

(Kasai *et al.*, 1992). The γδ hopping method (Berg *et al.*, 1993), is particularly useful for sequencing small clones (see Chapter 8). The nested deletion method for sequencing DNAs cloned in the pDUAL cosmid/plasmid vector has also been developed (Wang *et al.*, 1993b). A derivative of pDUAL, called 'pDELTA-1' (Wang *et al.*, 1993a), has been constructed to accommodate a two-step deletion approach: 'anchoring' deletions can be isolated as described above, and then gaps filled by generating nested deletions *in vitro* (for gap filling) from sites between the $\gamma\delta$ ends that are retained after *in vivo* deletion formation. This will be particularly valuable for accessing larger deletion plasmids (with small deletions), which are difficult to size accurately. pDELTA-1 is being marketed by GIBCO/BRL. Additional derivatives of pDUAL with a low copy number replication origin, different multiple cloning sites and other features are under development or testing.

ACKNOWLEDGMENTS

This work was supported by grants from the Department of Energy (DE-FGO2-89ER-60862 and DE-FGO2-90ER-610), the National Institutes of Health (HG-000563), the Ministry of Education of Japan (01655005, 01880029, 02237105 and 03221105), Life Technologies Incorporated, and the University of Connecticut Research Foundation.

REFERENCES

Ahmed, A. (1984) *J. Mol. Biol.* **178**, 941–948.

Ahmed, A. (1987) *Methods Enzymol.* **155**, 177–204.

Berg, C.M., Berg, D.E. & Groisman, E.A. (1989) In *Mobile DNA*, D.E. Berg & M.H. Howe (eds). American Society of Microbiology, Washington, DC, pp. 879–925.

Berg, C.M., Vartak, N.B., Wang, G., Xu, X., Liu, L., MacNeil, D.J., Gewain, K.M., Waiter, L.A. & Berg, D.E. (1992) *Gene* **113**, 9–16.

Berg, C.M., Wang, G., Strausbaugh, L.D. & Berg, D.E. (1993) *Methods Enzymol.* **218**, 279–306.

Berg, D.E. (1989) Transposon Tn*5*. In Mobile DNA (D.E. Berg & M.M. Howe, eds). American Society of Microbiology, Washington, DC, pp. 185–210.

Church, G.M. & Kieffer-Higgins, S. (1988) *Science* **240**, 185–188.

Kasai, H., Isono, S., Kitakawa, M., Mineno, J., Akiyama, H., Kurnit, D.M., Berg, D.E. & Isono, K. (1992) *Nucleic Acids Res.* **20**, 6509–6515.

Kersulyte, D., Krishnan, B.R. & Berg, D.E. (1992) *Gene* **114**, 91–96.

Kohara, Y., Akiyama, K. & Isono, K. (1987) *Cell* **50**, 495–508.

Krishnan, B.R., Kersulyte, D., Brikun, I., Berg, C.M. & Berg, D.E. (1991) *Nucleic Acids Res.* **19**, 6177–6182.

Krishnan, B.R., Kersulyte, D., Brikun, I., Huang, H.V., Berg, C.M. & Berg, D.E. (1993) *Methods Enzymol.* **218**, 258–279.

Kurnit, D.M. & Seed, B. (1990) *Proc. Natl. Acad. Sci. U.S.A.* **87**, 3166–3169.

Liu, L., Whalen, W., Das, A. & Berg, C.M. (1987) *Nucleic Acids Res.* **15**, 9461–9469.

Phadnis, S.H., Huang, H.V. & Berg, D.E. (1989) *Proc. Natl. Acad. Sci. U.S.A.* **86**, 5908–5912.

Shapiro, J. (1979) *Proc. Natl. Acad. Sci. U.S.A.* **76**, 1933–1937.

Sherratt, D. (1989) In *Mobile DNA* (D.E. Berg & M.M. Howe, eds) American Society of Microbiology, Washington, pp. 163–184.

Strausbaugh, L.D., Bourke, M.T., Sommer, M.T., Coon, M.E. & Berg, C.M. (1990) *Proc. Natl. Acad. Sci. U.S.A.* **87**, 6213–6217.

Tomcsanyi, T., Berg, C.M., Phadnis, S.H. & Berg, D.E. (1990) *J. Bacteriol.* **172**, 6348–6354.

Tsai, M.-M., Wong, R.Y.-P., Hoang, A.T. & Deonier, R.C. (1987) *J. Bacteriol.* **169**, 5556–5562.

Wang, G., Berg, C.M., Chen, J., Young, A.C., Blakesley, R.W., Lee, L.Y. & Berg, D.E. (1993a) *Focus* **15**, 47–49.

Wang, G., Blakesley, R., Berg, D.E. & Berg, C.M. (1993b) pDUAL: a transposon-based cosmid cloning vector for generating nested deletions and DNA sequencing templates in vivo *Proc. Natl. Acad. Sci. U.S.A.* **90**, 7874–7878.

Wang, G., Xu, X., Chen, J., Berg, D.E. & Berg, C.M. (1994) *J. Bacteriol.* (in press).

Yanisch-Perron, C., Vieira, J. & Messing, J. (1985) *Gene* **33**, 103–119.

Ye, S.Y. & Hong, G.F. (1987) *Sci. Sin.* **30**, 503–506.

CHAPTER EIGHT

Transposon-facilitated Sequencing: an Effective Set of Procedures to Sequence DNA Fragments Smaller than 4 kb

C.H. MARTIN,[1,2] C.A. MAYEDA,[1,2] C.A. DAVIS,[1,2] M.P. STRATHMANN[1,2] & M.J. PALAZZOLO[1–3]

[1] Human Genome Center, Lawrence Berkeley Laboratory, Berkeley, CA 94720, USA
[2] Drosophila Genome Center, University of California at Berkeley, Berkeley, CA 94720, USA
[3] Department of Molecular and Cell Biology, University of California at Berkeley, Berkeley, CA 94720, USA

8.1 INTRODUCTION

One of the costly and rate-limiting requirements in DNA sequencing is the need for sequencing priming sites every 300–400 bp. We have developed and are using a transposon-facilitated DNA sequencing scheme (Strathmann *et al.*, 1991) to satisfy this need. In this strategy, mobile priming sites are provided by the transposon termini. This methodology is based on a suggestion by Guyer (1983) that the bacterial γδ transposon system would be useful for sequencing. The F-factor of *Escherichia coli* is capable of mobilizing plasmids from a donor cell to a recipient. This transferred plasmid, under appropriate mating conditions, always contains an insertion of the γδ transposon. Thus, it is possible to select for the insertion of γδ into a plasmid that encodes an antibiotic resistance by selecting for the transfer of that resistance factor from a donor cell into a recipient cell. This recipient cell must contain an independent selectable marker. Each cell recovered after double antibiotic selection is a recipient cell (from the mating mixture) that contains a target plasmid with a γδ insertion. There is no need in this system for a selective antibiotic resistance marker on the transposon.

The set of procedures that we are using has the following steps. The DNA fragment to be sequenced is subcloned into a minimal plasmid. This construct is then introduced into a donor strain that contains an F-factor carrying a copy of the γδ transposon. This transposon has been modified so that the well-characterized '-21' and 'reverse' primers from M13 are positioned immediately adjacent to the inverted repeats so that they can be used as sequencing priming sites. The plasmid-bearing donor cells are then mated to a recipient strain that carries a gene that encodes a product which confers kanamycin resistance. The mating mixture is plated on a doubly selective antibiotic agar plate. Surviving cells contain target plasmids, each carrying a γδ insertion. These clones are selected and grown as overnight cultures. The sites of transposon insertion are then mapped by polymerase chain reaction (PCR) using a pair of reactions for each insert. In these reactions, which use overnight cultures as templates, the distance between the inverted repeats of the transposon and the plasmid regions that flank each side of the cloned insert are measured. This allows the identification of a minimal set of sequencing templates that can be used to provide the substrates for an ordered and efficient determination of the sequence of the target DNA fragment.

Our system, when introduced, differed from others

in several ways (Nag *et al.*, 1987; Strausbaugh *et al.*, 1990). First, the DNA fragment to be sequenced is subcloned into a minimal plasmid so that the transposon is forced into the target. Next, the sites of transposon insertion are mapped by PCR (Saiki *et al.*, 1985) so that a minimal set of sequencing templates can be rapidly obtained. Finally, the target fragment is kept under 4 kb. In this manner the site of insertion can be mapped from sites that flank both sides of the subcloned insert. We require the sum of the two mapping PCR assays to match the size of the cloned insert. This requirement for redundant information eliminates most of the PCR artifacts that can obscure accurate transposon mapping information.

8.2 EXPERIMENTAL PROTOCOLS

8.2.1 Subcloning

8.2.1.1 Key aspects of vector selection and preparation

As mentioned in the Introduction, the use of a minimal plasmid as a vector makes the set of procedures more efficient because the transposon is forced into the target sequence. Insertion events into the antibiotic resistance gene or the origin of replication are not recovered using this scheme. In our initial experiments we found that both the f1 origin sequences and β-galactosidase sequences found in many commercially available plasmids and phagemids attracted numerous transposon insertions. For this reason, the miniplasmid pMOB, composed of little more than a polylinker, an origin of replication, and an ampicillin resistance gene, was constructed (Strathmann *et al.*, 1991). We have used both this miniplasmid and the commercially available pSP72 (Promega) in transposon targeting experiments.

8.2.1.2 Preparation of vector DNA

Since the minimal vector does not contain a simple mechanism for scoring or selecting those clones that contain inserts, it is important that steps be taken to minimize the production of clones that contain copies of the vector plasmid without an insert. Thus, in our experiments we try to maximize the possibility that the vector is digested with an appropriate restriction enzyme prior to cloning. The addition of spermidine-HCl (5 mM) to the digestion facilitates more complete cutting. In addition, calf intestinal alkaline phosphatase is used to minimize vector recircularization-without-insert ligation.

8.2.1.3 Ligation and transformation

The phosphatased vector and blunt-ended insert molecules are ligated overnight and introduced into *E. coli*. The ligation reactions consist of 0.15 μg of phosphatased vector, an approximately equal amount of insert DNA, 66 mM Tris-HCl (pH 7.5), 5 mM $MgCl_2$, 1 mM dithioerythritol, 1 mM ATP, and 0.5 units of T4 DNA ligase (Boehringer Mannheim). The ligation reaction is then incubated at room temperature for at least 4 h. The ligated products must then be introduced into the appropriate *E. coli* strain. We have used both chemical transformation and electroporation using standard procedures. The most straightforward approach is to introduce the subclones directly into the donor strain that will be used in the transposition procedures (pOX38::gdUR HB101).

8.2.2 Transposition

The molecular mechanisms of the transposon events have been described by others (Guyer, 1983). Briefly, in the donor cell, prior to conjugation, the F-factor and target sequences are joined by two copies of the γδ transposon in an intermediate cointegrate structure. This cointegrate is capable of being transferred to the recipient cell; after transfer the cointegrate is resolved. One copy of the γδ remains in its original position on the F-factor. The second copy remains in the target plasmid at the point of the initial transposition. Double selection for the antibiotic resistance gene on the target plasmid (ampicillin resistance) and for the resistance factor (kanamycin resistance) in the recipient cell identify target plasmids that have been transferred by conjugation and therefore contain a copy of the γδ transposon.

8.2.2.1 Key aspects of the mating strains

There are two important features of the donor strain. First, it contains an F-factor that is capable of mobilizing a plasmid to a recipient strain via a γδ-based mechanism. Second, the strain is $RecA^-$. This prevents multimerization of the target plasmids. Monomers are important to the strategy – otherwise the transposons could insert into the duplicated (and hence nonessential) antibiotic resistance gene and origin of replication.

We have been using *E. coli* strain pOX38::gdUR HB101 as the donor strain in our transposon sequencing experiments. In addition to the requirements described in Section 8.2.2.2, this strain is notable in that the γδ has been modified in such a way that the '-21' and 'rev' sequences have been introduced immediately internal to the 38 bp inverted repeats at the ends of the transposon.

The recipient strain must not contain an F-factor

and must contain a selective marker that is different from the one contained on the target plasmid. The only other important criterion is that it be possible to prepare reliable sequencing templates from the recipient strain in a straightforward fashion. We have been using a strain called 'JGM', which is kanamycin resistant (Strathmann *et al.*, 1991).

8.2.2.2 Key aspects of bacterial mating

As mentioned above, in most of our experiments we have transferred plasmids with ampicillin resistance to recipient cells with kanamycin resistance. In these circumstances significant amounts of β-lactamase are present in the mating mixture from the donor cells. For this reason, it is essential to plate the mating mixture on carbenecillin plates to avoid the satellite colonies that would be present if the mixture is plated on ampicillin plates. The exact number of cells containing transferred plasmids tends to vary somewhat from mating to mating. Since it is desirable to do the mating only once, yet still isolate enough events to generate a sufficient number of sequencing starting points, it is best to plate several dilutions of the mating mixture.

8.2.2.3 Protocol for bacterial mating and plating

Overnight cultures of both the donor and recipient (0.1 ml) are mixed with 2 ml of L-broth (without antibiotics) in a 15 ml sterile culture tube. This tube is then incubated on a roller wheel for 3 h. In our experience using the wheel for mixing and aeration is far superior to rotary shaking. After 3 h, 1 μl is added to 1 ml of fresh LB. A series of four two-fold dilutions are then made and 0.2 ml of each dilution are plated on LB agar plates that contain 100 μg l^{-1} of carbenecillin and 50 μg l^{-1} of kanamycin. The plates are then incubated upside-down overnight at 37°C. The day after the mating, 96 clones are picked into individual tubes in a titertube rack. A titertube rack is a collection of 1.2-ml tubes that are arrayed in an 8×12 array. This rack is incubated at 37°C overnight with shaking.

8.2.3 Mapping sites of transposon insertion

8.2.3.1 Key aspects of PCR reactions on bacterial overnight cultures

The PCR mapping reactions are performed for two reasons. The first reason, of course, is the use of PCR to map the point of insertion of the transposon into the target DNA fragment to be sequenced. The second is to eliminate the transposon events that do not promote the sequencing of the target. The largest class of 'nonhelpful' events are those in which transposon insertion occurs in a nonlethal site in the plasmid (even the minimal plasmids that we have constructed have a few sites into which the transposon can insert without preventing recovery in the donor strain). A particularly useful screen against nontarget insertions is the mapping of the insertion site from both sides of the plasmid that flank the subcloned target. Insertions in the target fragment can be readily identified as those clones that support PCR amplification in both of the pair of reactions, in which the sum of the amplified fragments from each side adds up to the size of the insert.

The PCR-based transposon mapping assays are performed on dilutions of overnight bacterial cultures. Care must be taken in making the dilutions, as too large a volume of overnight culture can inhibit the Taq polymerase, while enough template must be provided to readily support detection of the PCR amplified product. The original overnight cultures are kept as stocks from which a minimal subset of clones can be regrown for sequencing template preparations.

8.2.3.2 Protocol for transposon mapping in target sequences

Two PCR reactions are performed on each potential sequencing template in order to map the transposon insertion site relative to each end of the cloned insert. The template for each PCR reaction is a 1:30 dilution of an overnight bacterial culture. In one PCR reaction a primer is used that matches a sequence near the β-lactamase side of the cloning site. In the second PCR reaction another primer is used that hybridizes to the region that flanks the cloned insert, nearest the origin of replication. For the plasmid pSP72, we call these oligos PM001 and PM002. The sequence of PM001 is 5′ CGTTAGAACGCGGCTACAAT 3′. The sequence of PM002 is 5′ GCCGATTCATTAATGCAGGT 3′. Since the transposon has a 38 bp terminal inverted repeat at each end, the same primer that matches part of this sequence can be used to map the site of transposon insertion relative to each of the regions that flank the cloning site. The sequence of the primer NGDIR-1 that matches the inverted repeat is 5′ GTTCCATTGGCCCTCAAAC 3′. Thus, one PCR reaction contains the primers PM001 and NGDIR-1, while the second reaction contains PM002 and NGDIR-1.

The PCR reaction mixture includes: the template which is 1 μl of a 1:30 dilution in water of a bacterial culture overnight, each PCR primer at a concentration of 0.25 μM, the four nucleotide triphosphates each at a concentration of 200 μM, 50 mM KCl, 10 mM Tris-Cl (pH 8.3), 1.5 mM $MgCl_2$, and 0.01% gelatin. The PCR amplifications are performed in a Perkin–Elmer Model 9600. Thirty cycles are used for each mapping experiment. The cycles include a 15-s denaturation step at 94°C, followed by a 15-s annealing step at 55°C, and finally an extension step for 1 min at 72°C. All of the mapping experiments are performed using a 96-well format.

8.2.4 DNA sequencing

8.2.4.1 Analysis and selection of a minimal set of γδ transposon-bearing templates for sequencing

The results of the PCR amplifications to determine the locations of the transposon insertions can then be analyzed by agarose gel electrophoresis. We currently size the PCR reaction products on 1.4% agarose gels run in 1× TBE buffer. Each of the bands from the same transposon event are then summed. Only those pairs whose size when summed matches the size of the insert are placed on the transposon map and considered for template selection. For each 3 kb clone 96 transposon events are mapped. In those cases where the insert was cloned into pSP72 about 60% of the transposon events met this criterion and were considered for template selection. The most likely explanation for the clones that failed this test is that the transposon inserted into one of the nonessential plasmid regions.

Once the transposon map has been generated it is straightforward to select the minimal set of about 10–12 transposition events that will provide a minimal set of templates. It is true that some of the targets we have sequenced have regions in which a mapped transposon cannot be identified after the mapping of 96 individual transposon-bearing clones. Based on results of sequencing over 30 approximately 3 kb subclones using these methods, we have found that about 93% of the sequence is determined using the mapped γδ transposons; the remaining 7% of sequence is determined using custom oligonucleotide primers with the ABI dye terminator kit. On average, each 3 kb subclone required two custom oligonucleotides in order to obtain double-stranded coverage at every base (i.e. the 7% missing typically represented two distinct single-stranded gaps).

8.2.4.2 Template preparations for sequencing

As described above, 10–12 clones, with transposons inserted about every 300 bp, are selected and 3 ml cultures are grown overnight at 37°C. Then, 1.5 ml of the culture is pelleted in a microfuge at 14 000 r.p.m. for 1–2 min. The supernatant is decanted and discarded. The pellet is microfuged again for 1 min and the remaining supernatant is withdrawn with a pipetman. Then, 300 μl of STET (8% sucrose, 5% Triton X-100, 50 mM EDTA, 50 mM Tris-HCl, pH 8.0) is added to the pellet and resuspended using a pipetman. The mixture is then vortexed at maximum speed until the pellet is completely resuspended. Twenty microliters of a Tris-lysozyme solution (50 mM Tris-HCl (pH 8.0), 10 mg ml^{-1} lysozyme) is added. The tube is inverted several times and allowed to incubate at room temperature for 5–10 min. The solution is then heated to 100°C for 2 min. This is done by placing the tube into a heat block that is positioned over a Bunsen burner. The Bunsen burner is held under the heat block so that the water boils lightly and continuously for the entire 2 min. Alternatively, the tubes can be placed in a boiling water bath in a floating rack designed for this purpose (Nalgene, Inc.). The boiled preparation is then immediately microfuged at 14 000 r.p.m. for 5 min. The accumulated bacterial debris is then removed using a toothpick. It is important that all of the debris is removed at this point even if a second toothpick has to be used. DNA is precipitated by the addition of 280 μl of an ammonium acetate–isopropanol solution (75% isopropanol, 2.5 M ammonium acetate). This mixture is vortexed for 1–5 s at the maximum setting. The plasmid is pelleted by microfuging at 14 000 r.p.m. for 5 min immediately after vortexing. The supernatant is removed and discarded. Then, 100 μl of 70% ethanol is used to rinse the DNA. Again the pellet is vortexed for a few seconds. The pellet is recentrifuged at 14 000 r.p.m. for 5 min. The supernatant is decanted and the pellet is microfuged again for a few seconds. The remaining small volume of ethanol is removed with a pipetman. The template is then air dried for about 5 min until all the moisture has disappeared. The pellet is resuspended in 50 μl of sterile double-distilled water containing 0.1 μg of RNase (DNase-free, Boehringer Mannheim). Prior to sequencing the templates are drop dialyzed. This is accomplished by floating a 100 000 Da molecular weight cutoff polysulfone membrane (Millipore PTHK0MS10; small squares, approximately 2 cm on a side, can be cut from these sheets) on 5–15 ml of sterile water. The resuspended DNA pellet (15–50 μl) is placed on the floating dialysis membrane for 2 h. The DNA solution is then carefully recovered from the dialysis membrane using a pipetman and placed into a clean tube. Typically, a 1:1 or 1:2 dilution of this dialyzed template is made prior to use in sequencing. Use of undiluted material often results in strong peaks for the first 100 bases of an ABI 373A sequencing chromatogram, but the intensity of the peaks declines rapidly with increasing distance from the primer site, which results in significantly reduced read lengths.

8.2.4.3 Protocol for sequencing reactions

The cycle sequencing reactions are performed using the protocol that is provided with the dye primer kit from Applied Biosystems. After trimming away the small region of the γδ inverted repeat that is present at the beginning of each sequencing run (the last 21 nucleotides of the γδ inverted repeat sequence, as seen on the ABI sequencing chromatogram with either -21 or reverse primer reactions, is 5′ TTC CAT TGG CCC TCA AAC CCC 3′) (Reed *et al.*, 1979); the next and subsequent bases read are derived from the insert region (adjacent to the transposon insertion site) and, determining the point at which data quality becomes unacceptable, we typically obtain between 360 and 420 nucleotides of useful data. One feature worth mentioning is that there is typically a 5 bp duplication created at the site of the γδ insertion. This results in the -21 primed and the reverse primed sequences overlapping for 5 nucleotides at their 5′-ends, on opposite strands, with each continuing in opposite directions away from the transposon insertion point.

8.3 DISCUSSION

We have found this transposon-facilitated DNA sequencing strategy to be a highly cost-effective approach for sequencing DNA fragments smaller than 4 kb. This size range makes it ideal for the complete double-stranded elucidation of the sequences of cDNA clones. Furthermore, it is useful for the determination of larger genomic sequences when a set of overlapping clones of appropriate size is available.

The transposon-facilitated scheme that we describe greatly reduces the number of oligonucleotides required to determine the complete double-stranded sequence while avoiding time-consuming *in vitro* manipulations and additional subcloning steps. The most difficult of the molecular manipulations are performed *in vivo* by the bacterial cell. While a relatively large number of transposon insertion events have to be mapped, these reactions can be performed in a 96-well format using diluted bacterial overnight cultures. Thus the number of DNA preparations for sequencing and the number of sequencing reactions are kept to a minimum. Also, the use of expensive automated sequencers is made more efficient by using a near-minimal number of sequencing templates.

Our continuing development of this system will involve the creation of improved miniplasmids towards the goal of increasing the proportion of mappable (and thus potentially useful) γδ insertions. Another area involves exploring additional transposons that may have a complementary set of insertion site biases, thus reducing the need for custom oligonucleotides. Additionally, our group is incorporating these methods into a strategy designed to efficiently determine the sequence of large genomic regions.

REFERENCES

Guyer, M. (1983) *Methods Enzymol.* **101**, 362–369.

Nag, D.K., Huang, H.V. & Berg, D.E. (1987) *Gene* **64**, 135–145.

Reed, R., Young, R., Steitz, J.A., Grindley, N. & Guyer, M. (1979) *Proc. Natl. Acad. Sci. U.S.A.* **76**, 4882–4886.

Saiki, R.K., Scharf, S., Faloona, F., Mullis, K.B., Horn, G.T., Erlich, H.A. & Arnheim, N. (1985) *Science* **230**, 1350–1354.

Strathmann, M.P., Hamilton, B.A., Mayeda, C.A., Simon, M.I., Meyerowitz, E.M. & Palazzolo, M.J. (1991) *Proc. Natl. Acad. Sci. U.S.A.* **88**, 1247–1250.

Strausbaugh, L.D., Bourke, M.T., Sommer, M.T., Coon, M.E. & Berg, C.M. (1990) *Proc. Natl. Acad. Sci. U.S.A.* **87**, 6213–6217.

CHAPTER NINE

Construction of Exonuclease III Generated Nested Deletion Sets for Rapid DNA Sequencing

L.-I. LIU & R.D. FLEISCHMANN
The Institute for Genomic Research, Gaithersburg, MD 20878, USA

9.1 INTRODUCTION

Advances in techniques in DNA sequencing and the assembly of large-scale DNA sequencing projects have revolutionized the study of gene structure. Several approaches have been developed to rapidly and efficiently obtain overlapping DNA fragments for sequencing. These approaches include primer walking (Strauss *et al.*, 1986), transposon insertion (Ahmed, 1984, 1985), shotgun cloning (Anderson, 1981; Deininger, 1983; Messing, 1983), and the construction of nested deletion sets (Henikoff, 1987; McCombie *et al.*, 1991). The construction of nested deletion sets using the enzyme exonuclease III (Exo III) is one of the most popular strategies used for generating progressive unidirectional deletions of any double-stranded DNA insert up to about 7 kb, providing a set of plasmids with progressive overlapping unidirectional deletions into the target sequence. The major advantages of this approach are:

(1) the stability of the reaction rate of Exo III makes the reaction conditions and results very reproducible for producing progressive deletions of 200–250 bp;
(2) sequencing of the digested products generated from the original clone does not require subcloning into another vector;
(3) the approach provides a single vector-based primer site progressively adjacent to the sequence of interest (particularly useful in dye-primer based fluorescent sequencing); and
(4) it provides sequence redundancy which should allow greater than 99.8% sequence accuracy.

The successful construction of nested deletion sets is based on the enzymatic characteristics of Exo III. Exo III removes nucleotides from blunt ends or 5′ overhangs of restriction cut double-stranded DNA and inefficiently removes nucleotides from DNA ends with four-base 3′ overhangs (Henikoff, 1984). The non-progressive nature of Exo III allows one to generate a set of plasmids deleted to the same extent from each time point at a given temperature. Therefore, the rate

Table 9.1 Exonuclease III deletion rate (20 units of exonuclease III for each 1 μg of DNA).

Temperature (°C)	*Deletion rate (bp/min)*
37	400
34	375
30	230
23	125

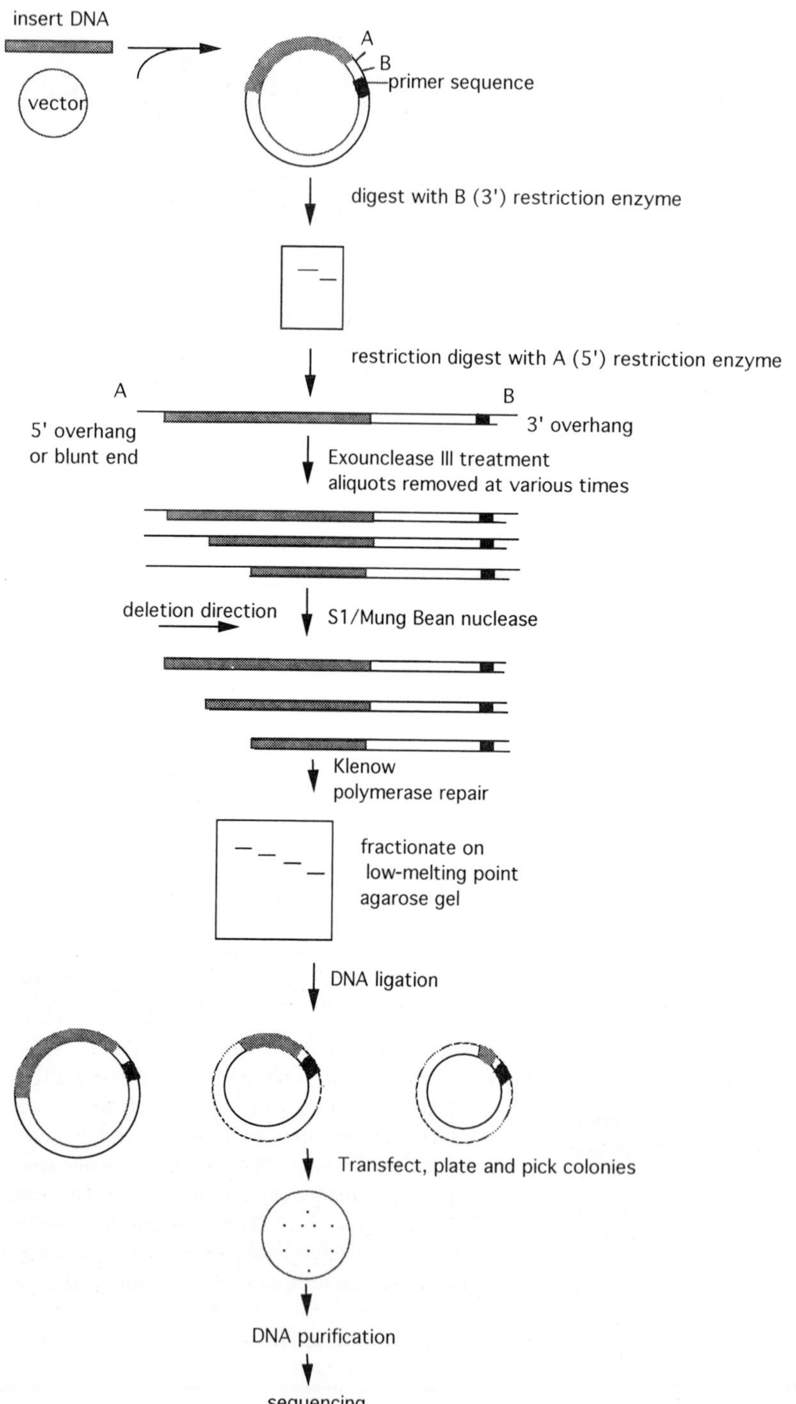

Figure 9.1 Schematic diagram of the steps in constructing a nested deletion set generated by exonuclease III digestion.

and range of deletions can be controlled by the reaction temperature and the incubation time of each time point (Table 9.1).

The method detailed below can be used in the following applications:

(1) sequencing of cDNA or genomic DNA clones of up to approximately 7 kb length;
(2) conformation of repetitive regions of cosmids;
(3) and filling gaps between assembled contigs in large-scale sequencing projects by the shotgun method.

9.2 PROTOCOL OVERVIEW

If the target insert is not already cloned into a vector containing the appropriate restriction sites, then the insert should be subcloned into a vector system such as pBluescript, pGEM, or equivalent vector by standard recombinant DNA techniques. These vectors will provide the appropriate restriction sites for generating Exo III protected and targeted ends.

In order to produce a set of clean nested deletions it is important to start with highly purified supercoiled DNA. Because Exo III targets both linear and circular double-stranded DNAs containing nicks or gaps (Zasloff *et al.*, 1978; Barnes *et al.*, 1983), pre-existing nicks in the DNA will lead to nonspecific deletions. In general, the plasmid DNA produced by a number of commercially available products using an alkaline lysis followed by a chromatography purification (Qiagen Tip 100 or Promega Magic Maxiprep) will produce DNA of sufficient quality. In addition, nicked and linear DNA can be selectively removed from supercoiled DNA by an acid-phenol extraction protocol (Zasloff *et al.*, 1978).

The plasmid DNA must first be linearized by digestion with the appropriate restriction enzymes prior to Exo III treatment. As illustrated in Fig. 9.1, there must be a restriction site (B site) between the insert DNA and the sequencing primer site that generates a four-base 3′ overhang end that is protected from Exo III digestion. It has been reported that 3′ overhangs generated by digestion with Apa I, Pst I, Pvu I, Hha I, or Sac II do not protect against digestion with Exo III (Titus, 1991). If no proper restriction site is present to generate a 3′ overhang (B site), a 5′ overhang adjacent to the primer site can be blocked from exonuclease digestion by incorporation of α-thio nucleotide analogs (Putney *et al.*, 1981; Henikoff, 1987). A second restriction site (site A) between the insert DNA and site B is digested to produce either a 5′ overhang or a blunt end which acts as the target site for the Exo III deletion activity.

One unit of Exo III will remove about 200 nucleotides from each 3′ recessed end of 1 μg of a 5000 bp linear double-stranded DNA template in 10 min at 37°C (Titus, 1991). Progressive deletions can be obtained in a single reaction by removing aliquots at successive time points. Each time point is then treated with S1 or mung bean nuclease (Vogt, 1973; Stratagene Protocol, 1990) and Klenow DNA polymerase to produce blunt ends for ligation. The success of the Exo III digestion can be monitored on a low melting point 0.8% agarose gel. The product from each time point is loaded onto a separate lane and the success of the digestion is monitored by the array of linear bands (Fig. 9.2). This step also acts as

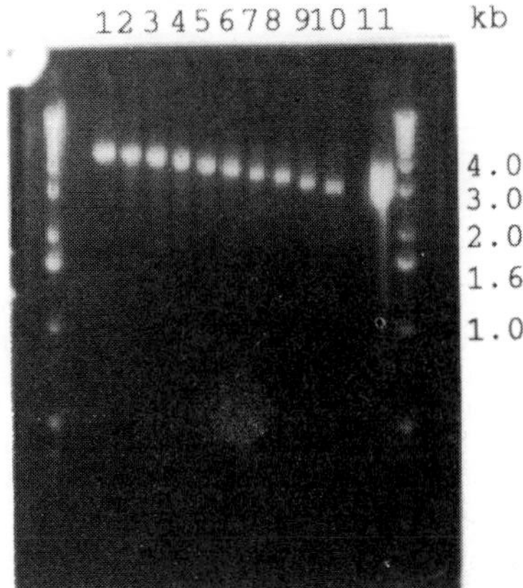

Figure 9.2 LMP agarose gel of a nested deletion set generated by exonuclease III digestion – 10 μg of plasmid DNA was digested with Not I and Sst I; 5μg of linear plasmid was digested with 300 U of exonuclease III and 10 time points (lanes 1–10) were taken at 30-s intervals. Each time point was treated with S1 nuclease, Klenow DNA polymerase and loaded onto a separate gel lane. (Lane 11) Plasmid DNA was digested as described above, but all time points were placed in the same S1 nuclease tube. Following Klenow DNA polymerase treatment and precipitation the sample was loaded onto a single lane on the gel.

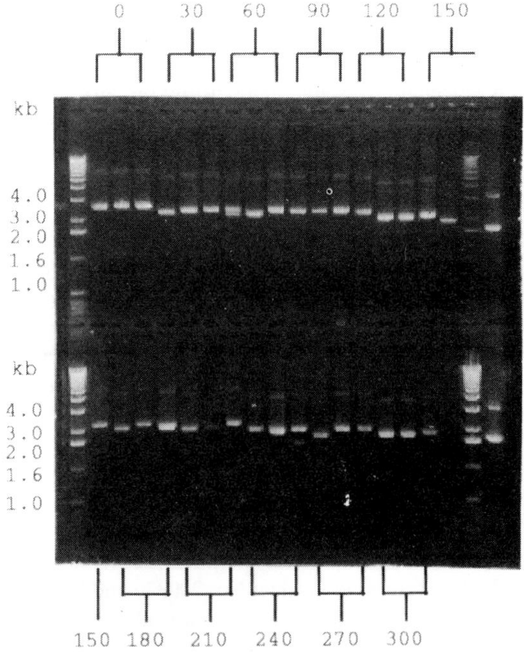

Figure 9.3 Agarose gel of supercoiled plasmid DNA prepared from exonuclease III time points. Three bacterial colonies were picked from each of 11 LB/Amp plates, each representative of an Exo III time point. Mini-prep DNA was prepared and a 1 μl aliquot was run on each lane. The gel illustrates the range of plasmid size recovered from each time point of the Exo III digestion.

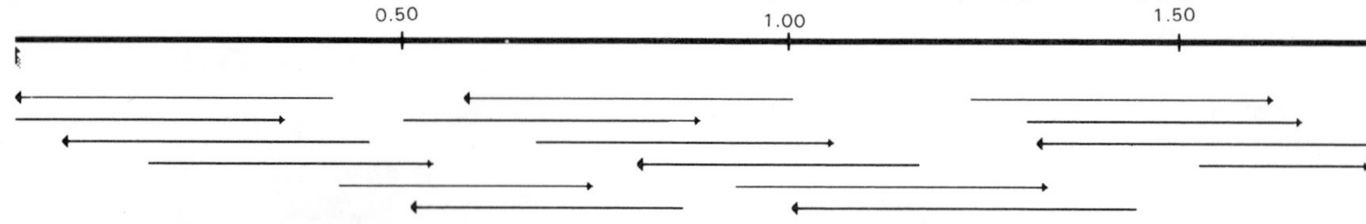

Figure 9.4 Schematic diagram of overlapping DNA sequences generated from two exonuclease III nested deletion sets. A 1745 bp insert cloned into pBluescript SK was digested with either Kpn I and Xho I or Sac I and Xba I. Plasmids prepared from each set of nested clones were primed with either fluorescent M13 reverse primers (Sac I/Xba I, left to right arrows) or fluorescent M13 forward primers (Kpn I/Xho I, right to left arrows). Cycle sequencing reactions were performed on an ABI catalyst 800 Robot and analyzed on an ABI 373A Fluorescent Automated DNA Sequencer. The resulting data were edited, assembled, and a contiguous unambiguous sequence derived using the ABI INHERIT software package.

a purification step by cutting out the band from each time point. The isolation of the desired band at each time point avoids unwanted smaller deletion products which can be generated. Alternatively, all the Exo III time points can be pooled into one tube, precipitated, and run on a single lane on the low-melting-point agarose gel (Fig. 9.2). The smear that is obtained can be conveniently divided into three segments. In both cases the isolated gel product is subjected to an in-gel ligation, transformed into X1-1 Blue or other appropriate bacterial host cell line, and plated onto LB/Amp plates (either one plate per time point or one plate per segment). Colonies are then picked from each plate, minipreps are prepared, and an aliquot of the supercoiled DNA is analyzed on a 0.8% agarose gel (Fig. 9.3). The appropriate array of plasmids are then selected for sequencing using the primer site adjacent to the deletions. The DNA sequence overlap generated by two Exo III deletion sets (one from each direction) of a cDNA insert cloned into pBluescript SK is schematically illustrated in Fig. 9.4. Ten overlapping DNA sequences primed with the M13 reverse primer (left to right arrows) and nine overlapping sequences primed with the M13 forward primer (right to left arrows) provide a contiguous 1745 bp DNA with no ambiguities. The individual sequences were assembled and the contiguous sequence of the cDNA clone was derived using the ABI INHERIT software package.

9.2.1 Protocol

A number of commercial exonuclease III kits are available (Stratagene Product No. 200330 and Promega Product No. E5850). The protocol outlined below is our modification of the Promega Erase-a-Base system.

Reagents

Exo III 10× buffer:	660 mM Tris-HC1, pH 8.0 6.6 mM $MgCl_2$
S1 7.4× buffer:	0.3 M potassium acetate, pH 4.0 2.5 M NaCl 10.0 mM $ZnSO_4$ 50% glycerol
S1 nuclease mix:	172 µl deionized water 27 µl S1 7.4× buffer 60 U S1 nuclease
S1 stop buffer:	0.3 M Tris base 0.05 M EDTA
Klenow buffer:	20 mM Tris-HCl, pH 8.0 100 mM $MgCl_2$
Klenow mix:	30 µl Klenow buffer 3–5 U Klenow DNA polymerase
dNTP mix:	0.125 mM each dNTP
SOC medium (1 l)	10 g bacto-tryptone 5 g bacto-yeast extract 5 g NaCl 10 mM $NaCl_2$ 10 mM $MgCl_2$ 10 mM glucose Adjust pH to 7.0 with NaOH and autoclave

(1) Once the DNA to be sequenced has been subcloned into an appropriate vector (pBluescript, pGEM, or equivalent), prepare a sufficient quantity of supercoiled plasmid using a Qiagen Tip 100 or Promega Magic Maxi prep. Both procedures prepare plasmid DNA of adequate quality.

(2) Do test restriction digests to determine which restriction enzymes cut only in the vector. We use Sst I and Not I or Xba I from one side of the insert and Kpn I and Xho I or Hind III from the other side of the insert.

(3) Choose the appropriate restriction enzymes and do a preparative double digest of your plasmid using 10 µg of plasmid DNA. Clean with Promega Magic DNA Clean up, or equivalent between digests and after the final digest. You

should recover 5–8 μg of linear DNA in 50 μl water. Monitor the complete digestion by agarose gel electrophoresis.

(4) Add 7.5 μl of S1 nuclease mix to each of up to 24 Eppendorf tubes and place on ice. We generally do 12 time points including the zero time point for up to a 3 kb insert.

(5) Dilute 5 μg of linear DNA into 60 μl of Exo III 1× buffer.

(6) Warm the tube to 37°C in a water bath.

(7) Before adding the Exo III to the tube, remove a 2.5 μl aliquot and add to the first (zero time point) S1 tube on ice.

(8a) Add 300–500 U of Exo III to the tube to begin the digestion. Remove 2.5 μl aliquots at 30-s intervals and place in successive S1 tubes that are on ice. (1 unit of Promega Exonuclease III (catalogue No. M1811) is equivalent to the amount of enzyme required to produce 1 nmol of acid-soluble nucleotide from [^{3}H]DNA duplex in 30 min at 37°C in 50 mM Tris-HCl, pH 7.6, 1 mM $MgCl_2$, 1 mM DTT, and 10 nmol of [^{3}H]DNA.)

(8b) Alternatively, to run all the time points in a single lane on the agarose gel, remove 2.5 μl aliquots at 30-s intervals into a single tube on ice containing 90 μl of S1 nuclease mix (for 12 time points).

(9) Move the S1 tubes to room temperature for 30 min.

(10) Add 1 μl of S1 stop solution to each tube and heat at 70°C for 10 min. (12 μl of S1 stop solution for the single-tube protocol.)

(11) Transfer the tubes to 37°C and add 2 μl of Klenow mix to each tube. (12 μl of Klenow mix for the single-tube protocol.) Incubate for 3 min and add 1 μl of dNTP mix. (12 μl of dNTP mix for the single-tube protocol.) Incubate for an additional 5 min at 37°C.

(12) For the single-tube protocol:
  (a) add 0.1 volumes of 2 M NaCl to the tube;
  (b) add 2 volumes of ice cold 100% ethanol;
  (c) centrifuge in a microfuge for 15 min;
  (d) aspirate off supernatant and wash DNA pellet with 250 μl of 70% ethanol;
  (e) centrifuge in a microfuge for 10 min; and
  (f) aspirate off supernatant and resuspend DNA pellet in 20 μl H_2O.

(13) Add 2 μl of 10× loading buffer to each tube and heat at 68°C for 3 min.

(14) Load each tube onto a lane of an 0.8% LMP minigel and separate for approximately 2.0 h at 60 V.

(15) At the completion of electrophoresis a ladder of linear fragments should be visible in successive lanes on the gel. Carefully slice the bands from each lane and place each one into an individual Eppendorf tube. For the single-lane protocol slice the smear into three individual bands.

(16) Melt the agarose slices at 65°C for 5–10 min. Prepare a ligation tube for each time point containing: 4 μl of 5× BRL ligation buffer, 10 μl of water, and 1 μl of BRL T4 DNA ligase.

(17) Place the melted gel slices at 37°C. Mix well and add 5 μl from each time point to the appropriate ligation tube.

(18) Incubate overnight at 12–16°C.

(19) Prepare $CaCl_2$ competent cells (Xl-1 Blue from Stratagene Cloning Systems, or equivalent).

(20) Put aliquots (100 μl) of competent cells into Falcon 2059 tubes on ice.

(21) Re-melt the ligations and add 5 μl from each ligation to 100 μl of competent cells.

(22) Incubate on ice for 30 min.

(23) Heat shock at 42°C for 45 s, and then place on ice for 2 min.

(24) Add 200 μl of SOC and incubate for 1 h at 37°C.

(25) Plate the entire transformation on LB/Amp (50 μg ml^{-1}) plate and incubate overnight at 37°C.

(26) Prepare minipreps from at least three colonies per time point. For the single-tube protocol prepare at least 10 colonies from each of the three plates.

(27) Run an aliquot of each plasmid on an agarose gel and select a ladder of plasmids for sequencing.

9.3 DISCUSSION

The protocols described in this chapter are used routinely in our laboratory to construct Exo III generated nested deletion sets. A nested deletion set provides a group of plasmids for sequencing that gives sufficient sequence overlap for efficient assembly and redundancy for sequence accuracy. In order to produce successful nested deletion sets it is particularly important to invest the time in the up-front preparative processes, in particular the choice of vector, the preparation of the supercoiled plasmid DNA, and the double restriction digestion of the plasmid. Phagemid vectors such as Stratagene pBluescript and the Promega pGEM series of vectors are particularly useful since they can be used for library construction and provide multiple cloning sites with the appropriate flanking restriction sites. We have found that preparation of plasmid DNA with the Qiagen Tip 100 reagent provides supercoiled plasmid DNA of sufficient quantity and quality for the exonuclease protocol. It is critical that the plasmid DNA is digested to completion by both restriction endonucleases. Monitoring the digestion in an agarose gel can help in diagnosing problems farther down the line.

In the Exo III protocol some titration of the particular Exo III lot may be necessary to get an appropriate and reproducible rate of digestion. The digestion of the ssDNA overhang generated by the Exo III digestion can be accomplished using either S1 nuclease or mung bean nuclease (Vogt, 1973; Stratagene Protocol, 1990). The S1 nuclease protocol is used successfully routinely in our laboratory and is convenient to use in the context of the exonuclease protocol described in this chapter. However, certain properties of mung bean nuclease (Kroeker & Kowalski, 1978) can make it the enzyme of choice, especially in the presence of nicked DNA. Mung bean nuclease can then be substituted, modifying the protocol with additional purification and precipitation steps (Stratagene Protocol, 1990).

Separation of the DNA product from each time point by electrophoresis on an LMP agarose gel is useful not only as an analytical step but also as a step for the purification of the DNA product. The purification step provides a more homogeneous representation of each time point without interference from smaller DNA contaminants which can be generated during the Exo III protocol. In-gel ligation overcomes the need for extraction of the DNA fragment from the LMP gel. Preliminary results indicate that a room temperature ligation of 2–3 h may obviate the need for an overnight ligation.

In general, choosing three colonies from each plate gives sufficient representation of the product produced from each time point. With experience, examination of the supercoiled plasmid on an agarose gel allows the investigator to choose an array of clones that cover the insert of interest. If gaps occur in the process of assembly, the investigator can return to the appropriate time point choosing additional clones to fill in the gap.

We have recently begun to examine the use of the single-lane protocol presented here. The major advantage of this modification is that it allows an individual to get the same amount of information from one-quarter the amount of ligations, transformations, and plates. This should increase the number of clones an individual can process at once and increase the overall throughput and efficiency of the laboratory.

REFERENCES

Ahmed, A. (1984) *J. Mol. Biol.* **178**, 941–948.

Ahmed, A. (1985) *Gene* **39**, 305–310.

Anderson, S. (1981) *Nucleic Acids Res.* **9**, 3015–3027.

Barnes, W.M., Bevan, M. & Son, P.H. (1983) *Methods Enzymol.* **101**, 98–122.

Deininger, P.L. (1983) *Anal. Biochem.* **129**, 216–223.

Henikoff, S. (1984) *Gene* **28**, 351–359.

Henikoff, S. (1987) *Methods Enzymol.* **155**, 156–165.

Kroeker, W.D. & Kowalski, D. (1978) *Biochemistry* **17**, 3236–3243.

McCombie, W.R., Kirkness, E., Fleming, J.T., Kerlavage, A.R., Iovannisci, D.M. & Martin-Gallardo, A. (1991) *Methods: Companion Methods Enzymol.* **3**, 33–40.

Messing, J. (1983) *Methods Enzymol.* **101**, 20–78.

Putney, S.D., Benkovic, S.J. & Schimmel, P.R. (1981) *Proc. Natl. Acad. Sci. U.S.A.* **78**, 7350–7354.

Stratagene Protocol (1990) *Exo III/Mung Bean Nuclease Deletions*. pp. 1–8.

Strauss, E.C., Kobori, J.A., Siu, G. & Hood, L.E. (1986) *Anal. Biochem.* **154**, 353–360.

Titus, D.E. (ed.) (1991) *Promega Protocols and Applications Guide*, 2nd edn. pp. 90–98.

Vogt, V.M. (1973) *Eur. J. Biochem.* **33**, 92–200.

Zasloff, M., Ginder, G.D. & Felsenfels, G. (1978) *Nucleic Acids Res.* **5**, 1139–1152.

CHAPTER TEN

Expressed Sequence Tags as Tools for Physiology and Genomics

M.D. ADAMS

The Institute for Genomic Research, 932 Clopper Road, Gaithersburg, MD 20878, USA

10.1 INTRODUCTION

DNA sequence data provide the highest level of identification of gene structure. Determination of the distribution of mRNA species in a cell provides an analysis of the physiological activity of that cell. Combining the technology of automated sequencing with randomly selected cDNA clones from libraries that represent mRNA distributions allows one to construct a very detailed picture of the transcriptional activity of a cell or tissue that includes not only the identification of transcribed genes, but also abundance level, and ultimately the degree of overlap in gene expression among various tissues. Partial cDNA sequences or expressed sequence tags (ESTs) offer the fastest way to obtain gene-specific sequence data for a large number of independent cDNA clones. By designing libraries and sequencing strategies carefully, the maximum amount of information can be derived from any particular mRNA population. EST sequence data on over 12 000 cDNA clones from several organisms have been published (Table 10.1).

The two key aspects of an EST project are the choice and construction of the library, and the facilities for accumulating and correlating the sequence analysis data. In most cases, the decisions to be made are independent of the organism, tissue, cell type, or developmental stage to be studied. An oligo-dT-primed, directionally cloned library, in which the 5′ and 3′ ends of the mRNA are oriented consistently with respect to the vector, is the most desirable. This type of library permits 3′ anchored sequencing, where each cDNA is sequenced from the extreme 3′ end, thus permitting accurate analysis of the mRNA distribution. Sequencing clones from the 5′ end of a directional library is also valuable because it is much more likely to reveal the coding sequence in the cDNA clone, especially among higher vertebrates where 3′ untranslated sequences tend to be longer than the standard sequencing reaction.

Sequence data are accumulated at a very rapid pace in EST projects. A single sequencing reaction gives all of the information necessary for extensive analysis of a partial gene sequence. If one is working with a good library, every clone sequenced has the potential to be an interesting new gene. Coping with the quantity of sequence analysis data while preserving their quality depends on maintaining adequate database facilities, but also on devoting sufficient time to analysis by scientists familiar with the biochemistry, physiology, and genetics of the organisms being studied.

Table 10.1 Data from EST projects.

Organism	*Tissue*	*No. of EST clones*	*Reference*
Human	Brain	7593	Adams *et al.* (1991, 1992, 1993a,b)
Human	Brain	1024	Khan *et al.* (1992)
Human	HepG2 cells	982	Okubo *et al.* (1992)
Human	Pancreatic islets	1000	Takeda *et al.* (1993)
Human	Retina	58	Gieser and Swaroop (1992)
Caenorhabditis elegans	Mixed stage	585	McCombie *et al.* (1992)
Caenorhabditis elegans	Mixed stage	1517	Waterston *et al.* (1992)
Mouse	Testis	171	Höög *et al.* (1991)
Rabbit	Muscle	178	Putney *et al.* (1983)

10.2 cDNA LIBRARIES

Chapters 15, 16 and 17 address technical issues of cDNA library construction. Key factors for construction of a library that is optimally useful for EST sequencing are presented briefly here. A directionally cloned library is usually primed with oligo-dT coupled to an adaptor/linker. A different adaptor/linker is ligated to the 5′ end of the cDNA, permitting asymmetric cloning into a vector disgested with two different restriction enzymes. Completely directional cloning is important, as noted above, for knowing *a priori*, the location of the 5′ and 3′ ends of the cDNA clone. For 3′ end anchored sequencing to be practical, the length of the poly(A) tail in a cDNA clone must be less than about 30 bp. Otherwise, sequencing reactions through the tail will stall, ruining further extension. Finally, all clones should in fact have poly(A) tails; if cDNA synthesis was initiated with the adaptor portion of the primer rather than the oligo-dT portion, the library will still be directional, but will not be anchored at the 3′ end.

Sequencing redundancy initially provides useful information regarding the abundance of clones in a library. Most cell-types are believed to express between 5000 and 20 000 different genes. In general, only a few dozen will represent more than 0.5% of the mRNA population (10–20% of the total mRNA). It is interesting to know the identity of these mRNA species that might be present at a rate of 5 or more per 1000 cDNA clones. After sequencing 1000 clones, however, repeated resequencing does not give additional information on abundant clones. One of the most effective ways of eliminating abundant clones is to screen or subtract with total cDNA (after blocking with unlabelled genomic DNA in the case of mammalian libraries – see Chapter 24). Use of total cDNA from the tissue of origin as a probe means that knowledge of the sequence of abundant clones ahead of time is not necessary. An alternative approach to reducing the number of abundant clones is to use a defined composition probe consisting of EST clones known to represent abundant messages. While this approach requires more effort to produce the probe, there is a lower probability of detecting false positives during hybridization or loss of clones nonspecifically during subtraction. In general, clones representing <0.5% of the mRNAs do not need to be subtracted as the effort expended provides diminishing returns. For more thorough characterization of a library, and to approach closure in identifying all cDNAs from a library, subtraction and normalization are necessary, since it is not possible to construct a hybridization probe of sufficient complexity to eliminate all the moderately abundant clones.

10.3 ESTs: MAXIMIZING SEQUENCING THROUGHPUT

Expressed sequence tags (ESTs) represent the largest amount of information possible per raw base sequenced. Single-pass sequencing, while less accurate than highly redundant contigs of overlapping sequences (see Section 10.4), is accurate enough for very sensitive similarity searches. Without redundancy, a large number of independent clones can be analyzed, rather than analyzing tens to hundreds of overlapping clones to determine the sequence of a single gene. For instance, over 3000 sequencing reactions were run to determine the 106 000 base sequence of three cosmids from chromosome 19 (Martin-Gallardo *et al.*, 1992); these cosmids probably contain five genes. Application of these same sequencing reactions to independent cDNA clones would result in the identification of well over 2000 different genes, even allowing for extensive resequencing of abundant mRNAs. A basic outline of our protocol for EST sequencing follows (Figure 10.1).

cDNA libraries are often made in λ-ZAP, using the directional cloning system with EcoRI/XhoI arms. cDNA is ligated and packaged essentially according to the manufacturer's (Stratagene) instructions in the Uni-ZAP and Gigapack gold kits. Other protocols are

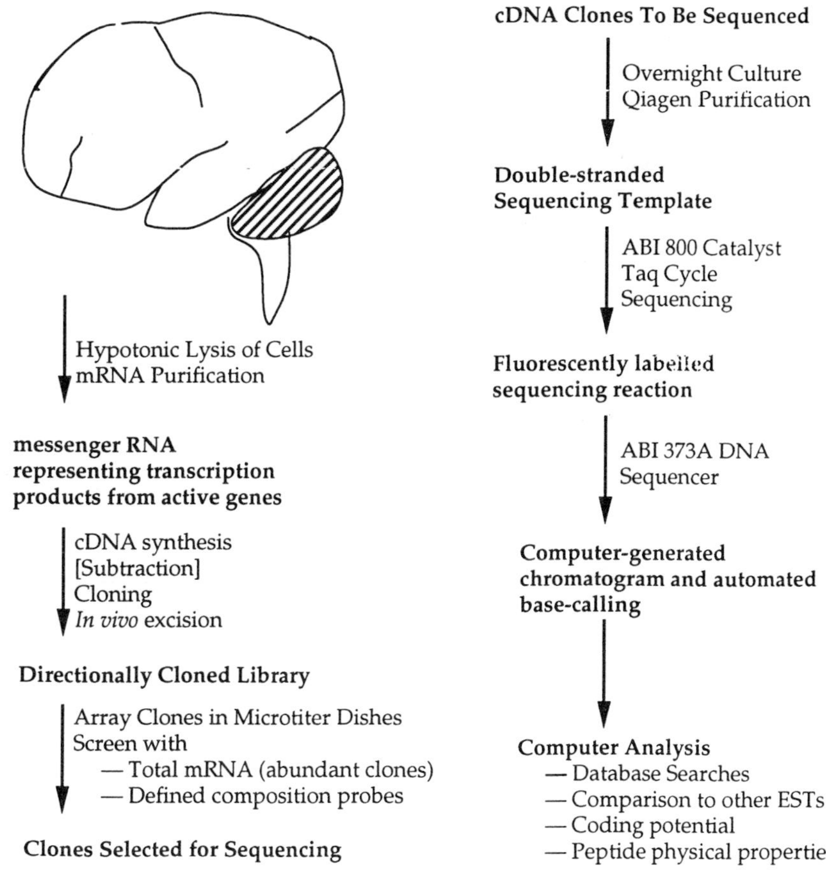

Figure 10.1 The building blocks of an EST project are listed with key decision points. At each step, relevant lab techniques including mRNA purification, cDNA library construction, clone selection, and automated sequencing are shown. Many variations on this theme are possible, depending upon the specific application, such as subtraction or screening of the library to bias sequencing toward particular groups of clones. Computional analysis may also take many forms based around sequence similarity searching and compositional characteristics.

presented in chapters in this book. The packaged library is amplified in *Escherichia coli* SURE cells. The amplified library is rescued *en masse* by *in vivo* excision into pBlue script packaged as single-stranded circles in M13 phage heads. While the standard protocols from Stratagene are used, great care is given to determining the optimum ratio of host cells, helper phage and cDNA-containing lambda phage. A successful *in vivo* excision generally gives a high titer of packaged pBlue script and only a minimal increase in the percentage of blue colonies compared with blue plaques from the amplified phage library. Colonies obtained from plating of the rescued library are analysed for insert size and compared to insert sizes from phage plaques obtained by polymerase chain reaction (PCR) to be certain that small clones were not preferentially amplified in the rescue. The supernatant from the *in vivo* excision is stable for at least a year at 4°C without loss of titer.

Fresh colonies are plated weekly for preparation of plasmid DNA templates for sequencing. An aliquot of the packaged pBlue script is mixed with host cells (generally XL1–Blue) and plated on LB plates supplemented with 50 μg ml^{-1} ampicillin. The host strain chosen depends in part on the plasmid preparation method used. XL1-Blue is far superior for Qiagen preps, for instance, than DH5 or JM109 strains. Overnight cultures (5 ml) are split into two 500 μl fractions for storage at −70°C and a 4-ml aliquot for plasmid DNA purification. We have tried many different plasmid preparation protocols, but find that Qiagen consistently provides the cleanest DNA template for automated sequencing. The standard Qiawell-8 protocol using strips of eight minicolumns in microtiter format is used, except an extra spin is added to clear the lysate and five washes of the column are used prior to elution rather than the suggested two washes. Determination and adjustment of the concentration of the template is crucial to successful automated sequencing. Concentration of the plasmid template is adjusted to 250 ng μl^{-1} with water based on the absorbance at 260 nm. A relatively narrow concentration range (200–300 ng μl^{-1}) of plasmid template provides the highest quality sequence data.

DNA sequencing reactions are performed as suggested in the Prism ReadyReaction cycle sequencing kits from Applied Biosystems, Inc. (ABI) on a Catalyst LabStation from ABI, and electrophoresed on an ABI 373 DNA Sequencer. Chromatogram files are stored on a Unix partition mounted remotely on Macintoshes via Columbia AppleTalk protocol. Chromatograms are edited and then archived on 8-mm tape. Sequence data and analysis results are entered into a specially designed Sybase database (Kerlavage *et al.*, 1993). Information on cDNA library, template preparation, sequencing gel, and chromatogram editing for each EST is stored in the database along with results of computational analyses of the sequences. The name of the EST encodes information on the cDNA library from which it is derived and the plate and well location of the clone in a microtiter plate.

10.4 ASSESSING THE ACCURACY OF EST SEQUENCES

Rapid throughput in EST sequencing is obtained through the use of single sequencing runs on many different clones. The low sequencing redundancy means that ambiguous base calls are not generally confirmed by a second reaction or by sequencing the other strand. It is important, therefore, to estimate the accuracy of the sequence to provide users of the data with some idea of the confidence that can be placed in it.

A number of factors affect the accuracy of sequence data including the quality of the template DNA, the fidelity of the sequencing reaction, and sequence-dependent characteristics such as poly(A) stretches and G/C compressions. Accuracy must be determined for each set of template/reaction conditions separately. Furthermore, in assigning an accuracy value for a data set, it is important to use consistent data editing procedures so that sequence ends are trimmed at comparable breakpoints between 'reliable' and 'unreliable' base calls.

The simplest way to estimate the accuracy of a group of sequences is to take advantage of matches between ESTs and previously sequenced genes from the organism being studied. For human studies, the ribosomal RNAs, mitochondrial genome, and several housekeeping genes have been useful for determining accuracy based on sequence alignments. Specifically, known sequences were extracted from GenBank (such as the human mitochondrial genome sequence HUMMTCG) and used to query a local EST database using FASTA, which inserts gaps to achieve optimal alignment. Each alignment can then be examined for the number of ambiguous and miscalled bases, insertions, and deletions in the EST compared to the known query sequence. Since each of these error types occurs at different rates and for different reasons with respect to the sequencing chemistry, each should be calculated separately. The Applied Biosystems 373A sequencer should be able to achieve a consistent accuracy rate of at least 98% correct base calls in the region between 30 and 400 bp from the sequencing primer using dye-primer chemistry. In our laboratory, ESTs shorter than 150 bp or with more than 3% ambiguous base calls on the 373A are rejected.

10.5 ANALYSIS OF EST DATA

Computational analyses can be applied to several thousand sequences in a fraction of the time required to perform a genetic, biological or chemical analysis on a single corresponding clone or protein. It is therefore important to extract as much information and as many clues about the identity and characteristics of new genes as possible from their sequences alone. An integrated sequence analysis and database system has been developed to do this using the Sybase database management system (Kerlavage *et al.*, 1993). A series of Unix and Macintosh programs has been written to facilitate data entry and retrieval from the database. The relational structure permits grouping of data and links based on common features across data sets. Other groups have used object-oriented database designs to store information. What all of these systems have in common is the ability to store analysis results and combine information from several sources for asking more detailed biological questions about the data sets as a whole. Several chapters in this volume address issues of database functionality for sequence information integration.

New ESTs are edited to remove polylinker sequence from the vector and trailing sequence of low quality at the end of the sequencing run. ESTs are then pre-searched against a defined-composition database to filter out vector sequence, ribosomal and mitochondrial sequences, and common repetitive elements (which are difficult to name accurately in searches of the complete databases). ESTs that have no matches in this prescreen database are searched against the public nucleotide and peptide (with hypothetical six-frame translations) databases using the BLAST (Altschul *et al.*, 1990) network server at the National Center for Biotechnology Information (NCBI), National Library of Medicine. A pair of comprehensive, nonredundant databases composed of entries from GenBank and EMBL (nucleotide) and GenPept, SWISS-PROT, and PIR (peptide) are used for the searches. Nucleotide searches distinguish between exact and nonexact matches to previously sequenced human genes and find matches in 5′ and 3′ untranslated regions. Peptide searches are generally more sensitive than nucleotide searches and are used to find more distant evolutionary or motif-related matches. Where possible, putative identifications are made based on the database matches. In general, BLAST scores of 300 for nucleotide and 85 for peptide matches are the minimum necessary to assign a putative identification.

Other computational methods have been used, especially in cases where BLAST is not able to provide a putative identification. The GRAIL E-mail server (grail@ornl.gov) is used to predict protein-coding ESTs. These can be used to search nucleotide databases (including other ESTs) with TBLASTN to find matches to unannotated protein-coding regions. Approximately 5% of ESTs without other matches in the public databases can be identified in this way. More sensitive searching algorithms such as FASTA or BLAZE may also identify additional significant matches, depending on the search criteria and threshhold for acceptable matches.

When ESTs are derived from the anchored 3′ end of a cDNA clone (e.g. by reading through the poly(A) tail), it is possible to accurately calculate the frequency of the clone in the library, and by extension the abundance of the mRNA, if the library is in fact representative. Furthermore, comparison of EST frequencies between tissues permits a detailed examination of tissue specificity and the number of genes that are in fact ubiquitously expressed.

10.6 CHARACTERIZATION OF TISSUE PHYSIOLOGY

Another aspect of EST sequencing is that relative levels of expression can be determined for a cell-line or tissue and compared to similar information from other tissues. Efficient storage of sequence analysis data, including database match, tissue of origin, and any categorization information is essential to facilitate making comparisons between EST data sets. A large number of ESTs from human brain and hepatoma cell lines have been published (Table 10.1) so that comparisons of these tissues at the gene expression level is possible. Information derived from database matches was used to describe the transcriptional activity in broad categories.

While a number of categorization schemes are probably equally valid, we have used a system that attempts to describe the physiological activity of the tissue by grouping genes by biochemical processes such as energy metabolism, other anabolic and catabolic metabolism, macromolecular synthesis, several categories of regulatory processes, etc., rather than by subcellular location or other catalog.

The differences between the two samples are striking. Most notably, secreted proteins and components of the transcription and translation apparatus are much more abundant in HepG2. Conversely, structural and regulatory proteins are more abundant in brain. The causes and relevance of these differences are worth considering. Albumin is the most abundant secreted protein in the hepatoma set, indicative of the tissue of origin. The increase in transcription and translation components and metabolic enzymes, however, could simply be due to the fact that the hepatoma cells are a rapidly dividing, transformed cell line. The brain cDNA libraries were derived from a tissue comprising a diverse set of cell types. The greater number of protein-modifying enzymes, receptors, transcription factors, and other signal transduction proteins indicates a much higher proportion of the biochemical activity of the tissue is directed toward signalling and regulatory processes. Components of the cytoskeleton and other structural proteins that are more prevalent in brain than hepatoma could be explained by the highly ordered structure of the axonal portions of neurons and other morphological characteristics of brain tissue.

In many cases, the annotations of sequences matched by ESTs in GenBank or other databases do not give a sufficient explanation to adequately classify the function of the encoded protein. We have used the program Entrez from the National Center for Biotechnology Information to link GenBank (and other database) sequences matched by ESTs to the abstracts from the papers in which they were published. Entrez's 'neighbouring' scheme allows related journal articles and sequences to be retrieved as well. It is, of course, often necessary to refer to the full text of an article in order to obtain sufficient information for adequate functional classification.

Undoubtedly, technological advances will appear over the next several years to bring the complete human genome sequence into our scientific lifetimes. In the meantime, the existing automated sequencing technology, when applied to partial sequencing of cDNAs, provides an extraordinary amount of information on the content of the genome and in the process, a glimpse at the transcriptional activity and diversity of different tissues and developmental stages.

ACKNOWLEDGMENTS

Tony Kerlavage, Chris Fields, and Mark Dubnick were instrumental in developing the EST database and analysis methods. Craig Venter provided laboratory facilities and intellectual vision at all stages of the development of the EST program.

REFERENCES

Adams, M.D., Kelley, J.M., Gocayne, J.D., Dubnick, M., Polymeropoulos, M.H., Hong, X., Merril, C.R., Wu, A., Olde, B., Moreno, R.F., Kerlavage, A.R., McCombie, W.R., Venter, J.C. (1991) *Science* **252**, 1651–1656.

Adams, M.D., Dubnick, M., Kerlavage, A.R., Moreno, R., Kelley, J.M., Utterback, T.R., Nagle, J.W., Fields, C. and Venter, J.C. (1992), *Nature* **355**, 632–634.

Adams, M.D., Kerlavage, A.R., Fields, C. and Venter, J.C. (1993) *Nature Genet.* **4**, 256–267.

Adams, M.D., Soares, M.B., Kerlavage, A.R., Fields, C., Venter, J.C. (1993) *Nature Genet.* **4**, 373–380.

Altschul, S.F., Gish, W., Miller, W., Myers, E.W. and Lipman, D.J. (1991) *J. Mol. Biol.* **215**, 403–410.

Gieser, L. and Swaroop, A. (1992) *Genomics* **13**, 873–876.

Hoog, C. (1991) *Nucleic Acids Res.* **19**, 6123–6127.

Kerlavage, A.R., Adams, M.D., Kelley, J.C., Dubnick, M., Powell, J., Shanmugam, P., Venter, J.C., Fields, C. (1993) *Proceedings of the Twenty-Sixth Annual Hawaii International Conference on Systems Sciences* 585–594 (IEEE Computer Science Press, New York.

Khan, A.S., Wilcox, A.S., Polymeropoulos, M.H., Hopkins, J.A., Stevens, T.J., Robinson, M., Orpana, A.K. and Sikela, J.M. (1992) *Nature Genet.* **2**, 180–185.

Martin-Gallardo, A., McCombie, W.R., Gocayne, J.D., FitzGerald, M.G., Wallace, S., Lee, B.M.B., Lamerdin, J., Trapp, S., Kelley, J.M., Lui, L.-I., Dubnick, M., Johnston-Dow, Kerlavage, A.R., de Jong, P., Carrano, A., Fields, C. and Venter, J.C. (1992) *Nature Genet.* **1**, 34–39.

McCombie, W.R., Adams, M.D., Kelley, J.M., FitzGerald, M.G., Utterback, T.R, Khan, M., Dubnick, M., Kerlavage, A.R., Venter, J.C., Fields, C. (1992) *Nature Genet.* **1**, 124–131.

Okubo, K., Hori, N., Matoba, R., Niiyama, T., Fukushima, A., Kojima, Y. and Matsubara (1992) *Nature Genet.* **2**, 173–179.

Putney, S.D., Herlihy, W.C. and Schimmel, P. (1983) *Nature* **302**, 718–721.

Takeda, J., Yano, H., Eng, S., Eng, Y. and Bell, G. (1993) *Hum. Molec. Genet.* **2**, 1793–1798.

Waterston, R., Martin, C., Craxton, M., Huynh, C., Coulson, A., Hillier, L., Durbin, P., Green, P., Shownkeen, R., Halloran, N., Metzstein, M., Hawkins, T., Wilson, R., Berks, M., Du, Z., Thomas, K., Theirry-Meig, J., Sulston, J. (1992) *Nature Genet.* **1**, 114–123.

CHAPTER ELEVEN

The Use of Automated DNA Sequencing in the Analysis of cDNAs of Model Organisms

W.R. McCOMBIE

Cold Spring Harbor Laboratory, Cold Spring Harbor, NY 11724, USA

11.1 INTRODUCTION

Experimental model systems have played a crucial role in the progress of biological research. The selection of a particular model system is a compromise between biological generality and ease of experimental manipulation in answering specific experimental questions. More recently developed model systems, such as *Caenorhabditis elegans*, have been selected based on their ability to answer questions in complex areas such as development and neural function at the molecular level (for a review see Wood, 1988). It would be extremely valuable to provide a link between various model systems (and humans) at the molecular level that could be accessed by computer searches.

Since the invention of automated DNA sequencers in 1986 (Smith *et al.*, 1986), there has been outstanding progress towards increasing the throughput of these instruments. Current Applied Biosystems DNA sequencers can sequence up to 72 templates per day (with two runs per day). This means that a single instrument can sequence 10 000–15 000 templates per year. One purpose to which this capacity has been applied is the rapid generation of expressed sequence tags (ESTs). The importance of sequencing expressed genes and using them as mapping probes to develop a high-resolution physical map of the human genome was suggested by Brenner (1990). This strategy can also be fruitfully applied to the expressed genes of model organisms. In these procedures large numbers of cDNA clones are partially sequenced by a single-pass sequencing run on an automated DNA sequencer, and the resulting sequence is analyzed and stored in the public databases (Adams *et al.*, 1991, 1992; Wilcox *et al.*, 1991; McCombie *et al.*, 1992; Waterston *et al.*, 1992; Kahn *et al.*, 1992; Okubo *et al.*, 1992). Since each template can represent a unique clone, up to 15 000 clones can be tagged per year with a single sequencer. Even if a fairly high percentage of the clones do not represent unique genes, these partial sequences allow a large number of gene identifications to be made in a short period of time.

In this chapter I describe strategies for applying these techniques to model systems and how this will impact on the future use of these organisms.

11.2 *CAENORHABDITIS ELEGANS*

C. elegans is a free-living nematode whose nutritional requirements, life-cycle and anatomy make it a valuable organism for many studies of the development of multicellular organisms and the development and function of the nervous system (for a review see: Wood, 1988). The genome of this organism, which is about 100 Mb and encodes about 10 000–20 000 genes, is currently being sequenced (Sulston *et al.*, 1992). The genome has already been extensively mapped and most of it has been cloned into a set of overlapping yeast artificial chromosomes (YACs) which can be used for rapid physical mapping (Coulson *et al.*, 1986, 1988). A set of overlapping cosmids comprising most of the genome has been ordered, mapped to these YACs (Coulson *et al.*, 1986, 1988), and calibrated against the genetic map of the animal (Edgley & Riddle, 1990). This allows a rapid interface between a cloned gene (or gene fragment), which can be physically mapped, and a testable hypothesis to be proposed regarding the phenotype of that gene based on the related genetic map and mutational analysis.

At the time we began our work, only a small percentage of the genes in *C. elegans* were known at the sequence level. When a gene from another species was identified whose function might be best examined in *C. elegans*, the *C. elegans* homolog first had to be cloned and sequenced. Often this was straightforward, but still required a few months. In less fortunate examples various factors could make the cloning of a gene from this organism problematic. It seemed likely that the availability of sequence information from a large percentage of the genes from this organism would radically change the way in which it was used as a model system. If a simple computer search would allow the identification of a potential *C. elegans* homolog (or homologs) of a human gene, the entire process of examining homologs in this organism would be made more rapid and efficient. The initial clone identified by partial sequencing would be sufficient for physical mapping or isolation of corresponding genomic clones from the cosmid set. The genomic clones could then be used for marker rescue to determine the phenotype of the gene, or for regulatory studies using gene fusions (Fire, 1986).

11.3 LIBRARY SELECTION

There are several factors to consider before a library is chosen for EST sequencing. The first is library quality. It must be remembered that this is essentially an operational definition of suitability for the intended application. As a result, the rather unusual application under discussion here places different restraints on the quality of a library than do more traditional applications such as screening for a particular gene.

In the case of random cDNA sequencing, several generally desired traits of cDNA libraries can be identified (see also Chapter 10). A good library for this purpose should have a large number of full-length clones. While not important for sequencing *per se*, it will minimize the additional work if a clone of interest needs to be studied through the analysis of its full-length cDNA. This is less important in *C. elegans* than in some systems because of the availability of the overlapping set of mapped cosmid clones comprising the genome of the organism. This asset makes it relatively easy to isolate genomic fragments containing the gene in question which can be used for many of the follow-up studies. Secondly, the library should contain a minimal amount of clones representing nonmessenger RNA populations. In a similar way, the library should contain as low a percentage as possible of inserts derived from bacterial host or phage vector DNA. This is not usually as great a concern when screening a library for a desired gene since the undesired host genes would not usually be selected. However, this becomes an important issue when sequencing cDNAs. If the gene has previously been identified as being from phage or *E. coli*, it can be removed from the database, but this is not as simple with contaminating DNA from an uncharacterized source. Lastly, an ideal library should contain no chimeric clones. This will facilitate mapping and eliminate confusion in the use of the clones for characterization of the structure and function of the genes they represent.

The next characteristic to consider is whether or not the library was directionally cloned. A directional library allows the investigator to choose which end of the cDNAs to sequence based on the advantage of sequencing from that end and the goals of the project (see Chapter 10). Such fine tuning is not possible without a directionally cloned library.

The final criterion to enter into library selection is the specific origin of the library. In our initial study of *C. elegans* cDNAs, the choice was made to use a library derived from mixed-stage animals (McCombie *et al.*, 1992). The mixed-stage library provides the widest cross-section of genes from *C. elegans* and provides a basis for future studies that might employ more specific libraries. Developmental stage is probably the most important variable to consider in *C. elegans* library selection, and is an important consideration in other developmentally complex organisms. In larger animals that are amenable to dissection, the choice of what tissues or organs from which to make the library is also important. The developmental window and level of gene expression for some genes in one tissue may be more advantageous for cloning than in the actual tissue of interest. The difficulty in getting sufficient material to make high-quality libraries from some sources also needs to be considered.

The important point is that in more complex systems the decisions become increasingly complex as well.

A related area that needs to be evaluated in library selection is the relationship of the organism to its environment at the time the library was made. This is clearly seen in the most simple systems where this relationship constitutes a major component of variable gene expression. For instance, we have been sequencing cDNAs from the fission yeast *Schizosaccharomyces pombe* (T. Matsumoto, N. Kaplan and W.R. McCombie, unpublished data). Owing to the simplicity of this organism, the growth stage and medium that the yeasts are growing on are the principal determinants of which genes are expressed in the library. The extent to which environmental effects modulate gene expression in more complex organisms is much less well understood. It seems likely, however, that such effects will eventually be noticed in random EST projects on complex eukaryotes.

11.4 ANALYSIS OF *C. ELEGANS* PARTIAL cDNA SEQUENCES

The analysis of cDNA sequence data consists of two phases. First is a series of computer-based manipulations that might include homology searches, analysis of open-reading frames, searches for DNA or protein coding motifs and compositional analysis that aid in classifying the clones. This serves as a starting point for the experimental analysis of the clones. With *C. elegans* partial cDNAs the primary computer analyses were homology searches (McCombie *et al.*, 1992; Waterston *et al.*, 1992). This identifies two types of potential relationship between the *C. elegans* cDNAs and sequences in the databases, These are exact matches to previously characterized *C. elegans* genes or genes similar to, but not identical to, those found in *C. elegans* or other species. In the case of similarity to *C. elegans* genes it should be noted that there is somewhat of a grey area between being identical to a previously characterized *C. elegans* gene and similar to it in cases where it cannot be ascertained if the differences are real or the result of sequencing errors. If this is important in a particular case it would need to be resolved with further experimentation.

The identification of partial sequences of genes similar to those found in humans or other higher organisms is the *raison d'être* of doing partial cDNA sequencing from a simple model system such as *C. elegans*. The computer searches should be targeted towards identifying *C. elegans* genes that can be used to elucidate the function of genes in higher organisms. As a result we have reported on only those genes showing a fairly high degree of similarity to a sequence in the protein databases as indicated by a Blast (Altschul *et al.*, 1990) score of 95 or above (McCombie *et al.*, 1992). Once the initial database matches have been found, more complete analysis can be carried out by gathering more experimental data. This can proceed along several different paths, depending on the nature of the original database match.

11.5 CONCLUSION

The rapid sequencing of partial cDNAs from easy-to-manipulate experimental organisms will greatly facilitate our ability to identify potential homologs of these new human genes from model organisms in which their normal function can be elucidated. The availability of cDNA sequence databases from a wide range of model systems will let the investigator rapidly move to characterize a new gene in the most advantageous model system. The work to characterize the gene can begin immediately since the sequence and clones are already available. In addition, this will greatly facilitate studies of the comparative biology of genes in different model systems. These can be tied rapidly to genetic studies in other organisms at the level of the individual gene. This combination of rapid sequencing and mapping in model organisms will change the way in which the normal function of human genes can be determined.

REFERENCES

Adams, M.D., Kelley, J.M., Gocayne, J.D., Dubnick, M., Polymeropoulos, M.H., Xiao, H., Merril, C.R., Wu, A., Olde, B., Moreno, R.F., Kerlavage, A.R., McCombie, W.R. & Venter, J.C. (1991) *Science* **252**, 1651–1656.

Adams, M.D., Dubnick, M., Kerlavage, A.R., Moreno, R., Kelley, J.M., Utterback, T.R., Nagle, J.W., Fields, C. & Venter, J.C. (1992) *Nature* **355**, 632–634.

Altschul, S.F., Gish, W., Miller, W., Myers, E.W. & Lipman, D.J. (1990) *J. Mol. Biol.* **215**, 403–410.

Brenner, S. (1990) *Ciba Foundation Symp.* **149**, 6–12.

Coulson, A., Sulston, J., Brenner, S. & Karn, J. (1986) *Proc. Natl. Acad. Sci. U.S.A.* **83**, 7821–7825.

Coulson, A., Waterston, R., Kiff, J., Sulston, J. & Kohara, Y. (1988) *Nature* **335**, 184–186.

Edgley, M.L. & Riddle, D.L. (1990) *Genetic Maps* **5**, 3.111–3.133.

Fire, A. (1986) *EMBO J.* **5**, 2673–2680.

Kahn, A.S., Wilcox, A.S., Polymeropoulos, M.H., Hopkins, J.A., Stevens, T.J., Robinson, M., Orpana, A.K. & Sikela, J.M. (1992) *Nature Genet.* **2**, 180–185.

McCombie, W.R., Adams, M., Kelley, J.M., Fitzgerald, M.G., Utterback, T., Khan, M., Dubnick, M., Kerlavage, A., Venter, J.C. & Fields, C. 1992. *Nature Genet.* **1**, 124–131.

Okubo, K., Hori, N., Matoba, R., Niiyama, T., Fukushima, A., Kojima, Y. & Matsubara, K. (1992) *Nature Genet.* **2**, 173–179.

Smith, L.M., Sanders, J.Z., Kaiser, R.J., Hughes, P., Dodd, C., Connell, C.R., Heiner, C., Kent, S.B. & Hood, L.E. (1986) *Nature* **321**, 674–679.

Sulston, J., Du, Z., Thomas, K., Wilson, R., Hillier, L., Staden, R., Halloran, N., Green, P., Thierry-Mieg, J., Qiu, L., Dear, S., Coulson, A., Craxton, M., Durbin, R., Berks, M., Metzstein, M., Hawkins, T., Ainscough, R. & Waterston, R. (1992) *Nature* **356**, 37–41.

Waterston, R., Martin, C., Craxton, M., Huynh, C., Coulson, A., Hillier, L., Durbin, R., Green, P., Shownkeen, R., Halloran, N., Metzstein, M., Hawkins, T., Wilson, R., Berks, M., Du, Z., Thomas, K., Thierry-Mieg, J. & Sulston, J. (1992) *Nature Genet.* **1**, 114–123.

Wilcox, A.S., Khan, A.S., Hopkins, J.A., Sikela, J.M. (1991) *Nucleic Acids Res.* **19**, 1837–1843.

Wood, W.B. (ed.) (1988) *The Nematode Caenorhabditis elegans*. Cold Spring Harbour Press, Cold Spring Harbor, NY.

PART II

Sample Preparation and Sequencing Methods

A. LIBRARIES

CHAPTER TWELVE

Applications of Cosmid Libraries in Genome Mapping and Sequencing Efforts

R.L. STALLINGS,[1] N.A. DOGGETT,[2] A. FORD,[2] J. LONGMIRE,[2] C.E. HILDEBRAND,[2] L.L. DEAVEN[2] & R.K. MOYZIS

[1] Department of Human Genetics, University of Pittsburgh, Pittsburgh, PA 15261, USA
[2] Life Sciences Division and Center for Human Genome Studies, Los Alamos National Laboratory, Los Alamos, NM 87544, USA

12.1 INTRODUCTION

Chromosome specific recombinant DNA libraries derived from flow sorted chromosomes are useful reagents for the rapid generation of chromosome specific genetic markers (Donis-Keller *et al.*, 1987), sequence tagged sites (Green *et al.*, 1991b) and fluorescent *in situ* hybridization probes for painting individual human chromosomes (Collins *et al.*, 1991). In addition, chromosome 16 and chromosome 19 specific cosmid libraries have also been used to develop contig maps that cover substantial regions of these human chromosomes (Stallings *et al.*, 1992; Trask *et al.*, 1992). The chromosome 16 specific cosmid library has allowed the contig map of the metallothionein gene cluster to be closed (West *et al.*, 1990) and has permitted a chromosome 16p breakpoint found in Rubenstein–Taybi disease to be cloned (Breuning *et al.*, 1992). Emphasis has now been placed on the use of yeast artificial chromosomes (YACs) to close gaps in the chromosome 16 cosmid contig map.

In general, overall emphasis on physical mapping of the human genome has shifted away from cosmids to the use of YAC vectors because YACs provide greater mapping efficiency (Bellanne-Chantelot *et al.*, 1992). The second-generation YAC libraries now being constructed have average insert sizes of 800 kb (Bellanne-Chantelot *et al.*, 1992). Unfortunately, all of the presently available YAC libraries have a high frequency (40–60%) of chimeric clones (clones containing co-ligated DNA segments from noncontiguous regions of the genome) that greatly complicate their use (Green *et al.*, 1991a). Furthermore, YACs are somewhat difficult to use in molecular biology experiments and do not provide the finer level of resolution provided by cosmid contig maps. Thus, cosmid contig maps are highly desirable. Cosmid contig maps are already providing the starting material for sequencing the entire *Caenorhabditis elegans* genome (Sulston *et al.*, 1992). For these reasons, it is highly advantageous to have procedures in place for the rapid cloning of YAC inserts into cosmid vectors and for the ordering of cosmid subclones into contigs.

In this chapter, we convey some of the successes and some of the problems that we have encountered in the construction and analysis of cosmid libraries from YAC clones containing human DNA inserts. Methods for the efficient and rapid subcloning of YACs into cosmid vectors should greatly increase the utility of YAC based contig maps.

12.2 LIBRARY CONSTRUCTION FROM YAC CLONES

We have used the cosmid vector sCos-1 (Evans *et al.*, 1989) to construct libraries from both flow sorted human chromosomes and YAC clones. The BamH1 cloning site of this vector is flanked by both Notl and EcoRI sites that allow liberation of the insert from the vector. T3 and T7 RNA promoter sequences flank the BamH1 cloning site and allow for the synthesis of riboprobes from the ends of inserts. sCos-1 also contains the ColE1 origin of replication, the ampicillin resistance gene, and SV2–neo for selection of bacteria on kanomycin and eukaryotic cells on neomycin.

For subcloning YACs into cosmids, target DNA can be obtained from either the total yeast clone DNA or from only YAC insert DNA isolated by preparative pulsed-field gel electrophoresis. If DNA from the total yeast clone is used as target DNA, the resulting library requires screening for cosmids containing human inserts. Isolating YAC inserts on pulsed-field gels does not eliminate the necessity of screening for clones with human inserts since some level of yeast contamination inevitably occurs. This is particularly true when YAC inserts are of similar size to some of the yeast chromosomes. Since yeast have relatively small genomes (about 15 Mb), screening for cosmids containing human inserts is not too large a task when DNA from the entire yeast clone is subcloned. We screen for human insert bearing clones by hybridization of ^{32}P labeled Cot1 fractionated human DNA to cosmid colony lifts.

12.2.1 Procedure 1: subcloning of yeast clone DNA into sCos-1

The following cloning procedure has been successfully employed for subcloning human chromosome 16 YACs and telomere bearing YACs (half YACs) into cosmids (Riethman *et al.*, 1989).

(1) Quantify high molecular weight DNA obtained from yeast clone.
(2) Digest 1 μg yeast DNA with three dilutions of Sau3A1 (0.05 U μ g^{-1} DNA, 0.01 U μg^{-1} DNA and 0.05 Uμg^{-1} DNA) in 100–200 μl total volume for 10 min at 37°C.
(3) Inactivate enzyme by heating to 70°C for 10 min. Store at 4°C.
(4) Examine some of the DNA on gels. Different lots of enzymes may give different results. In addition, different YAC DNA preparations may digest at different rates. Select sample that has evidence of digestion (smearing), but that has large amount of DNA that is still high molecular weight (>40 kb).
(5) Dephosphorylate 500 ng partially digested DNA sample with calf intestinal phosphatase using reaction conditions recommended by the supplier (Boehringer Mannheim). This usually involves adding X μl partially digested DNA (for 500 ng DNA), 20 μl 10× calf intestinal phosphatase (CIP) buffer, 2 μl CIP and X μl 1× TE to bring the volume up to 200 μl. After the dephosphorylation step, add 180 μl 1× TE, 20 μl STE to the reaction mixture and incubate at 60°C for 10 min. Then add 2 μl 20% sodium dodecyl sulfate (SDS) and incubate at 68°C for 10 min. Extract the DNA in an equal volume of phenol/chloroform for 30 min on a rotator and then extract the aqueous phase in chloroform for 30 min on the rotator. Following extraction, add 1 μg of sCos1 arms prepared as described by Evans & Lewis (1989) to the partially digested, dephosphorylated DNA preparation. 1 μg of sCos1 arms can also be added to the partially digested DNA that is still phosphorylated, which can be used as a control. Both preparations are then microdialyzed twice for 30 min in 1× TE buffer. Both samples are then concentrated to a 10–30 μl volume with sec-butyl alcohol and dialyzed 15 min in 1× TE.
(6) Perform ligation of dephosphorylated target DNA to vector arms at 12°C overnight. The two reactions contain all of the concentrated DNA from above and one-tenth volume 10× ligation buffer with ATP, and 1 μl T4 DNA ligase.
(7) Package the ligation mix *in vitro* into bacteriophage lambda using Gigapak II gold from Strategene using methods described by the supplier. Store at 4°C (total volume 550 μl).
(8) Test plate 5 μl of reaction mix from both dephosphorylated and phosphorylated DNA to determine cloning efficiency. Add 5 μl of packaged reaction mix to 45 μl SM without gelatin and 50 μl optimized HB101 bacteria and incubate for 20 min at 37°C. Pellet cells at 2500 r.p.m. for 5 min and resuspend cells in 100 μl NZYDT media. Plate different dilutions of the reaction mix on NZYDT + kanamycin plates and incubate at 37°C overnight. We typically get 10^5 c.f.u. μg^{-1} target DNA.
(9) Plate approximately 100–150 c.f.u. per 100 mm petri dish. The yeast genome is approximately 15 Mb in size, so about 2000 clones are needed for a 5x representation.
(10) Make colony lifts (Grunstein & Hogness, 1975). Colony lifts are hybridized to ^{32}P labeled Cot1 fractionated DNA (see Section 12.3) and positive clones are picked and stored individually in microtiter dish wells in 15% glycerol/LB media at −70°C.

Procedures for the preparation of standard buffers discussed above may be found in Sambrook *et al.* (1989).

12.3 ANALYSIS OF COSMID SUBCLONES AND CONTIG ASSEMBLY

Cosmids clones can be analyzed and assembled into contigs by a number of different methods. We have previously used repetitive sequence fingerprinting to identify overlap between clones (Stallings *et al.*, 1990). This method allows contigs to be assembled very accurately because the extent of overlap between clones can be ascertained (Stallings *et al.*, 1992). In addition, repetitive sequence fingerprinting provides a great amount of information on the distribution of repetitive and nonrepetitive DNA sequences within the contig (Stallings *et al.*, 1991). Alternatively, one could make riboprobes from the ends of inserts and hybridize the clones to gridded arrays of cosmids in order to identify overlap between clones (Evans & Lewis, 1989). Here, we present the technical details of repetitive sequence fingerprinting, as well as the details of our cosmid gridding procedure.

The fingerprinting protocol described here was designed for the fingerprinting of cosmid clones obtained from a chromosome 16 cosmid library. Over 4000 clones were fingerprinted. Since contigs were being constructed from cosmids derived from an entire chromosome (90 Mb), great care was taken to obtain highly standardized conditions. In these early experiments, reproducibility and accuracy of restriction fragment sizes was critical. All of the protocols described here work well with cosmids that were made from YAC inserts. It should be realized, however, that YAC inserts represent relatively small genomic regions (200–1000 Mb) and that shortcuts can be taken in the fingerprinting of cosmids from YAC inserts. For example, one might be able to reduce the number of restriction digests or the number of repetitive sequence probes that are hybridized and still be able to assemble clones into contigs. Other laboratories have employed variations of this approach to fingerprinting cosmid subclones from YAC clones (Bellanne-Chantelot *et al.*, 1991).

12.3.1 Procedure 2: technical details of cosmid fingerprinting

(1) Large gel forms, 20 cm wide × 25 cm long, are used in order to increase the resolution of restriction fragments obtainable by electrophoresis. Allowing the smallest restriction fragments to migrate about 20 cm allows for both greater resolution and reproducibility of fragment size determinations. Thirty well (1 mm thick) combs permit more samples to be run on a gel. To ensure accuracy and reproducibility in the determination of restriction fragment sizes, we load size standards into lanes 1, 8, 15, 22 and 29. Six samples digested with EcoRI, HindIII and EcoR/HindIII from two cosmid clones are loaded into the six wells between size standards. We have utilized a rapid boiling method to isolate cosmid DNA (Holmes & Quigley, 1981).

(2) Our size markers include: HindIII cut λ (23.1, 9.4, 6.6, 4.4, 2.3, and 2.0 kb) EagI cut λ (36.7, 27.5, 19.9, 16.7 and 11.8 kb) and HaeIII cut ϕX174 (1.35, 1.1, 0.9, and 0.6 kb). The EagI cut λ markers are critical for accurate sizing between the 23.1 and 9.4 kb HindIII λ markers. Restriction fragment sizes are determined from digitized images taken off negatives by a Visage 110. We have found that this system allows the relative error in sizing restriction fragments (between 0.6 and 12 kb) to be reduced to 0.4%. Figure 12.1 shows some representative fingerprints of cosmids.

(3) Gels are 0.5 cm thick (obtained by pouring 250 ml of 0.6% agarose in 1× TA buffer (Sambrook *et al.*, 1989) with 0.5 μg ethidium bromide/ml. A 0.6% agarose gel allows adequate resolution of larger DNA fragments and resolution of smaller fragments down to about 0.6 kb. Gels are run for 27 h at 40 V with a recirculated buffer. Polaroid Type 55 +/− film is used for photography.

(4) Following photography, gels are treated with 0.25 M HCl for 3 min, washed with distilled water and then denatured with 0.4 M NaOH for 30 min. Two sandwich blots are made from each gel by placing a nylon membrane (Zetabind) and blotting pads above and below the gel. A reservoir is not used and transfer of DNA is accomplished with liquid contained within the gel. After 6 h the blots are disassembled and membranes are neutralized in 0.5 M Tris pH 7.5, 1.5 M NaCl for 15 min on a shaker. After baking for 1 h at 80°C, membranes are washed in 0.5% SDS and 1× SSC. Membranes can then be hybridized to any probe or stored indefinitely in sealed plastic bags at 4°C. These membranes provide a resource for any further analysis of the fingerprinted clones.

(5) Southern blot information is associated with specific restriction fragments by using SCORE, a program for computer-assisted scoring of Southern blots (Cannon *et al.*, 1991). Identification of overlap between cosmid clones based on similarity of fingerprints is obtained by previously described algorithms (Balding & Torney, 1991). An example of a cosmid contig developed by fingerprinting cosmid subclones from a YAC is presented in Fig. 12.2.

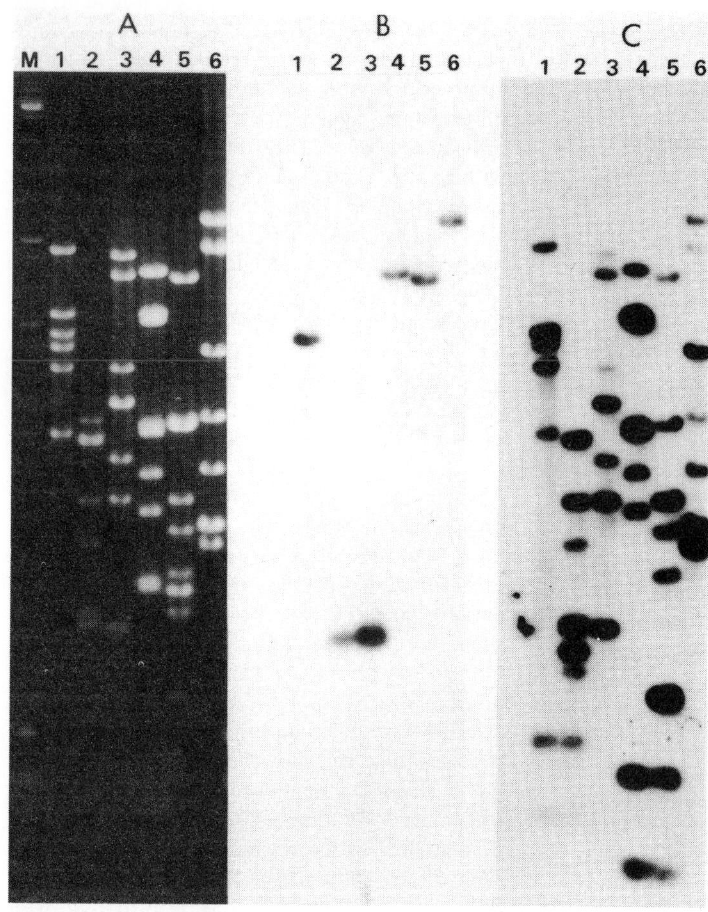

Figure 12.1 Examples of repetitive sequence fingerprints. (A) Ethidium bromide stained gel containing restriction digested DNA from clones 26C7 and 26C8. Lanes 1, 2 and 3 contain DNA from cosmid 26C7 digested with EcoRI, EcoRI/HindIII and HindIII, respectively, while lanes 4, 5 and 6 contain DNA from cosmid 26C8 digested with the same series of enzymes, respectively. In (A), lane M contains size standards (HindIII and EagI digested λ and HaeIII digested φX174). (B and C) Autoradiograms of Southern blots from the ethidium bromide stained gel in (A) hybridized to ^{32}P labeled GT (B) and ^{32}P labeled Cot1 fractionated human DNA (C).

12.4 GRIDDING OF COSMID CLONES ONTO NYLON MEMBRANES USING A BIOMEK AUTOMATED WORKSTATION

Gridded arrays of cosmids on nylon membranes provide a rapid means of identifying cosmids of interest by hybridization based protocols. Cosmids arrayed in microtiter dishes can be very easily transferred to nylon membranes for hybridization using commercially available automated workstations. The Beckmen Biomek has proven excellent for this purpose (Longmire *et al.*, 1991).

12.4.1 Protocol 3: automated gridding of cosmids

(1) The replicating tool should be presterilized with 95% alcohol and completely dried with a hair-dryer. We have found that it is occasionally necessary to degrease the replicating tool in a hood by immersion in chloroform followed by drying in a vacuum oven at 80°C for 5 min.

(2) Preparation of plates for gridding involves pouring 75 ml media with 11 g agar per litre and kanamycin (30 mg l^{-1}) into the lid of a microtiter dish. The agar medium is allowed to solidify. Nylon membranes are cut to fit over the agar layer in the lid. The membranes are pre-wet in sterile distilled H_2O and then soaked in LB media for 30 min before they are placed on top of the agar layer.

(3) We use the following format on the Biomek to sterilize the gridding tool: chlorox, 5 s; water, 10 s; 95% ethanol, 10 s; fan dry, 45 s.

(4) We are able to grid clones from 16 microtiter dishes (1536 clones) on a 8 cm × 12 cm membrane. This density of clones allows for unambiguous scoring of positive clones following hybridization (see Fig. 12.3).

(5) After gridding is complete, we cover the lid and place it in a humidified 37°C chamber for 22 h.

(6) Following incubation, the membrane is treated for 5 min in 0.5 M NaOH, 1.5 M NaCl; 5 min in 0.5 M Tris, 1.5 M NaCl; and 5 min in 2× SSC. Make sure the clone side of membrane is face up. Do not rub or disturb clones. Allow membrane to air dry and then bake in a vacuum oven at 80°C for

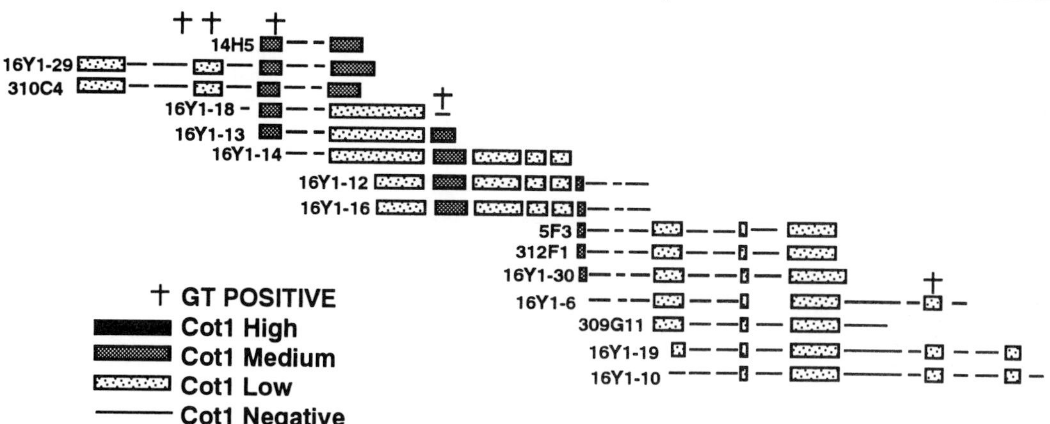

Figure 12.2 Example of a contig formed by subcloning a 150 kb YAC (16Y1) and repetitive sequence fingerprinting the resulting cosmid clones. Using this method, it is possible to obtain information on the distribution of unique and repetitive DNA sequences.

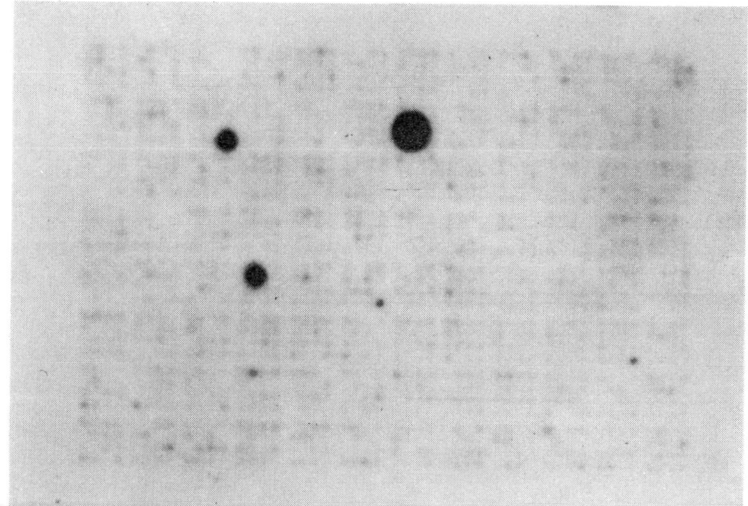

Figure 12.3 Example of a grid hybridization of a unique sequence probe to gridded arrays of cosmid clones.

1 h. The fixation of bacterial DNA on membranes is a procedure originally developed by Grunstein & Hogness (1975).

12.4.2 Procedure 4: hybridization of cosmid grids

(1) Wet membranes in sterile, distilled H_2O. Wash membranes in 0.1× SSC, 0.5% SDS at 65°C for 1 h. Ultraviolet crosslink for 30 s.
(2) Prehybridize membranes for 1 h at 65°C with: 6× SSC, 10 mM EDTA, pH 8.0, 1× Denhardt's solution, 1% SDS, 0.1 mg ml^{-1} denatured, sonicated salmon sperm DNA.
(3) Hybridize probe in the same prewash solution at 65°C overnight. Probes can be prepared by standard random primer labelling protocols (Feinberg & Vogelstein, 1983) or by other procedures.
(4) Membranes are given a quick rinse in 2× SSC/0.1% SDS at room temperature and then washed for 15 min in 2× SSC/0.1% SDS at room temperature; 15 min in 0.1× SSC/0.1% SDS at room temperature and then twice in 0.1× SSC/0.1% SDS at 50°C.

12.5 DISCUSSION

The protocols discussed here allow for the conversion of YAC clones into cosmids which are more amenable for molecular biology experiments. In addition to

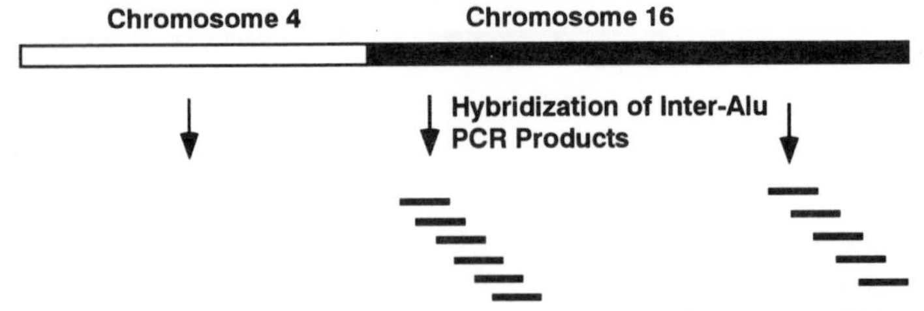

Figure 12.4 Strategy to obtain cosmid clones derived from only one insert of a chimeric YAC. If inter-Alu PCR products from the chimeric YAC are hybridized to chromosome specific cosmid libraries, only cosmids from the library of interest will be positive.

subcloning YACs into cosmids, it is also possible to hybridize YACs to gridded arrays of cosmids to identify cosmids that overlap regions of the YAC of interest. The use of inter-Alu PCR products generated from YACs provides a rapid and simple method of generating probes from YACs (Nelson *et al.*, 1989). The number of clones that must be screened can be greatly reduced if filters containing chromosome specific cosmid libraries are utilized. In fact, chromosome specific cosmids can be used to 'filter' chimeric YACs derived from genomic libraries, as illustrated in Fig. 12.4. The majority of chimeras from total genomic libraries will come from different human chromosomes. When inter-Alu PCR products from such YACs are hybridized to chromosome specific libraries, only cosmids from the chromosome of interest will be positive. Thus, although the original YAC clone was chimeric, one can very quickly derive nonchimeric cosmid probes that cover portions of only one of the YAC inserts.

REFERENCES

Balding, D.J. & Torney, D.C. (1991) *Bull. Math. Biol.* **53,** 853–879.

Bellanne-Chantelot, C., Barillot, E., Lacroix, B., Le Paslier, D. & Cohen, D. (1991) *Nucleic Acids Res.* **19,** 505–510.

Bellanne-Chantelot, C., Lacroix, B., Ougen, P., Billault, A., Beaufils, S., Bertrand, S., Georges, I., Gilbert, F., Gros, I., Lucotte, G., Susini, L., Codani, J.J. Gesnouin, P., Pook, S., Vaysseix, G., Lu, K., Ried, T., Ward, D., Chumakov, I., Le Paslier, D., Barillot, E. & Cohen, D. (1992) *Cell* **70,** 1059–1068.

Breuning, M.H., Dauwerse, H.G., Fugazza, G., Saris, J.J., Spruit, L., Wijnen, H., Tommerup, N., van der Hagen, C.B., Imaizumi, K., Kuroki, Y., van der Boogaard, M.J., de Pater, J.M., Mariman, E., Hamel, B., Himmelbauer, H., Frischauf, A.M., Stallings, R.L., van Ommen, G. & Hennekam, R.C.M. (1992) *Am. J. Hum. Genet.* **52,** 249–254.

Cannon, T.M., Koskela, R.J., Burks, C., Stallings, R.L., Ford, A.M. & Fickett, J.W. (1991) *Biotechniques* **10,** 764–767.

Collins, C., Kuo, W.L., Segraves, R., Fuscoe, J., Pinkel, D. & Gray, J.W. (1991) *Genomics* **11,** 997–1006.

Donis-Keller, H., Green, P., Helms, C., Cartinhour, S., Weiffenbach, B., Stephens, K., Keith, T.P., Bowden, D.W., Smith, D.R. & Lander, E. (1987) *Cell* **51,** 319–337.

Evans, G.A. & Lewis, K.A. (1989) *Proc. Natl. Acad. Sci. U.S.A.* **86,** 5030–5034.

Evans, G.A., Lewis, K. & Rothenberg, B.E. (1989) *Gene* **79,** 9–20.

Feinberg, A.P. & Vogelstein, B. (1983) *Anal. Biochem.* **132,** 6–13.

Green, E.D., Riethman, H.C., Dutchik, J.E. & Olson, M.V. (1991a) *Genomics* **11,** 658–669.

Green, E.D., Mohr, R.M., Idol, J.R., Jones, M., Buckingham, J.M., Deaven, L.L., Moyzis, R.K. & Olson, M.V. (1991b) *Genomics* **11**, 548–564.

Grunstein, M. & Hogness, O. (1975) *Proc. Natl. Acad. Sci. U.S.A.* **72,** 3961–3965.

Holmes, D.S. & Quigley, M. (1981) *Anal. Biochem.* **114,** 193–197.

Longmire, J.L., Brown, N.C., Ford, A.F., Naranjo, C., Ratliff, R.L., Hildebrand, C.E., Stallings, R.L., Costa, A.K. & Deaven, L.L. (1991) *Lab. Robotics Automation* **3,** 195–198.

Nelson, D.L., Ledbetter, S.A., Corbo, L., Victoria, M.F., Ramirez-Solis, R., Webster, T.D., Ledbetter, D.H. & Caskey, C.T. (1989) *Proc. Natl. Acad. Sci. U.S.A.* **86,** 6686–6690.

Riethman, H.C., Moyzis, R.K., Meyne, J., Burke, D.T. & Olson, M.V. (1989) *Proc. Natl. Acad. Sci. U.S.A.* **86,** 6240–6244.

Sambrook, J., Fritsch, E.F. & Maniatis, T. (1989) *Molecular Cloning*. Cold Spring Harbor Laboratory Press, Cold Spring Harbor, NY.

Stallings, R.L., Hildebrand, C.E., Torney, D.C., Longmire, J.L., Deaven, L.L., Jett, J.H., Doggett, N.A. & Moyzis, R.K. (1990) *Proc. Natl. Acad. Sci. U.S.A.* **87,** 6218–6222.

Stallings, R.L., Ford, A.F., Nelson, D., Torney, D.C., Hildebrand, C.E. & Moyzis, R.K. (1991) *Genomics* **10,** 807–815.

Stallings, R.L., Doggett, N.A., Callen, D., Apostolou, S., Chen, L.Z., Nancarrow, J.K., Whitmore, S.A., Harris, P., Michison, H., Breuning, M., Sarich, J., Fickett, J., Cinkosky, M., Torney, D.C., Hildebrand, C.E. & Moyzis, R.K. (1992) *Genomics* **13,** 1031–1039.

Sulston, J., Du, Z., Thomas, K., Wilson, R., Hillier, L., Staden, R., Halloran, N., Green, P., Thierry-Mieg, J., Qui, L., Dear, S., Coulson, A., Craxton, M., Durbin, R., Banks, M., Metzstein, M., Hawkins, T., Ainscough, R. & Waterston, R. (1992) *Nature* **356,** 37–41.

Trask, B., Christensen, M., Fertitta, A., Bergmann, A., Ashworth, L., Branscomb, E., Carrano, A. & van den Engh, G. (1992) *Genomics* **14,** 162–167.

West, A.K., Stallings, R.L., Hildebrand, C.E., Chiu, R., Karin, M. & Richards, R.I. (1990) *Genomics* **8,** 513–518.

CHAPTER THIRTEEN

A Role for the P1 Cloning System in Genome Analysis

D.A. SMOLLER,[1] W.J. KIMMERLY,[2,3] O. HUBBARD,[2,3] C. ERICSSON,[2,3] C.H. MARTIN[2,3] & M.J. PALAZZOLO[2,3,4]

[1] Genome Systems Inc., 7166 Manchester Rd, St Louis, MO 63143, USA
[2] Drosophila Genome Center
[3] Human Genome Center, Lawrence Berkeley Laboratory
[4] Department of Molecular and Cell Biology, University of California, Berkeley, CA 94720, USA

13.1 INTRODUCTION

One of the major recent advances in molecular genetics has been the development of the physical map. Such a map is a cloned, ordered representation of nearly the entire genome of a target organism. The early physical maps representing *S. cerevisiae* (Olson *et al.*, 1986) and *Caenorhabditis elegans* (Coulson *et al.*, 1986) were generated by determining the overlap relationships between DNA fragments cloned into phage λ and cosmid vectors, respectively. These overlaps were identified and characterized by fingerprint analysis. It is interesting to note that, even after years of fingerprinting, the average sizes of the contigs were still only small multiples of the size of the individual cloned inserts upon which the map is based. Thus, a major limitation to the physical mapping of an organism with a large genome has been the relatively small size of inserts that could be packaged using phage λ *in vitro* packaging extracts.

The insert size limitation was alleviated by the development of yeast artificial chromosomes (YACs) (Burke *et al.*, 1987). The average size of the cloned inserts in the initial YAC libraries was as large as 200–400 kb. Recently, the average size of the inserts has been surpassed by the development of libraries that ontain inserts approaching 1 Mb (Bellanne-Chantelot *et al.*, 1992; Chumakov *et al.*, 1992). YAC cloning technology has been used in physical mapping procedures in two ways. In some experiments the relatively small cosmid contigs built by fingerprint analysis have been joined by bridging adjacent contigs with YACs (Coulson *et al.*, 1988). Recently, other groups have used YAC libraries to initiate physical maps. In this second approach to mapping, YACs have been used to build large contigs without the arduous preliminary fingerprinting. The *Drosophila* map was constructed by large-scale *in situ* hybridization of individual YAC clones to the polytene chromosomes of the salivary glands (Garza *et al.*, 1989). Maps of the human Y chromosome and chromosome 21 (Chumakov *et al.*, 1992) were generated by PCR-based sequence tagged site (STS) content mapping (Olson *et al.*, 1989).

Despite the great recent successes in physical mapping, there are still limitations to the YAC cloning system. First, in most libraries chimeric clones exist in which nonadjacent genomic regions are artifactually joined. The typical level of such co-cloning artifacts is as high as 50% (Green *et al.*, 1991). In addition, the human mapping efforts show that a significant fraction of YAC clones carry deletions. Furthermore, it is difficult to purify large amounts of the artificial chromosome from the endogenous yeast chromosomes. Some investigators also find YACs difficult to work with, at least one reason being the large size of the individual clones.

The bacteriophage P1 is the foundation of a new cloning strategy developed by Sternberg (1990) which minimizes many of the difficulties inherent in working with cosmid and YAC clones but is still capable of carrying inserts that are relatively large (about 90 kb). The P1 cloning system is somewhat analogous to cosmid technology. Large genomic inserts are ligated to a plasmid containing an origin of replication, a selectable marker, and *cis*-acting signals that are recognized by an *in vitro* packaging system that mimics the lytic viral life cycle. The P1 system differs from cosmid cloning in two major ways. The cloned inserts in the P1 vector are twice as large a cosmid inserts because the system takes advantage of the fact that the P1 virus is more than twice as large as the phage lambda. Secondly, the origin of replication allows a single copy of the plasmid per cell. There is a growing body of evidence that single copy replicative mechanisms provide greater stability for large eukaryotic cloned inserts (Smoller *et al.*, 1991). P1 DNA is more manipulable and accessible than YAC DNA because the inserts are smaller in size and standard techniques have been developed for bulk preparation procedures in which the cloned vector can be purified from the endogenous host genome in a straightforward fashion.

13.2 AN OVERVIEW OF THE P1 CLONING SYSTEM

The P1 cloning system has three key elements: a plasmid with *cis*-acting *in vitro* packaging signals, an *in vitro* packaging system, and a set of host strains that can be used to rescue recombinant clones as plasmids. Packaging is a two-stage 'headful' P1 phage assembly reaction in which entry of the linear vector plus insert molecule into the phage head initiates at the *in vitro* pac cleaved end. Once packaged, an *Escherichia coli* strain containing an expressed *cre* gene is infected by the recombinant phage. The *cre* recombinase acts on the linear recombinant DNA at the loxP sites generating a stable, circular self-replicating molecule. This plasmid replicates at one copy per host cell, the copy control being maintained by the P1 plasmid replicon. When production of the P1 DNA is desired, the P1 plasmid copy level can be elevated by a *lac* promoter-controlled P1 lytic replicon capable of generating about 20 copies per cell. This amplification is performed by simply adding isopropyl β-D-thiogalactopyranoside (IPTG) prior to plasmid isolation. Several micrograms of plasmid DNA can be isolated from 10 ml of stationary phase *E. coli* cells.

13.3 PREPARATION OF P1 VECTOR DNA

Two different P1 vectors are available for the construction of libraries (Sternberg, 1990). The first-generation vector, pNS582tetAd10, contains all the sequences necessary for packaging into a P1 phage and viability in an *E. coli* host after infection. The cloning site of this vector, including *Bam*HI and *Sal*I restriction sites, lies within the tetracycline gene. Insert bearing clones, therefore, generate kanamycin resistant and tetracycline sensitivity colonies. Clones containing inserts can be scored by tetracycline sensitivity after survival on kanamycin. The second-generation vector, pAd10 sacBII, is a positive selection vector that contains a cloning site within the *Bacillus amyloliquefaciens* sacB gene (Pierce *et al.*, 1992a). The expression of this gene results in lethality on media containing 5% sucrose. It was constructed so that clones without inserts could be selected against. Since clones without inserts have been shown to overgrow insert-bearing clones in pooled libraries, this feature can be used to minimize the need for arrayed libraries. The sacBII vector was also engineered to contain SP6 and T7 promoters, and the rare-cutting restriction enzyme sites, NotI and SfiI, flanking the BamHI cloning site.

A major requirement of the cloning procedures is that steps must be taken to ensure that the two vector arms will not undergo ligation to generate concatamers that might function as substrates by the P1 packaging extract. The avoidance of vector concatamers can be accomplished in the following way. Vector arms are created by digesting to completion 1 μg of commercially available CsCl purified vector (NEN) with ScaI (New England Biolabs), treating these ends with 10-fold excess of calf intestinal alkaline phosphatase (Boehringer Mannheim) in 50 μl of 50 mM Tris-HCl, pH 8.0 at 37°C for 45 min heating for 10 min at 68°C, followed by phenol/chloroform extraction and ethanol precipitation. The resuspended DNA is then digested to completion with BamHI and again treated with calf intestinal phosphatase at a level to eliminate any detectable religation of the vector. The DNA is finally extracted, precipitated, and resuspended in TE (10 mM Tris-HCl, 1 mM EDTA, pH 8.0).

13.4 PREPARATION OF INSERT DNA

Genomic DNA in the size range 75–100 kb is required to clone into the P1 vector system. In order to generate enough cloneable DNA fragments in this size range by partial restriction enzymatic digestion the initial source DNA must be above 500 kb. Once high molecular weight DNA has been isolated, the DNA can be partially digested with *Sau*3A to generate the highest quantity of DNA fragments in the size range 75–100 kb fragments. The appropriate fragments are size selected on a 10–40% sucrose gradient, dialyzed, concentrated, and finally redialyzed in preparation for cloning.

Although the maximum headful size of DNA is 115 kb, occasionally smaller clones may be generated by headfuls containing smaller fragments. Proper sizing is necessary to remove any of the small inserts that might become substrates for packaging. Careful size selection of insert DNA prior to ligation also tends to minimize the possibility of artifactual co-ligation events that would be of packageable size. Thus size selection, used in conjunction with packaging extracts of the P1 cloning system, can provide a biological constraint against the creation of chimeric clones.

High molecular weight DNA has been isolated in a number of ways from different organisms. Preparations of high molecular weight DNA from both the rat and *Drosophila* can be generated by CsCl density gradient centrifugation of genomic DNA from nuclei isolated from total tissue (Smoller *et al.*, 1991; M. Southerd-Smith, unpublished). In the case of *Drosophila*, nuclei are prepared from 0.5 g of flies. A nuclear pellet is formed by homogenization of the flies in NIB (10 mM Tris, 10 mM NaCl, 10 mM EDTA, 0.15 mM spermine, 0.15 spermidine, 0.5% Triton X-100, pH 8.5). The pellet is washed twice with NIB and then lysed in NIB containing 3% sarkosyl. The DNA is then purified using a CsCl density gradient by the addition of 30 g of CsCl to the lysed nuclei and centrifugation for 24–36 h at 45 000 r.p.m. in a VTi50 rotor. The viscous DNA band is removed from the gradient and dialyzed three times against TE (10 mM Tris-HCl, 1 mM EDTA, pH 8.0).

DNA fragments used for ligation with the P1 vector can be obtained by partial digestion of genomic DNA with *Sau*3A. Initially, digestion reactions using different amounts of *Sau*3A are analyzed by CHEF gel electrophoresis to identify the conditions that gave the largest molar quantity of DNA fragments in the size range 70–100 kb. These conditions are then scaled up in preparative reactions. The DNA digestion products are then size fractionated on a 10–40% sucrose gradient. Aliquots of the fractions are analyzed by CHEF gel electrophoresis to identify the appropriate size fraction. Prior to ligation and packaging, the insert DNA is concentrated. Several methods have been used for this concentration, including careful butanol extractions or packing dialysis tubing containing the genomic DNA in dry sucrose for several hours (Smoller *et al.*, 1991; Pierce *et al.*, 1992b; Gorman *et al.*, 1992). Ethanol precipitation as a form of concentration is a viable approach but has been shown to reduce cloning efficiencies by 30–50% (Smoller *et al.*, 1991). Solvent exchange can be achieved by either drop dialysis on floating filters (Smoller *et al.*, 1991; Pierce & Sternberg, 1992) or using dialysis tubing (Gorman *et al.*, 1992).

13.5 LIGATION AND PACKAGING

Size-selected genomic DNA, prepared as described above, is ligated to an equal molar ratio of prepared vector arms in the presence of T4 DNA ligase (Gibco-BRL) in ligation buffer (0.05 M Tris-HCl, 0.01 M dithiothreitol, 1 mM ATP, 0.01 M $MgCl_2$). This ligation reaction is subsequently packaged as described in Sternberg (1990). Either host *E. coli* strain, NS3145 or NS3529, is the infected with the recombinant infectious phage particles. Both *E. coli* strains contain expressed *lacI*^{q} and cre recombinase. The cre acts upon the LoxP sites within the vector arms to circularize the incoming linear molecule. The *lacI*^{q} repressor protein maintains the plasmid at single copy until P1 lytic replication is derepressed by the addition of IPTG, generating about 20 copies of plasmid per cell. The initial host strain developed was NS3145. The *lacI*^{q} gene in this strain is contained on an F1 plasmid. P1 DNA prepared from clones grown in these strains tend to copurify with the F1 plasmid. NS3529 was constructed with a λ lysogen containing *lacI*^{q}, thereby eliminating the need for the *lacI*^{q} containing F1 plasmid. The host strains are also *recA*, and *mcrAB*. These features give the molecule the protection from intrahomologous recombination and loss of methylated insert DNA.

The packaging reactions are limited by their transformation efficiencies and the volume that can be packaged. Commercial extracts are available but are variable in their transformation efficiencies. The range of variability is from 10^3 to 10^4 per microgram of source DNA (Smoller *et al.*, 1991; Gorman *et al.*, 1992; Pierce *et al.*, 1992a). The extracts allow the packaging of only DNA in volumes of 9 μl or less. This volume restriction leads inevitably to many rounds of packaging and plating. The mouse library for example, required 15 packaging reactions to generate 127 000 independent clones (Pierce *et al.*, 1992b).

13.6 ARRAYING AND CONSTRUCTION OF A VARIETY OF GENOMIC LIBRARIES FROM DIFFERENT SPECIES

Preliminary characterization of the first-generation vector was carried out using human DNA (Sternberg,

1990) and *Drosophila* DNA (Smoller *et al.*, 1991). The initial results proved the utility of this cloning system and at the same time demonstrated the need, when using the pNS582tetAd10 vector, to dispense and array the clones into single colony stocks. The initial human P1 library was collected into pools. Analysis of the library revealed a number of clones (10–20%) without any insert. These clones grew significantly faster than those with insert. Amplification of this library for screening purposes resulted in pools containing up to 80% of the clones without insert.

The small size of the *Drosophila* genome allowed the generation of a hand arrayed library. Taking advantage of the fact that the cloning site in this vector is within the tetracycline gene, over 4000 clones were analyzed for the likely presence of an insert by tetracycline sensitivity. A total of 3840 independent tetracycline sensitive clones were grown and placed into single wells of 40 96-well microtiter plates.

The second-generation vector, pAd10-sacBII, allows for the positive selection of insert-containing plasmids. This vector has been used for the construction of rat, mouse, human and pine libraries. The rat, mouse and pine libraries were arrayed into several hundred pools containing 400–500 colonies. Because the libraries can be pooled they are less laborious to construct. The human library was, however, arrayed into individual wells of a microtiter plate. While pooled libraries are easier to construct, the screening for individual clones is much more expedient if the clones are individually arrayed and maintained. The identification of clones in arrayed libraries will be discussed in more detail below.

13.6.1 Screening P1 libraries by PCR

Polymerase chain reaction (PCR) can be used to screen large genomic libraries in order to identify individual clones that represent a given genomic region for which at least partial sequence (STS) data exist (Green & Olson, 1990). There are three important experimental operations that must be performed if PCR-based STS screening is going to be used to recover P1 clones that represent specific regions. These are: (1) the growth of individual clones prior to pooling, (2) the selection and implementation of an efficient pooling strategy, and (3) DNA preparations to isolate the templates for the PCR reactions. A selection of suitable protocols for these procedures is described in order below.

13.6.1.1 Growth of individual clones in 8 × 12 formats

PCR-based STS screening is most economically performed if any of a variety of pooling schemes is used to rapidly sort through the large number of clones. It is clear from a variety of data that cloned DNA frequently induces variable growth of the host cell; it is best to grow the cells individually and pool them after they have reached saturation. Of course, it is also obvious that if a greater cell density can be achieved at the stage of individual growth, then more DNA can be isolated when the clones are subsequently pooled. Because of the large number of clones in each of the libraries, it is essential to develop protocols that can be used to grow the clones in either microtiter plates or titer tubes.

We performed several experiments to test the rate and endpoint of the growth of P1 clones in different microtiter (8 × 12) formats. When grown without shaking in microtiter plates, in 150 µl of media, the cell density achieved at saturation as five-fold below that which was attained when the same clones were grown in a shaking flask or on a culture tube placed on a bacterial wheel. In contrast, cells grown in titer tubes in 400 µl of culture media on a rotary shaker at 300 r.p.m. reached the same cell density as clones grown when shaken in larger flasks or tubes. In most of our pooling experiments we grew individual P1s (or small collections of clones when the libraries have been pooled) with overnight incubation and vigorous shaking in titer tube racks in terrific broth (TB) media (Tartof & Hobbs, 1987).

13.6.1.2 Pooling strategies for screening by PCR

There are a variety of pooling strategies that have been described for screening large genomic libraries. Each method has its own advantages as well as its own drawbacks. Others have presented, both in practical and theoretical terms, comparison between different strategies (Green & Olson, 1990; Amemya *et al.*, 1992).

At the Lawrence Berkely Laboratory we are using a two-stage screening procedure. The first stage is to screen a set of simple plate pools in which each pool represents the 96 clones in a single microtiter plate. The second stage is based on the 8 × 12 row-column matrix. In other words there are eight pools that represent the 12 clones in each row and there are 12 pools that contain DNA that corresponds to the eight clones in each column. Twenty PCR reactions can be used to identify the individual clone in any plate that contains a positive clone.

13.6.1.3 DNA preps for generating templates for screening by PCR

We have used two different DNA preparation protocols to generate templates in our STS content mapping experiments. One set of procedures is based on alkaline lysis plasmid purification followed by ion exchange using commercially available Qiagen columns. As the Qiagen protocols and reagents are available commercially, we will not describe them further here. The second set of procedures is based on the lysozyme-boiling plasmid purification scheme. Our version of the lysozyme-boiling plasmid DNA preparation protocol is as described below.

13.6.1.4 *Preparation of DNA pools and PCR-based screening*

To prepare the plate, row, and column pools, clones are grown up independently in the titer tube format. Each titer tube contains 0.4 ml terrific broth (TB) plus kanamycin. The P1 DNA pools are prepared using a protocol involving lysozyme treatment followed by boiling. After pooling the clone cultures from appropriate tubes, the cells are harvested by centrifugation. To prepare the P1 DNA plate pools, the cells from the 96 clones of a single plate are suspended in 15 ml STET buffer (8% (w/v) sucrose, 5% (v/v) Triton X-100, 50 mM Tris, pH 7.5, 50 mM EDTA). Lysozyme is added to 0.1 mg ml^{-1} and incubated at room temperature for 5 min, then the samples are boiled in a water bath for 5 min. The samples are centrifuged at 12 000 r.p.m. for 20 min to remove chromosomal DNA and other particulate matter. The supernatant is recovered by decanting and P1 DNA is precipitated by the addition of an equal volume of 2.5 M NH_4OAc/75% isopropanol. The resulting pellet is suspended in 0.5 ml TE buffer with 0.5 µg ml^{-1} RNase A (Boehringer Mannheim) and incubated at 37°C for 1 h. Plate pool DNA preparations are treated further by a single extraction with a phenol/chloroform (1:1) mixture then precipitated with ethanol. After rinsing the pellets with 70% ethanol and drying under a vacuum, the DNA pools are finally suspended in 0.5 ml TE buffer and stored at −80°C. The row and column pools are prepared in a similar fashion except the phenol/chloroform extraction is omitted. All DNA pool preparations are diluted 1:100 in sterile water before use in PCR.

Individual PCR reactions are performed in 25 µl reaction volumes containing 1 µM each forward and reverse primers, 1:100 diluted DNA pool, and 0.5 units Taq DNA polymerase (Perkin–Elmer/Cetus). Using the Perkin–Elmer 9600 Thermocycler, the samples are initially denatured for 4 min at 95°C, then subjected to 30 cycles of 96°C for 15 s, 58°C for 15 s, and 72°C for 30 s. Aliquots of each sample are analyzed by agarose gel electrophoresis (2% agarose gel in TBE buffer, 89 mM Tris-HCl, 89 mM borate, 1 mM EDTA, pH 8.3).

13.6.2 Screening by colony filter hybridization

The final stage of arrayed library screening can be accomplished by PCR. In contrast, the final stage of screening of pooled libraries is typically performed by colony filter hybridization (Grunstein & Hogness, 1975). To identify an individual clone, 1500 clones are screened by hybridization to filter replicas made by plating portions of the amplified pools. Such a membrane is allowed to incubate at 37°C for 5 h on LB agar plates containing 1 mM isopropyl-1-thio β-D-galactopyroside (IPTG) and 25 µg ml^{-1} kanamycin. Filters can then be placed on chromatography paper (Whatman 3MM) 3% sodium dodecyl sulfate (SDS) to lyse cells, then denatured on 3 MM paper saturated with 0.5 M NaOH and 1.5 M NaCl, neutralized on 3 MM paper saturated with 0.5 M Tris-HCl and 1.5 M NaCl, pH 7.5 and finally washed on 3 MM paper saturated with 2 × SSC (1 × SSC is 0.15 M NaCl, 0.015 M sodium citrate). The filters are then baked at 80°C for 30 min. Filters then can be hybridized with ^{32}P labeled probe in high phosphate buffer (0.5 M NaCl, 0.1 M Na_2HPO_4, 0.005 M EDTA, 1% Sarkosyl) at 65°C. After 16–18 h of hybridization, filters were washed once in 1% Sarkosyl and 1 mM Tris-HCl, pH 8.0. Before autoradiography the filters are washed four times in 1 mM Tris-HCl, pH 8.0.

13.7 DNA SEQUENCING DIRECTLY FROM THE CLONE ENDS

Taq cycle sequencing protocols have been used to directly sequence the ends of the cloned inserts. The primers in these sequencing reactions are derived from the vector sequences that flank the cloning site. Pierce *et al.* (1992a) used radioactive cycle sequencing on P1 templates and have been able to elucidate the sequence of the first 100 bp adjacent to the priming site. In contrast, we prepare double-stranded P1 sequencing template DNA using a modified alkaline lysis preparation followed by CsCl density gradient centrifugation, PEG precipitation, and spot dialysis using Millipore Minitan S paper (Kimmerly *et al.*, 1992). Templates prepared in this fashion are analyzed by fluorescent automated DNA sequencing using an ABI 373A automated sequencer, the ABI DyeDeoxy Taq Cycle Sequencing kit, and adapted methodology. Using custom-made SP6 and T7 primers, we routinely obtain about 400 bases of sequence at each end of a cloned insert. The DNA sequence information obtained from the end of P1 cloned inserts is used to design primers for PCR.

13.8 CONTIG BUILDING IN A *DROSOPHILA* P1 LIBRARY

We have used STS content mapping and directed P1-end DNA sequencing to construct a contig that covers the *Drosophila* Bithorax Complex (BX-C). The pooling strategy that we used was the two-tiered screening strategy described above. Specifically, we screened 9216 clones that were distributed in 96 microtiter plates. The first stage of the screening consisted in PCR-based assays using pools of clones maintained in 96 plates. Subsequently, the secondary screens of the positive plates were done by screening the DNA in 20 smaller pools comprising the eight 12-member rows and the 12 eight-member columns.

The initial round of STS markers were designed using published sequences from some of the expressed regions in the BX-C. One pair of oligos was designed based on the sequence from the 5′-most Ubx exon, a second pair from the sequence of iab-4, a third from abd-A, and a fourth from abd-B. This screen identified a set of P1 clones that covered the bulk of the complex. However, a small gap remained in the region between the abd-A and the Ubx markers. To bridge this gap, a P1 clone that was identified by both iab-4 and abd-A was purified by alkaline lysis followed by cesium chloride density gradient centrifugation, then sequenced using the ABI 373A DNA sequencer, fluorescently labeled dideoxynucleotide chain terminators, and the SP6 and T7 sequencing primers. More than 350 bases of DNA sequence were elucidated from each end of the cloned insert. This DNA sequence provided a pair of primers from each end of the insert which were then used to screen the library again. This set of procedures identified a set of four overlapping P1 clones that span nearly 300 kb of the BX-C, and also closed the apparent gap which resulted from the initial round of screening.

13.9 PREPARATION OF THE DNA FROM INDIVIDUAL CLONES

Isolation of cloned supercoiled DNA from bacterial cultures is accomplished in the following way. The lytic origin of replication can be induced with IPTG by simply diluting an overnight culture containing a P1 clone 1:100 in LB media and growing this culture to an OD550 of 0.05–0.1, about 30 min at 37°C. Following this incubation IPTG is added to a final concentration of 1 mM. After this addition the culture is allowed to proceed at 37°C for 3–5 h. This culture can then be isolated with any standard plasmid preparation. DNA isolated as described above can be digested with restriction enzymes and analyzed by agarose gel electrophoresis or pulse field gel electrophoresis using standard protocols.

13.10 DISCUSSION

Our experience in working with P1 clones provides an example whereby STS markers taken from both the literature and physical mapping experiments can be used to convert a map based on YACs into a set of higher resolution P1 contigs. Gaps can be filled in a straightforward fashion by direct P1 sequence analysis from the ends of the cloned P1 inserts. In addition to the ease of PCR-based screening and the ability to sequence the ends, a body of preliminary data suggest that P1 clones are relatively free of chimeras and are more stable than YACs or cosmids. Finally, DNA from the P1 plasmid can be purified in relatively large amounts. These fragments can be subcloned using straightforward techniques, thus providing an excellent intermediate source of templates to facilitate large-scale genomic sequencing efforts.

REFERENCES

Amemya, C.T., Algeria-Hartman, M.J, Aslandidis, C., Chen, C., Nikolic, J., Gingrich, J. & de Jong, P. (1992) *Nucleic Acids Res.* **20,** 2559–2563.

Bellanne-Chantelot, C., Lacroix, B., Ougen, P., Billaut, A., Beaufils, S., Bertrand, S., Georges, I., Gilbert, F., Gros, I., Lucotte, G., Susini, L., Codani, J., Gesnouin, P., Pook, S., Vaysseix, G., Lu-Kuo, J., Le Paslier, D., Barrilot, E. & Cohen, D. (1992) *Cell* **70**, 1059–1068.

Burke, D.T., Carle, G.F. & Olson, M.V. (1987) *Science* **236**, 806–812.

Chumakov, I., Rigault, P., Guillou, S., Ougen, P., Billaut, A., Gausconi, G., Gervy, P., LeGall, I., Soularue, P., Grinas, L., Bougueleret, L., Bellanne-Chantelot, C., Lacroix, B., Barillot, E., Gesnouin, P., Pook, S., Vaysseix, G., Frelat, G., Schmitz, A., Sambucy, J., Bosch, A., Estivill, X., Weissenbach, J., Vignal, A., Riethman, H., Cox, D., Patterson, D., Gardiner, K., Hattori, M., Sakaki, Y., Ichikawa, H., Ohki, M., Paslier, D., Heilig, R., Antonarkis, S. & Cohen, D (1992) *Nature* **359**, 380–387.

Coulson, A.R., Sulston, J., Brenner, S. & Karn, J. (1986) *Proc. Natl. Acad. Sci. U.S.A.* 7821–7825.

Coulson, A.R., Waterston, J., Kiff J., Sulston, J. & Kohara, Y. (1988) *Nature* **335**, 184–186.

Garza, D., Ajioka, J.W., Burke, D.T. & Hartl, D.L. (1989) *Science* **246**, 641–646.

Gorman, S.W., Roberts-Oehlshlager, S.L., Cullis, C.A. & Teasdale, R.D. (1992) *Biotechniques* **12**, 722–726.

Green, E.D. & Olson, M.V. (1990) *Proc. Natl. Acad. Sci. U.S.A.* **87**, 1213–1217.

Green, E.D., Riethman, H.C., Dutchik, J.E. & Olson, M.V. (1991) *Genomics* **11**, 658–669.

Grunstein, M. & Hogness, D.S. (1975) *Proc. Natl. Acad. Sci. U.S.A.* **72**, 3961–3965.

Olson, M.V., Dutchik, J.E., Graham, M.Y., Brodeur, G.M., Helms, C., Frank, M., MacCollin, M., Scheinman, R. & Frank, T. (1986) *Proc. Natl. Acad. Sci. U.S.A.* **83**, 7826–7830.

Olson, M.V., Hood, L., Cantor, C. & Botstein, D. (1989) *Science* **245**, 1434–1440.

Pierce, J.C. & Sternberg, N. (1992) *Methods Enzymol.* **216**, in press.

Pierce, J.C., Sauer, B. & Sternberg, N. (1992a) *Proc. Natl. Acad. Sci. U.S.A.* **89**, 2056–2060.

Pierce, J.C., Sternberg, N. & Sauer, B. (1992b) *Mammalian Genome* in press.

Smoller, D.A., Petrov, D. & Hartl, D.L. (1991) *Chromosoma* **100**, 487–494.

Sternberg, N. (1990) *Proc. Natl. Acad. Sci. U.S.A.* **87**, 103–107.

Tartof, K.D. & Hobbs, C.A. (1987) *Bethesda Res. Lab. Focus* **9**, 12.

CHAPTER FOURTEEN

Generation and Mapping of Chromosome Specific Sequence-tagged Sites

N.A. DOGGETT, D.L. GRADY, J.L. LONGMIRE & L.L. DEAVEN
Life Sciences Division and Center for Human Genome Studies, Los Alamos National Library, Los Alamos, NM 87545, USA

14.1 INTRODUCTION

Sequence-tagged sites (STSs) (Olsen *et al.*, 1989) are unique sequence-based landmarks in the genome that are detectable by the polymerase chain reaction (PCR). The generation of STSs has become an integral part of strategies designed for mapping of specific chromosomes (Green & Olson, 1990; Green *et al.*, 1991). The procedure known as 'STS content mapping' (Green & Olson, 1990), involves the identification of cloned DNA fragments in a library (usually yeast artificial chromosome (YAC) clones) and the establishment of their position along the chromosome through regions of known sequence (STSs). One can now generate a 1 Mb STS framework map directly from two sets of resources, namely a library from flow sorted human chromosomes and a panel of somatic cell hybrids containing varying lengths of a single human chromosome (resulting from chromosomal deletions and translocations, or radiation-induced breakage). This type of map can provide a basis from which YAC contigs can be assembled and ordered. These combined approaches are being used to generate a low-resolution map of chromosome 5 with STS markers at intervals of 0.5–1 Mb. We anticipate that at least 300 STS markers will be needed, given random statistics, to generate this 1 Mb map. A similar approach involving STSs and YACs is being used to close the gaps in the cosmid contig map of chromosome 16 (Stallings *et al.*, 1990, 1992).

The process of STS generation depends upon the preparation and sequencing of the sample DNA, the establishment of an effective PCR assay, and the localization of the STS along a chromosome (Fig. 14.1). The strategy utilizes flow sorting technology developed at Los Alamos National Laboratory (Deaven *et al.*, 1986). These flow-sorting capabilities were used to construct three chromosome 5 libraries in M13 and a chromosome 16 specific cosmid library. The M13 clones from chromosome 5 can be directly sequenced on an automated DNA sequencer. The chromosome 16 cosmid clones were previously fingerprinted and assembled into cosmid contigs covering 60% of the chromosome (Stallings *et al.*, 1992). The end cosmid members of these contigs were subcloned into M13 prior to sequencing. In both these cases, the sequence was analyzed for the presence of repetitive elements, appropriate PCR primers were synthesized to 'tag' the site, and these primers were tested for functionality in the polymerase chain reaction.

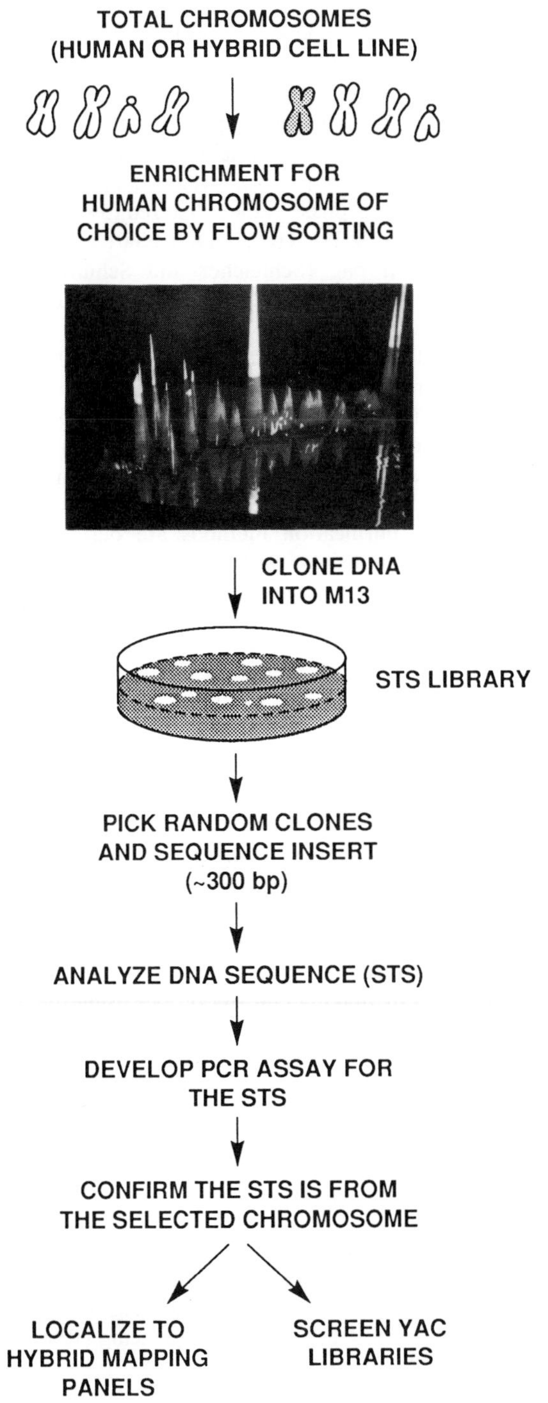

Figure 14.1 Schematic representation of the overall process for generating STSs from flow-sorted human chromosomes.

14.2 PROTOCOLS

14.2.1 Chromosome sorting

For the construction of the first complete digest chromosome specific recombinant DNA libraries, diploid human fibroblasts or Chinese hamster–human hybrid cells were used as the source of human chromosomes (Deaven *et al.*, 1986). In more recent work, human chromosomes are sorted almost exclusively from rodent–human hybrids (Bartholdi *et al.*, 1987; Van Dilla & Deaven, 1990). The cells are cultured using conventional techniques and chromosomes are isolated in the presence of polyamines as described (Van Dilla & Deaven, 1990). Chromosomes are sorted by bivariate fluorescence-activated flow sorting on a Coulter Epics V sorter. The purity of sorted chromosomes is critical to the construction of chromosome specific DNA libraries, thus a significant amount of effort goes into purity determinations of sorter outputs. In general, purity information is obtained by analysis of the histograms during sorter operation and by analysis of samples of sorted chromosomes using *in situ* hybridization techniques (Van Dilla & Deaven, 1990). Rodent chromosomes can easily be differentiated from human chromosomes using total genomic human DNA as an *in situ* probe because of differences in repetitive DNA between these species (Moyzis *et al.*, 1989). The construction of chromosome specific M13 total digest libraries requires significantly less DNA than is needed for partial digest cosmid libraries. As a result, it is practical to use more stringent sort window settings to insure higher chromosome purity in these libraries (greater than 90–95% versus 85–95%). Thus, flow sorting results in libraries with a 20- to 50-fold enrichment of the desired chromosome. The contaminating DNA when human chromosomes are sorted from hybrid cells is, of course, rodent DNA. If desired, clones containing rodent DNA can be detected with rodent repetitive DNA and eliminated from a library to achieve even higher levels of chromosomal purity. The chromosome 16 specific cosmid library was prescreened with mouse repetitive DNA to avoid the fingerprinting of cosmid clones containing mouse inserts (Stallings *et al.*, 1990).

14.2.2 Library constructions

14.2.2.1 Chromosome 5 specific single space total digest M13 STS libraries

Three distinct chromosome 5 libraries have been constructed in M13 to date. Two *Bam* HI/*Hind* III libraries were constructed with flow-sorted chromosomes 5 from the hamster/human somatic cell hybrid,

Q826-20 (C. Jones), and a single *Bam*HI/*Hin*dIII library was made with flow-sorted chromosomes 5 from normal human lymphoblastoid cells, GM130A. For each of these M13 libraries, approximately 200 000 chromosomes 5 (about 70 ng DNA) were sorted. The chromosomal DNA was purified (see below) and digested to completion with *Bam*HI and *Hin*dIII. The digested DNA (50 ng) was ligated to a two-fold excess of (*Bam*HI and *Hin*dIII digested and alkaline phosphatase treated) M13mp18 replicative-form DNA. Transformation of *Escherichia coli* strain DH5αF (BRL) yielded approximately 10^3 recombinants per nanogram of DNA. The average insert size of these clones was 2 kb, and 55% of the plaques hybridized to human Cot_{50} DNA which is consistent with distribution of repetitive sequences in human DNA of this length (Moyzis *et al.*, 1989). The use of four base cutters such as Sau3A for the construction of STS libraries is not recommended because the fragments cloned are frequently too small to yield sufficient sequence for an STS unless size selection is used prior to cloning. We have found that double digestion with *Bam*HI and *Hin*dIII produces clones approximately 2 kb in length which are appropriate for sequencing without the need for size selection.

14.2.2.2 Chromosome 16 specific partial digest cosmid library

For construction of a chromosome 16 cosmid library, 5–6 million chromosomes 16 (about 600 ng) were sorted from the mouse/human somatic cell hybrid CY18 (Callen *et al.*,1986) into siliconized glass tubes. After sorting, chromosomes are pelleted at 1500**g** for 30 min, and chromosomal DNA is purified (see below). Purified chromosomal DNA is partially digested in 100-ng aliquots with *Sau*3AI (0.005 to 0.01 U/μg DNA) for 15 min at 37°C. Aliquots (10 ng) of each of these partial digestions should be analyzed on a 0.4% agarose gel to examine the extent of the digestion. Partial digestions in the appropriate size range are then pooled, and extracted and dialyzed as described below. The partially digested chromosomal DNA is then dephosphorylated with calf intestinal phosphatase (Boehringer Mannheim) using conditions recommended by the supplier. For the chromosome 16 cosmid library construction, approximately 350 ng of partially digested and dephosphorylated chromosomal DNA was ligated to 1 μg of cloning arms from the cosmid vector sCos1 (Evans *et al.*, 1989). *In vitro* packaging and infection of *E. coli* host strain HB101 yielded 1.75 × 10^5 independent recombinants (67-fold representation). A 10-fold representation (26 000 clones) was selected as individual colonies and archived in 96-well microtiter dishes.

14.2.2.3 Purification of sorted chromosomal DNA for library constructions

Pelleted flow sorted chromosomes can be stored at 4°C. For DNA purification, 100 μl lysis buffer (0.1 M Tris-HCl, pH 8.0, 0.1 M EDTA, 0.01 M NaCl) is added. Sodium dodecyl sulfate and proteinase K are then added to final concentrations of 0.5% and 100 μg ml^{-1}, respectively, and the sample is incubated overnight at 37°C. After incubation, the DNA is gently extracted with an equal volume of phenol:chloroform (1:1) and then chloroform/isoamyl alcohol (24:1). The extracted aqueous phase is then dialyzed (3× 15 min) against TE (0.01 M Tris-HCl, pH 8.0, 0.001 M EDTA) in a collodion bag (Schleicher and Schull). The chromosomal DNA is then ready for restriction enzyme digestions, or can be stored at 4°C. After restriction digestion, the sample is heated to 70°C for 10 min, extracted gently with phenol/chloroform 1:1, chloroform/isoamyl alcohol 24:1, and dialyzed against TE (3× 15 min) (as before). The DNA is then ready for ligation, packaging and infection if cloning in cosmids or ligation and transformation if cloning in M13. These purification methods are designed to produce high molecular weight chromosomal DNA which is required for cosmid cloning. Gentle handling of DNA is not critical when cloning smaller fragments in M13.

14.2.3 Subcloning of cosmids into M13

There are several considerations worth discussing for the development of an STS from a cosmid clone. In the chromosome 16 mapping effort, cosmids are selected for sequencing because they are the end-members of cosmid contigs; however, the orientation of these clones (outward or inward) is unknown. In order to sequence the outward end of a cosmid in a contig, we would first have to determine which end was outward by making end riboprobes using the T3 and T7 promoters in the sCOS1 vector and hybridizing these back to the next overlapping cosmid. The probe which hybridized would come from the inward side of the cosmid and the probe which did not hybridize would be the end which extends outward. A variety of approaches could then be used to sequence the outward end including:

(1) direct sequencing of the cosmid using the appropriate T3 or T7 primer (Dugaiczyk *et al.*, 1992);
(2) amplification of the outward end using vectorette or bubble PCR followed by sequencing (Riley *et al.*, 1990); or
(3) subcloning of the cosmid into M13, identification of the outward end fragment by screening with the appropriate T3 or T7 oligonucleotide, and sequencing.

In our work it has not been necessary for STSs to come from the end-most region of contigs and we have avoided this additional work. (The STSs are used to localize contigs to a hybrid breakpoint map of chromosome 16, to confirm large contigs, and to close gaps between contigs with YAC clones – none of which requires STSs from the extreme ends of contigs.) Instead, we have chosen to digest cosmid DNA fully with *Eco*RI, shotgun subclone into M13mp18, and randomly sequence one (or more if necessary) of these subclones. *Eco*RI was chosen because it leaves the entire sCOS1 vector intact and this does not clone in M13 (because an M13-sCOS1 recombinant would contain two origins of replication which is fatal). As a result, all the M13 subclones are derived from the human insert contained in each cosmid. (We have sequenced over 400 of these random subclones, and we have not yet found a vector fragment.) Because the cosmids are the end clones of contigs, a random sequence from one of these clones is at most 35 kb from the end of a contig.

14.2.4 DNA sequencing

Propagation of individual M13 clones and isolation of single-stranded DNA are carried out using standard protocols. DNA sequencing can be performed using the dideoxy chain termination method (Sanger *et al.*, 1977) and universal M13 sequencing primers. For the chromosome 5 STS project M13 clones were chosen at random from the three libraries, and 250–450 bp of DNA sequence immediately adjacent to the vector was determined using a Dupont Genesis 2000 automated sequencer. For the chromosome 16 STS generation effort, M13 subclones of cosmids were sequenced (average of 300 bp) on either a manual sequencing apparatus (BRL) or on an ABI 373A automated DNA sequencer.

14.2.5 Development of unique PCR assays

14.2.5.1 Sequence analysis and primer selection

The chromosome specific DNA sequences generated were subjected to a number of computer-based analyses. Initially, each sequence was analyzed for similarity to repeats occurring in the human genome, specifically, Alu (Bains, 1986), L1 (Scott *et al.*, 1987), THE (Fields *et al.*, 1992) and alphoid (Wayne & Willard, 1987) motifs, and more recently to the consensus repeat database of Jurka *et al.* (1992), which contains these repeats and others. Sequences found to contain significant amounts of homology to any of these repetitive elements are eliminated from further analysis, unless it is possible to select PCR primers that flank the repetitive regions or that lay to one side of them. Each sequence is also analyzed using the GRAIL program (Uberbaucher & Mural, 1991) because we have found that primers synthesized in coding regions often yield poor results due to cross homology with rodent DNA, and amplification within multigene families. Finally, PCR primers are chosen and primer melting temperatures are determined using a computer program that calculates melting temperature based on the nearest neighbor frequency of nucleotides (R. Ratliff, personal communication). We have recently started using a computer program, ASTS, which fully automates the selection of PCR primers from DNA sequences and predicts primer melting temperatures based on the nearest neighbor equations (see Chapter 38).

14.2.5.2 PCR assays

We have synthesized oligodeoxynucleotide primers on Applied Biosystems Model 391 or 394 DNA synthesizers and purified the primers using NEN reverse phase chromatography. PCR is performed in 20 or 50 μl reactions containing 1.5 mM $MgCl_2$, 50 mM KCl, 10 mM Tris-HCl (pH 8.3), 0.001% gelatin, 200 μM dNTPs, 1 μM primers, 30–50 ng template DNA, and 1–2 U Amplitaq polymerase. We have used Perkin-Elmer/Cetus thermal cyclers (Models 450 or 9600) for cycling reactions with satisfactory results. Thermal cycling is performed for 30 cycles of 92–94°C to denature, 10°C below the average predicted primer melting temperature for primer annealing (usually 50–60°C), and 68–72°C for extension. The cycle times for denaturation, annealing, and extension are 1 min, 45 s and 5 min for the Perkin–Elmer/Cetus 450 and 15 s, 15 s and 65 s for the Perkin–Elmer/Cetus p600, respectively. PCR products are analyzed for appropriate size by either polyacrylamide gel electrophoresis or agarose gel electrophoresis (NuSieve GTG, FMC Bioproducts) and detected by ethidium bromide staining. An example of the successful generation of chromosome 5 STSs is shown in Fig. 14.2.

14.3 SUMMARY AND CONCLUSIONS

Chromosome specific STSs can be efficiently generated from flow-sorted human chromosomal DNA. Sorter purity levels of over 90% result in a 50-fold enrichment of any human chromosome. DNA obtained from the sorted chromosomes can be used to construct recombinant DNA libraries with varying insert sizes (plasmid through YAC). Direct cloning of sorted DNA into DNA sequencing vectors provides the most rapid means for generating chromosome specific STSs. The STSs generated from these libraries can be used in combination with somatic cell hybrid deletion panels and total genomic YAC libraries to

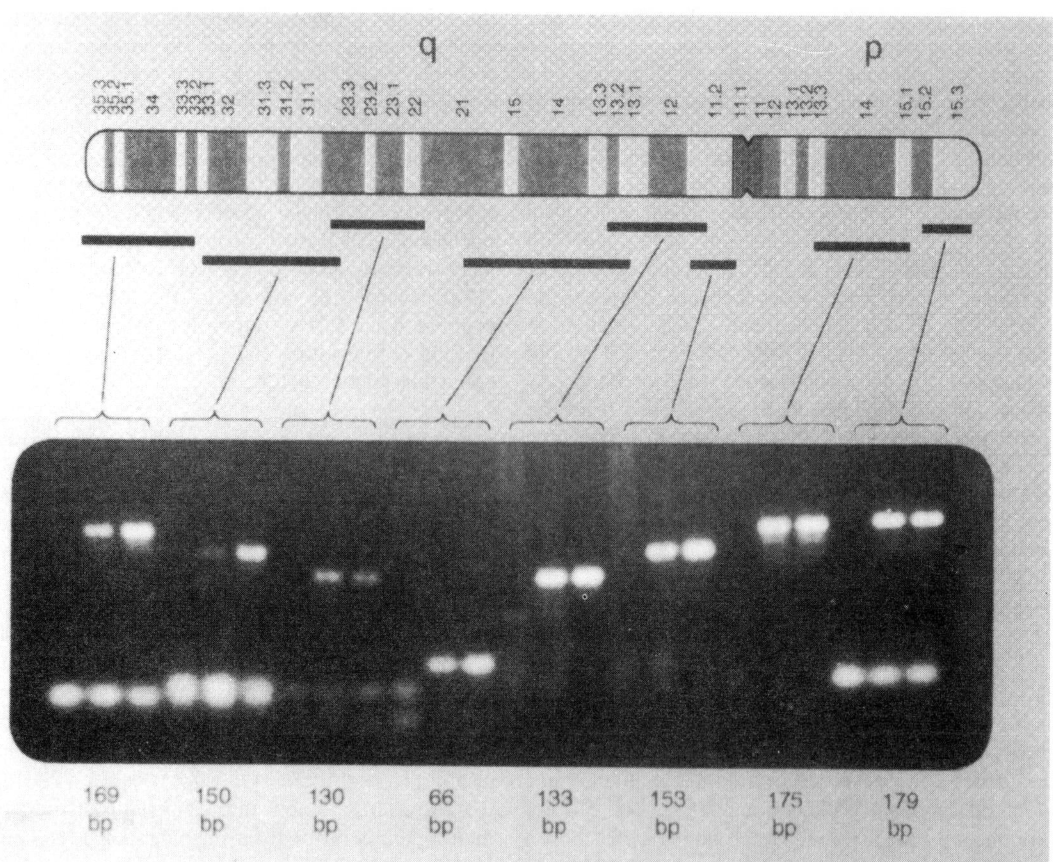

Figure 14.2 Example of STS generation and regional localization. The upper portion shows the regional localization along chromosome 5 of eight STSs generated from a chromosome 5 specific M13 library. The lower photograph shows the actual PCR products separated by agarose gel electrophoresis. The three gel lanes for each STS are PCR products generated from total-genomic human DNA (right lanes), chromosome 5 DNA (middle lanes), and total genomic hamster DNA (left lanes). The size of each STS product is given at the bottom of the figure.

rapidly construct low resolution ordered clone maps of entire chromosomes.

ACKNOWLEDGMENTS

The authors wish to thank Lynne Duesing, Donna Robinson, and Judy Tesmer for their excellent technical assistance as well as Bob Moyzis, and Bob Ratliff for helpful discussions. Supported by the US Department of Energy grant 005063 (F137) under contract W-7405-ENG-36.

REFERENCES

Bains, J. (1986) *J. Mol. Evol.* **23**, 189–199.

Bartholdi, M.F., Meyne, J., Albright, K., Luedemann, M., Campbell, E., Chritton, D., Deaven, L., Van Dilla, M. & Cram, L.S. (1987) *Methods Enzymol.* **151**, 252–267.

Callen, D.F. (1986) *Ann. Genet.* **29**, 235–239.

Deaven, L.L., Van Dilla, M.A., Bartholdi, M.F., Carrano, A.V., Cram, L.S., Fuscoe, J.C., Gray, J.W., Hildebrand, C.E., Moyzis, R.K. & Perlman, J. (1986). *Cold Spring Harbor Symp. Quant. Biol.* **51**, 159–168.

Dugaiczyk, A., Goold, R., Disibio, G. & Myers, R.M. (1992) *Nucleic Acids Res.* **20**, 6421–6422.

Evans, G., Lewis, K. & Rothenberg, B.E. (1989) *Gene* **79**, 9–20.

Fields, C.A., Grady, D.L. & Moyzis, R.K. (1992) *Genomics* **13**, 431–439.

Green, E.D. & Olson, M.V. (1990) *Science* **250**, 94–98.

Green, E.D., Mohr, R.M., Idol, J.R., Jones, M., Buckingham, J.M., Deaven, L.L., Moyzis, R.K. & Olson, M.V. (1991) *Genomics* **11**, 548–564.

Jurka, J., Walichiewicz, J. & Milosavljevic, A. (1992) *J. Mol. Evol.* **35**, 286–291.

Moyzis, R.K., Torney, D.C., Meyne, J., Buckingham, J.M., Wu, J.-R., Burks, C., Sirotkin, K.M. & Goad, W.B. (1989) *Genomics* **4**, 273–289.

Olsen, M., Hood, L., Cantor, C. & Bostein, D. (1989) *Science* **245**, 1434–1440.

Riley, J., Butler, R., Ogilvie, D., Finniear, R., Jenner, D., Powell, S., Anand, R., Smith, J.C. & Markham, A.F. (1990) *Nucleic Acids Res.* **18**, 2887–2890.

Sanger, F., Nicklen, S. & Coulson, A.R. (1977) *Proc. Natl. Acad. Sci. U.S.A.* **74**, 5463–5467.

Scott, A.F., Schmeckpeper, B.J., Abdelrazik, M., Comey, C.T., Ohara, B., Rossiter, J.P., Cooley, T., Heath, P., Smith, K.D. & Margolet, L. (1987) *Genomics* **1**, 113–125.

Stallings, R.L., Torney, D.C., Hildebrand, C.E., Longmire, J.L., Deaven, L.L., Jett, J.H., Doggett, N.A. & Moyzis, R.K. (1990) *Proc. Natl. Acad. Sci. U.S.A.* **87**, 6218–6222.

Stallings, R.L., Doggett, N.A., Callen, D., Apostolou, S., Harris, P., Michison, H., Breuning, M., Sarich, J., Hildebrand, C.E. & Moyzis, R.K. (1992) *Genomics* **13**, 1031–1039.

Uberbaucher, E.C. & Mural, R.J. (1991) *Proc. Natl. Acad. Sci. U.S.A.* **88**, 11261–11265.

Van Dilla, M.A. & Deaven, L.L. (1990) *Cytometry* **11**, 208–218.

Wayne, J.S. & Willard, H.F. (1987) *Nucleic Acids Res.* **15**, 7549–7569.

CHAPTER FIFTEEN

Construction of cDNA Libraries

R. F. MORENO-PALANQUES & R. A. FULDNER
The Institute for Genomic Research, Gaithersburg, MD 20878, USA

15.1 INTRODUCTION

Traditionally, cDNA libraries have been constructed in order to find one or a few particular gene(s) of interest. However, with the introduction of the expressed sequence tag (EST) studies related to the Human Genome Project, the interest in cDNA libraries has been expanded to finding all the genes which are expressed in a particular cell or tissue (Adams *et al.*, 1991, 1992; Khan *et al.*, 1992; McCombie *et al.*, 1992; Okubo *et al.*, 1992). The various EST studies have revealed that single-pass automated sequencing of randomly chosen clones from cDNA libraries is an efficient strategy for identifying unknown genes, some of which have interesting homologies to known human and non-human genes, including those of evolutionarily distant organisms (Adams *et al.*, 1991, 1992). The utility of data generated in an EST study is greatly dependent upon the quality of the cDNA library. Ideally, a cDNA library dedicated to a study of this nature should be:

(1) representative, containing all sequences present in the initial poly(A)$^+$ population in the same relative frequencies;
(2) unidirectionally cloned so that the orientation of each cDNA is known, facilitating subsequent sequence analysis;
(3) composed of a high proportion of long or full-length inserts;
(4) uncontaminated with genomic, mitochondrial or ribosomal RNA inserts; and
(5) composed of a large proportion of inserts with short poly(A) tails.

15.2 PREPARATION OF cDNA LIBRARIES

15.2.1 Preparation of mRNA

There are various methods for isolating total RNA from unfractionated cells and tissues using guanidinium salts (Chirgwin *et al.*, 1979; Davis *et al.*, 1986; Okayama *et al.*, 1987; McDonald *et al.*, 1987; Chomczyniski & Sacchi, 1987). RNA can also be isolated from subcellular fractions: nuclei (Nevins, 1987); cytoplasm, using sucrose gradients and ribonucleoside–vanadyl complexes (Berger, 1987b); and from membrane-bound or free polysomes (Mechler, 1987; Lynch, 1987; Berger, 1987b). The use of guanidinium salts to facilitate the purification of intact, functional RNA is common because of their ability to rapidly inhibit endogenous ribonucleases (McDonald *et al.*, 1987). However, this method does not allow subcellular fractionation because it is based on a complete disruption of the cells or tissues.

Although there is a small, but distinct subclass of mRNA which lacks poly(A) (Greenberg & Perry, 1972; Adesnik *et al.*, 1972; Van Ness *et al.*, 1979), most mRNAs are post-transcriptionally polyadenylated. Mature mRNAs usually display a poly(A) tract of 40–200 adenylate residues, with an average length of 40–65 residues (Jacobson, 1987). Selection of poly(A)$^+$ RNA after total RNA isolation is necessary in order to eliminate ribosomal RNA which could serve as a template for reverse transcription. The isolation of mRNA from total RNA is typically performed with one of the following techniques: an oligo(dT) cellulose or a poly(U)–sepharose matrix (Aviv & Leder, 1972; Davis *et al.*, 1986; Jacobson, 1987); oligo(dT) covalently linked to latex particles (Hara *et al.*, 1991); or oligo(dT)-coated magnetic beads (Hornes & Korsnes, 1990). New commercially available kits and protocols even allow the direct isolation of poly(A)$^+$ RNA from tissues or cells using oligo(dT) cellulose matrix (Fastrack, Invitrogen) or with oligo(dT)-coated magnetic beads (Jakobsen *et al.*, 1990). These methods rely on base pairing between the poly(A)$^+$ residues at the 3′-end of mRNAs and the oligo(dT) residues coupled to the solid support. Non-poly(A)$^+$ species are washed off, and bound mRNA is subsequently eluted with a low salt buffer. Purification of poly(A) mRNA using oligo(dT) probes attached to paramagnetic beads is a rapid method not requiring centrifugation or filtration steps thereby reducing the risk of physical or enzymatic degradation. In addition, it is possible to isolate mRNA (both polyadenylated and nonpolyadenylated) with benzoylated cellulose chromatography for the production of random hexamer-primed cDNA libraries (Van Ness *et al.*, 1979).

Further assessment of the quality of the purified mRNA is recommended before proceeding with library construction. This is commonly achieved by an electrophoretic fractionation of mRNA in denaturing gels, Northern blotting and hybridization with a probe of known size common to most cells, such as actin or glyceraldehyde 3-phosphate dehydrogenase. mRNA molecules should be distributed between 0.6 and 4 kb in size, and the mRNA is considered intact if the mRNA homologous to the probe appears as a distinct band without a large amount of degraded product (Berger, 1987a).

15.2.2 Preparation of cDNA

During the construction of a cDNA library, polyadenylated mRNA is converted to cloneable ds cDNA by using a series of enzyme-catalyzed reactions. The oligonucleotides used to prime the synthesis of a complementary DNA with reverse transcriptase can be oligo(dT) primers which bind to the 3′ poly(A) tail thereby potentially generating a full-length copy (Krug & Berger, 1987); random hexamer primers (Dudley *et al.*, 1978; Sargent, 1987); or primers specific for a particular gene. The yield of complementary cDNAs containing the full coding sequence depends upon the quality of the mRNA and the degree of secondary structure present in mRNA transcripts which can interfere with complete extension of the primer. Random primers, usually DNA hexamers, ensure equal representation of all mRNA sequences in the cDNA library and increase the chances of obtaining coding and 5′-untranslated sequence. However, random primers allow neither unidirectional cloning nor synthesis of full-length cDNA copies. Some attempts to make unidirectionally cloned random-primed libraries by adding a restriction site tail to the random primers have not been successful (L. Staudt, personal communication). The use of unidirectional libraries in an EST study allows the possibility to choose between getting information from the 3′ untranslated sequence or from mostly coding sequence from the 5′-end which can be useful in identifying homology and similarity to sequences already described in the nucleotide and peptide databases. The sequence information which has been gained from the 3′ untranslated region has been found very useful in the designing of primers for the assignment of ESTs to specific chromosomes using PCR and panels of hybrid DNA (Polymeropoulos *et al.*, 1992). Cloning of poly(A) tails is a consequence of oligo(dT) priming and although long poly(A) tails do not ordinarily present a problem in conventional cloning applications, they have proven to be a difficulty in automated sequencing. It is difficult to sequence a length of poly(A) exceeding 30–40 bases because the nucleotides in the reaction mix get depleted and create an imbalance, introducing errors in the following sequence. A library containing a high proportion of short poly(A) tails can be achieved by using an excess of oligo(dT) primer during the first strand cDNA synthesis, thereby saturating the poly(A) tails of the mRNAs. Since the reverse transcriptase enzyme cannot strand displace, the primer located at the most 5′-position of the poly(A) tail will serve as the origin of the cDNA resulting in clones containing short poly(A) tails.

To protect the cDNA from restriction enzymes some protocols use 5′-methyl dCTP instead of dCTP in the nucleotide mixture for the first strand synthesis. The presence of 5′-methyl dCTP will create an hemimethylated cDNA. Other protocols protect EcoRI internal sites by treating the double-stranded DNA with EcoRI methylase (Wu *et al.*, 1987). To reduce the extensive secondary structure present in some mRNAs, some protocols treat RNA solutions with strong denaturants or methylmercury hydroxide to destroy base pairing (Payvar & Schimke, 1979).

Efficient synthesis of the cDNA second strand is accomplished with RNase H which nicks the RNA of the RNA–cDNA hybrid. This results in the production of multiple fragments that act as primers for DNA polymerase I (Gubler, 1987b). Other methods use the transient hairpin loop at the 3′-end of the newly synthesized first strand as a primer to generate the second strand after the RNA fragments are removed with alkali (Gubler, 1987a). After the synthesis of the second strand, the procedures in common use diverge considerably. For a comprehensive review of cDNA techniques, see Kimmel & Berger (1987) and Sambrook *et al.* (1989).

There are two general approaches for unidirectional cDNA cloning. In one, poly(A)$^+$ RNA is primed from an adaptor primer which generates a unique restriction site at one end of the cDNA. After second-strand synthesis, adaptor oligonucleotides with a cohesive end different from the restriction site within the adaptor primer are ligated onto both ends of the cDNA (Helfman *et al.*, 1987; Han *et al.*, 1987; Meissner *et al.*, 1987). The second established method for unidirectional cDNA synthesis is to use the vector as the primer (Okayama & Berg, 1982; Alexander, 1987; Deininger, 1987; Heidecker & Messing, 1987; Okayama *et al.*, 1987; Eun & Yoon, 1989; Rubenstein *et al.*, 1990).

The use of λ phage vectors for unidirectional cloning of cDNA inserts commonly requires the filling of uneven termini of the double-stranded cDNA with T4 DNA polymerase in order to increase the efficiency of ligation to adaptors. The digestion of the construct with the appropriate restriction enzyme, using a restriction site incorporated to the 5′-end of the adaptor-oligo(dT) primer, releases the adaptor from the 3′-end of the cDNA. This digestion creates two different and specific ends that allow unidirectional cloning. The cDNAs are commonly size fractionated with commercially available Sephacryl columns to discard very short inserts of less than 500 bp. The size-selected cDNA is ligated to lambda vector arms or to a vector of choice which has been previously digested with the appropriate restriction enzymes, dephosphorylated, and purified. High-efficiency commercially available packaging extracts facilitate the last step before plating the library (Kretz *et al.*, 1989) (Gigapack, Stratagene, La Jolla, CA 92037, USA). Cloning and packaging can be carried out as needed, and can be followed by a conversion to phagemid form by *in vivo* excision without prior amplification of the library. Amplification of a phage library that is to be used in an EST study should be avoided as it may result in a change in the composition of the library as a result of differences in growth rates between clones (Ausubel *et al.*, 1990).

For first-strand cDNA synthesis and subsequent unidirectional cloning, other strategies use an oligo(dT) linker ligated to a plasmid vector to create vector primer (Okayama & Berg, 1982; Alexander, 1987; Deininger, 1987; Heidecker & Messing, 1987; Okayama *et al.*, 1987; Eun & Yoon, 1989; Rubenstein *et al.*, 1990). In this case, the linearized vector is utilized to prime the polyadenylated tail of the mRNA and to synthesize the first strand of the cDNA. After the synthesis of the second strand following standard methods, the construct is blunt ended, size selected and then ligated to itself. In doing so, a bimolecular ligation is converted to a more efficient cyclization reaction. Another advantage of the vector-primer method is that the cDNA is never exposed to a restriction enzyme. The resulting plasmid library is used to transform competent cells. A modification of this method makes use of the so-called 'primer-restriction end adaptors' (Coleclough, 1987; Gubler, 1987b; Rubenstein *et al.*, 1990) that serve both as primers and as ligation substrates. Some of these methods of insertion involve a strong selection for both termini of the original cDNA molecules, favoring the production of full-length clones.

15.2.3 Vectors for cDNA library construction

Although multiple types of vector may be used to construct a cDNA library, most of the large-scale cDNA sequencing studies are based on λ or phagemid libraries. Most of the current automated sequencing technology has been developed using M13 priming sites. Any vector, phage or plasmid with these sites and the ability to be amplified in prokaryotes should be suitable for this purpose. It is important to have a highly representative library for an EST study, and therefore, the cloning efficiency of the vector system chosen is of central importance. It has been estimated that cloning in λ is approximately 30-fold more efficient than cloning in plasmids (Welsh *et al.*, 1990). Efficiency for plasmid cloning, however, has increased with the introduction of vector-primer methods and electroporation methods to introduce DNA into competent cells (Okayama & Berg, 1982; Alexander, 1987; Deininger, 1987; Heidecker & Messing, 1987; Okayama *et al.*, 1987; Eun & Yoon, 1989; Rubenstein *et al.*, 1990). Finally, if additional clones need to be obtained for further studies, screening a high density of plaques from the phage library with DNA or antibody probes is more easily accomplished. Clones obtained from λ cDNA libraries may either be subcloned into a plasmid vector prior to sequencing or sequenced directly using a polymerase chain reaction (PCR) to generate an adequate amount of insert DNA (Mason, 1992; Tracy & Mulcahy, 1991). Some characteristics of the new λ vectors allow *in vivo* excision of the cDNA insert. A vector that is particularly good for these purposes is λ ZAP (Stratagene, La Jolla, CA 92037, USA). During the *in vivo* excision process, the insert and flanking fragments of the λ-phage are converted *in vivo* into a phagemid that facilitates subsequent sequencing. λ ZAP vectors also allow identification of nonrecombinant phages, and, since an EST study involves the random picking of clones, this is an important feature. One vector of this series, Uni-ZAP XR (Stratagene, La Jolla, CA 92037, USA), can accommodate DNA inserts from 0 to 10 kb in length. The polylinker is flanked by T3 and T7 promoters and it has a choice of six different primer sites for DNA sequencing. Transcripts made from the T3 and T7 promoters generate riboprobes useful in Southern and Northern blotting. The phagemid form, generated by *in vivo* excision, has the bacteriophage f1 origin of replication, allowing rescue of single-stranded DNA which can be used for various subtraction procedures, as described below. Finally, unidirectional deletions can be made with exonuclease III and mung bean nuclease to facilitate further sequencing of selected clones and the lacZ promoter may be used to drive expression of fusion proteins suitable for Western blot analysis or protein purification.

Plasmids or phagemids can be efficiently used for construction of a cDNA library by vector-priming methods (Okayama & Berg, 1982; Heidecker & Messing, 1987; Deininger, 1987; Alexander, 1987; Okayama *et al.*, 1987; Eun & Yoon, 1989; Rubenstein *et al.*, 1990) or restriction end adapter-priming methods (Coleclough, 1987; Gubler, 1987b; Rubenstein *et al.*, 1990). These methods have much higher efficiencies than do the older methods of cloning double-stranded cDNA into plasmids, and with the combination of electroporation of plasmids into bacteria, these methods rival the efficiencies obtained with phage cDNA cloning. They also have high recovery of full-length cDNAs, are simple and fast once the vector is optimized, and subcloning is not required for sequencing. However, this method has some disadvantages:

(1) sizing can be done with vector priming, but the length of the attached vector makes it more difficult to resolve small differences in cDNA lengths;
(2) screening by colony hybridization is not as simple as phage plaque based hybridization methods; and
(3) the modification of the multiple cloning site that takes place during the construction of vector primers (Rubenstein *et al.*, 1990) make it difficult to distinguish between recombinant and non-recombinant clones (R.F. Moreno-Palanques, unpublished).

In summary, of the current vectors available, only λ or plasmids used as vector primers (including primer-restriction end adapter methods) allow a high-efficiency cloning of long or full-length cDNAs.

15.2.4 Preparation of a cDNA library from limited amounts of RNA

In general, after standard cDNA synthesis procedures are followed, only about 10% of the input mRNA is actually converted to cloneable double-stranded cDNA. In addition, using *in vitro* phage packaging systems or transformation of competent bacteria with plasmid vectors, the amount of ligated cDNA that is introduced into cells is even further reduced. Therefore, often it is not possible to construct a representative cDNA library when the amount of RNA is limiting, such as in studies involving small numbers of cells, embryonic tissues or surgically removed tissue samples. Many researchers are turning to novel techniques that combine the amplification power of PCR with recent advances in solid phase RNA capture systems to produce cloneable amounts of cDNA from small amounts of starting material. For example, a representative library was constructed from RNA extracted from only 50 mouse ova (Welsh *et al.*, 1990). The cloning strategy relies on the synthesis of first strand cDNA from mRNA using an oligo(dT) primer followed by the addition of a homopolymer tail to the 3′-end of the newly synthesized first strand with terminal transferase (Welsh *et al.*, 1990). The second-strand cDNA is synthesized with the use of a return primer complementary to the homopolymeric tail. The resulting DNA duplex is then amplified with PCR prior to the subcloning of the cDNA into the vector of choice. However, it has been found that many artifacts are produced with this *in vitro* amplification strategy primarily as a result of nonspecific annealing of the reverse primer or from primer tailing of products, and that the average size of inserts obtained is only 0.6 kb. Other investigators have modified this general scheme in an attempt to optimize the production of representative libraries. Using a solid-phase RNA capture system, Lambert & Williamson (1993) have developed a method to construct a cDNA library from as little as 5 ng of poly(A)$^+$ RNA. In this method, first-strand synthesis is initiated from an oligo(dT) primer bound to a magnetic bead creating a solid-phase library that can serve as a template for second-strand synthesis prior to the amplification of the duplex with complementary primers. In another modification, a universal buffer system (KGB buffer) has recently been described which allows for the use of a single buffer during cDNA synthesis and ligation to specific linkers, thereby reducing yield losses due to multiple extractions and reprecipitations between enzymatic steps of cDNA synthesis (Don *et al.*, 1993). Additional amplification of the cDNA can also be carried out with primers complementary to the linkers if necessary. In order to circumvent the drawbacks associated with *in vitro* amplification of cDNA using nonspecific primers, a novel method has been introduced which tags the 3′-end of single-stranded cDNA (sscDNA). This method is based on the ligation of a specific oligonucleotide to the 3′-end of sscDNA using T4 RNA ligase (Edwards, 1991) prior to the second-strand synthesis of cDNA and the subsequent PCR amplification of the cDNA with nested primers.

15.3 MODIFIED cDNA LIBRARIES

15.3.1 Subtracted cDNA libraries

Subtractive hybridization has been applied successfully to a wide range of biological problems to identify differentially expressed genes of unknown sequence. The conventional approach has consisted of isolating clones with a labeled subtractive cDNA probe designed to detect cDNA clones which are present in one cell type but absent or expressed at significantly lower levels in another cell type (Owekamp & Firtel, 1980; Schutzbank *et al.*, 1982; Scott *et al.*, 1983; Sargent, 1987). Subtractive hybridization typically involves hybridizing tens of micrograms of poly(A)$^+$ RNA (driver) at a 10-fold excess with cDNA (target) prepared from mRNA from another cell type. RNA and cDNA are annealed to high R_0t values (approximately 3000 mol s^{-1}) (Britten *et al.*, 1974; Kohne *et al.*, 1977). The double-stranded RNA–DNA hybrids represent sequences which are present in both cell types and can be removed (subtracted) from the unhybridized single-stranded cDNA preparation using hydroxyapatite chromatography (Sargent, 1987; Schneider *et al.*, 1988; Timblin *et al.*, 1990; Kriegler, 1991). The remaining unhybridized cDNA can be used as a probe to screen libraries or used to generate a subtracted cDNA library if enough product is available. Two recent uses of subtractive probe technology were the identification of a 60 novel cDNA clones induced in resting human peripheral blood T cells after mitogen activation (Zipfel *et al.*, 1989) and the identification of a class of candidate tumor suppressor genes (Lee *et al.*, 1991). Another protocol dependent upon a low-ratio hybridization subtraction between driver RNA and target cDNA has been developed recently which was found to enrich for cDNAs representing low-abundance transcripts which are induced only several-fold over the base level. The hybridization reaction was carried out in a phenol emulsion which altered the kinetics such that the rates of association were increased markedly. Therefore, this subtraction technique allows for the isolation of clones which are quantitatively as well as qualitatively different between cell types (Fargnoli *et al.*, 1990).

One of the most common problems with subtraction techniques is that specific mRNAs or cDNAs from the target tissue are subtracted away by common sequences or repetitive elements (e.g. Alu repetitive elements) yielding very low amounts of cloneable material. Another major disadvantage of these procedures is that they require large amounts of driver mRNA which may be difficult to obtain for some studies in which tissues are limited. Various methods have been developed recently which combine PCR and solid-phase capture technologies to generate subtracted libraries. Several investigators have published techniques which base the subtractive hybridizations on ds cDNA which has been prepared from small amounts of mRNA and amplified with primers complementary to oligonucleotide linkers (Wang & Brown, 1991; Murakami *et al.*, 1992). Both these procedures developed methods to photobiotinylate the driver DNA in order to improve the efficiency of removal of hybridized DNA sequences. A novel strategy for preparing a subtracted library using an oligo(dT) primer covalently linked to latex particles has also been recently introduced (Hara *et al.*, 1991). Driver cDNA is synthesized on latex beads using mRNA template from one tissue source. After removal of the mRNA template by heat denaturation and centrifugation, the first strand remains bound to the solid support where it can serve as a subtractive probe for the target mRNA for several rounds of subtraction. The unhybridized subtracted mRNA can be recovered by centrifugation and converted to cDNA. Following a typical subtraction, only a few nanograms of target sequences remain and therefore the ds cDNA is first ligated to linkers, amplified with PCR and then subcloned into an appropriate vector. Another technique that generates tissue-specific or stage-specific subtracted probes using magnetic oligo (dT) beads has been recently described (Rodriquez, 1991). In this method, cDNA is synthesized on the magnetic beads from total RNA (target) as well as from driver mRNA. A random primed radiolabeled probe is synthesized from the target cDNA and subtraction hybridization is performed with an excess of driver cDNA bound to the solid support. The radiolabeled probe which remains can then be used to screen a library.

Subtraction procedures based on the commercially available λ ZapII phage vector system have been widely used to isolate rare mRNA which are differentially expressed in various tissues. The advantages of this system are that cDNA libraries can be established with high efficiency in phage particles from small amounts of mRNA using either directionally cloned or random primed cDNA fragments. During the rescue of a covalently closed circular plasmid library from a bacteriophage λ ZapII library, there is an intermediate step which involves the generation of single-stranded phagemid particles. These phagemids form the basis for a variety of subtraction procedures allowing for the facile isolation and characterization of rare mRNA (less than 0.2% abundance) enriched more than 100-fold after one or more rounds of subtraction (Duguid *et al.*, 1988, 1989). The original method of phagemids subtractions were based on random-primed cDNA libraries and was modified by the introduction of unidirectional insertion of cDNAs into single-stranded phagemids with modified noncomplementary multiple cloning steps (Rubenstein *et al.*, 1990). This technique has been applied in studies to enrich for sequences overexpressed in neoplastic tissue as compared with normal (Schweinfest *et al.*, 1990; Schweinfest & Papas, 1992). The procedure described recently by Houge (1993) is a different modification which is based on the production of biotinylated phagemids with asymmetric PCR which can be used as the driver DNA to subtract out complementary sequences from the phagemids produced from another phage library. In contrast to previous procedures, this method uses standard, unmodified, commercially available λ Zap II vector systems and therefore can be applied to already existing libraries.

Many techniques have been developed recently for the selection of cDNAs by hybridization with fragments of genomic DNA immobilized on a solid support (Parimoo *et al.*, 1991; Lovett *et al.*, 1991; Morgan *et al.*, 1992). Low-abundance cDNAs encoded by large genomic clones can be isolated, amplified and subcloned. In the simplest of these techniques, biotinylated cloned genomic DNA is hybridized in solution with amplifiable cDNAs (Morgan *et al.*, 1992). The genomic clones and hybridized cDNAs are captured on streptavidin-coated magnetic beads, the cDNAs are eluted, amplified with PCR and subcloned. All of the selected cDNAs which were initially present at very low abundance were found to be increased as much as 100 000-fold after two rounds of enrichment. The advantage of this technique is the potential application to the isolation of candidate disease loci as well as to the derivation of detailed transcription maps across large genomic regions.

15.3.2 Normalized cDNA libraries

The abundance of different mRNA varies characteristically within a cellular population, as well as during different stages of development. An individual class of transcript can represent from 1 in 10^6 to $\geqslant$10% of the total message population. In absolute numbers, it has been estimated that about 10 000 genes are expressed in a given human cell type at levels that may vary from less than one transcript to 200 000 transcripts per cell, with approximately one-third of the genes being expressed at 1–10 molecules per cell (Galau *et al.*, 1977). The variation in abundance classes in a given cell type implies that several hundred thousand clones from a unmodified and representative library would have to be sequenced to have a reasonable chance of finding a particular rare transcript (Sambrook *et al.*, 1989). The situation is even more complex with a tissue which comprises numerous cell types (Ohlsson, 1989).

Normalization is the process by which the abundant messages are reduced in number and rare messages are increased relative to other clones within a cDNA library. The number of different cDNA clones that would constitute a normalized library from a given cell type has not been accurately determined, although it has been estimated to be in the range of 0.5×10^4 to 25×10^4 (Weissman, 1987).

The elimination of highly abundant clones by selection by hybridization, subtraction or normalization would certainly avoid the redundant sequencing of these repetitive elements but would also result in the loss of the information regarding the accompanying coding sequence. Normalization procedures attempt to reduce the amount of redundant sequencing and are useful when a complex tissue is the source for cDNA. Two approaches have been proposed to obtain normalized cDNA libraries (Weissman, 1987). One approach is dependent upon hybridization with genomic DNA in order that the relative abundance of cDNAs is made proportional to the abundance of genes complementary to those cDNAs in genomic DNA. The other approach is based on the reannealing of double-stranded DNA in solution following second-order kinetics (Galau *et al.*, 1977). Using this kinetic approach, cDNA libraries (Patanjali *et al.*, 1991) have been constructed containing an approximately equal representation of all sequences present in the initial preparation of poly(A)$^+$ RNA. In this process, the randomly primed cDNA fragments of a selected size range were cloned into λ phage vector. Inserts were amplified by PCR using primers for the flanking sites of the EcoR I site in λ phage vector, then denatured and self-reannealed at different C_0t (24, 48, 72 h, etc.). Single-stranded cDNAs were separated from double-stranded cDNA by hydroxyapatite chromatography and were amplified by PCR. The 500–2000 bp fraction was size-selected and cloned into a λ gt10 vector. The sizes of ss DNA and ds DNA fractions decreased progressively with the increase in reassociation time, being the most pronounced after 120 h of reassociation. This was attributed to thermal degradation. The actual hybridization behavior was, however, more complex than predicted by second-order kinetics of hybridization (Patanjali *et al.*, 1991). The most abundant species hybridized more rapidly than predicted and, in some of the initial experiments, was actually found to be present in the single-stranded fraction at a level considerably less than that of scarcer species. To overcome this problem the investigators suggested that two different library fractions should be prepared with different annealing kinetics, and combined in an appropriate ratio.

Theoretically, removal of the more abundant species from a library prepared from a single cell type would enrich the rare species by only three-fold or less because only one-third of the mRNA in a single cell type is made up of species present at 1–10 copies per cell (Galau *et al.*, 1977). A single cell type may contain representatives of 10 000 or more mRNAs. If each mRNA contributes 2–3 cDNAs to the normalized cDNA library, any single cDNA fragment would be present at about one copy per 30 000. This representation would go down further when the source of cDNA is a complex tissue. In an hypothetical normalized library derived from all mRNAs expressed at any time in any tissue, each fragment would be present at an abundance of less than 1 part per 3×10^5 or less. This estimate would be even lower for rare transcripts and nonnormalized libraries. In the EST studies performed to date, we have found it is possible to identify the highly abundant transcripts present in a library by sequencing as few as 300 randomly selected clones. It is possible to selectively remove these identified sequences from a library in the phagemid form by subtractive hybridization with complementary biotinylated probes prior to additional sequencing, thereby enhancing the number of low-abundance transcripts sequenced. Therefore, by normalizing libraries prepared from different organs or tissues at different stages of development, one can approach the goal of isolating and characterizing the majority of the human genes.

REFERENCES

Adams, M.D., Kelley, J.M., Gocayne, J.D., Dubnick, M., Polymeropoulos, M.H., Xiao, H., Merril, C.R., Wu, A., Olde, B., Moreno, R.F., Kerlavage, A.R., McCombie, W.R. & Venter, J.C. (1991) *Science* **252**, 1651–1656.

Adams, M.D., Dubnick, M., Kerlavage, A.R., Moreno, R., Kelley, J.M., Utterback, T.R., Nagle, J.W., Fields, C. & Venter, J. C. (1992) *Nature* **355**, 632–634.

Adesnik, M., Salditt, M., Thomas, W. & Darnell, J.E. (1972) *J. Mol. Biol.* **71**, 21–30.

Alexander, D.C. (1987) In *Methods in Enzymology, Recombinant DNA, Part E*, R. Wu & L. Grossman (eds). Academic Press, San Diego, pp. 41–64.

Ausubel, F.M., Brent, R., Kingston, R.E., Moore, D.D., Seidman, J.G., Smith, J.A. & Struhl, K. (1990) *Current Protocols in Molecular Biology*. Greene Publishing Associates/Wiley-Interscience, New York.

Aviv, H. & Leder, P. (1972) *Proc. Natl. Acad. Sci. U.S.A.* **69**, 1408–1412.

Berger, S.L. (1987a) In *Methods in Enzymology, Guide to Molecular Cloning Techniques*, S.L. Berger & A.R. Kimmel (eds). Academic Press, San Diego, pp. 215–219.

Berger, S.L. (1987b) In *Methods in Enzymology, Guide to Molecular Cloning Techniques*, S.L. Berger & A.R. Kimmel (eds). Academic Press, San Diego, pp. 227–234.

Britten, R.J., Graham, D.E. & Neufeld, B.R. (1974) In *Methods in Enzymology, Nucleic Acids and Protein Synthesis, Part E*, L. Grossman & K. Moldave (eds). Academic Press, San Diego, pp. 363–418.

Chirgwin, J.M., Przybyla, A.E., MacDonals, R.J. & Rutter, W.J. (1979) *Biochemistry* **18**, 5294–5299.

Chomczyniski, P. & Sacchi, N. (1987) *Anal. Biochem.* **162**, 156–159.

Coleclough, C. (1987) In *Methods in Enzymology, Recombinant DNA, Part E*, R. Wu & L. Grossman (eds). Academic Press, San Diego, pp. 64–83.

Davis, L.G., Dibner, M.D. & Battey, J.F. (1986) *Basic Methods in Molecular Biology*. Elsevier, New York.

Deininger, P. (1987) In *Methods in Enzymology, Guide to Molecular Cloning Techniques*, S.L. Berger & A.R. Kimmel (eds). Academic Press, San Diego, pp. 371–389.

Don, R.H., Cox, P.T. & Mattick, J.S. (1993) *Nucleic Acids Res.* **21**, 783.

Dudley, J.P., Butel, J.S., Socher, S.H. & J.M.R., (1978) *J. Virol.* **28**, 743–752.

Duguid, J.R., Rohwer, R.G. & Seed, B. (1988) *Proc. Natl. Acad. Sci. U.S.A.* **85**, 5738–5742.

Duguid, J.R., Bohmon, T.C.W., Liu, N. & Tourtellotte, W.W. (1989) *Proc. Natl. Acad. Sci. U.S.A.* **86**, 7260–7264.

Edwards, J.B., Delort, J. & Mallet, J. (1992) *Nucleic Acids Res.* **19**, 5227–5232.

Eun, H.-M. & Yoon, J.-W. (1989) *BioTechniques* **7**, 992–997.

Fargnoli, J., Holbrook, N.J. & Fornace, A.J. (1990) *Anal. Biochem.* **187**, 364–373.

Galau, G.A., Klein, W.H., Britten, R.J. & Davidson, E.H. (1977) *Arch. Biochem. Biophys.* **179**, 584–599.

Greenberg, J.R. & Perry, R.P. (1972) *J. Mol. Biol.* **72**, 91–98.

Gubler, U. (1987a) In *Methods in Enzymology, Guide to Molecular Cloning Techniques*, S.L. Berger & A.R. Kimmel (eds). Academic Press, San Diego, pp. 325–329.

Gubler, U. (1987b) In *Methods in Enzymology, Guide to Molecular Cloning Techniques*, S.L. Berger & A.R. Kimmel (eds). Academic Press, San Diego, pp. 330–335.

Han, J.H., Stratowa, C. & Rutter, W.J. (1987) *Biochemistry* **26**, 1617–1625.

Hara, E., Kato, T., Nakada, S., Sekiya, S. & Oda, K. (1991) *Nucleic Acids Res.* **19**, 7097–7104.

Heidecker, G. & Messing, J. (1987) In *Methods in Enzymology,*

Recombinant DNA, Part E, R. Wu & L. Grossman (eds). Academic Press, San Diego, pp. 28–41.

Helfman, D.M., Fiddes, J.C. & Hanahan, D. (1987) In *Methods in Enzymology, Guide to Molecular Cloning Techniques*, S.L. Berger & A.R. Kimmel (eds). Academic Press, San Diego, pp. 349–359.

Hornes, E. & Korsnes, L. (1990) *Genet. Anal. Technol. Appl.* **7**, 145–150.

Houge, G. (1993) In *PCR Methods and Applications*. Cold Spring Harbor Laboratory Press, Cold Spring Harbor, New York, pp. 204–209.

Jacobson, A. (1987) In *Methods in Enzymology, Guide to Molecular Cloning Techniques*, S.L. Berger & A.R. Kimmel (eds). Academic Press, San Diego, pp. 254–261.

Jakobsen, K.S., Breivold, E. & Hornes, E. (1990) *Nucleic Acids Res.* **18**, 3669.

Khan, A.S., Wilcox, A.S., Polymeropoulos, M.H., Hopkins, J.A., Stevens, T.J., Robinson, M., A.K., O. & J.M., S. (1992) *Nature Genet.* **2**, 180–185.

Kimmel, A.R. & Berger, S.L. (1987) *Preparation of cDNA and the generation of cDNA libraries: Overview*. Academic Press, San Diego.

Kohne, D.E., Levison, S.A. & Byers, M.J. (1977) *Biochemistry* **17**, 5329–5341.

Kretz, P.L., Reid, C.H., Greener, A. & Short, J.M. (1989) *Nucleic Acids Res.* **17**, 5409.

Kriegler, M. (1991) *Gene Transfer and Expression. A Laboratory Manual*. W.H. Freeman, New York.

Krug, M.S. & Berger, S.L. (1987) In *Methods in Enzymology, Guide to Molecular Cloning Techniques*, S.L. Berger & A.R. Kimmel (eds). Academic Press, San Diego, pp. 316–325.

Lambert, K.N. & Williamson, V.M. (1993) *Nucleic Acids Res.* **21**, 775–776.

Lee, S.W., Tomasetto, C. & Sager, R. (1991) *Proc. Natl. Acad. Sci. U.S.A.* **88**, 2825–2829.

Lovett, M., Kere, J. & Hinton, L.M. (1991) *Proc. Natl. Acad. Sci. U.S.A.* **88**, 9628–9632.

Lynch, D.C. (1987) In *Methods in Enzymology, Guide to Molecular Cloning Techniques*, S.L. Berger & A.R. Kimmel (eds). Academic Press, San Diego, pp. 248–253.

Mason, I.J. (1992) *BioTechniques* **12**, 60.

McCombie, W.R., Adams, M.D., Kelley, J.M., FitzGerald, M.G., Utterback, T.R., Khan, M., Dubnick, M., Kerlavage, A.R., Venter, J.C. & Fields, C. (1992) *Nature Genet.* **1**, 124–130.

McDonald, R.J., Swift, G., Przybyla, A.E. & Chirgwin, J.M. (1987) In *Methods in Enzymology, Guide to Molecular Cloning Techniques*, S.L. Berger & A.R. Kimmel (eds). Academic Press, San Diego, pp. 219–227.

Mechler, B.M. (1987) In *Methods in Enzymology, Guide to Molecular Cloning Techniques*, S.L. Berger & A.R. Kimmel (eds). Academic Press, San Diego, pp. 241–248.

Meissner, P.S., Sisk, W.P. & Berman, M.L. (1987) *Proc. Natl. Acad. Sci. U.S.A.* **84**, 4171–4175.

Miller, F.D., Naus, C.C.G., Higgins, G., Bloom, F.E. & Milner, R.J. (1987) *J. Neurosci.* **7**, 2433–2444.

Morgan, J.G., Dolganov, G.M., Robbins, S.E., Hinton, L.M. & Lowett, M. (1992) *Nucleic Acids Res.* **20**, 5173–5179.

Murakami, A., Yajima, T. & Inana, G. (1992) *Biochem. Biophys. Res. Commun.* **187**, 234–244.

Nevins, J.R. (1987) In *Methods in Enzymology, Guide to Molecular Cloning Techniques*, S.L. Berger & A.R. Kimmel (eds). Academic Press, San Diego, pp. 234–241.

Ohlsson, R. (1989) *Cell. Differ. Dev.* **28**, 1–16.

Okayama, H. & Berg, P. (1982) *Mol. Cell Biol.* **2**, 161–170.

Okayama, H., Kawaichi, M., Brownstein, M., Lee, F., Yakota, T. & Arai, K. (1987) In *Methods in Enzymology, Recombinant DNA, Part E*, R. Wu & L. Grossman (eds). Academic Press, San Diego, pp. 3–28.

Okubo, K., Hor, N., Matoba, R., Niiyama, T., Fukushima, A., Kojima, Y. & Matsubara, K. (1992) *Nature Genet.* **2**, 173–170.

Owekamp, W. & Firtel, R.A. (1980) *Dev. Biol.* **79**, 409–418.

Parimoo, S., Patanjali, S.R., Shukla, H., Cahplin, D.D. & Weissman, S.M. (1991) *Proc. Natl. Acad. Sci. U.S.A.* **88**, 9623–9627.

Patanjali, S.R., Parimoo, S. & Weissman, S.M. (1991) *Proc. Natl. Acad. Sci. U.S.A.* **88**, 1943–1947.

Payvar, F. & Schimke, R.T. (1979) *J. Biol. Chem.* **254**, 7636–7642.

Polymeropoulos, M.H., Xiao, H., Glodek, A., Gorski, M., Adams, M.D., Moreno, R.F., FitzGerald, M.G., Venter, J.C. & Merril, C.R. (1992) *Genomics* **12**, 492–496.

Rodriquez, I.R. & Chader, G.J. (1992) *Nucleic Acids Res.* **20**, 3528.

Rubenstein, J.L.R., Brice, A.E., Ciaranello, R.D., Denney, D., Porteus, M.H. & Usdin, T.B. (1990) *Nucleic Acids Res.* **18**, 4833–4842.

Sambrook, J., Fritsch, E.F. & Maniatis, T. (1989) *Molecular Cloning. A Laboratory Manual*. Cold Spring Harbor Laboratory Press, Cold Spring Harbor, New York.

Sargent, T.D. (1987) In *Methods in Enzymology, Guide to Molecular Cloning Techniques*, S.L. Berger & A.R. Kimmel (eds). Academic Press, San Diego, pp. 423–432.

Schneider, C., King, R.M. & Philipson, L. (1988) *Cell* **54**, 787–793.

Schutzbank, T., Robinson, R., Oren, M. & Levine, A.J. (1982) *Cell* **30**, 481–490.

Schweinfest, C.W. & Papas, T.S. (1992) *Int. J. Oncol.* **1**, 499–506.

Schweinfest, C.W., Henderson, K.W., Gu, J.R., Kottaridis, S.D., Besbeas, S., Panotopoulou, E. & Papas, T.S. (1990) *Genet. Anal. Technol. Appl.* **7**, 64–70.

Scott, M.R.D., Westphal, K.-H. & Rigby, P.W.J. (1983) *Cell* **34**, 557–567.

Timblin, C., Battey, J. & Kuehl, W.M. (1990) *Nucleic Acids Res.* **18**, 1587–1593.

Tracy, T.E. & Mulcahy, L.S. (1991) *BioTechniques* **11**, 68–75.

Van Ness, J., Maxwell, I.H. & Hahn, W.E. (1979) *Cell* **18**, 1341–1349.

Wang, Z. & Brown, D.D. (1991) *Proc. Natl. Acad. Sci. U.S.A.* **88**, 11505–11509.

Weissman, S.M. (1987) *Mol. Biol. Med.* **4**, 133–143.

Welsh, J., Liu, J.P. & Efstratiadis, A. (1990) *Genet. Anal. Technol. Appl.* **7**, 5–17.

Wieland, I., Bolger, G., Asouline, G. & Wigler, M. (1990) *Proc. Natl. Acad. Sci. U.S.A.* **87**, 2720–2724.

Wu, R., Wu, T. & Ray, A. (1987) In *Methods in Enzymology, Guide to Molecular Cloning Techniques*, S.L. Berger & A.R. Kimmel (eds). Academic Press, San Diego, pp. 343–349.

Zipfel, P.F., Irving, S.G., Kelly, K. & Siebenlist, U. (1989) *Mol. Cell. Biol.* **9**, 1041–1048.

CHAPTER SIXTEEN

Construction of Directionally Cloned cDNA Libraries in Phagemid Vectors

M.B. SOARES

Department of Psychiatry, College of Physicians and Surgeons of Columbia University and New York State Psychiatric Institute, 722 West 168th Street, Unit 41, New York, NY 10032, USA

16.1 INTRODUCTION

All of us who have gone through the exercise of cloning a gene and defining its transcriptional unit, have learned the hard way that no cDNA clones should be considered bona fide copies of their respective mRNAs without further scrutiny. For example, cDNA clones may be chimeric, i.e. they may be composed of two or more cDNAs derived from different mRNAs artificially joined into a single molecule by blunt-end ligation. These so-called 'cDNA cloning artifacts' can be particularly troublesome in instances when one is studying a transcription unit that can generate a family of transcripts by a combination of alternative splicing, differential polyadenylation, and utilization of more than one promoter, as in the case of the rat insulin-like growth factor II gene (Soares *et al.*, 1986).

The issue of the quality of a cDNA library becomes of paramount importance in the context of the expressed sequence tag (EST) sequencing program, in which thousands of cDNA clones are randomly picked from libraries and sequenced from one or both ends (Adams *et al.*, 1991, 1992; Okubo *et al.*, 1992; Khan *et al.*, 1992). Cloning artifacts are likely not to be easily detected due to the very nature of such a large-scale operation. Ultimately, therefore, the relevance of the data generated is to a great extent dependent upon the quality of the cDNA library that is utilized.

The purpose of this chapter is to address some of the problems that are commonly observed in cDNA libraries, to discuss some modifications that I have introduced in the existing methods (Okayama & Berg, 1982; Gubler & Hoffman, 1983; D'Alessio *et al.*, 1987) and to provide a detailed protocol for the construction of directionally cloned cDNA libraries in phagemid vectors.

I focus in this chapter on the details of constructing a directionally cloned cDNA library primed with NotI-$(dT)_{18}$. A human infant brain cDNA library made specifically for the production of ESTs with this method proved to be very useful for identifying new genes and describing the transcriptional activity of this tissue. The library was found to contain very low numbers of mitochondrial and ribosomal transcripts, a low percentage of chimeric clones, and a high percentage of short poly(A) tails, all of which served to increase the efficiency of EST sequencing from both the 5′- and 3′-ends of the clones (Adams *et al.*, 1993).

16.2 OUTLINE OF THE CONSTRUCTION OF DIRECTIONALLY CLONED cDNA LIBRARIES

The general scheme for construction of directionally cloned cDNA libraries can be outlined as follows:

(1) A NotI-$(dT)_{18}$ oligonucleotide is utilized as primer for first-strand cDNA synthesis with RNase H^- reverse transcriptase from Moloney murine leukemia virus (Gibco-BRL).

(2) 'One tube' first- and second-strand cDNA syntheses are performed essentially as described by D'Alessio *et al.*, (1987).

(3) Double-stranded cDNAs are polished with T4 DNA polymerase, size selected on a Bio-Gel A-50m column as described by Huynh *et al.* (1985) and ligated to a large excess of adaptor molecules (EcoRI adaptors, for example).

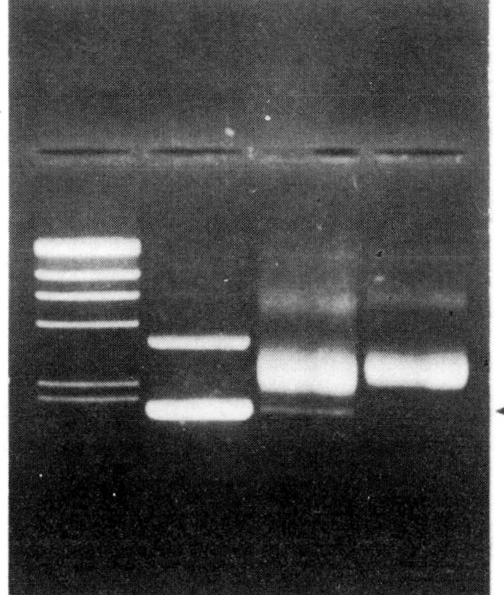

Figure 16.1 Agarose gel electrophoresis of total plasmid DNA from a directionally cloned cDNA library constructed in a phagemid vector. Total plasmid DNA from a human infant brain cDNA library directionally cloned into a phagemid vector (lafmid BA) was prepared using Qiagen columns according to the manufacturer's instructions. (1) λ DNA digested with HindIII. (2) cDNA cloning vector: lafmid BA. (3) Total plasmid DNA from the cDNA library before size selection (see step (5), Section 16.2). (4) Total plasmid DNA from the cDNA library after the final size selection procedure (see step (6), Section 16.2) which involved digestion of the plasmid DNA shown in lane 3 with NotI, purification of the recombinant molecules containing inserts larger than 500 bp by agarose gel electrophoresis, recircularization and electroporation into bacteria. As indicated by the arrow, the small amount of nonrecombinant (vector without an insert) clones present in the library before size selection (lane 3) can no longer be detected after size selection (lane 4).

(4) cDNAs are treated with T4 polynucleotide kinase to phosphorylate the adaptor ends (one of the two oligonucleotides of the adaptor molecule has a 5'-OH to prevent concatemerization of adaptors), digested with NotI, size selected again over a Bio-Gel A-50m column and ligated directionally into the NotI and Eco RI ends of a phagemid vector.

(5) The ligation mixture is electroporated into bacteria and propagated under appropriate antibiotic selection (see lane 3 in Fig. 16.1).

(6) To eliminate completely from the library all clones that contain inserts shorter than 500 bp, the nonrecombinants, and most existing chimeric clones, a plasmid preparation of the library is linearized with NotI, electrophoresed on an agarose gel and the linear recombinant molecules containing cDNAs larger than 500 bp are purified off the gel with β-agarase and recircularized in a large volume ligation reaction. The exact reaction conditions to promote recircularization rather than intermolecular ligations can be determined by the formula $3.3/\sqrt{kb}\ \mu g\ ml^{-1}$ as discussed by Smith *et al.* (1987).

(7) The ligation mixture is electroporated into bacteria and propagated under appropriate antibiotic selection to generate a cDNA library with an average size insert of 1.5 kb, no inserts shorter than 500 bp and a very low background of nonrecombinant clones (see lane 4 in Fig. 16.1).

16.3 PROBLEMS COMMONLY OBSERVED IN cDNA LIBRARIES

16.3.1 Long polyadenylate tails

Probably the most widely acknowledged problem of directionally cloned cDNA libraries is the presence, in high frequency, of clones that contain (or consist exclusively of) long polyadenylate tails. Because these clones contribute none or very limited sequence information they can seriously compromise the overall efficiency of production of 3' ESTs (Adams *et al.*, 1991).

I found that this problem can be practically eliminated by increasing the amount of the NotI-$(dT)_{18}$ oligonucleotide utilized to prime the synthesis of first-strand cDNA. The rationale behind this idea is that if the poly(A) tails of the mRNAs are completely saturated with primers only the most proximal primer can be extended to reverse transcribe the mRNA (reverse transcriptase cannot strand displace efficiently (Kornberg & Baker, 1992)). Extension of any other primer is limited to its distance to the next downstream primer, thus generating very small fragments that can be easily eliminated by an efficient size selection procedure.

Some precautions are necessary to avoid nonspecific priming at GC-rich regions of the mRNAs when using large amounts of the NotI-$(dT)_{18}$ primer for first-strand cDNA synthesis. Most importantly, the reaction mixture should be preincubated at 37°C before the addition of reverse transcriptase. I have observed that if the enzyme is added to the reaction mixture while it is at room temperature, an appreciable number of clones without tail can be obtained. For example, I have obtained clones for the mitochondrial 16S rRNA which resulted from priming events at two sites of the RNA sequence that differ from the recognition sequence of the NotI restriction enzyme by a single nucleotide. Presumably, if a GC-rich cluster is flanked by a few As located upstream on the RNA, the NotI sequence (GCGGCCGC) of the primer can anneal to it while most of the oligo(dT) tail loops out.

Presumably, this problem could be prevented by the utilization of a different (GC-less) primer for first-strand cDNA synthesis, in which the NotI site (GCGGCCGC) would have been replaced by the recognition sequence of the PacI restriction endonuclease (TTAATTAA), for example.

16.3.2 Chimerism

Chimeric clones often result from blunt-end ligation of cDNA molecules during the reaction in which adaptors are ligated to the cDNAs. To prevent formation of these cloning artifacts, adaptor molecules must be present in vast excess over cDNAs in this ligation reaction. Such conditions can be easily satisfied only if the cDNAs are efficiently size selected prior to ligation. Although time consuming, chromatography over a Bio-Gel A-50m column as described by Huynh *et al.* (1985) is a very reliable method for size selection of cDNAs.

Another step where chimeric clones can be generated is during ligation of the cDNAs to the cloning vector. This is less likely to occur, however, because the cDNAs have two different ends and three cDNA molecules must be joined together before they can be ligated to a vector molecule. Nonetheless, in order to minimize the probability of formation of chimeric clones during this ligation reaction, vector should be present in excess over cDNAs. Since dephosphorylation usually reduces cloning efficiencies, I favor the approach of not dephosphorylating the vector and using it in only a slight excess: a two-fold excess over cDNAs seems to be a good compromise. Under these conditions, chimeric clones are unlikely to be formed and the background of nonrecombinant clones still remains low. The percentage of chimeric clones can be easily detected by digestion of cDNA clones with NotI enzyme. A linear product should result from this digestion and release of a fragment would indicate a chimeric clone.

It should be emphasized that a significant percentage of chimeric clones is eliminated at the final size-selection step in which the library (as plasmid DNA), is digested with NotI and the linear recombinant molecules containing cDNA inserts larger than 500 nt are gel purified, recircularized and electroporated into bacteria (step (6) in Section 16.2).

16.4 PROTOCOL FOR CONSTRUCTING DIRECTIONALLY CLONED cDNA LIBRARIES

16.4.1 cDNA synthesis

'One-tube' first- and second-strand cDNA syntheses are essentially as described by D'Alessio *et al.* (1987).

16.4.1.1 First-strand synthesis

(1) Incubate 6.0 µl poly$(A)^+$ RNA in H_2O (total of 5 µg RNA) and 2.0 µl NotI-oligo(dT) (Pharmacia; 5 µg/µl; 5′ AACTGGAAGAATTCGCGGCCG-CAGGAA T_{18} 3′) at 65°C for 5 min and then at 37°C for 10 min.
(2) Add the following reagents (prewarmed to 37°C) to the reaction:

   4 µl 5× reverse transcriptase buffer (250 mM Tris-HC1, pH 8.3 at 22°C; 375 mM KCl; 15 mM $MgCl_2$),
   2 µl 0.1 M DTT, and
   1 µl 10 mM each dNTP.
   Incubate at 37°C for 5 min.

(3) Add 5 µl 200 U μl^{-1} RNase H^- reverse transcriptase (Superscript Reverse Transcriptase, Gibco/BRL). Incubate at 37°C for 1 h.

16.4.1.2 Second-strand synthesis

(1) Place the reaction on ice and add:

   86.4 µl H_20,
   32 µl 5× second-strand buffer (94 mM Tris-HCl, pH 6.9 at 22°C; 453 mM KCl; 23 mM $MgCl_2$; 50 mM $(NH_4)_2SO_4$),
   3.2 µl 7.5 mM βNAD,
   3 µl 10 mM each dNTP,
   6 µl 0.1 M DTT,
   2.5 µl 6 U μl^{-1} *Escherichia coli* DNA ligase (NEB),

4 µl 10 U/µl *E. coli* DNA polymerase I (NEB),
1.3 µl Pharmacia 1.1 U/µl RNase H (Pharmacia), and
1.6 µl α-[^{32}P]dCTP (10 µCi µl^{-1}).

(2) Incubate at 16°C for 1 h 30 min and then at 22°C for 30 min.
(3) Add 1 µl 3 U µl^{-1} T4 DNA polymerase (NEB) and incubate at 22°C for 10 min.
(4) On ice, add 6.5 µl 0.5 M EDTA and incubate at 65°C for 10 min.
(5) Add 340 µl TE (10 mM Tris-HCl, pH 8; 1 mM EDTA); save 1 µl for determination of total counts per minute (c.p.m.). Extract with one volume of phenol/Sevag (24:1 chloroform/isoamyl alcohol) and precipitate with two volumes of ethanol and 0.3 M f.c. sodium acetate pH 7. Resuspend DNA in TE and reprecipitate.

16.4.2 Size selection of double-stranded cDNAs

Resuspend DNA pellet in 8.2 µl TE, add 0.8 µl of 5M NaCl + 1 µl of 10× loading dye (20% Ficoll; 400 – 0.1 M EDTA (pH 8), 0.1% SDS, 0.25% bromophenol-blue, 0.25% xylene cyanol) and load on a Bio-Gel A-50m column equilibrated with 0.4 M NaCl in TE.

The Bio-Gel A-50m column is prepared in a piece of glass tubing 32 cm long with an inner diameter of 0.2 cm, essentially as described by Huynh *et al.* (1985). The column can be precalibrated by running radioactively labeled DNA restriction fragments of known length through it and assaying the column fractions by gel electrophoresis.

16.4.3 Addition of EcoRI adaptors to double-stranded cDNAs

(1) Combine fractions containing size-selected cDNAs, precipitate with two volumes of ethanol and resuspend the pellet in:

3 µl H_20 (<1 µg cDNA),
5 µl 0.4 µg µl^{-1} EcoRI adaptors —
5′OH AATTCGGCACGAG 3′OH and
3′OH GCCGTGCTC 5′p
1 µl 10× T4 DNA ligase buffer (500 mM Tris (pH 7.8), 100 mM $MgCl_2$, 100 mM DTT, 10 mM ATP, 250 µg ml^{-1} bovine serum albumin)
1 µl T4 DNA ligase (NEB, 400 U µl^{-1})
Incubate at 16°C overnight.

(2) Add 0.8 µl 0.5 M EDTA, incubate at 65°C for 20 min, add 90 µl TE, phenol/Sevag extract, and precipitate with two volumes of ethanol and 0.3 M f.c. sodium acetate pH 7.

16.4.4 Phosphorylation of ligated EcoRI adaptors

8.5 µl DNA in H_2O
1 µl 10× T4 DNA ligase buffer (500 mM Tris-HCl (pH 7.8), 100 mM $MgCl_2$, 100 mM DTT, 10 mM ATP, 250 µg ml^{-1} bovine serum albumin)
0.5 µl 10 U µl^{-1} T4 polynucleotide kinase (NEB)

Incubate the above mixture at 37°C for 30 min and then at 65°C for 10 min.

16.4.5 NotI digestion of cDNAs

(1) To the reaction above, add

10 µl 10× NotI buffer (supplied by the manufacturer)
78 µl H_2O
2 µl 10 U µl^{-1} NotI (Boehringer or Stratagene)

and incubate at 37°C for 1–2 h.

(2) Add 4 µl 0.5 M EDTA and incubate at 65°C for 10 min.
(3) Extract with phenol/Sevag and precipitate with two volumes of ethanol and 0.3 M f.c. sodium acetate pH 7.

16.4.6 Removal of excess EcoRI adaptors and size fractionation of cDNAs

Resuspend the DNA pellet in 8.2 µl TE, add 0.8 µl 5 M NaCl and 1 µl 10× loading dye and load on a Bio-Gel A-50m column equilibrated with 0.4 M NaCl in TE, as described before.

16.4.7 Ligation of cDNAs with EcoRI + NotI digested phagemid DNA

(1) Combine fractions containing size-selected cDNAs, take 1 µl for the determination of c.p.m. and calculation of cDNA mass, and precipitate the remaining cDNA with two volumes of ethanol. The vector/cDNA molar ratio in this ligation should be about 2:1.
(2) 1.5 µl vector
0.5 µl 10× T4 DNA ligase buffer (500 mM Tris-HCl (pH 7.8), 100 mM $MgCl_2$, 100 mM DTT, 10 mM ATP, 250 µg ml^{-1} bovine serum albumin)
2.5 µl cDNA (approximately 50 ng)
0.5 µl T4 DNA ligase (NEB)
Incubate at 16°C overnight.
(3) Add 0.5 µl 0.5 M EDTA. Incubate at 65°C for 10 min.
(4) Add 90 µl TE, extract with phenol/Sevag and precipitate with two volumes of ethanol and 0.3 M f.c. sodium acetate.
(5) Resuspend DNA pellet in 5 µl TE, electroporate

into bacteria and propagate under appropriate antibiotic selection. Electrocompetent cells should have an efficiency of 10^{10} colony forming units per microgram of control DNA (supercoiled plasmid DNA). Typically one should expect about 5–10 million clones from 50 ng cDNA.

(6) Plasmid DNA is then prepared from the culture.

16.4.8 NotI digestion of the library and size selection on agarose gel

Plasmid DNA from the library is digested with NotI and electrophoresed on a 1% agarose gel. One should expect to see a smear that represents the library and a band that corresponds to vector without an insert. The gel slice containing the DNA from the uppermost part of the smear, which represents the clones with the largest size inserts, is cast backwards into a 1% low-melting-point agarose gel and electrophoresed until all the DNA has run into the low-melting-point agarose gel. Library DNA is then purified with β-agarase (NEB) and recircularized in a large volume ligation reaction. The DNA concentration in this ligation is determined by the formula: $3.3/\sqrt{kb}$ μg ml^{-1}. For example, if the vector is 3.5 kb long and the average size insert 1.5 kb, the formula becomes: $3.3/\sqrt{5} = 1.48$ μg ml^{-1}. The ligation is performed at 16°C overnight.

After phenol/Sevag extraction and ethanol precipitation, the DNA pellet is resuspended in TE, electroporated into bacteria and propagated under appropriate antibiotic selection to generate a cDNA library with an average size insert of 1.5 kb. The library will contain no inserts shorter than 500 bp and essentially no background of nonrecombinant clones. Most importantly, the frequency of chimeric clones in the library should be very low. An example of such a library is shown in Fig. 16.1 (lane 4).

ACKNOWLEDGMENTS

I wish to thank Scott Zeitlin for most valuable discussions and critical reading of this manuscript. This work was supported by the W.M. Keck Foundation (grant No. 891033), the US Department of Energy (grant No. DE-FG02-91ER61233) and the National Institutes of Health (grant No. 1-R55-HD28422-01).

REFERENCES

Adams, M.D., Kelley, J.M., Gocayne, J.D., Dubnick, M., Polymeropoulos, M.H., Xiao, H., Merril, C.R., Wu, A., Olde, B., Moreno, R.F., Kerlavage, A.R., McCombie, W.R. & Venter, J.C. (1991) *Science* **252**, 1651–1656.

Adams, M.D., Dubnick, M., Kerlavage, A.R., Moreno, R., Kelley, J.M., Utterback, T.R., Nagle, J.W., Fields, C. & Venter, J.C. (1992) *Nature* **355**, 632–634.

Adams, M.D., Soares, M.B., Kerlavage, A.R., Fields, C. & Venter, J.C. (1993) *Nature Genet.* **4**, 373–380.

D'Alessio, J.M., Noon, M.C., Ley III, H.L. & Gerard, G.F. (1987) *Focus* **9**, 1–4.

Gubler, U. & Hoffman, B.J. (1983) *Gene* **25**, 263–269.

Huynh, T.V., Young, R.A. & Davis, R.W. (1985) In *DNA Cloning*, Vol. I, D.M. Glover (ed.). IRL Press, Oxford, pp. 49–78.

Khan, A.S., Wilcox, A.S., Polymeropoulos, M.H., Hopkins, J.A., Stevens, T.J., Robinson, M., Orpana, A.K. & Sikela, J.M. (1992) *Nature Genet.* **2**, 180–185.

Kornberg, A. & Baker, T.A. (1992) In *DNA Replication*, 2nd edn. W.H. Freeman, New York, pp. 217–222.

Okayama, H. & Berg, P. (1982) *Mol. Cell. Biol.* **2**, 161–170.

Okubo, K., Hori, N., Matoba, R., Niiyama, T., Fukushima, A., Kojima, Y. & Matsubara, K. (1992) *Nature Genet.* **2**, 173–179.

Smith, C.L., Lawrance, S.K., Gillespie, G.A., Cantor, C.R., Weissman, S.M. & Collins, F.S. (1987) *Methods Enzymol.* **151**, 461–489.

Soares, M.B., Turken, A., Ishii, D., Mills, L., Episkopou, V., Cotter, S., Zeitlin, S. & Efstratiadis, A. (1986) *J. Mol. Biol.* **192**, 737–752.

CHAPTER SEVENTEEN

Construction of Directional cDNA Libraries

A. SWAROOP

Departments of Ophthalmology and Human Genetics, Cellular and Molecular Biology Program, and Human Genome Center, University of Michigan, Ann Arbor, MI 48105, USA

17.1 INTRODUCTION

High cloning efficiency and ease in storage and screening makes λ phage the vector of choice for generating cDNA libraries. We have selected Charon BS λ vector (Swaroop & Weissman, 1988) for constructing directional cDNA libraries. The libraries or clones in this vector may be easily transferred to Bluescript KSM13 plasmid (Stratagene) by NotI digestion and subsequent religation under low DNA concentration. If poly(A)$^+$ RNA is not the limiting factor and high cloning efficiency is not required, Bluescript or a comparable plasmid vector may be used.

A detailed method for generating Charon BS libraries and their enrichment by subtraction cloning has been described recently (Swaroop, 1993). The protocol is based on the procedure of Gubler & Hoffman (1983), and uses directional cloning strategy described by Dorssers & Postmes (1987). The steps involved in cDNA library construction are illustrated in Fig. 17.1.

17.2 MATERIALS

17.2.1 Phage vector and host bacteria

Charon BS (+) or (−) vectors are available from the author. The vector arms (under the trade name *Lambda BlueMid*) can be purchased from Clontech Laboratories (Palo Alto, CA). *Escherichia coli* K802 rec A$^-$ may be obtained from American Type Culture Collection, Maryland.

17.2.2 Enzymes and reagents

Except for the following, all other enzymes are purchased from New England Biolabs: RNasin (Promega Biotech); MMLV reverse transcriptase and RNase H (G-BRL); and calf-intestine alkaline phosphatase (CIP) (Boehringer Mannheim). The packaging extracts are obtained from Stratagene. Other reagents are procured from Pharmacia, G-BRL or Sigma.

17.2.3 Solutions

(1) First-strand buffer (5×): 0.25 M Tris-Cl (pH 8.3) (at room temperature (RT)), 375 mM KCl, 15 mM $MgCl_2$, 50 mM dithiothreitol (DTT), 0.5 mg ml^{-1} BSA (nuclease free).
(2) Second-strand buffer (5×): 0.125 M Tris-Cl (pH 7.5), 0.5 M KCl, 25 mM $MgCl_2$, 25 mM DTT, 0.5 mg ml^{-1} BSA.
(3) T4 DNA polymerase buffer (10×): 0.67 M Tris-Cl (pH 8.7) (at RT), 67 mM $MgCl_2$, 100 mM β-mercaptoethanol, 166 mM $(NH_4)_2SO_4$, 67 μM (ethylene-dinitrilo)tetraacetic acid (EDTA).

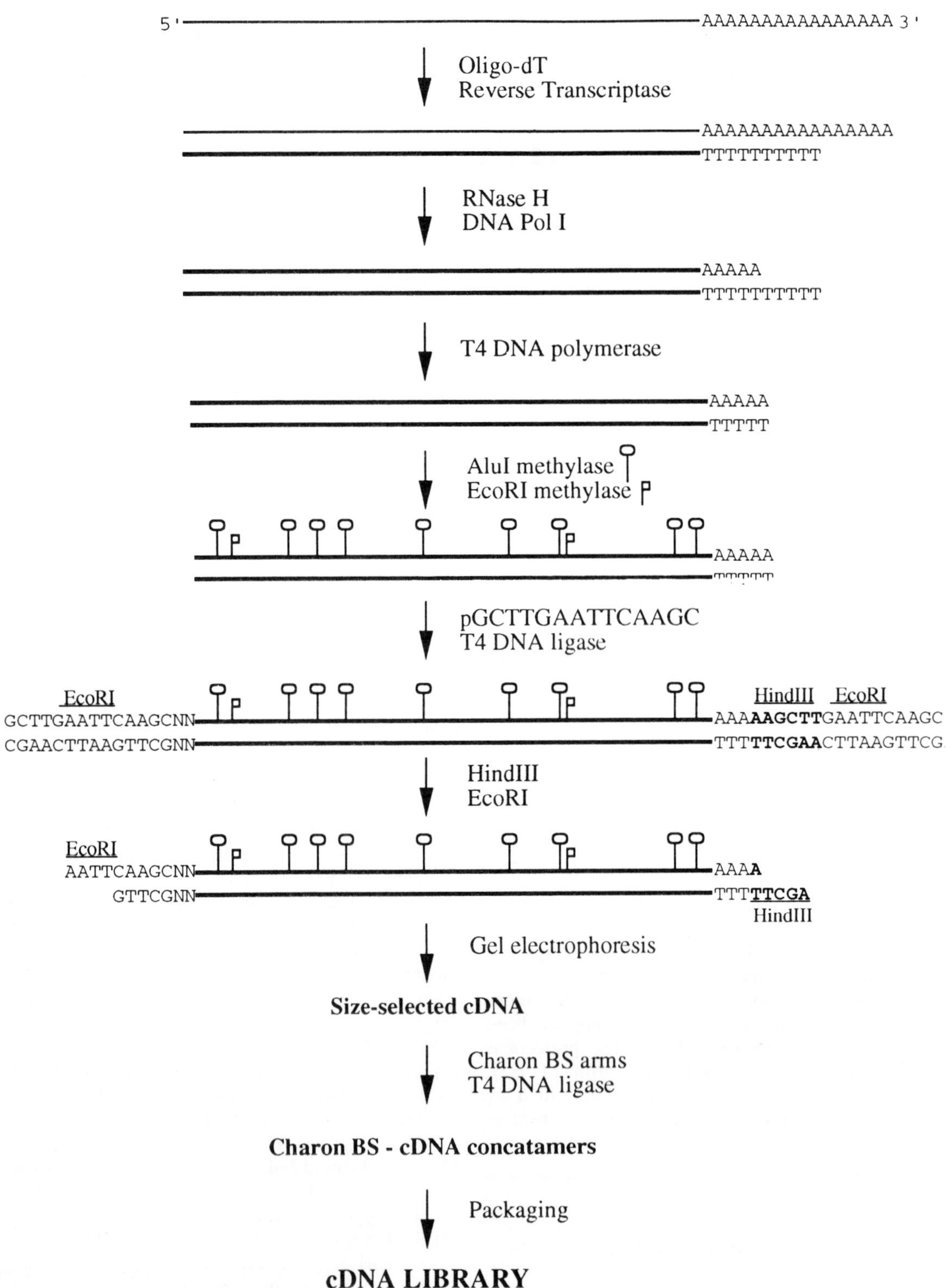

Figure 17.1 cDNA library construction.

(4) AluI methylase buffer (10×): 0.5 M Tris-Cl (pH 7.5), 0.1 M EDTA, 50 mM β-mercaptoethanol.
(5) TE: 10 mM Tris-Cl (pH 8.0), 1 mM EDTA.

All common solutions are prepared according to the cloning manual (Sambrook *et al.*, 1989). Precautions for working with RNA should be observed.

17.3 PREPARATION OF RNA

Total and poly(A)$^+$ RNA from tissues and cells can be prepared by any standard method. However, before starting the construction of a cDNA library, the quality of poly(A)$^+$ RNA should be assessed by Northern analysis (Sambrook *et al.*, 1989). We prefer the protocol of Chomczynski & Sacchi (1988) for preparing total RNA and oligo(dT) cellulose chromatography for poly(A)$^+$ RNA (Sambrook *et al.*, 1989).

17.4 PREPARATION OF VECTOR DNA

Charon BS phage DNA is digested with HindIII and EcoRI restriction enzymes for several hours, and subsequently treated with calf-intestinal alkaline phosphatase (Boehringer Mannheim). The phosphatase is inactivated by incubation at 65°C for 45 min in the presence of 1/10 volume of 0.5 M EGTA (ethylene-bis (oxyethylenenitrilo)tetraacetic acid). DNA is then purified by phenol extractions, followed by ethanol precipitation. The phage vector arms should be tested for the generation of background plaques (Sambrook *et al.*, 1989).

The plasmid vector (e.g. Bluescript) can also be prepared in a similar manner by HindIII and EcoRI digestion.

17.5 PREPARATION OF cDNA

17.5.1 First-strand synthesis

Place 2.5–5.0 μg poly(A)$^+$ RNA and 2.5 μg p(dT)$_{12-18}$ in a 0.5 ml eppendorf tube. Add sterile water to bring the final volume to 20 μl. Place the tube at 90°C for 2–3 min and chill quickly on ice. Add in the following order: 11 μl first-strand buffer (5×), 2.75 μl dNTP mix (10 mM each), 1.5 μl RNasin, 17.25 μl water and 2.5 μl MMLV-reverse transcriptase (200 U μl^{-1}). Remove 5 μl and place into another eppendorf tube containing 1–2 μCi of α-[^{32}P]dCTP (dried *in vacuo*). This is the test reaction. Both the first-strand and test reaction tubes are incubated at 37°C for 2 h. 1 μl 0.5 M EDTA (pH 8.0) and 44 μl of TE (pH 8.0) are then added to the test reaction tube which can be stored at −20°C until analysis at a later time. Place the first-strand synthesis reaction tube on ice.

17.5.2 Second-strand synthesis

The following are added to the 50 μl first-strand reaction mixture: 80 μl second-strand buffer (5×), 7.5 μl dNTP mix (10 mM each), 250 μl water, 10 μl of *Escherichia coli* DNA polymerase I (10 U μl^{-1}), and 1.75 μl *E. coli* RNase H (2 U μl^{-1}). Remove 40 μl of the second-strand synthesis reaction and place in another tube containing 5 μCi of α[^{32}P]dCTP. This serves as the test reaction for second-strand synthesis. Both tubes are incubated at 15°C for 4–5 h. After completion, add 2 μl 0.5 M EDTA and 8 μl TE (pH 8.0) to the test reaction tube, and store at −20°C for further analysis. Add 10 μl 0.5 M EDTA to the second-strand reaction tube, and extract once each with phenol/chloroform ($CHCl_3$) (1:1) and chloroform/isoamylalcohol (IAA) (24:1). Precipitate the cDNA by adding ammonium acetate (2 M, final concentration) and 2.5 volumes of ethanol. Store overnight at −20°C.

17.5.3 Quality and yield of cDNA

The quality of the cDNA synthesized is assessed by alkaline gel electrophoresis using a 10–20 μl aliquot from the labeled test reactions (Sambrook *et al.*, 1989). The product of second-strand reaction may also be analyzed by neutral agarose gel electrophoresis. The size of the cDNAs is estimated by using labeled markers during electrophoresis. The yields of the first- and second-strand cDNA synthesis are determined by the following method: 10 μl from each test reaction is spotted on two sets of glass fiber filters. The first set is dried in air or under an infrared lamp and counted in scintillation solution to obtain *total* counts per minute (c.p.m.). To measure incorporation in cDNA, a second set of filters is washed 3–4 times in a petri dish with chilled 10% trichloroacetic acid (each wash, 2–3 min). After rinsing with 95% ethanol, the filters are dried under an infrared lamp and counted. The yield is calculated as:

$$\text{First-strand yield } (\mu g) = \frac{\text{c.p.m. incorporated}}{\text{Total c.p.m.}} \times 33$$

$$\text{Second-strand yield } (\mu g) = \frac{\text{c.p.m. incorporated}}{\text{Total c.p.m.}} \times 132$$

The multiplication factors are based on the reaction conditions and test aliquots described in this protocol. The yield of first-strand cDNA should be 20–50% of the amount of poly(A)$^+$ RNA. The second-strand cDNA is usually >80% of the first-strand cDNA. The yield of second-strand is multiplied by 2 to determine the total amount of double-stranded cDNA obtained.

17.5.4 Generating blunt-ended cDNA

The double-stranded cDNA is pelleted by centrifugation, washed with 70% ethanol, and dried *in vacuo*. The pellet is dissolved in 42.5 μl water and 5 μl T4 DNA polymerase buffer (10×), 1 μl dNTP mix (10 mM each), 0.5 μl BSA (20 μg $μl^{-1}$), and 1 μl T4 DNA polymerase is added. The reaction is incubated at 37°C for 30 min. TE (pH 8.0) (150 μl) is then added and the reaction mixture is extracted once with phenol/ $CHCl_3$ (1:1) and once with $CHCl_3$/IAA (24:1). The cDNA is precipitated with 2.5 volumes of ethanol in the presence of 2 M ammonium acetate and placed either in dry ice for 30 min or at −70°C for 1–2 h before centrifugation.

17.5.5 Methylation

The internal HindIII and EcoRI sites in cDNA are protected by AluI and EcoRI methylases, respectively. The cDNA pellet is collected by microcentrifugation, washed once with 70% ethanol, dried in a speed-vac and dissolved in 20 μl of water. To this, add 2.5 μl 10× AluI methylase buffer, 1 μl (*S*)-adenosylmethionine (from a 2.5 mM solution, prepared fresh by diluting the stock), and 1.2 μl AluI methylase (5 U $μl^{-1}$). Incubate at 37°C for 60 min. Continue incubation for a further 1–2 h after adding 60 μl H_2O, 8.75 μl 1 M Tris-Cl (pH 8.0), 2 μl 5 M NaCl, 2.5 μl 2.5 mM *S*-adenosyl methionine, and 1.5 μl EcoRI methylase (40 U $μl^{-1}$). At this stage, treatment with Klenow DNA polymerase is recommended but optional. The cDNA is then extracted once each with phenol/$CHCl_3$ (1:1) and $CHCl_3$/IAA (24:1), and precipitated with ethanol in the presence of ammonium acetate.

17.5.6 Linker ligation and HindIII+ EcoRI digestion

The cDNA is pelleted and dissolved in 6 μl water. Add 0.5–2 μg kinased oligonucleotide pGCTTGAATTCAAGC and add sterile water to adjust the final volume to 10 μl. Incubate overnight at 15°C after adding 1.2 μl ligation buffer (10×) and 0.8 μl T4 DNA ligase. Ligation of the oligonucleotide will generate a HindIII site at the 3′-end of cDNA because of the presence of the poly(A) tail (Dorssers & Postmes, 1987). Inactivate the ligase by heating at 65°C for 15 min. cDNA is then digested with the restriction enzymes, HindIII and EcoRI, precipitated with ethanol and dissolved in 20 μl TE (pH 8.0).

17.5.7 Removal of unligated linkers

The unligated linkers are removed from cDNA by electrophoresis using 1% low-melting-point agarose gel. This also allows the selection of cDNAs in an appropriate size range. The gel containing cDNAs (e.g. >0.8 kb) is cut with a clean, sharp blade, and the cDNAs purified by phenol extraction using standard procedures (Sambrook *et al.*, 1989). Only long wavelength ultraviolet light should be used when viewing the agarose gel.

17.5.8 Ligation of cDNA to vector arms

The linker-ligated and size-selected cDNA is dissolved in 20 μl water. Addition of α-[^{32}P]dCTP during the second-strand reaction allows an estimation of the amount of cDNA at this stage. Approximately 1 μg phage arms is ligated in separate tubes to two different concentrations of cDNA. Similarly, plasmid vector can also be ligated to the cDNA.

17.5.9 Packaging

About 4 μl ligation reaction is packaged with Gigapack-Gold packaging extract (Stratagene) to obtain the phage cDNA library. The phage molecules are titrated by using *E. coli* K802recA$^-$ host bacteria to determine the complexity of the library. The remaining ligation reaction can also be packaged to increase the depth of the library. Several million independent plaques are needed for complete representation of all cDNAs in a library.

For generating plasmid libraries, the ligated DNA is used to transform the appropriate host *E. coli* cells. We prefer electrotransformation of bacterial cells (using electroporation apparatus from Bio-Rad or BTX-ECM 600) because of higher cloning efficiency.

17.6 AMPLIFICATION OF THE PHAGE cDNA LIBRARY

If desired, the unamplified library can be screened directly to isolate a particular cDNA clone. For amplification, the library is plated on 20–40 petri dishes (50 000–100 000 plaques per large NZ-agar petri dish) to obtain complete lysis, and the phages are collected in 10–15 ml SM per plate (Sambrook *et al.*, 1989). Aliquots of combined SM from all the plates serve as stocks of the library, which can be stored at 4°C without significant loss of titer.

17.7 TRANSFER OF CHARON BS PHAGE LIBRARY (OR CLONES) TO BLUESCRIPT PLASMID

Digest 10–20 μg phage library (or clone) DNA with NotI restriction enzyme for 2–6 h in 200 μl total volume. Incubate at 65°C for 30–60 min to inactivate NotI. Add 100 μl 10× ligation buffer and make up the volume to 1 ml. Add 2 μl T4 DNA ligate and incubate for several hours to overnight at 12°C. Precipitate the DNA with ethanol in the presence of 0.3 M sodium acetate (pH 5.0). Pellet the DNA, wash with 70% ethanol, dry and dissolve it in 10–50 μl water. Use 1 μl for transforming the competent host *E. coli* cells. Electrotransformation is preferred because of its higher efficiency.

17.8 SUMMARY

This protocol consistently provides high-fidelity directional cDNA libraries. The transfer of cDNAs from Charon BS phage to Bluescript KSM13(−) plasmid vector is efficient and generally yields untruncated clones. The plasmid or phage cDNA clones can be sequenced by using T3 or T7 promoter primers to obtain the 3′- or 5′-end sequence, respectively. The expressed sequence tags from independent cDNAs are valuable for physical and genetic mapping of large genomes (Adams *et al.*, 1991; Gieser & Swaroop, 1992; Khan *et al.*, 1992; Okubo *et al.*, 1992).

This protocol has also been used for the preparation of directional cDNA libraries in λSHK, a phage mammalian expression vector (Swaroop *et al.*, 1992b). A number of size-selected cDNA libraries from human tissues and cell lines have been constructed (A. Swaroop & J. Xu, 1993). Several of these libraries have been characterized extensively and used for the isolation of novel cDNA clones (Agarwal *et al.*, 1991; Ton *et al.*, 1991; Swaroop *et al.*, 1992a; Gieser & Swaroop, 1992).

ACKNOWLEDGMENTS

Research in the author's laboratory is supported by grants from the National Eye Institute (EY 07961), the National Retinitis Pigmentosa Foundation, and the George Gund Foundation. The construction of cDNA libraries at the Michigan Human Genome Center (Genome Technology and Genetic Disease, Director, Dr Francis Collins) is funded by NIH grant No. P30 HG00209.

REFERENCES

Adams, M.D., Kelley, J.M., Gocayne, J.D., Dubnick, M., Polymeropoulos, M.H., Xiao, H., Merril, C.R., Wu, A., Olde, B., Moreno, R.F., Kerlavage, A.R., McCombie, W.R. & Venter, J.C. (1991) *Science* **252**, 1651–1656.

Agarwal, N., Hsieh, C.-L., Sills, D., Swaroop, M., Desai, B., Francke, U. & Swaroop, A. (1991) *Exp. Eye Res.* **52**, 549–561.

Chomczynski, P. & Sacchi, N. (1988) *Anal. Biochem.* **262**, 156–159.

Dorssers, L. & Postmes, A.M.E.A. (1987) *Nucleic Acids Res.* **15**, 3629.

Gieser, L. & Swaroop, A. (1992) *Genomics* **13**, 873–876.

Gubler, U. & Hoffman, B.J. (1983) *Gene* **25**, 263–269.

Khan, A.S., Wilcox, A.S., Polymeropoulos, M.H., Hopkins, J.A., Stevens, T.J., Robinson, M., Orpana, A.K. & Sikela, J.M. (1992) *Nature Genet.* **2**, 180–185.

Okubo, K., Hori, N., Matoba, R., Niiyama, T., Fukushima, A., Kojima, Y. & Matsubara, K. (1992) *Nature Genet.* **2**, 173–179.

Sambrook, J., Fritsch, E.F. & Maniatis, T. (1989) *Molecular Cloning: A Laboratory Manual*, 2nd edn. Cold Spring Harbor Laboratory Press, Cold Spring Harbor, NY.

Swaroop, A. (1993a) In *Photorecepter Cells* P. Hargrave (ed.). Methods in Neurosciences, Academic Press, New York, pp. 285–300.

Swaroop, A. & Weissman, S.M. (1988) *Nucleic Acids Res.* **16**, 8739.

Swaroop, A. & Xu, J. (1993) *Cytogenet. Cell Genet.* **64**, 292–294.

Swaroop, A., Xu, J., Pawar, H., Jackson, A., Skolnick, C. & Agarwal, N. (1992a) *Proc. Natl. Acad. Sci. U.S.A.* **89**, 266–270.

Swaroop, A., Ganguly, S., Sarkar, S.N. & Vasavada, H.A. (1993b) *Gene* **123**, 287–288.

Ton, C.T., Hirvonen, H., Hiroshi, M., Weil, M.M., Monaghan, P., Jordan, T., van Heyningen, V., Hastie, N.D., Meijers- Heijboer, H., Drechsler, M., Royer-Pokora, B., Collins, F., Swaroop, A., Strong, L.C. & Saunders, G.F. (1991) *Cell* **67**, 1059–1074.

CHAPTER EIGHTEEN

Generation of Expressed Sequence Tags of *Brassica napus* by Single-run Partial Sequencing of Random cDNA Clones

J.M. KWAK & H.G. NAM

Department of Life Science, Pohang Institute of Science and Technology, P.O. Box 125, Pohang, Kyungbuk, 790–600, South Korea

18.1 INTRODUCTION

Plants provide a unique biological frontier to study many biological processes not found in other biological systems. Plants possess distinctive development strategies including modular development of organs, unique reproductive systems, indeterminate and determinate growth patterns, photomorphogenesis, and a high degree of developmental plasticity. As sessile organisms, plants developed many intricate environmental perception mechanisms such as photoreception, geotropism and plant–microbe interaction. In addition, plants are one of the major targets for genetic engineering to introduce more desirable breeding characteristics. For molecular interpretation of plant cellular processes and plant genetic engineering, development of plant gene resources is needed. Despite large investment in plant molecular biology in recent years, plant gene resources are still very limited. The result obtained in our laboratory shows that the recently developed expressed sequence tag (EST) approach (Adams *et al.*, 1991) is an efficient way of obtaining a vast amount of plant gene resources (Park *et al.*, unpublished).

In this chapter we describe detailed overall procedures, from isolation of RNA from plant tissues to preparation of templates for single-run partial sequencing. All the following procedures have been successfully used in this laboratory to produce ESTs of *Brassica napus* and should be applicable to most plant species.

18.2 RNA ISOLATION

Efficient isolation of pure and intact RNA from plant tissues has been hampered by the presence of high levels of cell-wall components and endogenous RNase. Use of a strong denaturant such as guanidium isothiocyanate and a strong detergent such as sodium laurylsarcosinate prevents RNA degradation in the following procedure. We routinely use this procedure for isolation of total or poly(A)$^+$ plant RNA and have been always successful in making good-quality cDNA libraries.

18.2.1 Isolation of total RNA from various plant tissues

The following protocol has been adapted from the method described by Cox & Goldberg (1988). The yield of RNA depends on the plant species and the tissue material. For *B. napus*, the approximate yield of RNA per gram of plant tissue by this procedure is 100 μg for root, 70 μg for leaf, 300 μg for stamen, 60 μg for stem, 700 μg for pistil, 450 μg for petal, and 300 μg for whole flower. This yield may not be exactly applicable to other plant species, but will give an idea of the amount of material to use in the beginning.

18.2.1.1 Procedures

(1) Freeze harvested plant material immediately in liquid nitrogen and store at −70°C until use.
(2) Weigh the amount of tissue to be used and then, using a mortar with pestle, grind the tissue to a fine powder in liquid nitrogen.
(3) Transfer the powder to a Potter S tissue-homogenizer (B. Braun) and allow the residual liquid nitrogen to evaporate for a moment. Fill the outer jacket of the homogenizer with ice before transferring the tissue powder.
(4) Add 6 ml of ice-cold guanidium isothiocyanate extraction buffer per gram of tissue and homogenize at 1100 r.p.m. for 5 min.
(5) Follow the procedure described by Cox & Goldberg (1988) from this point.
(6) Measure the yield of RNA by reading optical density of RNA solution at 260 nm (1 OD_{260} = 40.0 μg ml^{-1} RNA).

For an example of total RNA preparation, see Fig. 18.1.

18.2.2 Isolation of poly(A)$^+$ RNA

For isolation of poly(A)$^+$ RNA, we routinely use the commercially prepacked oligo(dT) cellulose column

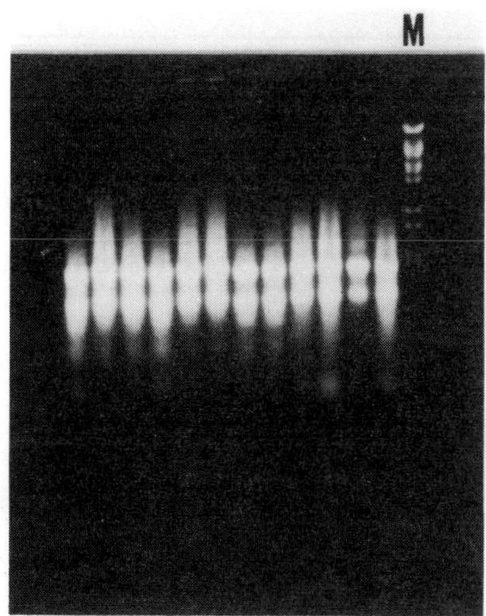

Figure 18.1 Agarose gel electrophoresis of total RNA preparation. M, λ/Bst E II.

from Stratagene and follow the instructions provided by the manufacturer with a minor modification. The usual yield of poly(A)$^+$ RNA by this procedure is 1–2% total RNA.

The materials required for the procedure are a sterile 20 ml luer-lock syringe and a poly(A) Quik mRNA purification kit (Stratagene).

It is important to make uniform column packing. Invert the column several times and place it upright for 10 min to settle the oligo(dT) cellulose. It is also important to push the buffer through the column constantly to obtain good yield. Use 400–500 μg of total RNA per column.

(1) Follow the manufacturer's instructions up to the RNA elution step and then follow the procedure described below.
(2) After elution of RNA from the column, add 0.1 vol. of 3 M sodium acetate (pH 5.2) to each 1.5 ml tube containing mRNA. Add 2 vol. of absolute ethanol. Store at −20°C overnight.
(3) Spin at 12 000 r.p.m. for 15 min in a microcentrifuge at 4°C.
(4) Pour off the supernatant and wash with 70% ethanol.
(5) Spin at 12 000 r.p.m. for 5 min in a microfuge at 4°C.
(6) Pour off the supernatant and vacuum dry for 10 min.
(7) Resuspend each pellet in 6 μl of DEPC-treated water and pool them in a sterile 1.5-ml tube.
(8) Electrophorese 2 μl of mRNA on 1% agarose gel to examine the quality and approximate amount of poly(A)$^+$ RNA. Store the remaining mRNA at −70°C until use.
(9) If the RNA contains too much rRNA, repeat the whole procedure with a new oligo(dT) column.

For an example of poly(A)$^+$ RNA preparation, see Fig. 18.2.

18.3 CONSTRUCTION OF cDNA LIBRARY

18.3.1 Procedure for random hexamer-primed cDNA synthesis

cDNA made by random hexamer primering would provide more expressed sequence tags with more sequences in the protein coding regions since the cDNA molecules are not synthesized from the non-coding poly(A) track sites but from random site in the transcripts (Haymerle *et al.*, 1986). This library will be useful if the main purpose of generating ESTs is to obtain related genes by database search. However, it should be emphasized that the cDNA library made

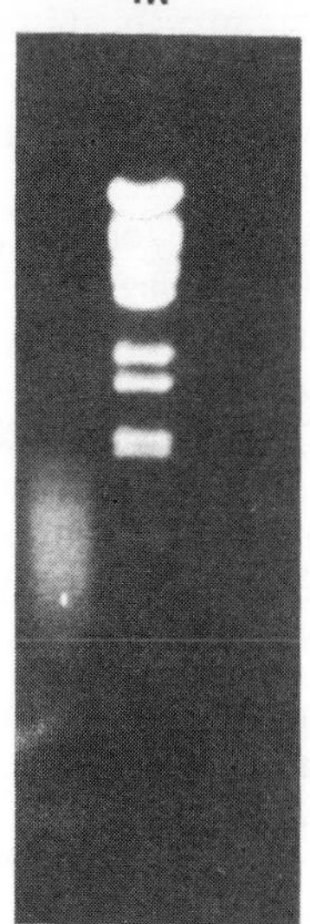

Figure 18.2 Agarose gel electrophoresis of poly(A)$^+$ RNA preparation. M, λ/Bst E II.

by this procedure is likely to contain many noncoding RNA or nonnuclear RNA sequences.

18.3.2 Procedure for insertion of cDNA into M13 bacteriophage vector

If the ESTs are to be generated mostly by manual sequencing, the M13 vector is preferable since the sequencing reaction conducted with the ssDNA template produces longer sequence readings per reaction and thus gives higher probability of database matching. Alternatively, the phagemid vector can be used for ssDNA sequencing. We prefer to use the M13 vector.

(1) Set up test ligation conditions, but use 135 ng μl^{-1} of RF M13 DNA instead of 50 ng μl^{-1} of plasmid DNA to find the best ligation conditions. Use ds M13 DNA that has been cut with appropriate restriction enzyme and has been dephosphorylated.
(2) Electroporate 1 μl of ligate into competent cell carrying F′ episome.

18.3.3 Procedure for insertion of cDNA into bacteriophage λ vector

(1) Find the best ligation conditions. According to the best ligation conditions, mix 4 μl of appropriate λ vector (500 $\mu g\ ml^{-1}$) and an appropriate amount of cDNA. (*Note:* be sure to convert the mole number of the test plasmid into the mole number of λ DNA and then calculate the amount of cDNA to use.)
(2) Bring it up to 19 μl by adding 1× ligation buffer.
(3) Add 0.1 vol. of 3 M sodium acetate and 2.5 vol. of cold absolute ethanol. Mix well and coprecipitate at −70°C for 15 min.
(4) Spin at 12 000 r.p.m., for 15 min in a microcentrifuge at 4°C.
(5) Pour off the supernatant and vacuum dry for 2 min.
(6) Dilute 1 μl of ATP solution with 73 μl of 1× ligation buffer.
(7) Resuspend the DNA pellet in 8 μl of 1× ligation buffer.
(8) Add 1 μl of diluted ATP solution and 1 μl of T4 DNA ligase.
(9) Mix well, spin briefly and incubate at 16°C for 30 min.
(10) Perform *in vitro* packaging on 4 μl of ligate using Gigapack II Gold Packaging Extract according to the manufacturer's instructions.
(11) Titer the library. Amplify the library according to the manufacturer's instructions.

18.4 SUMMARY

Brassica templates prepared using these procedures have been sequenced in our laboratory to produce ESTs as described by Adams *et al.* (1991). We have analyzed these sequences using both FASTDB (Brutlag *et al.*, 1990) and BLAST (Altschul *et al.*, 1990). We have also used the QUEST program of Abarbanel *et al.* (1984) for motif analysis. The *Brassica* EST sequences represent a large number of gene families, and significantly enhance the database of genes from *Brassica* napus (Park *et al.*, unpublished).

REFERENCES

Abarbanel, R.M., Wieneke, P.R., Mansfield, E., Jaffe, D.A. & Brutlag, D.C. (1984) *Nucleic Acids Res.* **12**, 263–280.

Adams, M.D., Kelley, J.M., Gocayne, J.D., Dubnick, M., Polymeropoulos, M.H., Xiao, H., Merril, C.R., Wu, A., Olde, B., Moreno, R.F., Kerlavage, A.R., McCombie W.R. & Venter J.C. (1991) *Science,* **252**, 1651–1656.

Altschul, S.F., Gish, W., Miller, W., Myers, E.W. & Lipman, D.J. (1990) *J. Mol. Biol.* **215**, 403–410.

Brutlag, D.L., Dautricourt, J.-P., Maulik, S. & Relph, J. (1990) *Comput. Appl. Biol. Sci.* **6**, 237–245.

Cox, K.H. & Goldberg, R.B. (1988) In *Plant Molecular Biology, A Practical Approach*, C.H. Shaw (ed.). IRL Press, Oxford, pp. 1–4.

Haymerle, H., Herz, J., Bressan, G.M., Frank, R. & Stanley, K.K. (1986) *Nucleic Acids Res.* **14**, 8615–8624.

CHAPTER NINETEEN

Abundance Screening of Human cDNA Libraries

J.M. SIKELA, T.J. STEVENS, J.A. HOPKINS, A.S. WILCOX, J. GLOD, A.S. KHAN & A.K. ORPANA

Department of Pharmacology, University of Colorado Health Sciences Center, Denver, CO 80262, USA

19.1 INTRODUCTION

Large-scale single-pass sequencing of cDNAs from human and other organisms has recently emerged as a valuable strategy for rapid gene identification, characterization and mapping (Adams *et al.*, 1991, 1992; Waterston *et al.*, 1992; McCombie *et al.*, 1992; Okubo *et al.*, 1992; Khan *et al.*, 1992; Polymeropoulos *et al.*, 1992; Wilcox *et al.*, 1991). This chapter describes one approach that we have found to reduce the selection of abundant clones.

19.2 LIBRARY PRESCREENING WITH TOTAL BRAIN cDNA

One consequence of using random selection of cDNA clones for sequencing is that highly represented cDNAs are repeatedly sequenced. If desired, clone redundancy can be minimized by prescreening the cDNA library with total cDNA, corresponding to the library tissue source, followed by selection of non-hybridizing clones (Wilcox *et al.*, 1991; Khan *et al.*, 1992). Under standard conditions, probe/target hybridization kinetics strongly favor interaction between abundant class cDNA species and their corresponding targets. Under such conditions, highly represented cDNAs typically will yield positive hybridization signals, while rarer species will produce no signal above background.

Prior to probing, cDNA libraries (in phage vectors) are plated at relatively low density (1000–2000 plaques per 15-cm plate) and plaque lifts are prepared on nylon filters by standard methods (Sambrook *et al.*, 1989). Prescreening is then carried out as follows.

(1) Reverse transcribe first-strand cDNA from total brain poly(A)$^+$ mRNA (e.g. 0.5–1.0 μg) using the cDNA Synthesis System Plus Kit (Amersham, or an equivalent) as described by the supplier. After first-strand synthesis with oligo(dT) priming, sodium hydroxide is added to 0.1 N and the mixture is heated at 68°C for 1 h to hydrolyze the RNA. The cDNA is then precipitated with sodium acetate/ethanol.
(2) Resuspend the pellet (approximately 100–200 ng cDNA) in 100 μl TE(pH 8.0).
(3) Incubate at 95°C for 5 min, and then place on ice.
(4) Label cDNA using the Multiprime Labelling Kit (Amersham) as directed by the supplier: 25 μl (50 ng) cDNA, 10 μl random primers (kit), 20 μl 5× buffer (kit), 31 μl ddH_2O, 10 μl α-[^{32}P]dCTP

(10 μCi μl^{-1}), 4 μl Klenow (4 units) (kit); 100 μl total.
(5) Mix and incubate at room temperature for 3.5 h.
(6) Add 1 μl 0.5 M EDTA, 4 μl tRNA (10 mg ml^{-1}), 73 μl 5 M ammonium acetate, and 363 μl cold 100% ethanol.
(7) Mix and incubate at −20°C for 20 min.
(8) Spin in a microfuge at 13 000 r.p.m. (12 000**g**) for 15 min.
(9) Aspirate the ethanol and rinse the pellet with 500 μl cold 70% ethanol.
(10) Spin at 13 000 r.p.m. for 5 min.
(11) Aspirate the ethanol and resuspend the pellet in 100 μl ddH_2O.
(12) Add 250 μl of sheared human genomic DNA (50 μg ml^{-1}) made by passing the DNA through a 21-gauge needle 20 times.
(13) Heat to 95°C for 5 min.
(14) Incubate at 37°C for 20 min.
(15) Add probe to blots that have been prehybridized at 68°C for at least 1 h in the following solution (Sambrook *et al.*, 1989): 10× Denhardt's, 4× SET, 0.001 μg ml^{-1} poly(A), and 50 μg ml^{-1} salmon sperm DNA (heated to 95°C for 5 min). (*Note:* typically 4–5 ml of solution is used per filter, with up to 20 filters per hybridization bag and probe at 10^5–10^6 c.p.m. ml^{-1}.)
(16) Shake overnight at 68°C.
(17) Wash at 68°C for 20 min in: 4× SET, 10× Denhardt's, and 0.1% SDS.
(18) Wash twice at 68°C for 20 min each in: 0.2× SET and 0.1% SDS.
(19) Wash at room temperature in 4× SET.
(20) Expose blots overnight at −70°C with an intensifying screen.
(21) Align autoradiograms with plates and select nonhybridizing plaques as described below.

19.3 PCR AMPLIFICATION FROM PLAQUES FOR AUTOMATED SEQUENCING

The ability to pick plaques corresponding to individual cDNA clones, to PCR amplify the inserts of each cDNA directly from plaque material and to use the resulting PCR products for automated sequencing provides a rapid and inexpensive approach to single-pass sequencing of cDNAs. This approach works very effectively for sequencing into the 5′-end of cDNA inserts, for random-primed cDNAs and for sequencing into the 3′-end of cDNAs if the poly(A) tail is short. The only major limitation we have encountered is that sequencing through the poly(A) tail of the cDNA using a Taq cycle sequencing protocol is much more difficult if the template is generated by PCR. We suspect that the slippage associated with PCR amplification of runs of mononucleotides (e.g. poly(A)) and dinucleotides is exacerbated when the templates are generated by PCR and then cycle sequencing is used. We have found that this problem can often be overcome if the templates are generated as conventional plasmid preparations instead of by PCR amplification. If PCR-generated templates are used, it is also possible to reduce this problem by using Sequenase (instead of Taq cycle sequencing). In our experience the most effective approaches to sequencing through poly(A) tails are to either use oligo(dT)-anchored primers (Khan *et al.*, 1991) or to use conventional plasmid preparations and sequence with Sequenase. However, the short poly(A) tails found in some libraries (e.g. those that have been made using saturating levels of oligo(dT) for cDNA synthesis (Khan *et al.*, 1992)) can be routinely sequenced by using plasmid preparations and cycle sequencing.

(1) Plate cDNA library at low density and choose well-isolated plaques. (*Note:* freshly plated plaques work best; plates that have been stored at 4°C for a few weeks can also be used effectively.)
(2) Using a Pasteur pipette tip, scoop up the plaque on the agarose layer, avoiding the agar layer beneath it.
(3) Place the plaque in 50 μl water.
(4) Incubate at 37°C for 30 min.
(5) Remove 25 μl of the supernate and add it to 25 μl 2× SM buffer (Sambrook *et al.*, 1989) containing 14% DMSO. This can be carried out using wells of microtiter plates or by using individual tubes. Store this phage stock at −70°C.
(6) To the remaining 25 μl containing the plaque, add 25 μl 2× Triton buffer (1 ml TritonX-100, 2 ml 2 M Tris pH 8.5, 0.4 ml 0.5 M EDTA pH 8.0, bring volume to 50 ml with water).
(7) Incubate at 90°C for 10 min.
(8) Use 10 μl of this Triton stock in a 100 μl PCR reaction. Store the remaining stock at −20°C for future use. (*Note:* optimal conditions for PCR amplification may vary between inserts of differing length. An important consideration for subsequent sequencing is the amount of primer used. Sequencing works well with 100 ng or less of each primer (30-mers).)
(9) The following conditions work well for amplification of DNA 100–2000 bp. Using the GeneAmp kit from Perkin–Elmer/Cetus: standard PCR buffer without Mg^{2+} (kit), 3 μl 100 mM Mg^{2+} (3 mM final concentration), 100 ng each vector primer, 200 μM each dNTP final concentration, 0.5 μl Taq (5 U μl^{-1}) (kit), 10 μl of template (from Triton stock), bring volume to 100 μl with ddH_2O. Cycling conditions (GeneAmp System

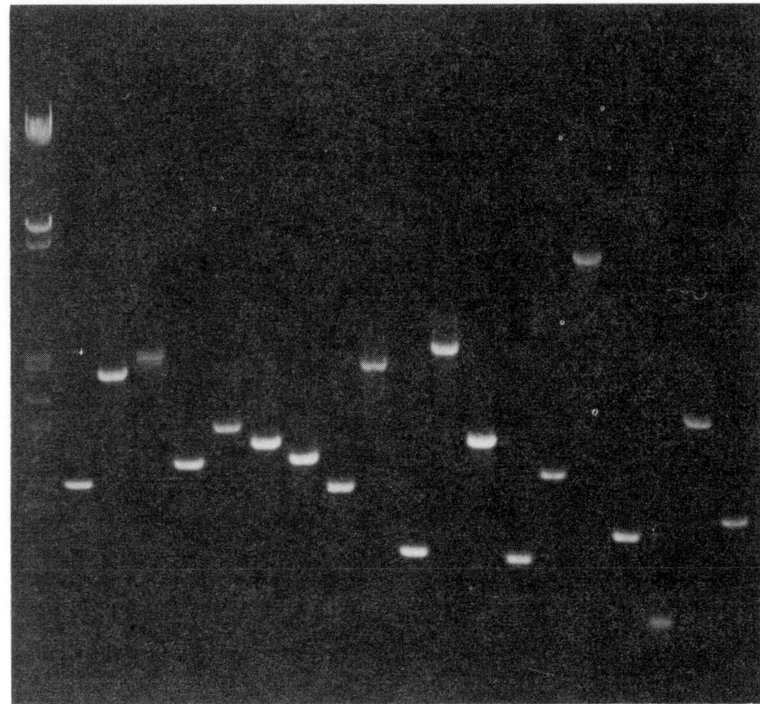

Figure 19.1 Ethidium/agarose gel analysis of PCR-generated cDNA inserts from a λZAPII human brain cDNA library. Individual plaques were selected and treated as described in the text. Inserts were amplified using M13 universal and reverse primers which flank the vector cloning site. The lane on the left is a λ/HindIII/EcoRI size marker.

9600 (Perkin–Elmer)): 95°C for 1 min; 15 cycles (95°C for 4 s, 55°C for 10 s, 70°C for 1 min); 15 cycles (95°C for 4 s, 70°C for 1 min); hold at 4°C. (*Note:* PCR amplifications can be quickly tested by running 10% of the reaction on a 1–2% agarose minigel (lower % for larger pieces) at 100 V for 1 h (see Fig. 19.1). We have found that a good rule of thumb is that if any product of the appropriate size can be detected on the ethidium/agarose gel there should be enough DNA to sequence using the Applied Biosystems, Inc. (ABI) 373A instrument protocol.)

(10) Purify PCR reactions using a Centricon 100 spin column (Amicon) as directed by the supplier. After the PCR reaction is run through the column in 1 ml of ddH_2O, the column is washed with 2 ml of ddH_2O. The retentate is then resuspended in 20 µl of ddH_2O. (*Note:* the completed PCR reactions can be stored at 4°C for several days with no substantial loss of sequenceability. In contrast, the template DNA should be Centricon-purified immediately before sequencing reactions are run. Regarding purification of PCR products for sequencing, we have found that gel elution works effectively as does ammonium acetate/ethanol precipitation. Our best results, however, have been obtained using Centricon columns.)

(11) For automated cycle sequencing 1 µl of resuspended product (from step (10)) is used for each A and C reaction and 2 µl for each G and T reaction. (*Note:* additional quantification of the DNA prior to sequencing is usually unnecessary.)

REFERENCES

Adams, M., Kelley, J., Gocayne, J., Dubnick, M., Polymeropoulos, M., Xiao, H., Merril, C., Wu, A., Olde, B., Moreno, R., Kerlavage, A., McCombie, W. & Venter, J.C. (1991) *Science* **252**, 1651–1656.

Adams, M., Dubnick, M., Kerlavage, A., Moreno, R., Kelley, J., Utterback, T., Nagle, J., Fields, C. & Venter, J.C. (1992) *Nature* **355**, 632–634.

Khan, A.S., Wilcox, A.S, Hopkins, J.A. & Sikela, J.M. (1991) *Nucleic Acids Res.* **19**, 1715.

Khan, A.S., Wilcox, A.S., Polymeropoulos, M., Hopkins, J.A., Stevens, T.J., Robinson, M., Orpana, A.K. & Sikela, J.M. (1992) *Nature Genet.* **2**, 180–186.

McCombie, W., Adams, M., Kelley, J., FitzGerald, M., Utterback, T., Khan, A., Dubnick, M., Kerlavage, A., Venter, J.C. & Fields, C. (1992) *Nature Genet.* **1**, 124–131.

Okubo, K., Hori, N., Matoba, R., Niiyama, T., Fukushima, A., Kojima, Y. & Matsubara, K. (1992) *Nature Genet.* **2**, 173–179.

Polymeropoulos, M., Xiao, H., Glodek, A., Gorski, M., Adams, M., Moreno, R., FitzGerald, M., Venter, J.C. & Merril, C. (1992) *Genomics* **12**, 492–496.

Sambrook, J., Fritsch, E. & Maniatis, T. (1989) *Molecular Cloning: A Laboratory Manual*, 2nd edn. Cold Spring Harbor Laboratory Press, Cold Spring Harbor, NY.

Waterston, R., Martin, C., Craxton, M., Huynh, C., Coulson, A., Hillier, L., Durbin, R., Green, P., Shownkeen, R., Halloran, N., Metzstein, M., Hawkins, T., Wilson, R., Berks, M., Du, Z., Thomas, K., Theirry-Mieg, J. & Sulston, J. (1992) *Nature Genet.* **1**, 114–123.

Wilcox, A.S., Khan, A.S., Hopkins, J.A. & Sikela, J.M. (1991) *Nucleic Acids Res.* **19**, 1837–1843.

CHAPTER TWENTY

High Density Grid Technologies

G. LENNON

Human Genome Center, L-452, Lawrence Livermore, National Laboratory, Livermore, CA 94550 USA

20.1 INTRODUCTION

Primarily as a result of the Human Genome Project, the number of individual clones to be analyzed on a per-laboratory basis has increased dramatically. Laboratories worldwide have turned to high-throughput processes where possible, and for techniques involving molecular hybridization, more and more laboratories are choosing to use arrayed clone libraries. In these libraries, each clone has its own 'address', which usually consists of the number of the plastic dish in which it resides along with the row and column information needed to specify the exact well within that dish. These arrayed libraries may consist of either genomic or complementary DNA (cDNA) clones.

Arrayed libraries may be readily shared, as copies of the library dishes may be made and sent to collaborating laboratories. These libraries have two main uses: first, since the clones are in a format amenable to multiple processing, many clones may be simultaneously prepared for amplification, sequencing, or other processes. Second, the clones (or products from them still in an arrayed format) may be placed in defined positions on suitably processed membranes for use in hybridization-based experiments. These experiments typically involve identifying the subset of clones that hybridize to a given probe, thereby identifying attributes of clones relevant to gene localization, isolation, sequence, or expression. This chapter focuses on the technology by which hybridization membranes can be made from arrayed libraries. For a review of the scientific use of arrayed cDNA libraries, see Lennon & Lehrach (1991). The focus will also be on the maximization of clone density on these filters, since for most of these types of experiment the odds of success increase with the number of clones screened, and because such maximization reduces handling time and cost.

High density gridding technology consists of three main components (Fig. 20.1): (1) the format of the arrayed library; (2) the means by which samples are transferred to membranes, either manually or robotically; and (3) the format of the membranes and their processing. These components are discussed in turn.

20.2 ARRAYED LIBRARIES

Clones spread at densities of about 3/cm^2 are individually picked manually or robotically into wells of plastic dishes. Although the tradition of picking clones with sterile toothpicks continues to this day in certain

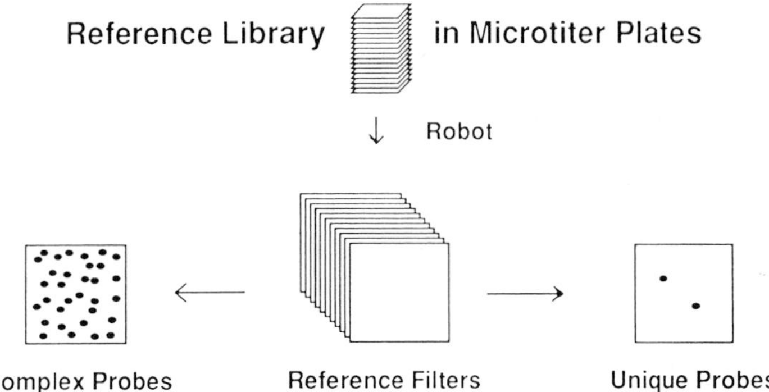

Figure 20.1 Basic scheme for production of high density filters.

laboratories, the use of a 'picking wheel' consisting of 12 spokes attached to a handle allows more efficient picking. Colony picking robots have also been made (J.M. Jaklevic & D.C. Uber, personal communication; Jones *et al.*, 1992), but none is currently in mass production.

The most popular plastic dish has been the 96-well polystyrene dish widely available since its debut in the 1950s as a dish for immunological assays. More recently, dishes with the same outer dimensions (8 cm×12 cm) but containing either 384 or 864 wells have been produced. While the 384-well plates retain compatibility with 96-well-oriented accessories, since their well center-to-center distance is precisely half that of their 96-well precursor, the 864 plates do not. Working volumes for these plates vary between 110 μl (96-well plates; numerous manufacturers), 75 μl (384-square-well plates; available from the author), 40 μl (384-round-well plates; USA Plastics, Ocala, FL), and 20 μl (864-well plates; Helix, San Diego, CA). Plastic plates with outside dimensions of 22 cm×18 cm, containing 1536 wells, are also available (D. Caput, Sanofi, Labège, France).

Once the picked clones have grown, multiple copies are made by replicating one dish at a time into one or more duplicate dishes. Common accessories for this include 96-pin tools (available from Sigma) and, more recently, 384- and 864-pin tools (Helix). This process has also been automated in a number of laboratories. Plates are generally stored at −80°C for long-term storage, requiring large amounts of freezer space. The trend to higher density dishes is certainly in part a response to the need for freezer-space saving. To aid in the conversion of arrayed libraries currently in 96-well dishes to 384-well dishes, at least one device (guide apparatus for replicating tools and hardware (GARTH)) has recently been used (B. Wong & P. de Jong, personal communication).

20.3 MANUAL GRIDDING SYSTEMS

Once libraries have been arrayed, either the clones themselves can be interleaved at high density onto membranes, or, for relatively short clones, their inserts (as polymerase chain reaction (PCR) amplified using common vector primer sequences). Various methods of transferring small amounts of liquids simultaneously from many samples have been proposed; in practice, only tools containing pin arrays are used. Although one can use pin tools manually without some sort of guide mechanism to transfer samples to membranes, the process quickly becomes untenable as spotting density increases. Two manual, but guided, systems are available.

One manual guide system mentioned previously, GARTH, can be used to array samples at four-fold density relative to the library format. If the library is in 96-well dishes, GARTH can spot filters at densities of 384 samples per 8 cm×12 cm membrane; if the library is in 384-well dishes, GARTH can spot 1536 samples per membrane.

A similar system, the Multiprep 96, is available commercially (Techne, Cambridge, UK). This transfer device spots at five-fold density relative to the library dish. The fixed pins are stainless steel, and transfer about 500 nl of liquid. Although marketed for use with 96-well dishes, and thus nominally spotting 480 samples per membrane, if the pin tool and dish are replaced by their 384-well counterparts this system should be able to spot 1920 samples per membrane.

Both these systems require multiple-pin tools for optimal efficiency, as each tool must be sterilized in between spotting from different dishes. Efficiencies can be improved by spotting numerous membrane copies in a cycle without sterilizing the pin tool in between. Spotting efficiencies with these systems are around 5000 samples per hour (if 96-well based) to

20 000 samples per hour (if 384-well based), although actual rates will vary depending on numerous factors (sterilization time per cycle, ease of dish swapping, ease of membrane support changing, etc.) These systems require continuous operator attention and concentration and are thus most suitable for use in laboratories where the total number of membranes needed is not high. Their overall utility, however, is great since they rarely require mechanical maintenance, are inexpensive, and take up little space.

20.4 AUTOMATED GRIDDING SYSTEMS

The desire to maximize filter production rates and clone densities while minimizing operator attention has spurred the development of both commercially available and custom robots suited to this task.

With over 100 systems in use worldwide, the Biomek 1000 and high density replicator tool (HDRT) system (Beckman, Fullerton, CA) is the most commonly used commercial robot for high density grid spotting. The Biomek 100 is a Cartesian robot, and typically has a platform that contains space for one microtiter dish source plate, two filter plates, and space for a rinsing and sterilization system. Platforms have been designed by a number of groups (Bentley *et al.*, 1992; Drmanac *et al.*, 1992) that increase the work area to 6–8 plate areas, but in general these modifications are incompatible (for both software and hardware reasons) with the use of a side-loader for automatically loading and unloading plates. The HDRT 96-pin replicator uses a floating-pin design that compensates for the uneven surface of the nylon membrane (see also Jaklevic *et al.*, 1991), and is available with two styles of pin. The first has 0.060 in. pin tips, and transfers about 300 nl, while the second has 0.015 in. tips and transfers about 10 nl. Biomek-compatible pin tools are also available with 384 or 864 pins per tool (Helix; these tools can also be used manually.) Additional software beyond the standard HDRSYS software includes the BioTest command language, which is available from Beckman at no cost should the user be interested in creating custom routines. Most laboratories use the 96-well format for dishes and pin tools, and spot at 4×4 density (Fig. 20.2, above diagonal line); however, more recently, a number of laboratories have begun routinely spotting at 6×6 density (Fig. 20.2, below diagonal line); for a recent detailed discussion of the use of this system at LLNL, see Olsen *et al.* (1993). The HDRSYS software does not have unlimited freedom to specify spacing and density parameters, and at least three groups have worked on programs to expand the Biomek's capabilities (S. Smith, Stanford Yeast Genome Group; S. Garner, Helix; and S. Reifel, Palo Alto). Perhaps the highest spotting densities have been achieved (Drmanac *et al.*, 1992) using software developed at Argonne National Laboratories in conjunction with custom pin tools.

With current sterilization routines, and attended by an operator to change plates as needed, the Biomek/HDRT system achieves a spotting rate of approximately 7000 clones per hour, being either four 8 cm×12 cm filters of 1536 clones (16 96-well dishes), or two filters of 3456 clones (at 6×6 density.) If the Biomek SL side-loader is used, the spotting rate is approximately halved, but the operator is not needed during the run.

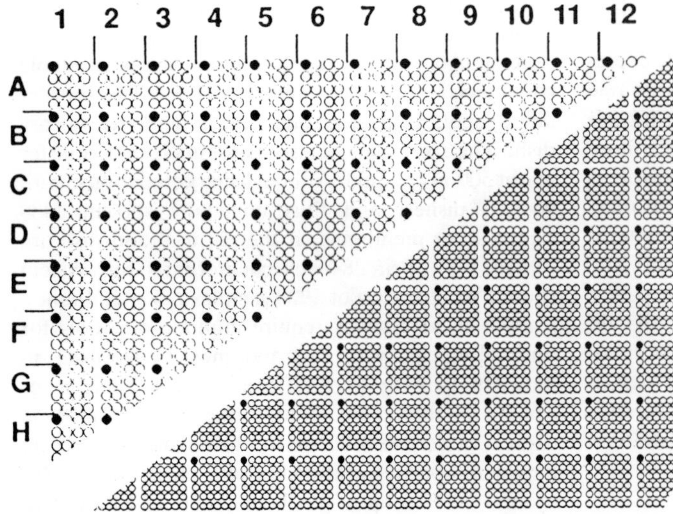

Figure 20.2 Common spotting densities. Above diagonal line, 4×4; below diagonal line, 6×6.

Another significant increase is achieved by using 384- or 864-well dishes; with 384-well dishes, even using the standard platform and side-loader leads to a spotting rate of over 12 000 clones per hour. With the side-loader, the Biomek can be run overnight, producing about 32 4×4 density filters.

Recently, Hewlett Packard (Palo Alto, CA) has released the Optimized Robot for Chemical Analysis (ORCA). This robot uses an articulated arm with six degrees of freedom, sliding along either a 1 or 2 m rail. The horizontal reach of the arm is 0.6 m, and an option ('flip') allows the arm to access both sides of the rail. Through Hewlett Packard's Methods Development Software (MDS), a Microsoft Windows based package, the user can program both simple and complex motions. A joystick teach pendant may also be used to conveniently specify positions and motions. This software does not limit the configurations available for spotting, and the system's open architecture is well suited to interfacing with other instruments. Accessories are becoming available through a number of vendors, and include microplate 'fingers' for moving microtiter dishes, and dish hotels and stackers (Scitec, Newark, DE). The same pin tools can be used with the ORCA as with the Biomek, but they must be modified slightly such that they can be held firmly by the grippers of the ORCA arm.

Throughput with the ORCA is difficult to evaluate at this time given the few systems currently in use. In our laboratory, where some optimization procedures are still underway, it appears that the spotting rate will be between 30 000 and 100 000 clones per hour, depending on whether one uses 96- or 384-well formats. The larger workspace can accommodate enough dishes such that unattended operation is feasible. Thus, a side-loader is not being used, but one can readily envisage how it could be integrated.

Custom robots have also been manufactured by groups aiming to increase throughput or decrease cost (or both). Perhaps the very first robot specifically designed to produce high density filters was developed by Hans Lehrach (Imperial Cancer Research Fund, London), and newer robots designed by this group continue to lead the field. The current robot is enclosed within a cabinet measuring 3 m×1.5 m, and is a Cartesian robot driven by custom software run on a Macintosh IICX. With one operator to change the 384-well source plate every 3 min or so, this robot takes 2 h to spot in duplicate onto 12 large-format (22 cm×22 cm) filters, for a throughput of over 200 000 clones per hour. Contrary to the commercial designs, the pin tools for this system use fixed pin positions. Although the filters are distributed worldwide (Nizetic *et al.*, 1991) and collected data are kept in the associated ICRF Reference Library Database, the hardware and software for this system are not distributed.

Another notable effort is the workcell system developed at the Human Genome Center at Los Alamos National Laboratory (Medvick *et al.*, 1991, 1992). This system also consists of a basic Gantry design, but has dish handling capability such that it can run unattended. About the size and height of a standard office desk, this system makes two large format (22 cm×22 cm) filters each containing 6144 clones. This is achieved in about 2 h, for a throughput of 6000 clones spotted per hour, but with minor modifications this rate can be doubled. Although this is significantly slower than the ICRF robot, it does not require constant attendance and can thus run overnight unattended; furthermore, the rate will increase if either 384- or 864-well formats are adopted. A DOS-based program runs this system, and the software is available. One other feature of this system is the ability to read barcodes affixed to the microtiter dishes.

Other custom or modified robots in use include systems at Généthon (C. Schmit, Evry, France), Lawrence Berkeley Labs (D. Uber & J. Jaklevic, Berkeley, CA), and the Salk Institute (K. Snider & G. Evans, La Jolla).

20.5 CONCLUSIONS

High density grid technology can no longer be considered in its infancy, as numerous options are available for manual, automated, commercial, or custom systems. Worldwide filter production, whether of clones or PCR products, easily exceeds 20 000 filters, and 100 million samples annually. High density filters are playing an essential role in genome biology, and are likely to continue evolving concomitantly with hybridization technology itself.

ACKNOWLEDGMENTS

This work was performed under the auspices of the US DOE by the Lawrence Livermore National Laboratory under contract No. W-7405-Eng-48. Ongoing discussions related to genome matters including robotics may be monitored on the BIOSCI electronic bulletin board 'chromosomes@net.bio.net'. An Email distribution list for those interested in genome automation is also maintained by Skip Garner (Helix), who can be reached by Email at garner@vaxd.gat.com.

REFERENCES

Bentley, D.R., Todd, C., Collins, J., Holland, J., Dunham, I., Hassock, S., Bankier, A. & Giannelli, F. (1992) *Genomics* **12**, 534–541.

Drmanac, R., Drmanac S., Labat, I., Crkvenjakov, R., Vicentic, A. & Gemmell, A. (1992) *Electrophoresis* **13**, 566–573.

Jaklevic, J.M., Hansen, A.D.A., Theil, E. & Uber, D.C. (1991) *Lab. Robot. Autom.* **3**, 161–168.

Jones, P., Watson, A., Davies, M. & Stubbings, S. (1992) *Nucleic Acids Res.* **20**, 4599–4606.

Lennon, G.G. & Lehrach, H. (1991) *Trends Genet.* **7**, 314–317.

Medvick, P.A., Hollen, R.M. & Roberts, R.S. (1991) *Lab. Robot. Autom.* **3**, 169–173.

Medvick, P.A., Hollen, R.M., Roberts, R.S., Trimmer, D. & Beugelsdijk, T.J. (1992) *Int. J. Genome Res.* **1**, 17–23.

Nizetic, D.N., Zehetner, G., Monaco, A.P., Gellen, L., Young, B.D. & Lehrach, H. (1991) *Proc. Natl. Acad. Sci. U.S.A.* **88**, 4123–4127.

Olsen, A.S., Combs, J., Garcia, E., Elliott, J., Amemiya, C., de Jong, P. & Threadgill, G. (1993) *BioTechniques* **14**, 116–122.

PART II

Sample Preparation and Sequencing Methods

B. AUTOMATED SAMPLE PREPARATION

CHAPTER TWENTY-ONE

Solid Phase Preparation of Sequencing Templates from PCR Products

D. R. SIBSON
HGMP Resource Centre, Harrow, Middx HA1 3UJ, UK.

21.1 INTRODUCTION

The polymerase chain reaction (PCR) has gained widespread application, allowing sequences of interest, from a complex sample of nucleic acid, to be amplified for study (Erlich, 1989; Innis *et al.*, 1990). Commonly, amplified material is sequenced. Templates for sequencing can advantageously be produced by PCR, even from cloned material, because growing clones and isolating DNA is avoided.

Solid phase methods of producing sequencing templates from PCR products for conventional, including fluorescent, methods of nucleic acid sequencing are gaining widespread use. They produce quality templates at a high rate, particularly when automated and avoid centrifugation, phenol extraction, ethanol precipitation, or column chromatography (Hultman *et al.*, 1989, 1991; Schofield *et al.*, 1989; Green *et al.*, 1990; Rosenthal & Jones, 1990; Kaneoka *et al.*, 1991). This chapter is concerned with the use of a streptavidin coated magnetic bead as the solid phase for sequencing PCR products, but the reader should refer to similar methods which potentially allow any source of nucleic acid to be used (Hawkins, 1992).

21.2 PRINCIPLE OF THE METHOD

Sequencing reactions are best performed using a given single-strand of DNA produced by a biological system, such as a filamentous bacteriophage (Messing, 1983), as template. PCR produces double-strands which, even when purified from other reaction components, are less satisfactory templates.

Strongly binding a single-strand from a PCR reaction to a solid phase allows the remainder of the reaction components to be removed by thorough washing (Fig. 21.1). Streptavidin-coated magnetic beads have been used as the solid phase to successfully pioneer the approach (Hultman *et al.*, 1989). Streptavidin has an extremely high affinity ($K_D = 10^{-15}$ mol^{-1}) and specificity for biotin (Pahler *et al.*, 1987). The strand to be immobilized on streptavidin-coated magnetic beads must therefore contain biotin, which is conveniently achieved by biotinylating its primer.

The beads, on which a strand has been immobilized, can be repeatedly washed. They are collected by a magnet, other reaction components aspirated, and then the beads resuspended in fresh liquid, following removal of the magnet.

Each strand of a PCR can be prepared separately for sequencing because in nondenaturing conditions

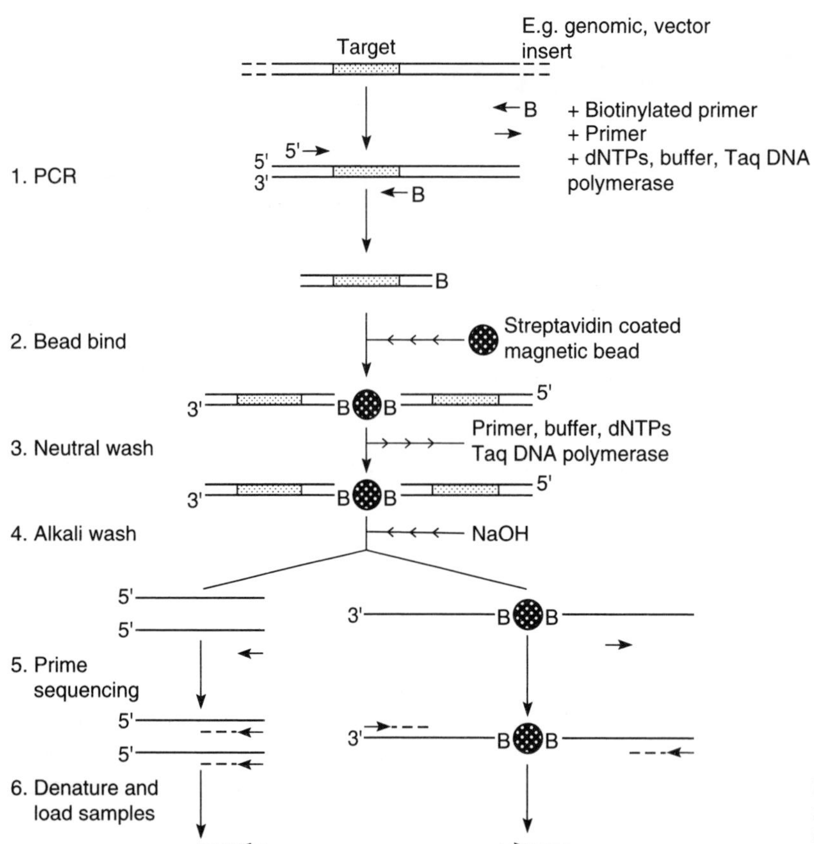

Figure 21.1 Outline of the approach used to isolate the sequencing template from either strand of a PCR product.

the nonbiotinylated strand remains hydrogen bonded to the immobilized strand but can be removed by strong alkaline denaturing conditions without affecting the biotin/streptavidin binding.

21.2.1 Preliminaries

The protocols described here concern the optimization of procedures for the routine use of a solid phase for producing sequencing template from PCR products, whether amplified from an essentially native source, such as genomic DNA, or from recombinant DNA, such as that found in a cloning vector. First, the primers should be selected and prepared and then the PCR and bead binding are optimized.

21.3 METHODS

21.3.1 Selection of the PCR primers

PCR primers must avoid nonspecific amplification products which, if bound to the beads and sequenced with the product of interest, lower sequence quality. Many good sets of PCR primers have already been published, especially those flanking the cloning sites of commonly used vectors. We use the forward and reverse sequencing primers 5′ GTAAAACGACGGC CAGT and 5′ AACAGCTATGACCATG as PCR primers for any of the common vectors which utilize a multiple cloning site with the lacZ(Δ)M15 gene color selection system (Messing, 1983), e.g. M13mp series bacteriophage, pUC series plasmids, pGEM, λZAP and pBluescript. The primer pairs 5′AGCAAGTTCA GCCTGGTTAAG, 5′ TTGACACCAGACCAACT GGTAATG, and 5′ ATTAACCCTCACTAAAG, 5′AATACGACTCACTATAG have worked well, respectively, for inserts in λ gt11 or between the T3 and T7 promoters, although 5′AGCAAGTTCAGCC TGGTTAAG and 5′ CTTATGAGTATTTCTTCCA GGGTA with λ gt10 are more difficult. If inserts have been cloned between two different adaptors (more than about 15 bases), for example when directionally cloning cDNA, these adaptors may be suitable as PCR primers.

Oligonucleotide design software, for example Primer (Whitehead Institute for Biomedical Research) can, by using a standard personal computer, predict pairs of primers which have a high probability (about 70%) of working. Primers that give the best results have to be finally determined empirically. Producing more than one oligonucleotide for each possible direction of priming most efficiently achieves several candidate sets because the number of possible primer combinations equals the product of the number of possible oligonucleotides for each direction. Nested primers, where an initial PCR with one pair of primers is followed by a second PCR with a second set of primers internal to the first set, and which may include one of the first primers, may give best results (Simmonds *et al.*, 1990). The size of the fragment to amplify should also be considered. Short fragments (<300 bases) will both amplify and bind to beads most efficiently but will not make full use of sequencing. The bases of most interest should also be considered, especially when low levels of nonspecific PCR products derived from short dimers or concatamers of primer may be present, since their sequences are likely to overlap the shorter sequences (up to about 50 bp) of the target.

Only one of a pair of primers should be biotinylated. Nonbiotinylated PCR primer may often also be used as the sequencing primer.

21.3.2 Synthesis of the oligonucleotide primers

Oligonucleotides are synthesized (for example using the ABI 380B, on a 1 μM scale synthesis), Trityl on, to aid the efficiency of the purifications which follow. Biotin is incorporated at the final (5′) position of the oligonucleotide during the synthesis with a biotin phosphoramidite having a spacer arm of at least six carbon atoms (for example, DMT-Biotin-C6-Phosphoramidite, Cambridge Research Biochemicals Incorporated), in which case oligonucleotides are made Trityl off. Biotinylation by transfer of biotin from Biotin-X-NHS-Ester to 5′-amino modified oligonucleotides, where the amino link is introduced by the use of Aminolink 2 during the automated synthesis, may be less efficient. Traces of unincorporated biotin will compete efficiently for binding to the beads; therefore, exercise caution.

21.3.3 Purification of the oligonucleotides

Full-length oligonucleotides are purified by reverse phase high performance liquid chromatography (HPLC) to achieve greater efficiency and scale. This also removes nonbiotinylated oligonucleotides from biotinylated oligonucleotides which otherwise would lower the overall yield of biotinylated product and compromise sequencing. Nonbiotinylated oligonucleotides can be synthesized Trityl off because they need not be purified by HPLC. However, less specific PCR may result and, in the case of sequencing primers, more ambiguities may occur.

21.3.3.1 Post-synthesis

(1) Deprotect all oligonucleotides at 55°C, in a water bath, for 8–16 h.
(2) Prepare flasks for rotary evaporation by soaking in 10% nitric acid overnight, rinsing in copious amounts of high quality water and drying in a warm air cabinet. Fresh flasks are required before and after HPLC.
(3) Add a drop of 3 M triethyl amine acetate to Trityl on oligonucleotides to protect the Trityl group.
(4) Dry, using a rotary evaporator at 50°C or 35°C for Trityl off or Trityl on deprotected oligonucleotides, respectively.
(5) Redissolve oligonucleotides in 0.5 ml HPLC grade water. Trityl off oligonucleotides, except biotinylated ones, are now ready for use.

21.3.3.2 HPLC purification (Trityl on and biotinylated oligonucleotides)

(1) Set up the HPLC at a flow rate of 4.7 ml min^{-1}, using buffer B (acetonitrile) and buffer A (100 mM triethylamine acetate, pH 7.0) with a reverse phase C18 semi-prep column 5μ(filter), 25 cm × 1 cm (Beckman Ultrasphere), according to Table 21.1.
(2) Filter the oligonucleotide using a 0.22 μM filter.
(3) Perform a test HPLC run by injecting 50 μl (10%) of the filtered oligonucleotide and running the appropriate gradient according to Table 21.1.
(4) The largest peak (eluting between about 9 and 11 min) should comprise the required oligonucleotide.
(5) Repeat HPLC using the remainder of the oligonucleotide, collecting the peak predicted from the test run.
(6) Dry the eluate in the rotary evaporator at 50°C.
(7) Redissolve the dried oligonucleotide in 200 μl of HPLC grade water. Biotinylated oligonucleotides are now ready for use.
(8) Detritylate Trityl on oligonucleotides by adding glacial acetic acid to 80% and incubating for 20 min at ambient temperature.
(9) Add an equal volume of absolute ethanol to the detritylated oligonucleotide and dry by rotary evaporation at 50°C.
(10) Redissolve in 400 μl of HPLC grade water, and precipitate for 30 min at room temperature by adding 40 μl of 3 M sodium acetate pH 5.4 and 1000 μl of absolute ethanol.
(11) Collect the pellet in a microfuge at full speed for 20 min.

Table 21.1 Conditions for reverse phase HPLC gradients used for purifying either Trityl on or biotinylated oligonucleotides.

Trityl on oligonucleotides		*Biotinylated oligonucleotides*	
% B	*Duration (min)*	*% B*	*Duration (min)*
15	Initial	10	Initial
15	3	10	2
15–40	5	10–20	5
40	8	20–21.2	7
40–95	2	21.2–90	2
95	2	90	1
95–15	1	90–10	3
15	End	10	End

(12) Carefully remove the supernatant and dry the pellet at 37°C.
(13) Redissolve the pellet in 200 µl of water. It is now ready for use.

21.3.4 Optimization of the PCR

Optimization of PCR aims to minimize use of the biotinylated primer which competes for bead binding. Prepare large batches of reagents, aliquot into convenient amounts and store at −20°C.

Thaw the reaction components and enzyme separately on ice; discard unused reagents. Use a standard PCR (for example, using the M13 forward and reverse primers) to amplify different serial dilutions of M13mp series single-stranded DNA, down to 0.1 pg per reaction, to check common reaction components.

Take suitable precautions to avoid cross-contaminating PCRs.

(1) Set up 54 trial PCR reactions using the range of pH 8.1–8.9 in increments of 0.1 with each magnesium chloride concentration between 1 and 3.5 mM in increments of 0.5, inclusive. Each reaction should contain:

 4 µl 10× dNTPs, (2mM each)
 4 µl appropriate 10× Tris-HCl buffer,
 4 µl appropriate 10× magnesium,
 20 pmol of each primer (one biotinylated), and
 either
 5–10 ng of test plasmid or phage DNA in 1–2 µl of appropriate 10 mM Tris-HCl, 1 mM EDTA buffer,
 or
 0.5 µg of test genomic DNA.

 Add water to 40 µl.

 Include negative controls, for example, omit the sample. Set up reactions on ice and add 0.2–1 units (actual amounts may have to be determined empirically) of Taq DNA polymerase (AmpliTaq, Perkin–Elmer) immediately before use. Perform heat denaturation at 95°C for 5 min, then 36 cycles of denaturation at 95°C for 30 s; annealing at the T_m for 1 min, extension at 72°C for 2 min; and follow cycling with 72°C for 10 min, where T_m is the estimated melting temperature of the oligonucleotides (Itakura *et al.*,1984). Time and temperature may have to be adjusted, depending on the length of the expected product and the type of thermocycler used.
(2) Analyse 10 µl of each reaction by agarose gel electrophoresis (Sambrook *et al.*, 1989). Note the reaction conditions giving exclusively the required product. Low levels of excess primers but not dimerized or concatermerized primers are acceptable.
(3) Repeat the optimum conditions of PCR determined above, but vary the amount of primer. A total of 25 reactions will test each possible combination of 1, 3, 9, 15 and 20 pmol of each of the primers.
(4) Examine the reaction products by agarose gel electrophoresis as above. Note the conditions which use the least amount of biotinylated primer to produce the specific product.

21.3.5 Optimization of the bead binding

It is also necessary to optimize the amount of beads used. Too many beads will inhibit sequencing reactions, while too few will not yield sufficient sequencing template.

21.3.5.1 Starting material

Perform five 40 µl PCR reactions according to the optimized conditions determined above.

21.3.5.2 Preparation of the beads

(1) Thoroughly resuspend streptavidin-coated magnetic beads in the shipping bottle by gentle agitation. Vortexing will damage the beads.
(2) Remove 100 µl of bead suspension to a 1.5 ml microfuge tube.
(3) Place the tube in the magnetic particle concentrator (MPC-E), Dynal. Within 30 s the beads will have sedimented.
(4) Carefully remove the liquid.
(5) Remove the tube from the MPC-E and gently resuspend the beads in 200 µl of binding/washing buffer 10 mM Tris HCl pH 7.5, 1 mM EDTA, 2M NaCl.
(6) Repeat steps (3) to (5) three more times to thoroughly wash the beads, with the exception that the final resuspension should be in 100 µl of buffer.

Traces of the bead's original storage buffer could inhibit the reactions which follow. Never allow the beads to dry out or their properties will be impaired.

21.3.5.3 Bead binding

If mineral oil has been used, the aqueous layer from each PCR should be transferred to a fresh microfuge tube. Traces of oil will not interfere with subsequent steps.

(1) Add 10 µl of bead suspension and 30 µl of binding/washing buffer to the first PCR reaction in a 1.5 ml microfuge tube. Repeat for each successive reaction, increasing the amount of beads by 10 µl and reducing the amount of buffer by 10 µl each time so that all reactions receive 40 µl of liquid but different amounts of beads. Include a control where 40 µl of binding/washing buffer is used but no beads are added.
(2) Bead bind at 28°C for 30 min. Use a finipipette to occasionally resuspend the beads. (*Note*: long PCR products may bind more efficiently at higher temperatures, for example 37°C.)
(3) Place the tubes in the MCP-E.
(4) Carefully transfer the buffer to a fresh tube.
(5) Remove the tubes from the MCP-E and gently resuspend the beads in 40 µl of fresh binding/washing buffer.
(6) Repeat steps (3) to (5) but resuspend the beads in 40 µl of 0.15 M freshly prepared NaOH. Incubate at 28°C for 10 min.
(7) Repeat steps (3) and (4) to transfer the NaOH to a fresh tube.
(8) Analyze by agarose gel electrophoresis, using TBE buffer (Sambrook *et al.*, 1989), 20 µl of the liquid removed at each stage.

Note the least amount of beads required to give the greatest yield of single-stranded DNA in the NaOH wash for future use. Examine the buffer removed immediately following bead binding to determine the proportion of PCR product which is bead bound – represented by the difference in intensity between a given sample and the one where no beads were added. Check that no DNA is present in the buffer used to wash the beads to confirm that the beads are binding strongly.

21.3.6 Preparation of sequencing template (96-well format)

(1) Perform the required number of PCR reactions in a 40 µl reaction volume in a 96-well format according to the optimum conditions determined in Section 21.3.4. Optimization of the PCR is done as described in Section 21.3.4. Suggested targets are: 0.5 µg of genomic DNA and 5–10 ng of plasmid or phage DNA. Purified DNA should be in TE (10 mM Tris, 1mM EDTA):
    (a) 2 µl of bacterial suspension prepared originally by resuspending a fresh plasmid containing colony in 10 µl of water; and
    (b) 2 µl of phage suspension prepared by resuspending a fresh plaque in 10 µl of TE.
(2) Prepare sufficient beads for all of the PCRs as described in Section 21.3.5.2. and determined in optimization bead binding as described in Section 21.3.5. Resuspend the beads in 40 µl per PCR of binding/washing buffer before use. The aqueous phase from PCR reactions with an oil overlay should be transferred to a fresh microtiter plate. The microtiter plate should have a geometry which is compatible with the MCP-96, (Dynal) for example Hi-Temp 96 (Techne). The MCP-96 is important because it draws the beads to the sides of the wells so that liquid can be withdrawn without disturbing the pellet (Fig. 21.2).
(3) Add 40 µl of bead suspension to each of the PCRs.
(4) Bead bind at 28°C for 30 min. Occasionally resuspend the beads. Beadprep 96 (Techne) resuspends the beads or keeps them in suspension, but works best with small volumes.
(5) Place the microtiter plate onto the MCP-96. The beads will sediment within 30 s.
(6) Carefully remove the buffer.
(7) Remove the microtiter plate from the MCP-96 and gently resuspend the beads in 40 µl of fresh binding/washing buffer.
(8) Repeat steps (5) to (7) but resuspend the beads

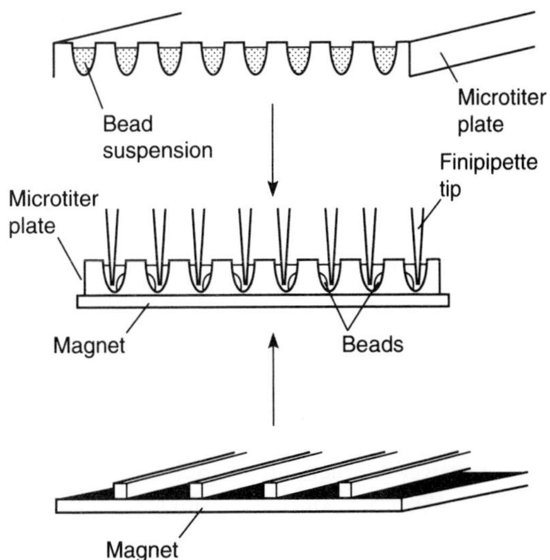

Figure 21.2 A microtiter plate and magnet, with complementary geometries, draw beads to well sides where they are undisturbed by pipetting.

in 40 µl of 0.15 M freshly prepared NaOH. Incubate at 28°C for 10 min.

(9) Repeat step (5).

(10) Carefully transfer the NaOH to the corresponding position in a fresh microtiter plate.

(11) Repeat step (7) but again resuspend the beads in 40 µl of 0.15 M NaOH. Incubate at 28°C for 10 min.

(12) Repeat step (5) and carefully remove the NaOH.

(13) Repeat step (7) but resuspend the beads in 40 µl of water.

(14) Repeat steps (5) to (7) twice more, but again using water to resuspend the beads. The final water wash can be replaced with × 1 reaction buffer for the sequencing reactions which follow.

Sequencing reactions should be performed directly on the beads as though they were pure DNA, according to standard sequencing protocols. The nonbiotinylated primer can be used as a sequencing primer, but a primer which is internal to this may be better, especially when genomic DNA has been amplified. Unless cycle sequencing has been performed, used sequencing reaction products can be removed by magnetic separation, in which case then add formamide directly to the beads before gel loading.

(15) To sequence from the strand removed by alkali, first neutralize by adding to the saved NaOH wash an equal volume of 0.15 M HCl and 1 µl of 1 M Tris (pH adjusted to match the sequencing buffer).

(16) To concentrate or change the buffer of the neutralized, NaOH released material, precipitate by adding 0.1 volumes of 3 M sodium acetate and 2.5 volumes of ethanol.

(17) Collect the precipitate by centrifugation and dry before redissolving in the buffer of choice. 96-Deep-well titer plates with a capacity of 96 × 1 ml can be obtained (Beckman). These can be centrifuged at low speed (3000**g**) which is sufficient over a long period (30 min) to pellet the precipitate.

21.4 AUTOMATION

All bead washing described in this paper can be performed by a conventional liquid handling robot which improves throughput and reliability (Uhlen *et al.*, 1992). We use a Beckman Biomek with a Side Arm Loader. The side arm loader transfers labware to any of four possible positions on an operations tablet. An MPC-96 is at one of the positions where the side arm transfers plates to determine, automatically, sedimentation of the beads. A frame support increases the handling stability of Hi-Temp 96 (Techne) microtiter plates. The magnet can, alternatively, be mounted on an elevating table, coming into position when required. Washing routines should entirely resuspend the beads.

ACKNOWLEDGMENTS

Staff at the Molecular Genetics Unit, Cambridge, are thanked for introducing me to the techniques described in this paper. The importance of the developments to the field, made by Professor M. Uhlen and his colleagues, at the Royal Institute of Technology, Stockholm, are acknowledged.

REFERENCES

Erlich, H.A. (ed.) (1989) *PCR Technology*. Stockton Press, New York.

Green, A., Roopra, A. & Vaudin, M. (1990) *Nucleic Acids Res.* **18**, 6163–6164.

Hawkins, T.L. (1992) *J. DNA Sequence Mapping* **3**, 65–70.

Hultman, T., Stahl, S., Hornes, E. & Uhlen, M. (1989) *Nucleic Acids Res.* **17**, 4937–4946.

Hultman, T., Bergh, S., Moks, T. & Uhlen, M. (1991) *BioTechniques* **10**, 84–93.

Innis, M.A., Gelfand, D.H., Sninsky, J.J. & White, T.J. (eds) (1990) *PCR Protocols*. Academic Press, New York.

Itakura, K., Rossi, J.J. & Wallace, R.B. (1984) *Annu. Rev. Biochem.* **53**, 323–323.

Kaneoka, H., Lee, D.R., Hsu, K.-C., Sharp, G.C. & Hoffman, R.W. (1991) *BioTechniques* **10**, 30–34.

Messing, J. (1983) *Methods Enzymol.* **101**, 20–46.

Pahler, A., Hendrickson, W.A., Gawinowicz Kolks, M.A., Argarana, C.E. & Cantor, C.R. (1987) *J. Biol. Chem.* **262**, 13933–13937.

Rosenthal, A. & Jones, D.S.C. (1990) *Nucleic Acids Res.* **18**, 3095–3096.

Sambrook, J., Fritsch, E.F. & Maniatis, T. (eds) (1989) *Molecular Cloning*. Cold Spring Harbor Laboratory Press, Cold Spring Harbor, NY.

Schofield, J.P., Vaudin, M., Kettle, S. & Jones, D.S.C. (1989) *Nucleic Acids Res.* **17**, 9498.

Simmonds, P., Balfe, P., Peutherer, J.F., Ludlam, C.A., Bishop, J.O. & Brown, A.J.L. (1990) *J. Virol.* **64**, 864–870.

Uhlen, M., Hultman, T., Wahlberg, J., Lundeberg, J., Bergh, S., Pettersson, B., Holmberg, A., Stahl, S. & Moks, T. (1992) *Trends Biotechnol.* **10**, 52–55.

CHAPTER TWENTY-TWO

Automatic Preparation of DNA Templates for Sequencing on the ABI Catalyst Robotic Workstation

A. HOLMBERG,[1] G. FRY[2] & M. UHLÉN[1]

[1] Department of Biochemistry and Biotechnology, Royal Institute of Technology, S-100 44 Stockholm, Sweden
[2] Applied Biosystems, 850 Lincoln Center Drive, Foster City, CA 94404, USA

22.1 INTRODUCTION

The need for automation of the various steps in DNA sequencing increases as the number of applications and genomic sequencing applications grow. It is therefore necessary to design robotic workstations for template preparation and sequencing reactions in order to allow high throughput of samples with reproducible results of high quality. Also, it is not economically feasible to use manual labor in such large-scale projects.

Several protocols based on paramagnetic microbeads for the recovery and purification of DNA templates to be used for sequencing have recently been described. These protocols involve both purification of polymerase chain reaction (PCR) products (Hultman *et al.*, 1989) and M13 templates (Fry *et al.*, 1992) and allow the isolation of specific single-stranded DNA templates. Protocols using both T7 DNA polymerase (Tabor & Richardson, 1989) and Taq cycle sequencing (Carothers *et al.*, 1989) have been investigated. The fact that pure single-stranded templates are obtained allows accurate sequencing of heterozygotes using T7 DNA polymerase (Gibbs *et al.*, 1989; Syvänen *et al.* 1991), mixed viral populations such as human immunodeficiency virus (HIV) (Wahlberg *et al.*, 1992a) and direct mitochondrial sequencing of hair shafts and semen (Hopgood *et al.*, 1992). The advantages of using magnetic beads as solid support include the possibility to perform manipulations such as strand melting and hybridization in a small volume with rapid kinetics (Uhlén, 1989). It is also important to point out that the magnetic beads can be transferred with a simple pipette tool, thus allowing immobilized DNA to be aliquoted and mixed within different wells or tubes within a robotic workstation. Thus, easy liquid handling is combined with rapid phase separation using a magnetic field. Recently, several automated protocols for solid phase sequencing have been described based on the Beckman Biomek-1000 laboratory workstation (Hultman *et al.*, 1991; Wahlberg *et al.*, 1992b).

Here, we describe an experimental model of a robotic workstation (ABI Catalyst) to allow for magnetic preparation of PCR products and/or M13 templates to be used directly for sequencing either with T7 DNA polymerase or Taq DNA polymerase. A novel user interface to allow for modifications of the protocols is also described. Thus, automatic protocols with flexible parameter settings are obtained, allowing an easy setup of integrated protocols for template preparations and sequencing reactions.

22.2 PRINCIPLE OF TEMPLATE PREPARATION

The principle of solid phase recovery of PCR products (Hultman *et al.*, 1991) and M13 (Fry *et al.*, 1992) has been described before. In summary, purification of PCR products relies on the stability of the biotin–streptavidin interaction, which allows for direct immobilization of double-stranded DNA fragments via a biotin introduced at the 5′ end of one of the two PCR primers. For inserts cloned in the standard plasmid vectors (i.e. pUC, pEMBL, Bluescript and pTZ), general PCR primers which are complementary to the sequences flanking the cloning site can be used. PCR may be carried out directly on a single bacterial colony of the strain carrying the phage or plasmid, thus eliminating the need to prepare purified DNA. The bacterial cells are simply lysed by the first thermocycling step. Colonies are 'toothpicked' into separate wells of a microtiter plate or a PCR tube, which is then placed in the PCR thermocycler.

For direct genomic sequencing (Fig. 22.1), specific PCR primers for the target region must be synthesized. To allow the use of universal sequencing primers, it is possible to introduce the sequence complementary to the primer annealing site by using a specific primer containing a 'handle' with this sequence (Wahlberg *et al.*, 1990). The annealing site of the universal sequencing primer is thus introduced into the amplified fragment and can subsequently be used during the sequencing reactions. Obviously, a specific sequencing primer can instead be used if a protocol involving labeled terminators is used (Lee *et al.*, 1992).

For the M13 protocol (Fig. 22.1), the general template purification scheme involves phage lysis followed by liquid hybridization with a synthetic oligonucleotide, complementary to the *lacZ* region on the M13, linked to paramagnetic particles (LacZBeads) (Fry *et al.*, 1992). The biotinylated oligonucleotide is

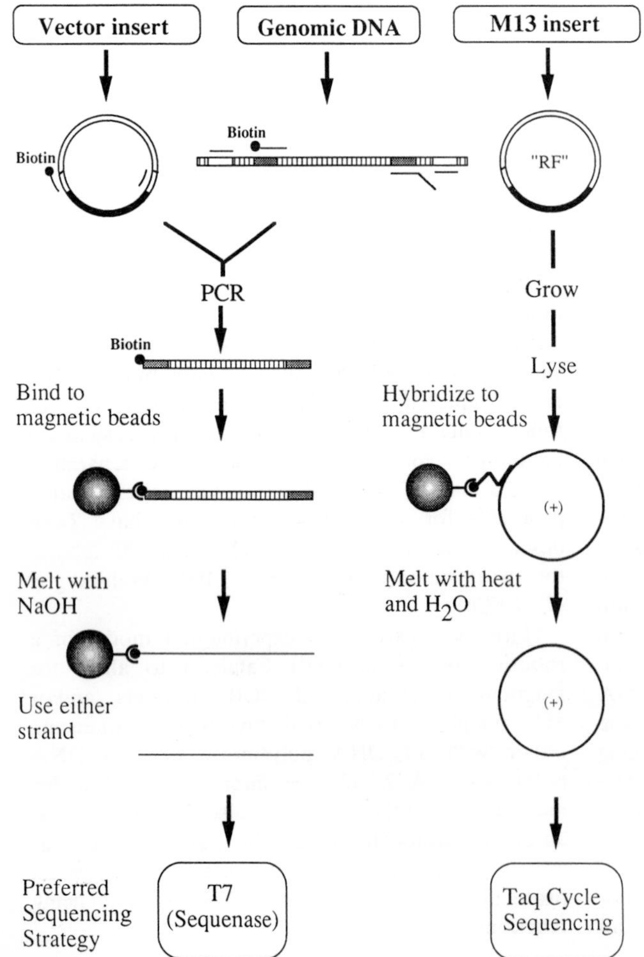

Figure 22.1 Schematic drawing of the principles for magnetic purification of vector inserts (PCR), genomic DNA (PCR) and M13.

coupled to the streptavidin coating the paramagnetic particles, thus forming the bead–DNA hybridization complex that can be magnetically captured from the solution.

22.3 ROBOTIC WORKSTATION

The ABI Catalyst consists of two parts connected by AppleTalk, the robot and a Macintosh IIsi computer running System 7 with 5 Mbyte Ram and an 80 Mbyte hard disk. The worksurface (Fig. 22.2) has: (i) two independently cooled storage positions down to 4°C, each with a capacity of 48 1.2-ml tubes; (ii) a position for six 30-ml glass bottles with heating capacity up to 80°C, (iii) two ambient temperature positions, one for 48 and one for 8 1.2-ml tubes; (iv) one magnetic station, with both cooling and heating capacity (4–95°C), fitted with either two adapters for 24 1.5-ml tubes each or one 96-well microtiter plate; (v) a thermocycler with a capacity of 96 samples capable of handling volumes up to 50 μl; and (vi) a wash station for the steel tip used in pipetting.

The arm is controlled by three separate stepmotors, (*X*, *Y*, and *Z* positions), with a positioning of about 0.02-mm accuracy each. The pipette tip is made of steel with a coating of an inert polymer and has a capacitance sensor for liquid and surface detection that can be turned on or off. Pipetting is controlled by two syringes (5 ml and 100 μl) using a positive liquid displacement fed from a large container with dH_2O that is also used for rinsing the tip. Pipetting with the 5-ml syringe is done in increments of 5 μl, and with the 100 μl syringe in increments of 0.1 μl. The syringes are connected to the different tubing by two three-way valves making it possible to switch from aspirations either through the tip or drawing liquid backwards from the dH_2O bottle or an ethanol bottle used for removal of gas that might collect in the tubing. Before each protocol begins, the robot performs a purge cycle with ethanol and water to ensure that any bubbles that have formed are removed.

The magnets in the magnetic module are also controlled by a stepmotor allowing them to move in and out of the separation area of interest. They consist of 18 vertical bars that each handle four or six samples in a cluster. Heating and cooling of the magnetic module are carried out in two ways; when using the large volume adapters (2×24), each adapter can be heated individually, but when using the microtiter plate adapter which is then fitted on top of the two others, it is heated by simultaneous control of the two heaters.

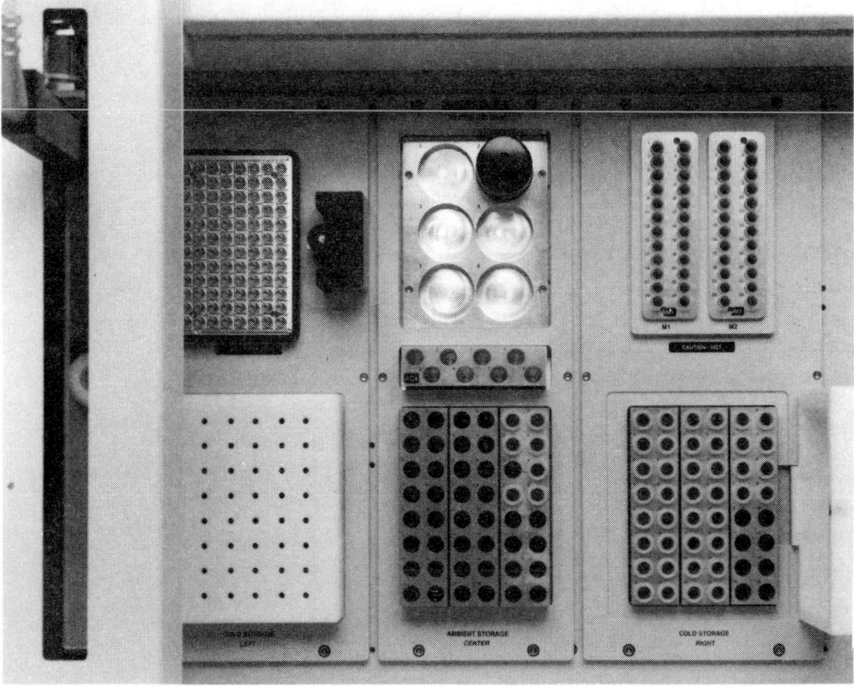

Figure 22.2 The ABI Catalyst robotic workstation equipped with a magnetic module (upper right) for 48 1.2-ml microcentrifuge tubes. The thermocycler unit can be seen (upper left) as can the heated storage area (upper middle) and the two cooled storage units (lower left and right).

The thermocycler consists of a fixed 96-well aluminium plate coated with an inert polymer. After a finished run, the block undergoes a wash cycle to regenerate a clean surface. No oil is used, instead it is fitted with a lid that has an independent heating capacity to prevent evaporation. As the lid is actually clamped down very tight it is possible to heat the thermocycler to +100°C without boiling.

22.4 SOFTWARE DEVELOPMENT

The programs were written in PolyFORTH and implemented using a special engine on the Catalyst, ABILink PolyFORTH v.6.03. This allowed us to create multiple tasks running independently, for instance heating and cooling in various positions while pipetting. By using a Hypercard application, loading and testing of new subroutines were facilitated. The same application was then used to create 'clickable' applications that the end-user can recognize and implement without further knowledge.

All user inputs are obtained by answering questions from the program. They are displayed on the screen using standard Macintosh formats with radio buttons for simple Yes/No type questions and selectable fields when a number input is required, e.g. volumes, temperatures, and incubation times (Fig. 22.3). In these protocols (ABI release version 2.0) it is also possible for the users to design their own sample preparation and sequencing schemes. This is done by stepping through the questions of importance and selecting desired conditions. The custom-designed method can thereafter be stored under an appropriate name. The method or parts of it can later be retrieved, selected and used for the sequencing schemes of interest.

22.5 M13 PURIFICATION

The principle of the M13 purification protocol is outlined in Fig. 22.4. Culture (1 ml) was incubated at 37°C for at least 5 h, and then centrifuged for 5 min in a microcentrifuge to remove the host cellular material. Cleared supernatant (600 μl) was transferred to a new 1.2-ml tube and inserted into the magnetic module. All the subsequent steps are performed by the robot. The phages were lysed by adding 35 μl 0.5 M Na_2EDTA (pH 8.5), 45 μl 10% SDS, mixed and incubated at 80°C for 10 min. Hybridization mixture (360 μl) and 100 μl of LacZBeads (Dynal, Norway) were added, and the mixture was incubated at 42°C for 30 min with intermittent mixing. After incubation, the beads were drawn to the side of the tube by the magnet. The supernatant was discarded and the beads washed in 300 μl of BeadRinse. Distilled water (35 μl) was then added if sequencing with dye-primers was

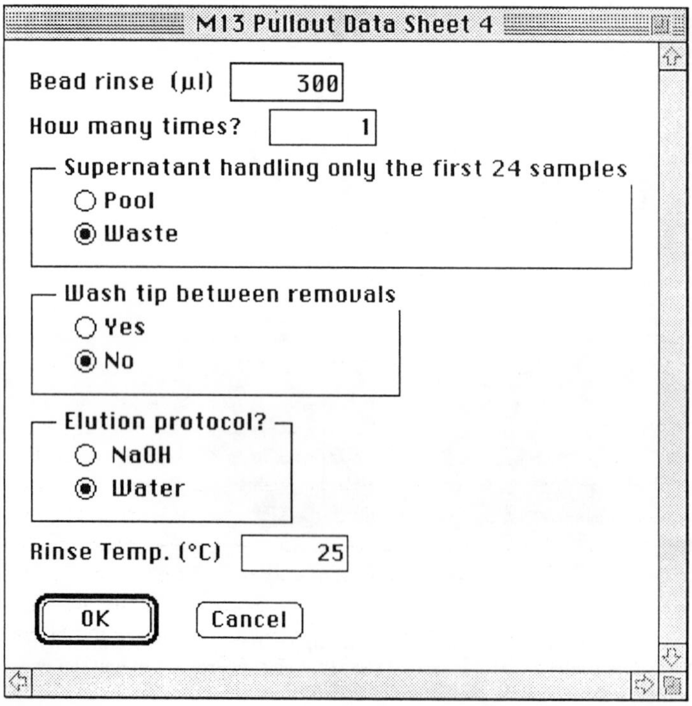

Figure 22.3 An example of the user interface of the ABI Catalyst software version 2.0.

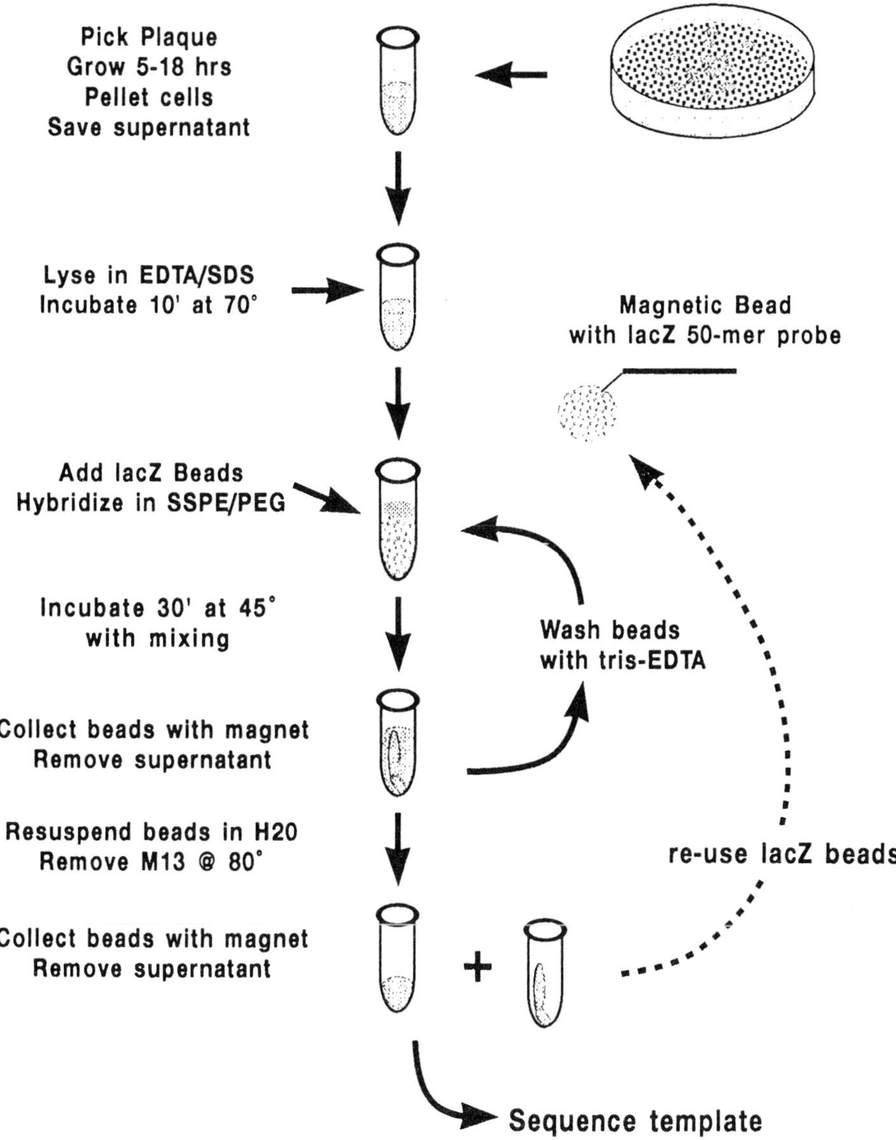

Figure 22.4 A schematic representation of procedure for M13 template purification.

later performed or 20 μl if dye-terminators were to be used, mixed with the beads and heated to 80°C for 2 min. The beads were pulled to the side and the supernatant containing the DNA collected and either stored for later use or transferred by the robot for immediate sequencing of the purified M13 template by either Taq dye deoxy terminator or Taq dye primer cycle sequencing (ABI, USA).

The sequencing reactions were performed as specified by the kit protocols, except for the template volumes. The template volume used for the dye-terminator reactions was 10 μl and for the dye-primer reactions, 5 μl for A and C and 10 μl for G and T. The difference in volume of water added to the beads to release the template and the actual volume used for sequencing is due to the amount of evaporation at 80°C. The beads are then recycled using a recondition solution as described by the manufacturer (Dynal, Norway) and reused in the next purification cycle. Upon completion, sequencing reactions were processed according to the sequencing protocols and loaded onto a 373 DNA sequencer for automated electrophoresis detection, data collection and sequence analysis.

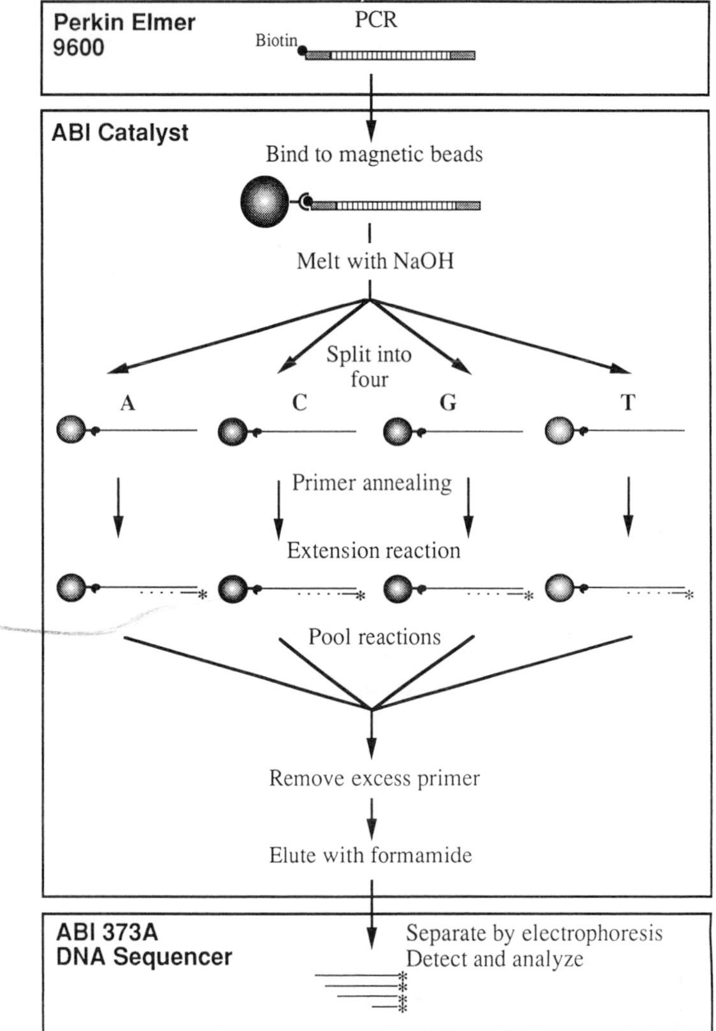

Figure 22.5 The principle of the solid phase protocol for sequencing PCR products using T7 dye-primers. Note that the PCR is performed in the Perkin–Elmer 9600 thermocycler and the detection is performed on the ABI 373A DNA sequencer.

22.6 SOLID PHASE DYE-PRIMER SEQUENCING OF PCR PRODUCTS USING T7 DNA POLYMERASE

The principle of the solid phase sequencing protocol is outlined in Fig. 22.5. The procedure involves three separate parts: (i) PCR; (ii) magnetic purification, separation and sequencing reactions; and (iii) automated electrophoresis and data collection. The PCR amplification of DNA was performed in a Perkin–Elmer 9600 thermocycler. The rack containing up to 24 samples was then transferred to the ABI Catalyst and the robot started.

The PCR product (50 μl) was transferred from the tubes to the 96-well microtiter plate positioned in the magnetic module and mixed with 20 μl streptavidin-coated paramagnetic bead suspension (Dynal, Norway) and 30 μl binding/washing solution, giving a final salt concentration of 1 M NaCl. The mixtures were incubated for 15 min at 30°C with intermittent mixing. After binding of the PCR fragments to the beads, the magnets in the magnetic module were raised and the paramagnetic beads pulled to the side of the wells. The beads were washed once with the binding/washing solution, with a mixing in between, and then secondly with a TE buffer to remove excess salt that otherwise would disturb the sequencing reactions. Before the first wash it is possible to save the supernatant for an agarose gel check of the binding efficiency, as described by the manufacturer (Dynal, Norway). The strands were then melted by mixing the bead pellet with 50 μl 0.1 M NaOH and incubating at room temperature for 5 min. Again, the magnets were raised and the supernatant containing the eluted strand transferred to wells further down on the microtiter plate, neutralized with HCl and stored for later processing. The magnetic beads containing the bound strand were then washed twice in 50 µl TE buffer, separated and finally resuspended in 26 µl annealing buffer (28 mM Tris-HCl (pH 7.5), 10 mM $MgCl_2$). Primer annealing was then performed as follows.

Each template was divided and transferred into four different positions in the thermocycler unit, 4 µl to the A and C position and 8 µl to the G and T position. Thereafter, 1 µl of primer A and C and 2 µl of primer G and T were pipetted to their respective positions and incubated at 65°C for 15 min. The lid was preheated to and kept at 80°C to prevent condensation. Finally, the temperature was lowered at a controlled rate of 5°C per minute down to 20°C and then rapidly cooled to 4°C. During the annealing of the four primers, the extension reactions were pipetted and mixed into four microcentrifuge tubes. For the A and C reactions, 1 unit of T7 DNA polymerase was mixed with 1.5 µl of deoxy/dideoxy mixture (3.65 µM specific ddNTP, 840 µM of each dNTP: c^7dGTP was used instead of dGTP), 1.5 µl of extension buffer (300 mM citric acid (pH 7.0), 318 mM DTT, 40 mM $MnCl_2$) and distilled water to a final volume of 5 µl. For the G and T reactions all volumes were doubled. The extension reaction mixes were then pipetted into the thermocycler and incubated at 37°C for 5 min with the lid kept at 60°C to prevent condensation. The extension reactions were stopped by rapidly cooling to 4°C and the addition of TE buffer to a final volume of 30 µl in each well on the thermocycler. The A, C, G, and T reactions were pooled and transferred back to the magnetic module and the supernatant containing excess primer and salt was removed. Thus, by using solid phase sequencing with magnetic beads, it was possible to completely eliminate ethanol precipitation and centrifugation procedures that otherwise would be necessary. Instead, both concentration and excess primer removal were obtained in one step. Finally, 5 µl of formamide was added and the magnetic module heated to 90°C for 2 min. The sequencing products were put on ice and loaded on the ABI 373A sequencer.

22.7 CONCLUDING REMARKS

Automated protocols for magnetic separation of DNA to be used for sequencing are suitable for automation of many routine applications in molecular biology, which can be performed by pipetting, incubation and magnetic separation. This might include solid phase cloning (Hornes *et al.*, 1990), solid-phase *in vitro* mutagenesis (Hultman *et al.*, 1990), solid phase synthesis of DNA (Stahl *et al.*, 1993) and various qualitative (Lundeberg *et al.*, 1990; Wahlberg *et al.*, 1990; Hedrum *et al.*, 1992) and quantitative (Lundeberg *et al.*, 1991) assays. Thus, magnetic separator modules might be important tools in the development of automated procedures in clinical, biological and biochemical research.

ACKNOWLEDGMENTS

This work was supported by grants from the Swedish Board for Technical Development and Dynal AS, Norway. We are grateful to Drs Erik Hornes and Rick Cathcart for helpful discussions and critical comments.

REFERENCES

Carothers, A.M., Urlaub, G., Mucha, J., Grunberger, D. & Chasin, L.A. (1989) *BioTechniques* **7**, 494–499.

Fry, G., Lachermeier, E., Mayrand, E., Giusti, B., Fisher, J., Johnson-Dow, L., Cathcart, R., Finne, E. & Kilaas, L. (1992) *BioTechniques* **13**, 124–131.

Gibbs, R.A., Nguyen, P.-N., McBride, L.J., Koepf, S.M. & Caskey, T. (1989) *Proc. Natl. Acad. Sci. U.S.A.* **86**, 1919–1923.

Hedrum, A., Lundeberg, J., Påhlson, C. & Uhlén, M. (1992). *PCR Methods Appl.* **2**, 167–171.

Hopgood, R., Sullivan, K.M. & Gill, P. (1992) *BioTechniques* **13**, 82–92.

Hornes, E., Hulman, T., Moks, T. & Uhlén, M. (1990) *BioTechniques* **9**, 730–737.

Hultman, T., Bergh, S., Moks ,T. & Uhlén, M. (1989) *Nucleic Acids Res.* **17**, 4937–4946.

Hultman, T., Murby, M., Ståhl, S., Hornes, E. & Uhlén, M. (1990) *Nucleic Acids Res.* **18**, 5107–5112.

Hultman, T., Ståhl, S., Hornes, E. & Uhlén, M. (1991) *BioTechniques* **10**, 84–93.

Lee, L.G., Connell, C.R., Woo, S.L., Cheng, R.D., McArdle, B.F., Fuller, C.W., Halloran, N. & Wilson, R.K. (1992) *Nucleic Acids Res.* **20**, 2471–2483.

Lundeberg, J., Wahlberg, J., Holmberg, M., Pettersson, U. & Uhlén, M. (1990) *DNA Cell Biol.* **9**, 287–292.

Lundeberg, J., Wahlberg, J. & Uhlén, M. (1991) *BioTechniques* **10**, 68–75.,

Ståhl, S., Hansson, M., Ahlborg, N., Nguyen, T.N., Liljeqvist, S., Lundeberg, J. & Uhlén, M. (1993) *BioTechniques* **14**, 424–434.

Syvänen, A.C., Hultman, T., Aalto, S.K., Söderlund, H. & Uhlén, M. (1991) *Genet. Anal.* **8**, 117–123.

Tabor, S. & Richardson, C.C. (1989) *Proc. Natl. Acad. Sci. U.S.A.* **86**, 4076–4080.

Uhlén, M. (1989) *Nature* **340**, 733–734.

Wahlberg, J., Holmberg, A., Bergh, S., Hultman, T. & Uhlén, M. (1990) *Proc. Natl. Acad. Sci. U.S.A.* **87**, 6569–6573.

Wahlberg, J., Albert, J., Lundeberg, J., Cox, S., Wahren, B. & Uhlén, M. (1992a) *FASEB J.* **6**, 2843–2847.

Wahlberg, J., Lundeberg, J., Hultman, T. & Uhlén, M. (1992b) *Electrophoreses* **13**, 547–551.

CHAPTER TWENTY-THREE

Custom Magnetic Particles: Their Use in DNA Purification

T. HAWKINS

MRC Laboratory of Molecular Biology, Hills Road, Cambridge, CB2 2QH, UK

23.1 INTRODUCTION

The introduction of automation in the field of DNA sequencing has had a dramatic effect on the potential for data production and collection. However, DNA sequencers only address one small area and, in real terms, shift the bottlenecks to upstream and downstream procedures. Now, more than ever, the emphasis is on the development of new and novel biochemical techniques that are faster, cheaper and provide products suitable for input into automated devices without significant changes or modifications.

The advantages of using magnetic particles in molecular and diagnostic biology have been described previously (Uhlen, 1989; Hultman *et al.*, 1989, 1991). The use of solid phase techniques has dramatically increased over the last few years as more biochemical methods have become adapted for use with magnetic particles. Recently, I described a method for M13 single-strand purification using a biotinylated probe and streptavidin-coated magnetic particles (Hawkins, 1992). This technique was specifically developed for the increased throughout required by the *Caenorhabditis elegans* sequencing project (Cambridge, UK) (Sulston *et al.*, 1992). In this chapter I describe an improved method for template production and describe the use of Promega magnetic particles for the purification of polymerase chain reaction (PCR) products either by direct biotin–streptavidin binding or by using a custom magnetic particle analogous to the M13 magnetic particle purification system.

23.2 MATERIALS AND METHODS

23.2.1 Manufacture of the magnetic particle/probe unit

The M13 specific probe was synthesized on an ABI 380B DNA synthesizer on a 1 μmol column. A biotin phosphoramidite was used as supplied from Amersham UK. The sequence of the probe was:

5′ TAT CGG CCT CAG GAA GAT CGC ACT CCA GCC AGC AAA AAA Biotin A 3′

Following cleavage from the column, the oligonucleotide was deprotected in ammonia at 55°C overnight. The probe was then purified using a NAP-10 column (Pharmacia PL) to remove the ammonia and the probe was eluted in water.

Promega nucleotide quality magnetic particles were used. Particles (100 ml) were probed in one go by adding the whole 1 μmol synthesis of the probe to the particles. The probe was mixed by shaking the particles and then allowed to bind overnight at 4°C. The 100 ml of particles was then divided into two 50-ml Falcon tubes and the particles separated using a magnet. The supernatant was poured off and 30 ml of wash buffer was added to the particles. The particles were resuspended by shaking. This wash procedure was repeated a total of eight times followed by one wash in water and then an incubation at 45°C for 10 min to remove any loose probe. Finally, the particles were washed once in re-use buffer to remove any unstable biotin. The particles were then washed in PBS and taken up in a total of 40 ml (20 ml per Falcon tube). This is the storage concentration.

23.2.2 Buffers

Lysis buffer:	15% sodium dodecyl sulfate.
Wash buffer:	0.1×SSC. Stock solution of 20× SSC, 87.7 g NaCl, 44.1 g sodium citrate, per 500 ml.
Re-use buffer:	0.15 M NaOH, 0.001% Tween20.
PBS:	8 g NaCl, 0.2 g KCl, 1.44 g Na_2HPO_4, 0.24 g KH_2PO_4, adjusted to pH 7.4, made up to 1 l with water.

23.2.3 Preparation of particles for template purification

Template preparations were carried out in batches of 96 using a total of 4 ml of particles per 96 reactions. Before use, 4 ml of particles was removed from the pool and placed in a 15-ml Falcon tube. The particles were then separated using a magnet and the storage buffer poured off. The particles were then washed once in water and finally taken up in 1 ml water ready for use.

23.2.4 Growth of M13 phage

A single M13 plaque was picked into 1.5 ml TY inoculated with a 1/100 dilution from a TG1 overnight culture and incubated at 37°C for 5–6 h with shaking at 300–400 c.p.m. The cells were spun down at 12 000**g** for 10 min. This whole procedure is best carried out in 2 ml Eppendorfs which can be loaded into the Eppendorf spin racks to allow easier handling for the growth stage and a faster spin throughput when using the appropriate centrifuge.

23.2.5 Lysis

To each of the 96 wells of a Falcon 3911 microtest plate 10 μl of lysis buffer was added. Supernatant (200 μl) of each of the clones was added to each of the wells in turn. The phage was lysed by placing the plate on an 80°C heat block for 5 min.

23.2.6 Hybridization and annealing of custom magnetic particles

The plate was removed from the heat block and 50 μl of the hybridization buffer added to each well using an Eppendorf multidispensing combitip. No mixing is required so the same tip was used for all the samples. Finally, 10 μl of the particles/probe were added to each well, again with the same tip as no mixing is required. The probe was then allowed to anneal to the target DNA by placing the plate on a 45°C heat block for at least 20 min.

23.2.7 Wash steps

The microtiter plate was removed from the block and placed on the dot magnet, as described above. The magnetic particles were drawn to the bottom of the wells together with the probe/template attached to their surface. After 1 min the particles had settled. The supernatant containing the nonspecific DNA and excess buffer was removed by aspiration. The particles were then washed by adding 100 μl of wash buffer to each well while the plate was off the dot magnet. The force of adding this liquid dispersed the particles and, therefore, removed any interspersed material. The particles were washed in this way at least twice.

23.2.8 DNA recovery

In order to recover the template, the particles were taken up in 60 μl water and then the plate was put onto a 80°C heat block to disassociate the template from the probe. After 3 min on the 80°C block, the plate was removed and placed on a magnetic unit such as the Dynal bar magnet (MPC-96) which pulls the particles to the side of each well. This enables the DNA to be removed to a new plate using a multichannel pipette without taking any particles. The DNA is then ready to be used for the sequencing process.

23.2.9 Re-use of magnetic particles

The particles were immediately covered with 50 μl of wash buffer to prevent drying out and then collected for recycling. The particles were placed into a 50-ml Falcon tube and this was then placed into a large magnet such as the Promega tube magnet which pulls all the particles to the side. The supernatant was then removed and the particles were taken up in re-use buffer for 1 min before being replaced. This procedure was repeated twice and the particles were then washed once in PBS and finally taken up in half their original volume. For this procedure, 4 ml of particles were used per 96 clones so that the particles were made back up into 2 ml PBS after recycling.

In order to prevent old or new particles from causing fluctuations in the yields, it is more effective to work from a mixed population of 50 ml of particles to which recycled particles are added after use and the volume is topped up periodically with fresh particles when needed; this is called the 'pool'.

The average yield from this type of M13 purification was approximately 600–800 ng per clone. The average yields and reproducibility were monitored by taking a random sample of 6 μl from each 96 template plate and testing by agarose gel electrophoresis.

23.3 POLYMERASE CHAIN REACTION

23.3.1 Buffers and solutions

Buffer:	400 mM Tris (pH 8.9), 100 mM ammonium sulfate, 25 mM $MgCl_2$.
NTP mix:	1.25 mM of each dNTP.
Taq polymerase:	2 U μl^{-1}.
Template:	M13 plaque, *or* M13 size – cosmid size 10 ng; genomic or YAC 25 ng.
Primers:	approximately 15 pmol μl^{-1}.
Cycler:	Techne PHC-3.
Magnetic particles:	preprobed with biotinylated custom probe and purified as described above.
PCR per well:	16 μl water, 3 μl buffer, 5 μl buffer, 2 μl primer A (about 15 pmol μl^{-1}), 2 μl primer B (about 15 pmol μl^{-1}), 1 μl template, 1 μl Taq polymerase (5 U μl^{-1})

Overlay each well with 20 μl light mineral oil. Cycle as follows:

30 cycles of: 95°C 30 s, 52°C 30 s, 72°C 2 min.

23.3.2 PCR purification (direct biotin–streptavidin system)

Buffers: Hybridization buffer 2.5 M NaCl, 20% PEG 8000; wash buffer 0.1× SSC.

Take up 80 μl of Promega magnetic particles and wash once in water. Resuspend in 20 μl water. To each PCR add 20 μl washed particles and 50 μl hybridization buffer. Incubate the plate at 45°C for 10 min before placing on a magnetic station (Dynal MPC-96).

After 3 min, aspirate off the liquid and then remove the plate from the magnet. Add 50 μl wash buffer to each well and replace the plate on the magnetic station. Repeat this wash procedure a total of three times.

Finally, resuspend the magnetic particles in 50 μl water and place the microtiter plate onto a 95°C heat block for 3 min. Remove the plate and put onto the magnetic station. Immediately remove the liquid to a new microtiter plate. Test the DNA recovery by agarose gel electrophoresis. This DNA will have to be diluted prior to DNA sequencing. Typically, the addition of 200 μl water will be sufficient to dilute the PCR fragments for DNA sequencing, as described above.

23.3.3 PCR purification (custom magnetic particle)

Buffers: as for M13 magnetic purification system (Section 23.4)

Replace the microtiter plate on the thermal cycler and heat to 95°C for 5 min. Add hybridization buffer and the custom magnetic particles as described for the M13 magnetic particle purification system (Section 23.4) and then follow the annealing and wash steps as described.

23.4 M13 PHAGE PURIFICATION USING MAGNETIC PARTICLES

The sequencing of the 100 Mb *C. elegans* genome will require an estimated 3 million DNA templates of consistent quality if the strategy outlined previously is pursued (Sulston *et al.*, 1992). This number of templates cannot be produced using the currently available methods. Automation of the processes described is currently under development (Watson *et al.*, 1993).

The traditional methods for M13 template production, such as PEG/phenol procedures (Bankier *et al.*, 1988), are designed to produce microgram quantities of DNA from milliliter volumes. These procedures are not easily scaled down and use centrifugation for the separation of phenol layers and the precipitation of DNA. Methods do exist that purify DNA on a smaller scale. However, these methods use either specialized filtration steps (Kristensen *et al.*, 1987) or PEG/SDS with multiple centrifugation steps (Smith *et al.*, 1990). A more recent method (Alderton *et al.*, 1992) uses paramagnetic particles to purify DNA from a PEG induced phage coagulation. However, this method is not template specific and cannot be scaled down to standard microtiter plate format.

The advantages of using magnetic particles are based on the ease with which DNA can be manipulated once

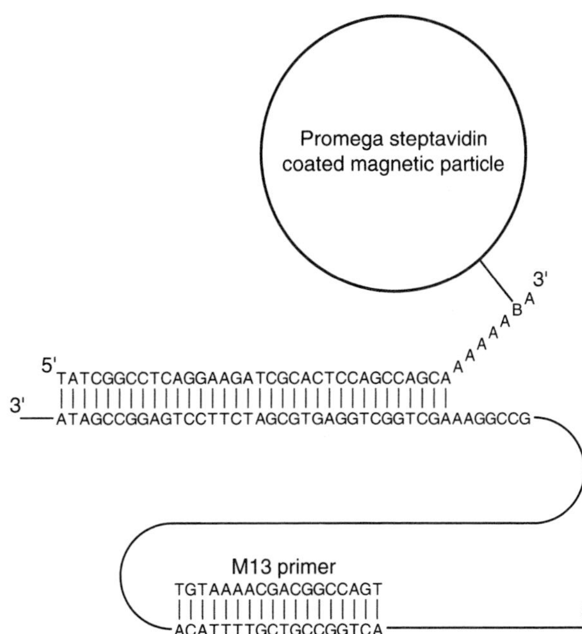

Figure 23.1 The M13 custom magnetic particle.

immobilized to the magnetic particles. In theory, the need for centrifugation stages can be removed from molecular biology if DNA fragments were fixed, temporarily or permanently to the magnetic solid phase.

23.4.1 Outline

The method uses an oligonucleotide probe which has been synthesized with a biotin group at the 3′-end. In the case of M13 purification, the probe was designed to be complementary to a region upstream from the M13–21 universal priming site. The probe was added to streptavidin-coated magnetic particles to allow the biotin to bind to the streptavidin coating, forming one unit, as shown in Fig. 23.1 and described in Section 23.3.

The procedure is based upon hybridization, during which the complementary probe binds to its target sequence on the DNA template. Once the probe has bound to the target sequence, the template/probe/magnetic particle can be treated as one stable unit and manipulated using the magnetic particle. It is then possible to wash away any nonspecific DNA or proteins which may otherwise contaminate the DNA template. Once this has been completed, the template can be removed from the probe/magnetic particle unit by heating.

23.4.2 Purification procedure for M13 single-stranded templates

M13 subclones are grown in culture, as previously described (Bankier *et al.*, 1988) and after a period of 5–6 h the cells can be harvested by centrifugation and 200 µl of the supernatant transferred to a Falcon microtiter plate. In practice, this small volume contains enough DNA for one thermally cycled sequencing reaction after DNA purification. To lyse the host cells, sodium dodecyl sulfate is used at a final concentration of 1%. The lysis stage is the most crucial step as access to the phage DNA is not possible without removal of the phage coat. To further encourage lysis, the phage are heated to 80°C for 5 min.

Once lysis is complete, the next stage is to anneal the probe/magnetic particle unit to the template. This stage is important as the highest annealing efficiency is required to reduce the quantity of magnetic particles used per reaction and thereby reduce costs. The addition of polyethylene glycol (PEG) to the hybridization buffer dramatically increases the overall yields obtained. The presence of PEG increases the concentration of the target DNA by excluding nucleic acids from the water hydrogen bonded to the polymer (Amanso, 1986). PEG also acts to increase the viscosity of the reaction mixture holding the magnetic particles in solution rather than allowing them to settle. Another component of the hybridization buffer is sodium chloride which effectively increases the ionic strength of the reaction during the annealing stage.

The most important reagent is the probe/magnetic particle unit which is stored in PBS/BSA, washed and resuspended in water before being added to the reaction. Once both the hybridization buffer and the probe/magnetic particle units have been added to the microtiter plate, it is placed at 45°C for 20 min to allow annealing. After 20 min, the annealing is complete. The magnetic particles are then pulled to the side of the tube using a magnet (Dynal MPC-96) and the supernatant removed by aspiration. The magnetic particles can now be washed by adding low ionic strength wash solution and moving the magnetic particles through the solution by moving the magnet from one side of the plate wells to another. This action of pulling the magnetic particles through solution was found to be a very effective way to remove nonspecific products. Also the wash solution has a lower ionic strength than the annealing conditions to help remove any contaminants linked to the probe or the magnetic particles. By experimentation, it was found that three washes are sufficient to remove nonspecific products.

Once the wash steps are complete, the final stage is to reclaim the template DNA from the probe/magnetic particle unit. This can either be carried out using alkali or heat to denature the template from the probe sequences. If alkali is used it will produce DNA which needs neutralization before DNA sequencing. Some enzymes such as T7 polymerase are very sensitive to pH fluctuations which leads to sequence failures or reduction in data quality. The use of heat and the resuspension of the magnetic particles in water after washing is found to give high yields and to produce the DNA template in an ideal solution to be sequenced. After heating to 80°C for 3 min, the probe/magnetic particle units can be pulled to the side of the wells and the DNA removed to another microtiter plate.

23.4.3 The design of the custom magnetic particle

During the purification procedure, the probe was found to shear from the magnetic particles; therefore, it is important that it should not interfere with the sequencing reactions by acting as a secondary sequencing primer, which is sufficient to dramatically reduce the quality of the dye-labeled reactions (results not shown). To prevent loose probe from acting in this way, the probe was designed with a linker arm at the 3′-end which was noncomplementary to the M13mp18 sequence, as shown in Fig. 23.1. The linker arm also reduces the steric hindrance between the probe and the 20 times larger magnetic particle. The biotin moiety is positioned at the 3′-end to further discourage probe to template 3′-hybridization. This strategy led to increased template yield and prevented loose probe from acting as a secondary sequencing primer.

23.4.4 The re-use of the magnetic particles

The probe/magnetic particle units can be reused if they are first rejuvenated to remove any excess template bound to the probe. It was found that 0.15 M sodium hydroxide is ideal to denature any remaining template while not stripping any streptavidin coating from the magnetic particles. When the magnetic particles are reused, the binding capacity is reduced. This is due to some of the probe shearing from the magnetic particles, as described earlier, but also to physical loss of particles when pipetted. However, the magnetic particles can be reused up to 50 times if they are treated carefully with limited heating and pipetting. Alternatively, to maintain a usable population of magnetic particles, a halving in volume of the particles every time they are reused will more than compensate for the loss of binding capacity and particle loss.

23.4.5 Small-scale DNA production

The procedure outlined above was specifically designed to produce sufficient DNA for one thermally cycled sequencing reaction to be electrophoresed on an Applied Biosystems 373A DNA sequencer. This requirement allows the purification procedure to be performed in microtiter format with one well being used per template purification. When using this volume of supernatant (200 μl), the average yield is 600–800 ng total. The microtiter format purification system is outlined in Fig. 23.2. The procedure has been set up to use multidispensing and multichannel pipettes without the need for mixing of reagents. This leads to one microtiter plate of 96 templates being typically completed in under 1 h.

The use of standard microtiter plates and other common molecular biology equipment should lead to this procedure being used widely for small-scale DNA purification for sequencing and for more general molecular biology applications.

23.4.6 Sequencing results from the M13 magnetic purified DNA system

The specification of this new template purification method was to produce high quality DNA sequence when a selection of random M13 subclones were used. Since January 1992, the microtiter format method has been introduced to the production sequencing for the *C. elegans* genome project (Cambridge, UK) and, by December 1992, over 30 000 M13 templates were produced. The majority of templates were used for Taq dye-primer reactions although some of the DNA was also used in Taq dye-terminator reactions with similar quality data produced. When T7 polymerase based reaction chemistries are used, more DNA is required such that two wells are needed per clone. When purifying M13 single-stranded DNA the average yield was 600–800 ng DNA, which was sufficient for one thermally cycled sequencing reaction plus

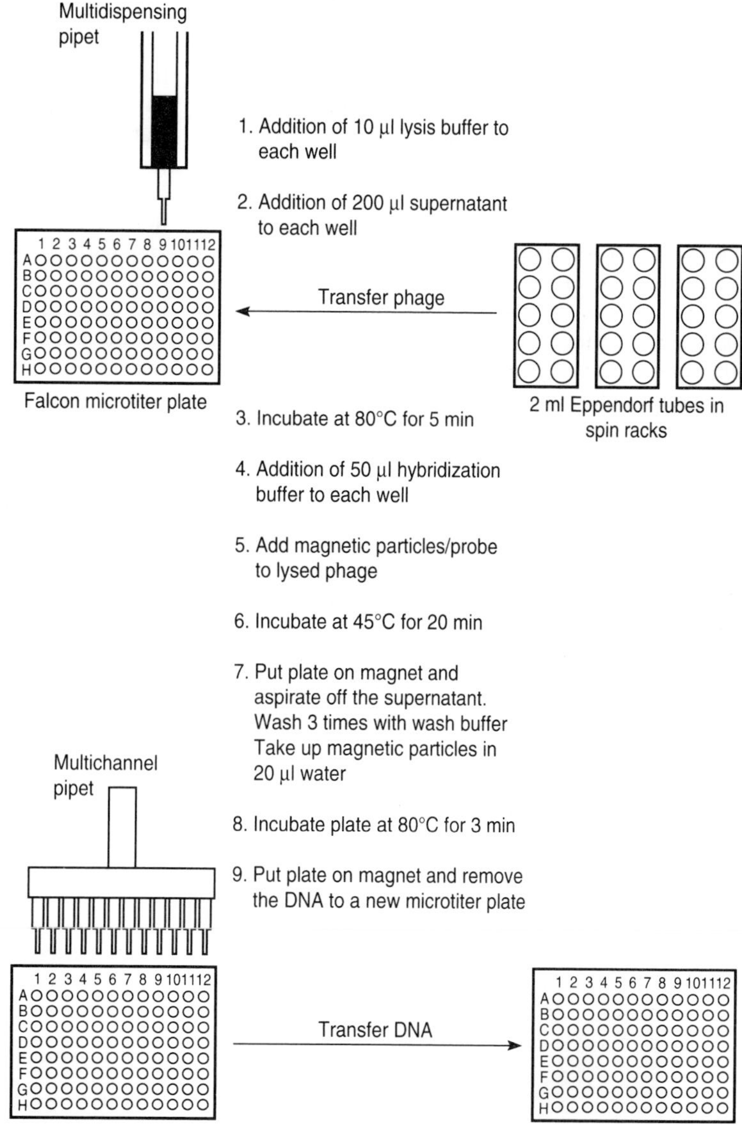

Multidispensing pipet
1. Addition of 10 μl lysis buffer to each well
2. Addition of 200 μl supernatant to each well
Transfer phage
Falcon microtiter plate
2 ml Eppendorf tubes in spin racks
3. Incubate at 80°C for 5 min
4. Addition of 50 μl hybridization buffer to each well
5. Add magnetic particles/probe to lysed phage
6. Incubate at 45°C for 20 min
7. Put plate on magnet and aspirate off the supernatant. Wash 3 times with wash buffer Take up magnetic particles in 20 μl water
Multichannel pipet
8. Incubate plate at 80°C for 3 min
9. Put plate on magnet and remove the DNA to a new microtiter plate
Transfer DNA

Figure 23.2 The microtiter M13 magnetic purification procedure.

remaining template for use in PCR amplifications, as described in the following sections.

23.5 THE PURIFICATION OF PCR PRODUCTS USING MAGNETIC PARTICLES: DIRECT BIOTIN–STREPTAVIDIN BINDING

23.5.1 Outline

Sequencing directly from PCR products has many advantages over subcloning; the ability to PCR directly from plaques or colonies removes the need for template preparation altogether. The main problem with this approach is the success rate of the PCR reaction and the DNA sequence quality obtained from the PCR product.

Performing PCR directly from M13 plaques was found to be dependent upon the amount of DNA available. In most cases, one M13 plaque provided too much DNA and led to little or no PCR product. However, the amount of DNA could be standardized by picking plaques into water then taking a small amount of this as the PCR template. Despite this, the success rate from PCR reactions using the M13 forward and reverse primers from M13 plaques was approximately 85%. Also, on average, a further 5% of clones gave more than one PCR product when amplified. This was due to other possible priming sites being present in the DNA insert.

Once a PCR product has been obtained, the next stage is purification prior to DNA sequencing. The sequence quality is directly proportional to the purity of the PCR template; if sequencing is performed directly after PCR without any purification, little or no usable sequence data are obtained. The most effective way to purify these fragments away from excess primer and nucleotides has been to use PEG/$MgCl_2$ (Paithanker & Prasad, 1991) to precipitate the larger fragments, thereby allowing the excess primers and nucleotides to be washed away. However, PEG inhibits the sequencing reactions and this procedure is time consuming and requires centrifugation to pellet the fragments. If PCR is to be used as a routine procedure, there must be some way of purifying the fragments without using PEG and centrifugation and providing PCR fragments which yield high quality sequence data.

The use of Dynal magnetic particles for the purification of PCR products has been described previously (Hultman *et al.*, 1991). The use of Promega particles presents some problems since the hydrophobic surface of these particles binds to the fluors used for the ABI sequencing chemistry and prevents the extension products from being visualized during electrophoresis. However, the lower cost of the Promega magnetic particles warrants their further investigation.

To tailor the reaction chemistry for use in a large-scale project, the PCR reactions are scaled down to work in microtiter plates with a total reaction volume of 30 μl. The PCR reactions are performed, as described in Section 23.3, with one biotin and one nonbiotinylated primer. When the PCR reactions are complete, hybridization buffer and magnetic particles are added to each sample. The double-stranded PCR fragments then bind to the streptavidin-coated magnetic particles which are then immobilized using a magnetic station (Dynal MPC 96). The magnetic particles are then washed to remove excess primer and nucleotides and the double-stranded PCR fragment is resuspended in water. On heating, the two PCR strands dissociate and, once the biotinylated strand has been immobilized using the magnet, the nonbiotinylated strand is removed and used for DNA sequencing.

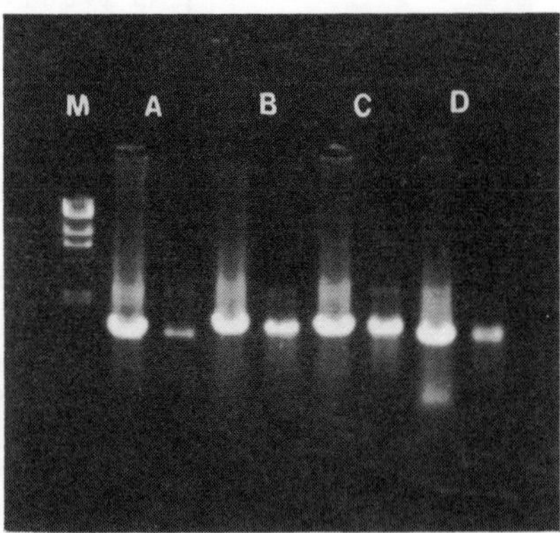

Figure 23.3 Agarose gel showing the relative amounts and quality of pre- and post-purified PCR products. Four PCR reactions (A, B, C & D) were set up and cycled as described in Section 23.3. One-tenth of the prepurified PCR product was set aside and the remainder purified using the direct biotin–streptavidin system as described in the text. The gel shows a 100 ng HindIII cut λ molecular marker (M) with each PCR set run with pre- and post-purified fragments (A, B, C & D).

The advantage of this system is that all the purification is performed while the PCR fragment is double-stranded which, when dissociated, yields a purified single-strand with no magnetic particles attached. Also, the concentration of PCR product recovered is dependent upon the amount of magnetic particles added to the PCR and in this way the fluctuations in PCR yields can be reduced. This procedure would also be useful in other methods in which the presence of a magnetic particle attached to the DNA would inhibit subsequent applications.

The gel photo in Fig. 23.3 shows the yield and purity before and after purification using this method. The DNA can be sequenced either using a dye-labeled primer or by the incorporation of dye-labeled terminators.

23.6 THE PURIFICATION OF DOUBLE-STRANDED PCR PRODUCTS USING MAGNETIC PARTICLES: ICE MAGNETIC PARTICLES

23.6.1 Outline

Fluorescence-based DNA sequencing is in many ways more demanding than the radioisotope methods (Hawkins *et al.*, 1992). The incorporation of a fluorescent label can either be in the form of a dye-primer or by the incorporation of dye-terminators into the extension products. For most purposes, the use of dye-terminators provides the only route for primer directed DNA sequencing as any primer can be used. This approach has the advantage of *in situ* labeling but requires significant postreaction purification to remove excess dye. For this reason, it would be an advantage to use dye-primer chemistry for primer walking in a similar way to the use of the M13–21 universal primer. The problems of using a dye-labeled chemistry for primer walking on the ABI 373A have been discussed previously (Hawkins *et al.*, 1992). This dye-primer approach can only be realistically used if the same primer site can be found in all samples; this could be done by synthesizing a primer with a specific sequencing tail at the 5′-end followed by the PCR specific region at the 3′-end. This would require the synthesis of a primer approximately 40 bp in length.

23.6.2 PCR purification using a custom magnetic particle

This technique uses custom PCR primer as usual but with the sequencing tail at the 5′-end. This primer can then be paired with the M13 reverse primer, on a suitable M13 subclone for example. The resulting fragment after amplification has the sequencing site at the 5′-end followed by the insert fragment. Following purification, this PCR fragment can be sequenced with the appropriately labeled sequencing primers to give the primer walk information.

The sequencing primer was designed as the Boehringer Omega-Nuclease recognition sequence (Monteilhet *et al.*, 1990) which was an 18 bp recognition site known to have few genome target sites and, therefore, reduced background-noise level due to secondary priming. This 'ICE' primer set was synthesized and labeled as described in Section 23.3. Once this new approach was set up, it became obvious that the purification step was critical to the data quality, as with the PCR sequencing described above.

The method for PCR purification outlined above would not be appropriate since each new custom primer made would have to be synthesized with a biotin group at the 5′-end. A much better system would use the hybridization of a probe analogous to the M13 magnetic template purification system. A primer was synthesized with the same sequence as the ICE sequencing primer but with a nonspecific linker at the 3′-end with the biotin linked to the penultimate base, as with the M13 specific probe (Fig. 23.4).

This probe was then attached to the Promega magnetic particles as described in Section 23.3. The procedure for purification was identical to the one used for the M13 magnetic template system with the exception of a denaturing period at 95°C for 5 min prior to the magnetic particle/probe unit being added. The whole procedure for the purification of 'ICE' PCR fragments is outlined in Fig. 23.5. The advantage of this approach is the specificity of the purification step and the quality of the DNA sequence produced. The procedure is fast and simple and yields a double-stranded PCR product which has the potential to be sequenced on either strand and has no magnetic particles attached. Also, unlike the direct biotin–streptavidin system, only full-length PCR products are recovered since the probe target site is only created in completed PCR fragments.

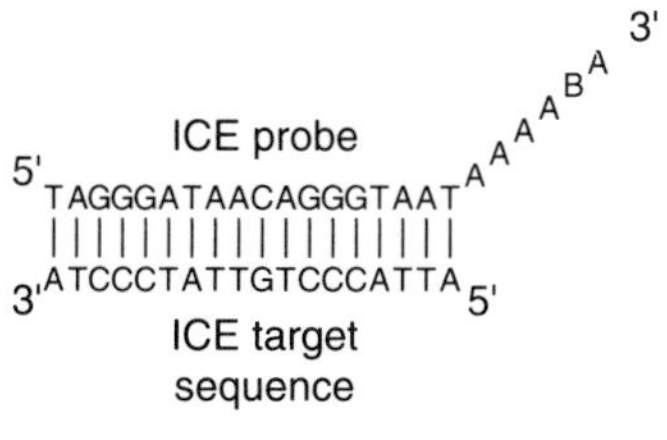

Figure 23.4 The 'ICE' probe sequence.

The extra cost of producing a custom primer 40 bp in length as opposed to 20 bp was somewhat offset by the quality of the data and the high success rate of the dye-primer chemistry. It should also be considered that the in-house manufacture of dye-labeled 'ICE' primer almost eliminates the cost of fluorescent label in each reaction.

23.7 DISCUSSION

The ability to use PCR products for large-scale DNA sequencing, using a procedure which is fast, cheap and, above all, produces high quality sequence data, could be very useful in any sequencing strategy. However, from the data currently available, it is not advisable to stop using the M13 magnetic particle purification system and to instead perform PCR reactions directly from the M13 subclones on a routine basis. The overall success rate of the PCR based approaches was 80% producing usable sequence, which was lower than the approximately 90% success rate found with the M13 magnetic template purification system.

When the three methods are compared (Table 23.1) it is found that the M13 magnetic purification process not only produces the highest quality DNA but also at the lowest cost per sample. This is due to the principal growth being from culture and not via PCR which requires Taq polymerase. With both PCR based methods, the cost of the Taq polymerase makes up a significant proportion of the final cost per sample.

The M13 magnetic purification system and the 'ICE' custom magnetic particle system are both based on specific template purification methods. In both cases, a custom magnetic particle has been used to selectively recover DNA fragments from a mixed population. This procedure, obviously, has many biochemical applications in situations where a DNA fragment with a specific motif or target sequence needs to be recovered, such as in selective cDNA capture or for the isolation of correctly recombined or inserted fragments into plasmid or genomic DNA.

The requirement for faster template purification systems is vital for effective use of automated systems

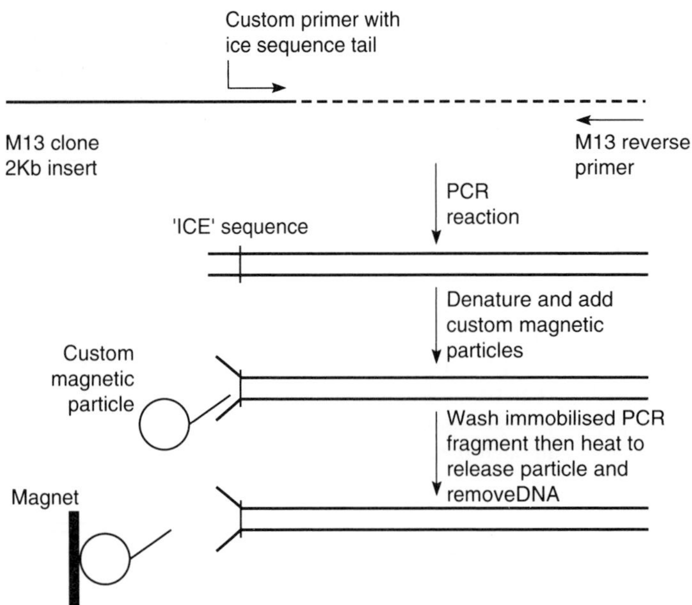

Figure 23.5 The 'ICE' custom magnetic particle system.

Table 23.1 Comparison of the three magnetic particle procedures.

	Time taken per 96 samples	*Cost per sample (£)*	*Template type*	*Application*
M13 magnetic purification system	6.5 h	0.45	ss/ds DNA max. size ≥40 kb	Specific DNA recovery
'ICE' PCR custom magnetic particle	PCR: 3 h purification: 1 h	0.75	ds DNA max. size ca. 3 kb	Specific DNA recovery
Direct biotin–streptavidin PCR system	PCR: 3 h purification: 40 min	0.60	ss DNA max. size ca. 3 kb	Recovery of biotinylated PCR products

such as the ABI 373A DNA sequencer. More importantly, it is vital that new biochemical techniques are easily automated, especially when considering the requirements of a large-scale genome project. To meet the increased throughput of the *C. elegans* genome project, a robotic system has been developed to automate the procedures outlined in this chapter (T. Hawkins & A. Watson, unpublished). This automated system is based upon the easy manipulation and isolation of DNA when using magnetic particles, as demonstrated here.

ACKNOWLEDGMENTS

This work has been supported by an MRC-HGMP grant. I would like to thank the members of the *C. elegans* mapping and sequencing group for their comments and discussions.

REFERENCES

Alderton, R., Eccleston, L., Howe, R., Read, C., Reeve, M. & Beck, S. (1992) *Anal. Biochem.* **201**, 166–169.

Amanso, R. (1986) *Anal. Biochem.* **152**, 304–307.

Bankier, A., Weston, K. & Barrell, B. (1988) *Methods Enzymol.* **155**, 52–93.

Hawkins, T. (1992) *J. DNA Sequence Mapping* **3**, 65–69.

Hawkins, T., Du, Z., Halloran, N. & Wilson, R. (1992) *Electrophoresis* **13**, 552–559.

Hultman, T., Stahl, S., Hornes, E. & Uhlen, M. (1989) *Nucleic Acids Res.* **17**, 4937–4946.

Hultman, T., Bergh, S., Moks, T. & Uhlen, M. (1991) *BioTechniques* **10**, 84–94.

Kristensen, T., Voss, H. & Ansorge, W. (1987) *Nucleic Acids Res.* **15**, 5507–5516.

Monteilhet, C., Perrin, A., Thierry, A., Colleaux, L. & Dujon, B. (1990) *Nucl. Acids. Res.* **18**, 1407–1413.

Paithanker, K. & Prasad, K. (1991) *Nucleic Acids Res.* **19**, 1346.

Smith, V., Brown, C., Bankier, A. & Barrell, B. (1990) *J. DNA Sequence Mapping* **1**, 73–78.

Sulston, J., Du, Z., Thomas, K., Wilson, R., Hillier, L., Staden, R., Halloran, N., Green, P., Thierry-Mieg, J., Qui, L., Dear, S., Coulson, A., Craxton, M., Durbin, R., Berks, M., Metzstein, M., Hawkins, T., Ainscough, R. & Waterston, R. (1992) *Nature* **356**, 37–41.

Uhlen, M. (1989) *Nature* **340**, 733–744.

Watson, A., Smalde, K., Lucke, R. & Hawkins, T. (1993) *Nature* **362**, 567–568.

PART II

Sample Preparation and Sequencing Methods

C. SEQUENCING METHODS

CHAPTER TWENTY-FOUR

Large-scale, Automated Sequencing of Human Chromosomal Regions

W. R. McCOMBIE[1] & A. MARTIN-GALLARDO[2]

[1] Cold Spring Harbor Laboratory, Cold Spring Harbor, NY 11724, USA

[2] Lawrence Livermore National Laboratory, Biomedical Sciences Division, Livermore, CA 94551, USA

24.1 INTRODUCTION

Reliable methods to sequence DNA were described in the mid-1970s (Sanger *et al.*, 1977b; Maxam & Gilbert, 1977). Owing to the time-consuming nature of manual DNA sequencing techniques, these have usually been applied to either cDNAs or genomic regions known to contain one or several genes of interest. Essentially, the primary application of these techniques has been to determine the complete sequence of a single gene or, more recently, the partial sequences of many genes (Adams *et al.*, 1991, 1992; Wilcox *et al.*, 1991; McCombie *et al.*, 1992a; Waterston *et al.*, 1992; Kahn *et al.*, 1992; Okubo *et al.*, 1992). Advances in the mid-1980s (Smith *et al.*, 1986) made feasible the sequencing of large regions of human chromosomal DNA for the purpose of discovering the genetic content in previously uncharacterized areas. This is fundamentally different from previous efforts which were directed by a goal of characterizing known coding regions. Once this can be accomplished in a systematic and routine manner, we will be able to obtain the data that will ultimately enable us to determine the structure of complex eukaryotic genomes.

Early sequencing efforts had in fact sequenced small genomes such as that of the bacteriophage φX174 (Sanger *et al.*, 1977a). These studies revealed interesting structural aspects of genomes such as the overlap of some of the genes in the φX174 genome (Sanger *et al.*, 1977a). A similar phenomenon has been observed in more complex genomes, such as the presence of genes within the introns of the factor VIII and NF1 genes (Levinson *et al.*, 1990; O'Connell *et al.*, 1990; Cawthon *et al.*, 1991). As more genomic regions are characterized as a result of sequencing or high-resolution mapping to find disease genes, it seems apparent that the coding regions of genes are found in a complex matrix of structural elements the function and, in many cases, identity of which have not yet been determined. Genomes are now best viewed as functioning entities with a level of organization that supersedes that observed by viewing collections of individual genes (for review of function at the genome level see Fedoroff & Botstein (1992)). It is this level of structural and functional organization that will have to be penetrated in order to ultimately understand the transmission of genetic information, the regulation of gene expression and the role of structural elements in the genome in organizing these activities.

The technological advances in DNA sequencing that have recently occurred make possible the generation of machine-readable sequence from either single- or double-stranded templates using automated instruments (Smith *et al.*, 1986). The remaining steps in DNA sequencing have not kept pace with these developments. There is, in fact, an ongoing shifting of the rate-limiting steps of the process in response to the changing technology. Perhaps the major challenge in organizing a sequencing project is to take the maximal advantage of current state of the art, automated sequencing instruments. The organizational issues in dealing with such a project can be divided into three main areas. First, the presequencing phase, which includes various steps, such as breaking the cloned DNA to be sequenced into usable size fragments and preparing the templates; second, the actual sequencing; and third, the assembly of the sequence data into their final form and the initial sequence analysis.

The goal is to carry out these steps in an efficient, coordinated manner that will cause the overall process to proceed as quickly as possible. A current ABI 373 can analyze 36 samples per run, resulting in 350–450 bases of sequence per sample. Two runs per day are possible, enabling 70 samples (plus one control on each run) to be sequenced per day. Assuming 800 of these random sequences are determined for each cosmid (a conservative assumption), this leads to an estimate of about 750 000 bp of finished DNA sequence per year per machine as a theoretical maximum output. While this is possible, very few if any groups are sequencing at this rate. Clearly, other steps in the process are rate limiting, not the running of sequencing gels and the acquisition of raw data. This simple observation impacts initial choices of strategy and procedures as well as the modification of these for optimal output. These decisions must be made with respect to the ability to scale-up, and not only with respect to performance on a small scale. This chapter addresses the issues of strategy, scale-up, project management and appropriate application of technology at each of the major stages in a genome sequencing project based on the knowledge we have obtained in managing two previously published large-scale sequencing projects (Martin-Gallardo *et al.*,1992; McCombie *et al.*, 1992c).

24.2 STRATEGIES

There are three basic strategies for DNA sequencing. These can be divided into two ordered strategies (primer walking and unidirectional deletions (Henikoff, 1987)) and one random strategy, shotgun sequencing (Messing *et al.*, 1981). The ordered strategies have the advantage of being more efficient than a shotgun approach. However, they have significant drawbacks as strategies for large-scale projects. Primer walking, as it was carried out when we began our projects in early 1990, was not well suited to automated sequencers. Initial chemistry for carrying out sequencing using custom primers on ABI 370 DNA sequencers gave inadequate results in our experience. In addition, primer walking requires the synthesis of a large number of oligonucleotides, one for each sequencing reaction. This results in adding about $30 to the cost of each reaction. Moreover, the results of each reaction must be obtained and analyzed before the next primer can be chosen and synthesized. This requires that a large number of start points must be initiated to scale-up the process and this in turn can be a burden on project management. The combined requirements of choosing a primer, coupling it with the correct template and carrying out that reaction makes automation less straightforward than with a random approach where many templates are used, but each is used in exactly the same way.

Ordered deletions are also an efficient strategy for sequencing. However, the use of this approach in large projects has serious disadvantages since it requires substantial effort to subclone the cosmid DNA into pieces that can readily be subjected to exonuclease III (ExoIII) deletions. This requires extensive labor from individuals with a fairly high skill level. Moreover, it is not clear how such procedures could be highly automated. While very effective for small fragments, this is basically an approach that does not scale-up well for a large-scale sequencing project which requires parallel operations. As a result of these factors we chose to implement a shotgun sequencing strategy for our two large-scale sequencing projects (Martin-Gallardo *et al.*, 1992; McCombie *et al.*, 1992c).

24.3 SUBCLONE GENERATION

In order to generate templates for sequencing it is necessary to subclone cosmid DNA into smaller fragments that can be readily isolated in sufficient yield to carry out sequencing reactions. This has historically been done by fractionating large DNA molecules into small sized fragments of 400–600 bases (Messing *et al.*, 1981). It was suggested subsequently that the use of much larger fragments (1.5–5 kb) would result in a considerable increase in the flexibility of gap filling procedures (Edwards *et al.*, 1990). We wanted to test this improvement and chose to use rather large insert sizes when subcloning into M13 (Martin-Gallardo *et al.*, 1992; McCombie *et al.*, 1992c). Approximate sonication times are established using salmon sperm DNA as a control. DNA is sonicated in short pulses while the sample is kept on ice. Following sonication for the estimated time, an aliquot of the sample is analyzed on an agarose gel to determine the average size of the sonication products. Details of the sonication procedure appear in Chapter 5 in this volume. Sonicated DNA is end-repaired and then size-selected on an agarose gel. DNA fragments in the size range 1.5–3.0 kb are recovered from the gels by electrophoresis onto NA45 paper (Schleicher and Schuell). The size-selected DNA is ligated into SmaI cut, dephosphorylated M13 DNA. Ligation mixtures should contain 100 ng of vector and varying amounts of insert DNA. It is best to carry out several ligation reactions with varying insert/vector ratios.

A small portion of each ligation is then used for pilot transformations. Ligation and transformation efficiency are indicated by the number of white plaques on medium with X-gal and IPTG. Libraries that are acceptable based on this criteria also may be tested with a polymerase chain reaction (PCR) in the following manner. Random plaques are chosen and amplified using conditions and primers described previously (Wilson *et al.*, 1990), except that primers are used in equimolar amounts as opposed to the asymmetric amplification described in that study. Following amplification, an aliquot of each reaction is analyzed by gel electrophoresis. These primers amplify a region of about 500 bases from M13 vectors with no insert (Wilson *et al.*, 1990). Hence, clones with no insert show a visible band of this size on an agarose gel, while larger PCR products correspond to clones containing an insert. Clones with very large inserts will show no band on the gel since amplification cannot proceed through the insert. Libraries with different vector:insert ratios can then be rapidly evaluated and the best library chosen for subsequent use based on the percentage of insert containing clones and the average size of the insert.

24.4 TEMPLATE PREPARATION AND ORGANIZATION

A very important point in the scaling-up process is the orderly use and storage of templates for sequencing. We find it is most convenient to organize template production in sets of 96 to readily allow their storage in commercially available 96-well plates. We typically grow M13 for template preparation in 5 ml cultures as sets of 96 clones. Following growth, the cultures are centrifuged to remove bacteria. If the clones are being used for DNA isolation immediately they are centrifuged again and lysis carried out. Single-stranded phage DNA isolation is carried out according to a slightly modified version of a procedure described previously, including the optional $CHCl_3$ extraction (Sambrook *et al.*, 1989). This is usually done in sets of 24 or 48 clones. The supernatants of the remaining cultures that are stored at 4°C after the first centrifugation are kept until they are needed for template preparation. This can be up to several days after they are grown. With these samples the second centrifugation step is carried out immediately before proceeding with phage lysis. Prior to lysis of all clones, aliquots of each culture are removed and stored at −70° in 96-well plates.

24.5 SEQUENCING REACTIONS

Cycle sequencing reactions carried out with the fluorescent universal primer (ABI, Inc.) on the ABI 800 robotic workstation or manually with a thermal cycler (according to the protocol supplied by ABI) (McCombie *et al.*, 1992b) give robust reactions that work well with the ABI373 sequencer. Following the addition of ethanol, which is also done automatically by the ABI 800, the products of all four base reactions of each sample are pooled, incubated on ice for at least 20 min and then centrifuged. The pellets are washed with 70% ethanol and dried under vacuum for 5–10 min. Alternatively, if the reactions are not to be loaded on a gel immediately, they are stored at −20°C until needed, usually as the ethanolic suspension prior to precipitation. The samples are resuspended in formamide/EDTA as recommended by ABI, heated to 95°C for 2 min and then quenched on ice prior to loading on a sequencing gel.

24.6 GEL ANALYSIS OF SEQUENCING REACTIONS, DATA TRANSFER, STORAGE AND PREPARATION FOR ASSEMBLY

Sequencing gels are run and analyzed using ABI DNA sequences as described previously (Gocayne *et al.*, 1987). The sequence files are transferred via a Macintosh network to a Macintosh server with a large disk drive. Data are backed up from that system to a mass storage device for long-term storage. Preferably this is a well-indexed duplicate back-up system to allow recovery of any data that may be lost as a result of disk failures. Prior to entry into an assembly project, the sequences need to be edited to remove vector and poor quality sequence at the end of the run. This can be done with the program SeqEd (Applied Biosystems). After vector and poor quality sequence has been removed, the modified sequence is exported in a format that can be read by the assembler. When we began working on these projects we took this opportunity to edit the sequences themselves (Martin-Gallardo *et al.*, 1992; McCombie *et al.*, 1992c). This can be done by looking at the sequence and trying to manually call bases where the software is unable to do so and has inserted an 'N'. In retrospect, this is not a good thing to do at this stage of a project. It is quite time consuming to edit each sequence. It is also not particularly useful, since the manually called bases are often wrong. We currently recommend resolving these ambiguities at the time of assembly, not by editing individual sequences. At the assembly stage more data are available to resolve ambiguities since the great majority of sequences are covered at least once on each strand and typically at a much higher level of redundancy. Given that the final sequence needs to be determined by comparing all the available sequences at a discrepancy, it seems redundant to attempt to do this on individual sequence files.

24.7 ASSEMBLY

We carried out sequence assembly with the Gel program of the Intelligenetics package on a Sun Microsystems workstation (Martin-Gallardo *et al.*, 1992; McCombie *et al.*, 1992c). In addition, the assembly process was greatly aided by adjunct programs that interacted with the IG data files (Dubnick, 1992). There are two other assembly packages in common use for large-scale projects: the Staden Xdap package (Dear & Staden, 1991) and DNASTAR (DNASTAR, Inc., Madison, WI). Sequence assembly is done as follows. Sequences are entered into an assembly file and, once several hundred sequences are available, the assembly is begun. We found that we were unable to use the automated assembly feature of the Gel program with our data and had to use the manual assembly mode (Martin-Gallardo *et al.*, 1992; McCombie *et al.*, 1992c). The requirement for manual intervention in the assembly process has also been observed by Iris *et al.* (1993) using different software. What these projects have in common is that the regions sequenced show a high Alu density (about 1 per kb) with some degree of Alu clustering (Martin-Gallardo *et al.*, 1992; McCombie *et al.*, 1992c; Iris *et al.*, 1993), while some other regions of the human genome do not (Wilson *et al.*, 1992; Koop *et al.*, 1992). This is a serious problem in the assembly of shotgun sequencing projects that may be overlooked with the success of some of the current software on assembling sequences from prokaryotic or simple eukaryotic sources. As a result of problems with assembly of Alu-rich sequences the average read length of each sequence is critical. The more sequence that flanks the 300 base repeat, the easier it is to exclude the repeat from a match and align the flanking sequence. Based on this, we feel that average read length is one of the most important parameters to improve in order to better the overall sequencing rate. Procedures that do not generate sequences with read lengths of at least 350–450 bases will be extremely difficult to assemble if used on human DNA, and hence of little value.

In order to use the assembly software manually we recommend the following strategy. The assembly parameters should be set to only accept matches of greater than 95%. A sequence is compared with all the other sequences in the project and the possible matches are presented. The operator then must decide which matches should be accepted. In determining whether to accept a match between two sequences, a definate mismatch at a single base, such as an A to a G, should be taken more seriously than an A to an ambiguity. In addition, limitations of the sequence should be taken into account. Two examples are at the ends of sequences. Typically, ABI machines will add extra bases to strings of bases at the end of sequence runs. An example of this would be to call the sequence ATTC as ATTTC. A mismatch generated by this type of predictable error should not be considered a serious problem if the alignment is between the beginning of the sequence which showed two Ts in the above example and the end of the sequence which showed three Ts. Also, in general, if a match is very good at the beginning of a sequence and errors start appearing near the end of a sequence, it may be assumed that one of the sequences was not trimmed aggressively enough to remove poor-quality data at the 3′-end. In this case, one need be careful that the mismatch is not due to unique sequences flanking two disparate repeats. This is usually not a problem since matches in long repeats are typically not greater than 95% and in such cases the mismatch will be quite abrupt and severe when the unique sequence is reached rather than the slow deterioration in the quality of the matches, as is observed with inadequately trimmed sequences. In addition, matches which terminate in repeat sequences, such as Alus, should not be accepted at early stages in the assembly.

After one sequence has been compared with all the other sequences in the project and all those that match have been included in the new contig, the process is repeated with the newly formed contig. This process is repeated until the growing contig matches no more sequences and the process then starts with the next single sequence. This is continued until all possible sequences are admitted into contigs. This process results in the contigs containing ambiguities where the multiple sequences that comprise the contig disagree. The ambiguities will eventually interfere with subsequent matches. Owing to this limitation, the contigs should be edited at this point to remove ambiguities wherever possible. This is done by consulting the primary data from the various sequences that contained the conflicting residues. In some cases, there will not be enough data at a given base to allow a decision to be made with confidence and the base should be left as an ambiguity until more data are included in the contig. In many cases, however, it is possible to confidently resolve an ambiguity. As an example, several sequences representing both strands might agree at a given position but one sequence, near the end of its run, might have an extra N or repeat base (as mentioned above) inserted in the sequence. This type of error could propagate by shifting the sequence and causing a large number of ambiguities that could be readily and confidently resolved by a few editing changes. Once this editing has been done on all the newly formed contigs, the entire process of generating new contigs is repeated until no more matches are formed. Then the resulting contigs are examined. Based on their individual sizes, aggregate size, and percentage coverage on both strands, as well as the number of sequences currently in the assembly project, a decision is made whether to sequence more random fragments or to begin filling in the gaps in an ordered fashion. If the decision is to do more random fragments, this is done and the sequences are added to the contigs as described above. Once the decision has been made that the random sequencing is at the point of diminishing marginal return, the ordered, gap-filling phase of the project can begin.

24.8 GAP FILLING AND REDUNDANCY ESTABLISHMENT

The next step in the sequencing process is filling the gaps in the sequence and the establishment of sufficient redundancy on both strands. Custom primers can extend sequences at the ends of contigs by walking along an already existing template. Double-stranded M13 clones can be sequenced using the reverse primer to determine the sequence at the other end of the clone from that already in the contig. By choosing several clones at varying distances from the end of the contig, it is possible to fill gaps in this way. Sequence data from either of these types of reactions are immediately merged into the contigs. Both these approaches are useful for ordering contigs. As was suggested by Edwards *et al.* (1990), the use of longer inserts greatly adds to the flexibility of gap-filling strategies using reverse sequencing and primer walking. With the reverse sequences there will still be a gap in some cases between contigs, but the matches from opposite ends of the same clones will order the contigs. The custom primer-derived sequences may fill the gaps but will be of lower quality than is obtained from universal or reverse primers. Sequencing on both strands will sometimes allow the gap to be filled with a sequence of adequate quality. In many cases, however, the gap will need to be sequenced again once the order of the contigs or the rough sequence in the gap has been established. This can be done by amplifying the region in the gap using PCR and cloning the amplification product into plasmid or M13 vectors. Several clones from each amplification should be sequenced using the M13 forward primer, M13 reverse primer or both. These data are then merged into the assembly. This should fill all the remaining gaps and areas of insufficient redundancy. Recent experience with the new generation of ABI dye terminators, which use T7 polymerase (Hawkins *et al.*, 1992; W.R. McCombie, unpublished), indicate that significant improvements in custom primer-based sequencing will be achieved. Since gap filling is a difficult, time-consuming step, this could significantly improve overall sequencing rates.

24.9 SEQUENCE ANALYSIS

The methods for analysis of long sequences such as this are still in their infancy. This problem is greatly compounded in human sequences by the presence of introns and long intragenic regions. This negates the value of such simple analyses as open-reading frame analysis to look for genes or evaluate error rates. The magnitude of this problem is not generally appreciated. The large amounts of data that must be analyzed and the need for combining multiple comparisons require extensive computational analysis. These computer predictions are required to guide the design of experimental strategies to characterize a region, yet computer predictions are of little value without experimental confirmation. For an interesting discussion of the dynamics of computer versus experimental analysis, see the recent article by Karlin & Brendal (1993).

In analyzing large, human genomic sequences it is necessary to try several approaches. The first should be to search for possible similarities between the genomic sequence and those in the protein and DNA sequence databases. The recent explosion of partial cDNA sequence data from humans (Adams *et al.*, 1991, 1992; Wilcox *et al.*, 1991; Kahn *et al.*, 1992; Okubo *et al.*, 1992) and *Caenorhabditis elegans* (McCombie *et al.*, 1992a; Waterston *et al.*, 1992) makes these comparisons much more likely to yield an informative result. Owing to the presence of repeats such as Alus in human sequences, it is necessary to either splice them from the query sequence or divide the query sequence into ordered, overlapping fragments and search using each of these. In addition, it is useful to determine all the open-reading frames and use these for query sequences. In both cases it is important to search both protein and DNA databases. When we did this on the sequence from the chromosome 4p16.3 region, no convincing matches were seen (McCombie *et al.*, 1992c). A region from the chromosome 19 sequence, however, showed a convincing similarity to a rat protein phosphatase gene (Martin-Gallardo *et al.*, 1992; Tamura *et al.*, 1989).

The next step in the computer phase of sequence analyses should be compositional analysis to look for regions that are likely to code for amino acids. There are two currently available approaches to accomplish this with human sequence. The first is the program Testcode, designed by Fickett (1982). This program found a large number of potential exons in the chromosome 4 sequence (McCombie *et al.*, 1992c). More recently, the program CRM was developed, which uses a number of different factors, including a Fickett analysis to predict potential coding regions (Uberbacher & Mural, 1991). CRM produced fewer predicted exons when used to analyze the genomic sequences from regions on chromosome 19 and 4 (Martin-Gallardo *et al.*, 1992; McCombie *et al.*, 1992c). CRM is probably the best currently available method for predicting exons from human sequence. Next, information from such sources as exon predictions, homology searches or the presence of CpG islands (Bird, 1986) must be correlated and used to design experiments to search for genes.

The first step in searching for genes is to use interexon PCR to amplify predicted exons from cDNA libraries. One difficulty with this approach is the problems that derive from deciding which cDNA library to use. Any choice is a compromise and will limit the investigator to a subset of possible genes present. However, PCR amplifications allow the investigator to rapidly and easily screen a large number of libraries once the genomic sequence is known. After amplification of an exon has been achieved, it is possible to go to the next step, i.e. the amplification across exon boundaries. In many cases this will in fact be the first step in the analysis since the predicted exons are often too small to readily amplify and visualize individually. Intraexon PCR accomplishes two things. Firstly, it minimizes or eliminates the possibility that the amplification is from contaminating genomic DNA in the cDNA library since the intron makes the amplification product noncolinear with the genomic DNA. Secondly, the intraexon PCR ties two exons together as part of a single gene. It is crucial to remember at this stage that CRM attempts to identify individual exons, not entire genes (Uberbacher & Mural, 1991). Hence, predicted exons must not only be tested but anchored to the neighboring exons for the final gene structure to be elucidated. Direct sequencing of some of the PCR products at this stage of the analysis is important. We used seminested PCR and sequenced the products with an ABI 373 sequencer (Martin-Gallardo *et al.*, 1992; McCombie *et al.*, 1992c), although other approaches are available. In this approach, the first amplification is carried out and an aliquot of that reaction is amplified again, using one of the original primers and a new primer 3′ to the original opposing primer. This third primer has a sequence identical to the M13 universal sequencing primer at its 5′-end. This allows the product of the second amplification to be sequenced directly (after purification on a Centricon column) using fluorescent M13 primer and an ABI373 DNA sequencer. New technology using fluorescent dideoxy nucleotides on ABI373 DNA sequencers promises to make the direct sequencing of PCR products more rapid and eliminate the need to tail on sites for universal sequencing primers.

We were able to identify several new genes in the genomic sequences we analyzed with these approaches (Martin-Gallardo *et al.*, 1992; McCombie *et al.*, 1992c). Interestingly, in one case a gene was found based on a two-tier computer analysis and confirmed with PCR (Martin-Gallardo *et al.*, 1992). A region of the chromosome 19 sequence where CRM showed a cluster of exons was analyzed by gm, which attempts to predict gene structure (Fields & Soderlund, 1990; Soderlund *et al.*, 1992; Martin-Gallardo *et al.*, 1992). Using these two programs, a gene was predicted that had a very high degree of similarity to the mouse *fosB* gene (Zerial *et al.*, 1989; Martin-Gallardo *et al.*, 1992). In the case of the gene designated HDA1-1, found in the sequence from chromosome 4, we were able to determine the entire gene structure based on the genomic sequence, CRM predictions, PCR reactions and use of the amplified products as probes to screen cDNA libraries (McCombie *et al.*, 1992c). The results of these experiments are encouraging for the future of genome sequence analysis. These studies demonstrate that information concerning the gene content of human genomic sequence can be determined based on an iterative process of computer predictions, PCR and sequencing. They also showed us the great distance that this approach is beginning to lag behind genome sequencing and the difficulties that the scale-up of human genome sequencing are placing on our ability to understand the meaning of these sequences.

24.10 FUTURE DIRECTIONS

There are a number of issues still challenging those trying to decipher the structure and function of complex genomes. Great strides have been made in automating the various individual steps in DNA sequencing. Two fundamental problems have arisen as a result of this work, however. First, these technical advances typically arise in isolation from related steps in the sequencing process. This means that in some cases they simply serve to shift the rate-limiting step in sequencing. Often this results in little or no net increase in the overall speed of the process. As an example, it is probably possible, using a variety of techniques, to generate short sequence fragments (200–300 bases) more rapidly than longer fragments (350–500 bases) can be generated. While this is useful in model organisms it will probably make no significant contribution to sequencing the human genome due to the problem of Alus as discussed above. What is gained in sequencing is more than lost in assembly. The need is for logistical or process control to combine currently available techniques into the most efficient overall process. In some cases the individual steps employed may not be the most rapid available but the steps used fit best with the other procedures in the process. In addition, more efficient interactive software tools are needed to help manage the resources of the project, track the data and prevent bottlenecks.

The second major problem is that the automation has been moving in the direction of automating very random sequencing strategies. While these approaches are the easiest to automate, they are also less efficient than the more ordered approaches. The challenge is to begin to develop more ordered, efficient approaches that can be automated to the same extent. The recent advances in using strings of short oligonucleotides to prime sequencing reactions (Kieleczawa *et al.*, 1992) may make primer walking viable in large-scale projects. This will accelerate the rate of DNA sequencing and make possible the complete determination of the genome of important model organisms in a cost-efficient manner. This information will in turn allow the sequencing of the human genome to be completed and the information therein to be used to its full advantage.

Great challenges remain with the analysis as well. Current methods for sequence analysis are for the most part based in the analysis of small regions where there is an exception of gene content available. These methods are inadequate for analyzing large regions, especially of totally unknown content. CRM (Uberbacher & Mural, 1991) is a valuable contribution to the analysis of human sequence as is the database expansion provided by human cDNA tagging projects (Adams *et al.*, 1991, 1992; Wilcox *et al.*, 1991, 1992; Kahn *et al.*, 1992; Okubo *et al.*, 1992). In addition, genome and cDNA tagging from model organisms such as the nematode (Sulston *et al.*, 1992; McCombie *et al.*, 1992a; Waterston *et al.*, 1992) will provide not only database expansion but in many cases a rapid way to determine the normal function of the newly discovered genes (see Chapter 11 in this volume). There is a great need to develop software to integrate the results of these various analyses into a common graphical interface that can aid in understanding the available data. This will greatly facilitate the use of available software aiding in the design of experiments to analyze the newly determined sequence.

REFERENCES

Adams, M.D., Kelley, J.M., Gocayne, J.D., Dubnick, M., Polymeropoulos, M.H., Xiao, H., Merril, C.R., Wu, A., Olde, B., Moreno, R.F., Kerlavage, A.R., McCombie, W.R. & Venter, J.C. (1991) *Science* **252**, 1651–1656.

Adams, M.D., Dubnick, M., Kerlavage, A.R., Moreno, R., Kelley, J.M., Utterback, T.R., Nagle, J.W., Fields, C. & Venter, J.C. (1992) *Nature* **355**, 632–634.

Bird, A.P. (1986) *Nature* **321**, 209–213.

Cawthon, R.M., Andersen, L.B., Buchberg, A.M., Xu, G.F., O'Connell, P., Viskochil, D., Weiss, R.B., Wallace, M.R., Marchuk, D.A., Culver, M., Stevens, J., Jenkins, N.A., Copeland, N.G., Collins, F.S. & White, R. (1991) *Genomics* **9**, 446–460.

Dear, S. & Staden, R. (1991) *Nucleic Acids Res.* **19**, 3907–3911.

Dubnick, M. (1992) *Comput. Appl. Biosci.* **8**, 291–292.

Edwards, A., Voss, H., Rice, P., Civitello, A., Stegemann, J., Schwager, C., Zimmermann, J., Erfle, H., Caskey, C.T. & Ansorge, W. (1990) *Genomics* **6**, 593–608.

Fedoroff, N. & Botstein, D. (1992) *The Dynamic Genome*. Cold Spring Harbor Laboratory Press, Cold Spring Harbor, NY.

Fickett, J.W. (1982) *Nucleic Acids Res.* **10**, 5303–5318.

Fields, C. & Soderlund, C. (1990) *Comput. Appl. Biosci.* **6**, 261–270.

Genetics Computer Group (1991) Program Manual for the GCG Package, Version 7, April 1991, 575 Science Drive, Madison, Wisconsin, USA 53711.

Gocayne, J., Robinson, D.A., FitzGerald, M.G., Chung, F.Z., Kerlavage, A.R., Lentes, K.U., Lai, J., Wang, C.D., Fraser, C.M. & Venter, J.C. (1987) *Proc. Natl. Acad. Sci. U.S.A.* **84**, 8296–8300.

Hawkins, T.L., Du, Z., Halloran, N.D. & Wilson, R.K. (1992) *Electrophoresis* **13**, 552–559.

Henikoff, S. (1987) *Methods Enzymol.* **155**, 156–165.

Iris, F.J.M., Bougueleret, L., Prieur, S., Caterina, D., Primas, G., Perrot, V. Jurka, J., Rodriguez-Tome, P., Claverie, J.M., Dausset, J. & Cohen, D. (1993) *Nature Genet.* **3**, 137–145.

Karlin, S. & Brendel, V. (1993) *Science* **259**, 677–680.

Kahn, A.S., Wilcox, A.S., Polymeropoulos, M.H., Hopkins, J.A., Stevens, T.J., Robinson, M., Orpana, A.K. & Sikela, J.M. (1992) *Nature Genet.* **2**, 180–185.

Kieleczawa, J., Dunn, J.J. & Studier, F.W. (1992) *Science* **258**, 1787–1791.

Koop, B.F., Wilson, R.K., Wang, K., Vernooij, B., Zallwer, D., Kuo, C.L., Seto, D., Toda, M. & Hood, L. (1992) *Genomics* **13**, 1209–1230.

Levinson, B., Kenwrick, S., Lakich, D., Hammonds, Jr., G. & Gitschier, J. (1990) *Genomics* 7, 1–11.

Martin-Gallardo, A., McCombie, W.R., Gocayne, J.D., FitzGerald, M.G., Wallace, S., Lee, B., Lamerdin, J., Trapp, S., Kelley, J.M., Liu, L.-I., Dubnick, M., Dow, L.A., Kerlavage, A.R., de Jong, P., Carrano, A.V., Fields, C. & Venter, J.C. (1992) *Nature Genet.* **1**, 34–39.

Maxam, A.M. & Gilbert, W.A. (1977) *Proc. Natl. Acad. Sci. U.S.A.* **74**, 560–564.

McCombie, W.R., Adams, M., Kelley, J.M., FitzGerald, M.G., Utterback, T., Kahn, M., Dubnick, M., Kerlavage, A., Venter, J.C. & Fields, C. (1992a) *Nature Genet.* **1**, 124–131.

McCombie, W.R., Heiner, C., Kelley, J.M., Fitzgerald, M.G. & Gocayne, J.D. (1992b) *DNA Sequencing* **2**, 289–296.

McCombie, W.R., Martin-Gallardo, A., Gocayne, J.D., FitzGerald, M., Dubnick, M., Kelley, J.M., Castilla, L., Liu, L.-I., Wallace, S., Trapp, S., Tagle, D., Whaley, W.L., Cheng, S., Gusella, J., Frischauf, A.-M., Poustka, A., Lehrach, H., Collins, F.S., Kerlavage, A.R., Fields, C. & Venter, J.C. (1992c) *Nature Genet.* **1**, 348–353.

Messing, J., Crea, R. & Seeburg, P.H. (1981) *Nucleic Acids Res.* **9**, 309–321.

O'Connell, P., Viskochil, D., Buchberg, A.M., Fountain, J., Cawthon, R.M., Culver, M., Stevens, J., Rich, D.C., Ledbetter, D.H., Wallace, M., Carey, J.C., Jenkins, N.A., Copeland, N.G., Collins, F.S. & White, R. (1990) *Genomics* **7**, 547–554.

Okubo, K., Hori, N., Matoba, R., Niiyama, T., Fukushima, A., Kojima, Y. & Matsubara, K. (1992) *Nature Genet.* **2**, 173–179.

Sambrook, J., Fritsch, E.F. & Maniatis, T. (1989) *Molecular Cloning: A Laboratory Manual*, 2nd edn. Cold Spring Harbor Laboratory Press, Cold Spring Harbor, NY.

Sanger, F., Air, G.M., Barrell, B.G., Brown, N.L., Coulson, A.R., Fiddes, C.A., Hutchison, C.A., Slocombe, P.M. & Smith, M. (1977a) *Nature* **265**, 687–695.

Sanger, F., Nicklen, S. & Coulson, A.R. (1977b) *Proc. Natl. Acad. Sci. U.S.A.* **74**, 5463–5467.

Smith, L.M., Sanders, J.Z., Kaiser, R.J., Hughes, P., Dodd, C., Connell, C.R., Heiner, C., Kent, S.B. & Hood, L.E. (1986) *Nature* **321**, 674–679.

Soderlund, C., Shanmugam, P., White, O. & Fields, C. (1992) *Proceedings of the 25th Hawaii International Conference on System Sciences (Hawaii International Conference on System Sciences 653–622)*, V. Milutinovic & B. Shriver (eds) IEEE Computer Society Press. Los Alamitos, CA, pp. 653–662.

Sulston, J., Du, Z., Thomas, K., Wilson, R., Hillier, L., Staden, R., Halloran, N., Green, P., Thierry-Mieg, J., Qiu, L., Dear, S., Coulson, A., Craxton, M., Durbin, R., Berks, M., Metzstein, M., Hawkins, T., Ainscough, R. & Waterston, R. (1992) *Nature* **356**, 37–41.

Tamura, S., Lynch, K.R., Larner, J., Fox, J., Yasui, A., Kikuchi, K., Suzuki, Y. & Tsuiki, S. (1989) *Proc. Natl. Acad. Sci. U.S.A.* **86**, 1796–1800.

Uberbacher, E.C. & Mural, R.J. (1991) *Proc. Natl. Acad. Sci. U.S.A.* **88**, 11261–11265.

Waterston, R., Martin, C., Craxton, M., Huynh, C., Coulson, A., Hillier, L., Durbin, R., Green, P., Shownkeen, R., Halloran, N., Metzstein, M., Hawkins, T., Wilson, R., Berks, M., Du, Z., Thomas, K., Thierry-Mieg, J. & Sulston, J. (1992) *Nature Genet.* **1**, 114–123.

Wilcox, A.S., Khan, A.S., Hopkins, J.A. & Sikela, J.M. (1991) *Nucleic Acids Res.* **19**, 1837–1843.

Wilson, R.K., Chen, C. & Hood, L. (1990) *BioTechniques* **8**, 184–189.

Wilson, R.K., Koop, B.F., Chen, C., Halloran, N., Sciammis, R. & Hood, L. (1992) *Genomics* **13**, 1198–1208.

Zerial, M., Toschi, L., Ryseck, R.P., Schuermann, M., Muller, R. & Bravo, R. (1989) *EMBO J.* **8**, 805–813.

CHAPTER TWENTY-FIVE

Zen and the Art of Large-scale Genomic Sequencing

L. ROWEN[1] & B. F. KOOP[2]

[1] Department of Molecular Biotechnology, University of Washington, Seattle, WA, USA

[2] Department of Biology, Center for Environmental Health, University of Victoria, Victoria, British Columbia, Canada

25.1 INTRODUCTION

Large scale genomic DNA sequencing requires commitment, courage, and curiosity. Implementing a sequencing operation that consistently maintains high throughput is what the work is partly about. It is also about incorporating technical improvements as they come along. Most importantly, however, it is about molecular archeology, the discovery of secrets within the sequence itself.

Genomic sequencing, that is, the sequence determination of whole genomes, chromosomes, or genetic loci, is a process that begins with mapping and subcloning the region of interest and ends with analyses of the sequence data produced. Out of these analyses, in conjunction with cDNA sequence information, come a delineation of gene organization and, occasionally, the discovery of new genes. Comparative genomic sequencing from different species may yield conserved sequence blocks that will produce insights into gene regulation or alternative functions of DNA sequence related to chromosomal structure and maintainance. We anticipate that long contiguous stretches of DNA sequence from homologous regions of closely and distantly related species will shed new light on the mechanisms of molecular evolution that have made human DNA the sequence it is today.

As a model system for large scale genomic sequencing we have chosen the human and mouse T-cell receptor loci. These multigene families are organized into three loci: alpha/delta (> 1 megabase); beta (600 kb) and gamma (150 kb). Each locus contains multiple copies of three or four types of gene segments ('variable,' 'diversity,' 'joining,' and 'constant') that contribute to the diversity of the T-cell receptor repertoire required for effective immune responses.

While project leaders in Dr. Leroy Hood's laboratory, we have overseen the sequencing of over a megabase of the T-cell receptor loci and have thus accumulated a body of experience regarding the joys and pitfalls of mammalian DNA. Our greatest challenge has been the presence of long homologous repeats that have rendered sequence assembly difficult. As a result of dealing with these repeats, we have concluded that high redundancy shotgun sequencing strategies are the most effective way to handle mammalian genomic DNA. Additionally, we have found cosmids to be a useful and manageable unit of sequencing. In what follows, we will discuss the rationale and strategy for shotgun sequencing in more depth.

25.2 SEQUENCING A COSMID

A typical cosmid contains between 33 and 45 kb of insert DNA and between 5 and 8 kb of vector DNA. Cosmids are easy to come by. Their sequence, however, is not. Two things need to be accomplished by a sequencing strategy, namely, the generation of a contig (contiguous stretch of DNA) that equals the cosmid insert in size, and the generation of a sequence that has a high likelihood of being correct at each nucleotide position.

Two tactical approaches to sequencing a cosmid present themselves: divide-and-conquer or slow-and-steady advance. These logically distinct tactics go by the names of 'shotgun' and 'directed' sequencing. In the shotgun approach, a cosmid is broken down randomly (by a nonspecific shearing method such as sonication or DNase I treatment) into small (around 1 kb) DNA fragments which are individually subcloned into an M13 or plasmid vector, sequenced with universal primer, and assembled en masse into longer contigs, the end point being a contig the size of the original cosmid insert. In the directed approach, the cosmid itself (or a limited set of subclones) is walked across by using primers made of pre-existing sequence as stepping stones for generating new sequence.

Each of these tactics has advantages and disadvantages. Acknowledging this, most large-scale sequencing groups use some combination of the two to accomplish the job. Groups differ mainly in the *proportion* of the cosmid sequencing done using each approach. Intuitively, it makes sense to use shotgun sequencing first, to generate a handful of contigs, and then to use directed sequencing to walk across the gaps, thus completing the cosmid. We believe that this is an inefficient strategy, and that it is better to aim at completing the cosmid via a shotgun approach right from the start.

25.3 MORE ABOUT SHOTGUN SEQUENCING

On the surface, shotgun sequencing a cosmid to completion appears to be a waste of resources in comparison with directed sequencing because at least four times the number of individual sequence reads are required. The reason is that no method of fragmenting a cosmid gives completely random results. To determine a consensus sequence for an entire cosmid in the first round of sequencing, we have found it necessary to obtain enough sequence reads to cover each nucleotide position an *average* of seven times. Of the 25 cosmids we have sequenced to an average redundancy of seven or higher, only five could not be assembled into a single contig from the initial shotgun batch of sequences. (As discussed below, gaps can usually be filled by sequencing the other end of PCR products generated from selected M13 clones.) Under normal conditions, it takes about 1000 reads to obtain a consensus sequence for a cosmid using the shotgun strategy. (The number goes down to 800 if the cosmid vector sequences are removed.)

The shotgun approach seems more cumbersome than directed sequencing because it requires the upfront steps of cosmid fragmentation, subcloning, and DNA template preparation, whereas walking strategies require only the generation of custom primers. To make matters worse, with shotgun strategies, a sophisticated assembly program for performing the contig merge is essential, whereas with directed sequencing, alignment of sequences is a straightforward task because new sequence data are always being added to sequences already in hand.

Given these drawbacks, why should a production-oriented group do primarily shotgun sequencing? Two answers: speed and precision. Shotgun sequencing is fast because gap-filling is largely avoided, editing is facilitated by the high redundancy of the sequence data, and sequencing procedures are becoming more automated and easier to do. The results are precise because of the high redundancy of sequencing.

25.4 GAP-FILLING

If a cosmid is sequenced to less than a seven-fold average redundancy using a shotgun approach, the individual sequence reads will assemble into a handful of nonoverlapping contigs. To join these into one, additional sequencing must be done to fill the gaps. If the gaps are small, custom primers can be made from the end sequences of each contig. These can be used in conjunction with dye-terminators to elongate the sequence read of M13 clones that extend into the gap. (The M13 clone contains an insert about 1 kb in size, only half of which is used to give readable sequence data in the shotgun phase.) Alternatively, a PCR product can be made from selected M13 clones by using primers that lie outside the M13 universal and reverse primer sites. This PCR product can then be sequenced with reverse primer to generate new data. These two strategies are most effective for gaps less than 200 bases. Bigger holes generally require that a new PCR product be made from the cosmid, subcloned, and sequenced, or that a portion of the cosmid itself be subcloned and sequenced. Alternatively, gaps can be filled by walking across the cosmid itself.

Gap-filling is slow for two reasons. One is that time is required to make the primers and do the subcloning and sequencing. Although these procedures are straightforward in theory, we have often encountered setbacks in their application. The other is that time may be required to resolve ambiguities. With a low redundancy sequencing strategy, such as dye-terminators in conjunction with a custom primer, one or more nucleotide positions may remain unresolved. This is so because of the variable signal strength obtained in cycle sequencing, or the difficulties Sequenase has with compressions. Higher redundancy strategies, such as sequencing several subclones of a PCR product, can generate sequence reads that are inconsistent with each other. If problems such as these arise, alternative sequencing strategies must be used to resolve ambiguities.

Thus there is a tradeoff between time spent doing the few hundred more shotgun sequences it takes to bring the redundancy of coverage up to an average of seven- or eight-fold, and time spent filling gaps. Shotgun sequencing has the advantage, though, of being more automatable than the process of gap-filling. When template preparations and sequencing reactions are going well, four fluorescent sequencers can produce about 100–120 reads a day. Shotgunning a cosmid to completion thus enables us to get a consensus sequence in weeks. Cosmids that are almost but not quite finished due to gaps or regions of ambiguity have a tendency to drag on for months because of the labor-intensive nature of the repair sequencing that has to be done. Thus, we try at the outset to prevent these problems from occurring.

25.5 ASSEMBLING SEQUENCE DATA

The enormous power of shotgun sequencing is revealed at the stage where we resolve ambiguities in the sequence data for the sake of obtaining a consensus sequence we have confidence in. To explain this point, a few words about the sequence assembly and editing strategy we employ are needed. With the shotgun approach, a 'contig merge' has to be performed on the hundreds of individual sequence reads that comprise the data for the cosmid. The merge can be done on batches of sequences as they are produced or alternatively one can wait until all the data are in before doing the assembly. Likewise, with editing, one can modify each chromatogram separately before feeding it into the assembly program. Modifications include a removal of the vector cloning site at the 5′-end of the read and the inaccurate sequence at the 3′-end. Alternatively, all of the end-clipping can be done in a batch step prior to assembly, and the base-calling problems can be addressed after the assembly has occurred.

We believe batch editing and assembly to be more efficient. With the programs we have, the end-clip editing of a thousand or so sequences takes about 3 min, and the contig merge takes about 18–24 h, depending on the complexity of the project. It can then take anywhere from 2 days to 3 weeks of computer work to resolve the resolvable ambiguities in the consensus sequence produced by a contig merge (more about this below). The remaining difficulties must be resolved by resequencing problematic regions. Once the ambiguities in the consensus sequence are resolved, the cosmid is finished. At our current rate of productivity, we are sequencing, assembling, editing, and finishing about 3 cosmids in 6 weeks.

We have learned that a trial assembly of the first 200 sequences obtained for a cosmid is helpful for three reasons. One is to ensure that the distribution of sequences is random, indicating that the fragmentation and subcloning of the cosmid worked well. The second is to get an indication as to whether repeats are going to complicate the contig merge. The third is to check that the cosmid being sequenced is the one intended. Regarding the first reason, we have learned the hard way that cosmid subcloning can sometimes go awry, and that it is better to discover this fact sooner rather than later. As for the second, several of the cosmids with DNA inserts from the human T-cell receptor beta locus have contained long (>10 kb) repeats of greater than 92% sequence similarity. In an assembly that allows a merge at a lower percentage match, these repeat sequences align into the same contig. Their presence is indicated by interspersed conflicts at single base positions in an otherwise clear consensus sequence. If we find these, we know to do the final assembly at a much higher match stringency. Finally, it is worth checking the sequence reads for known genes or known cosmid end sequences, as it is always possible for tubes of DNA to be unknowingly mixed up or mislabeled. Since a trial assembly can be done in about 2 h, it yields a high payoff for relatively little effort.

When we do the final assembly for a cosmid, we typically direct the program to add new sequence reads to a pre-existing contig if they match the consensus sequence by 85%. If the trial assembly indicated the presence of highly homologous repeats, we increase the stringency of match to 94%.

We sometimes solve problems with an assembly by redoing it at a different match stringency. If the match parameter is set at too low a percentage, there is a danger of merging sequences that do not belong together. On the other hand, if the match percentage is overly high, fewer long contigs are generated from the sequence data. Where the match stringency is best set is a function of the overall quality of the sequence data and the homology of repeats, if there are any.

The desired result of a contig merge is, of course, one contig containing all of the input sequences. In practice, this never happens. Inevitably, there are orphan sequences that do not align anywhere. These can be *Escherichia coli* sequences, artifacts of the subcloning strategy, or sequences from a different cosmid mistakenly put into the wrong subdirectory. They comprise less than 5% of our sequence reads. Along with these orphans are found a few large contigs that do not merge because the sequence reads at the ends fall below the percentage match criterion. So long as we have high redundancy, we can usually put together one contig from the cosmid-related sequences by an appropriate force joining of overlapping contigs, once the overlaps are discovered by examining the end sequences. This process is straightforward except in those cases where long homologous repeats occur. In the case of repeats, where we set the match stringency to 94%, we end up with scores of tiny contigs around 1–2 kb in size. Extreme care must then be taken not to join contigs in the wrong order. High redundancy among the sequence reads assists the sorting out process significantly, as it allows us to distinguish ambiguities due to sequencing errors from those that occur because the sequences are from different parts of the cosmid.

25.6 EDITING SEQUENCE DATA

To review, the high redundancy provided by the shotgun sequencing approach is useful for three reasons: (1) to assemble a cosmid's worth of sequence data into a single contig; (2) to help sort out highly homologous repeats; and (3) to provide sufficient data for the generation of a precise consensus sequence. The last benefit comes into play during the editing of an assembly project. By 'editing', we mean the resolution of conflicts and ambiguities in the sequence reads by the appropriate correction of base-calling mistakes.

When individual sequence reads are aligned, systematic difficulties with a particular sequencing method show up as a repetition of the same mistake in each read. Redundancy also helps to identify errors due to imprecise base-calling towards the end of a read (where G and C insertions and A and T deletions are fairly common). In general, redundancy gives hints as to when sequence data are apt to be mistaken. Discrepancies can usually be resolved by examining the original data or by selective resequencing using an alternative method that does not produce the same kind of systematic error. Once the editing is as complete as possible, we do the resequencing required to solve the remaining difficulties.

In essence, editing turns an initial consensus sequence determined by majority rule into a consensus sequence that is more accurate because serious efforts have been made to fix errors. In the interest of determining how much of a difference editing makes, we looked at the occurrence of miscalls, ambiguities, insertions, and deletions in the initial consensus sequence as compared with the final consensus sequence for two cosmids, one sequenced to a redundancy of 7, the other to 9.5. In one case the difference was 1.2% (428 discrepancies out of 34 619 nucleotides) and in the other case it was 1.0% (385 discrepancies out of 38 156 nucleotides). These numbers suggest that editing indeed makes a significant difference. This conclusion is supported further by a comparison of the final consensus sequence of three pairs of partially overlapping cosmids. The average redundancy for each cosmid was about eight-fold. In one case, there were 12 discrepancies over 20.5 kb. (Most of these occurred in a region where the redundancy of one of the cosmids was only two- to three-fold.) In another case, there was one discrepancy over 4 kb, and in the third there were no discrepancies over 9.5 kb of overlap. These results suggest an error rate of not more than one in 1500, with the average being significantly better than that.

It is, of course, debatable whether genomic DNA sequencing needs to give results that are this precise. Some argue that most of the DNA is junk and, consequently, that it is not worth the effort to solve problems with a consensus sequence unless there is good reason to believe that a particular region is interesting or important. Because we are interested in molecular evolution, and because we anticipate that DNA sequence information serves numerous functions not currently understood, we would like to have results that are as precise as is reasonably possible given prevailing sequencing technologies.

An alternative approach to consensus determination, one that strikes a middle ground between simple majority rule and careful editing, is favored by some computation experts. This approach involves developing a program that would calculate the confidence with which each base in a consensus sequence is called. Factors to be considered in the calculation would include redundancy (e.g. whether both strands are covered) and the likelihood of correctness of each base in the individual sequence read (functions of the systematic errors of the sequencing method and the position of the base in relation to the length of the overal read). Such an approach might ultimately replace the human judgment that editing now involves.

In summary, we believe that the gains in precision, along with the gains in speed, argue strongly in favor of a high redundancy shotgun approach to sequencing cosmids. We doubt that directed strategies can do as well on both fronts simultaneously. It may be that a directed strategy that employs ligated or unligated hexamers as primers rather than custom-made primers will effectively compete with the shotgun approach, once the technique is optimized and automated. Until this technique or some other viable alternative for doing directed sequencing rapidly is developed, though, we would encourage groups sequencing long DNA contigs from overlapping cosmids to set up a shotgun operation. Once the instrumentation and technical staff required are in place, shotgun sequencing operations can proceed quite smoothly.

The following sections discuss the procedures we use to sequence a cosmid to completion. As is typical for molecular biology research, the protocols we employ have been borrowed or adapted from other sources. There is nothing particularly original about what we do. Our strength lies more in the overall organization of our operation than it does in our specific procedures. For detailed protocols and general instructions, we recommend manuals such as those edited by Sambrook *et al.* (1989) and Ausuber *et al.* (1991). The instructions given below detail our specific adaptations of commonly available protocols.

More specifically, we briefly describe our procedures for cosmid growth, DNA fragmentation, subcloning of insert DNA into M13 vector, selection screens, DNA template preparation, sequencing, and finishing strategies.

25.7 COSMID GROWTH, FRAGMENTATION, AND SUBCLONING

To prepare cosmid DNA, we follow roughly the procedure detailed in Sambrook *et al.* (1989, pp. 1.38–1.39). Cells containing the cosmid are grown in 150 ml of L broth with the appropriate antibiotic (ampicillin or tetracycline) for 16–20 h at 37°C. They are spun and resuspended in 2.4 ml of solution I plus lysozyme. After 5 min at room temperature, 4.8 ml of solution II is added and the suspension is gently mixed by inversion. After 5 min, 4.0 ml of solution III is added, the suspension is mixed by shaking and, after 15 min is spun for 10 min. To the supernatant is added 7 ml (0.6 volume) of isopropanol. The solution is mixed, incubated at 15 min at room temperature, and spun. After removing all the supernatant, the pellet is resuspended in 500 μl Tris-EDTA (pH 7.5 or pH 8) containing 10 μg ml^{-1} RNase A and incubated at 37°C for 30 min. Phenol (350 μl) equilibrated in TE and 350 μl of chloroform/isoamyl alcohol (24:1) is added, the solution vortexed and spun, and 600 μl of the top phase transferred to an Eppendorf tube. Sodium chloride (125 μl,5 M) and polyethylene glycol (750 μl,13%) are added and, after mixing, the solution is stored at least an hour at 0°C. The material is spun for 15 min in a microfuge, washed twice with 200 μl 70% ethanol, dried, and resuspended in 150 μl water or TE.

This procedure yields roughly 100 μg to 1 mg of clean DNA, which is sufficient for subcloning purposes. Cosmids prepared by cesium chloride banding will also work, but the banding is not necessary if the above procedure is followed. We have had little difficulty with *E. coli* DNA contamination.

To prepare randomly generated 1 kb insert DNA fragments from a cosmid, we sonicate 10–15 μg of DNA in a final volume of 50 μl water. We use a Heat Systems-Ultrasonics Inc. cup horn sonicator with the following settings: output control, 4.5; % duty cycle, 100; pulse, continuous. Generally, 20–40 s of sonication is sufficient. We suggest that a time-course be done on each cosmid to ensure that an effective batch of insert fragments is obtained. After sonication, the fragment ends are repaired at 37°C for 30 min after adding 12 μl of water, 7 μl of 10× T4 polymerase buffer (500 mM Tris (pH 8.8), 150 mM ammonium sulfate, 65 mM magnesium chloride, 1 mM EDTA, 500 μg ml^{-1} BSA, 100 μM 2-mercaptoethanol), 1 μl 10 mM dNTPs, and 1.5 units of T4 DNA polymerase (Boehringer Mannheim Biochemicals). Glycerol (10 μl) is added to the repaired DNA and it, along with a 1 kb ladder, is loaded onto a 1.5% agarose gel (Tris–acetate–EDTA system). The gel is run at 100 V (10 V cm^{-1}) for 45 min. The portion of sonicated insert running between roughly 800 and 1500 bases is cut out of the gel and electrophoresed onto DEAE paper (Whatman DE 81 2.5 cm filters). The DNA is eluted from the DEAE filters with Tris-EDTA plus 1 M sodium chloride, ethanol precipitated, washed with 70% ethanol, and resuspended in 25 μl of Tris-EDTA, pH 7.5.

To subclone the insert DNA into M13 vector, we titrate the sonicated insert (generally, between 0.05 and 0.5 μl insert; around 5–50 ng) in a 20 μl ligation reaction containing 10–50 ng M13 vector, 2 μl 10× ligase buffer (500 mM Tric-Cl (pH 7.6), 100 mM magnesium chloride, 10 mM dithiothreitol, 1 mM ATP), and 1 unit of T4 DNA ligase (Boehringer Mannheim Biochemicals). The reaction is incubated overnight at room temperature. (For vector, we use M13 mp9 RF DNA (Boehringer Mannheim) that has been cut with *Hinc*II, and treated with calf intestinal alkaline phosphatase.) Ligation mixture (0.7 μl) is transformed into 85 μl of frozen competent DH5aF′ or DH5aF′IQ cells (BRL, Life Technologies, Inc.) in accordance with the manufacturer's instructions. We generally obtain 50–150 clear plaques per plate, at the optimum concentration of insert. Thus, one ligation tube usually suffices to generate enough clones to sequence a cosmid.

We have at times encountered two problems with this procedure, one being that the yield of plaques is extremely low, and the other being that a percentage of clear plaques contain no insert but are instead deletions of M13 (false positives). Unfortunately, we have not discovered the cause of either of these difficulties, nor have we devised truly effective strategies to overcome them. Resonication of the cosmid often solves the low yield problem, and fresh preparation of reagents for the repair and ligation steps often solves the false-positive problem. It may be that sonication is overly damaging to DNA and that fragmentation procedures being devised in other laboratories (for example, use of a nebulizer or partial restriction enzyme digests) will prove to be more routinely reproducible and effective than sonication.

25.8 SCREENS FOR CLONE SELECTION

We often subject M13 clones to two types of screen in order to enrich our percentage of clones yielding useful sequence data. The first type involves separating clones that contain foreign DNA insert from those that do not. The second involves identifying clones that contain cosmid vector. In order to perform either of these screens, we prepare a small volume of phage culture by adding a plaque to 200 μl of DH5aF'IQ cells (see Section 25.9) and growing the cells overnight in racks of 96 1-ml minitubes covered with parafilm.

To identify M13 clones containing insert, we use a sizing procedure, either of a PCR product or of total DNA. To generate PCR products from M13 clones we first release the coat protein from the DNA by adding 10 μl of phage culture to 100 μl 0.5% Tween 20, and heating at 95°C for 15 min. This material (5 μl) is added to 45 μl of a PCR reaction premix containing water, 5 μl 10× Taq buffer (100 mM Tris-Cl (pH 8.3), 500 mM potassium chloride, 15 mM magnesium chloride, 0.1% gelatin), 0.5 μl 10 mM dNTPs, 10 pmol of each primer, and 0.25 units of Taq polymerase (Amplitaq, Perkin–Elmer), and the PCR is performed at 95°C for 1 min, 50°C for 1 min, and 72°C for 1 min for a total of 25 cycles. The products are sized on a 1.5% agarose gel. The procedure above, and the choice of primers (PCR-1: GTTTTCCCAGTCACGACGTTG. PCR-2: GAATTGTGAGCGGATAACAAT), are adapted from the protocol developed by Applied Biosystems. These primers generate PCR products slightly larger than the size of the inserted DNA fragment.

A cheaper and faster but significantly less precise method of distinguishing M13 clones with insert from those without involves simply adding 50 μl of phenol to 50 μl of phage culture, vortexing, spinning, and running 10 μl of the top phase on a 0.7% agarose gel along with an M13 control as size standard. Phage DNA containing insert runs slightly slower than the M13 control. (Several bands will be seen on the gel, stemming from cellular background, but these do not interfere with the screen.)

To identify M13 clones containing cosmid vector, we use a hybridization screen. Dot–blot filters are prepared from 10 μl of the phage culture described above, and probe is prepared from 50 ng of cosmid vector DNA (PWE15A), using standard labeling procedures. Hybridization is performed at 37°C overnight, and excess radioactivity is removed from the filters by successive washings for 15 min at 65°C with: 2× SSC, 0.01% SDS; 1× SSC, 0.01% SDS (two times); 0.5× SSC, 0.01% SDS (two times); and 0.25× SSC, 0.01% SDS. Alternatively, a non-radioisotopic screen can be used for selecting or eliminating positives.

25.9 DNA TEMPLATE PREPARATION

Because we require at least 5 μg of purified template, we prepare DNA from 10 ml phage cultures, in batches of 84. To 1 l of 2× YT broth, we add 1 ml of frozen DH5aF'IQ cells and 1 ml of 10 mg ml^{-1} kanomycin. (Frozen cells are prepared by adding a scraping from the DH5aF'IQ lawn cells in a frozen competent cell kit (see above) to 500 ml of L broth or 2× YT broth containing 10 μg ml^{-1} kanamycin and then growing overnight at 37°C. Cells (1 ml) are added to 70 μl DMSO in Eppendorf tubes and stored at −70°C.) For phage growth, cells are distributed into 50 ml polypropylene centrifuge tubes (Corning), and a phage plaque or 4 μl of phage culture (see above) is added, for an overnight (12–16 h) incubation at 37°C. The cells are spun for 15 min at 4000 r.p.m. in a Beckman J6 centrifuge. After transferring the supernatant to 15 ml polypropylene centrifuge tubes (Corning) 2 ml of 20% polyethylene glycol (mol. wt 8000)/2.5 M sodium chloride are added. The tubes are capped, mixed by inversion, and incubated for at least 30 min at room temperature. The material is spun for 30 min at 3500 r.p.m. in the J6 centrifuge. The supernatant is poured off, and the remaining supernatant, after draining to the bottom of the tube, is removed by aspiration. It is important to remove as much PEG as possible. The phage pellets are resuspended in 250 μl of Tris-EDTA (pH 8) and transferred to Eppendorf tubes. Phenol (250 μl) equilibrated in Tris-EDTA is added. After vortexing and spinning, 200–220 μl of the top phase is transferred to a new Eppendorf tube. Chloroform/isoamyl alcohol

(24:1; 220 μl) is added, followed by vortexing and spinning. The top phase (140–170 ml) is transferred to a new Eppendorf tube. The DNA is precipitated with 0.1 volume of 3 M sodium acetate (pH 5.2) and 2 volumes of ethanol. After a 70% ethanol wash and vacuum drying, the DNA pellet is resuspended in 50–80 μl Tris-EDTA (pH 8). The DNA concentration is determined by diluting 2 μl of sample to 1 ml of water and measuring the absorbance at 260 nm. (At this dilution, OD 0.007 equals roughly 100 ng μl^{-1} DNA concentration.)

25.10 SEQUENCING

In order to improve the precision of our consensus sequence, we sequence a portion (10–30%) of our clones with Sequenase and the remainder with Taq polymerase (cycle sequencing). We perform cycle sequencing reactions either in the Catalyst sequencing robot (Applied Biosystems, Inc.) or on a 96-well thermocycler (Perkin–Elmer Gene Amp 9600) using the cycle sequencing kits and protocols developed by Applied Biosystems. The optimal DNA concentration for reactions run in the 96-well thermocycler is about four- to five-fold lower than it is for reactions run in the Catalyst.

For Sequenase reactions the following stocks are needed: 5× sequencing buffer (1 M Tris-Cl (pH 7.4), 1 M sodium chloride, 0.1 M dithiothreitol); Sequenase dilution buffer (2 M Tris-Cl (pH 7.5), 10 mM 2-mercaptoethanol, 1 mg ml^{-1} BSA); 1 M manganese chloride; dNTP mix (2 mM each of dATP, dCTP, dGTP, 3 mM dGTP); 50 mM ddATP; 50 mM ddCTP; 100 mM ddGTP; 50 mM ddTTP; 0.4 pmol μl^{-1} fluorescent dye-primers (Applied Biosystems); and Sequenase Version 1 (United States Biochemical Corporation). Four triphosphate brews, one for each dideoxynucleoside, are prepared by combining the dNTP mix and a dideoxynucleoside stock in a 12:1 ratio. We do 24 sets of sequencing reactions at a time in round-bottom 96-well plates (Costar) as follows. Divide the microtiter plate into three sets of four columns, designated A, C, G, and T. Add 3 μl of manganese chloride stock to 197 μl 5× sequencing buffer. Distribute 6 μl to each of the A wells. Add 2.5 to 3 μg template DNA and water to bring the total volume in the A wells to 25 μl. Add 1 μl of C primer to the C wells, 2 μl G primer to the G wells, and 2 μl T primer to the T wells. Distribute 4 μl of the template mix to the C wells, and 8 μl to each of the G and T wells. Add 1 μl A primer to the remaining template in the A wells. Cover the microtiter plate with parafilm and a tinfoil lid, and float on a 55°C water bath for 5 min. Let cool at room temperature for 15 min (annealing step). Prepare a premix of Sequenase by adding 28 μl of enzyme (13 U μl^{-1}) to 270 μl Sequenase dilution buffer. Prepare four premixes from the enzyme and triphosphates by adding 48 μl of the diluted enzyme to 60 μl of the dNTP/ddATP brew and 60 μl of the dNTP/ddCTP brew, and 96 μl of diluted enzyme to 120 μl of the dNTP/ddGTP brew and 120 μl of the dNTP/ddTTP brew. After annealing, add 3.5 μl of the appropriate enzyme/triphosphate premix to the A and C wells, and 7.0 μl of the enzyme/triphosphate premix to the G and T wells. Cover with parafilm and a lid, and float on a 37°C bath for 7 min. Terminate the reactions by adding 100 μl of sodium acetate/ethanol (150 μl 3 M sodium acetate (pH 5.2), 4.8 ml absolute ethanol) to the A wells. Using a multichannel pipette, pool the A, C, G, and T reactions horizontally. Transfer to Eppendorf tubes. Cool for at least 15 min, spin for 15 min in a microfuge, wash pellet with 200 μl 70% ethanol, dry, and resuspend in 4 μl gel loading buffer, as described in the Applied Biosystems 373A Sequencer manual.

For all sequencing protocols, the gel conditions recommended by Applied Biosystems for the 373A automated sequencer are followed.

25.11 FINISHING STRATEGIES

Our procedures for assembling and editing sequence data have been described earlier. Generally, some additional sequencing is required to complete a cosmid, either because gaps need to be filled or ambiguities need to be resolved. We fill gaps in two ways. Our first-choice strategy involves the generation of PCR products from M13 clones containing inserts that are likely to fill the gap. The PCR reaction is performed as described above for screening phage cultures, except that 2 μl of a 1:100-fold dilution of the purified template DNA is used in a 100 μl reaction. The PCR product is precipitated with 5 μl of 5 M ammonium acetate (pH 7.4) and 200 μl ethanol. After spinning, washing with 70% ethanol, and drying, the product is resuspended in 25 μl of water, and sequenced with reverse primer following the cycle sequencing protocol developed by Applied Biosystems. For sequencing reactions that work poorly, use of more or less DNA (depending on the signal strength and the quality of the read) frequently gives better results. (When too much DNA has been used in the sequencing reaction, most of the signal will be displayed at the beginning of the read.) If a gap cannot be filled in this manner, an alternative strategy involves generating a PCR product from the cosmid using custom primers, subcloning this product into M13, and sequencing M13 clones to the desirable level of redundancy.

As for ambiguities, when a nucleotide position in

the consensus sequence of a cosmid has been determined on the basis of both cycle sequencing and Sequenase reads, we resolve conflicts by accepting the results of the strategy that produces the clearest reads. (Cycle sequencing reads often drop bases that are clearly indicated in the Sequenase read. Sequenase reads often condense GC regions that are clearly separated on the cycle sequencing reads.) Thus, the easiest way to resolve ambiguities in a given read is to try a different sequencing protocol. If this approach fails, use of nucleoside analogs in the sequencing reaction, such as 7-deaza ITP in place of dGTP, will usually resolve the problem.

25.12 CONCLUSION

Techniques for large-scale genomic sequencing are changing fairly rapidly in the direction of automation and higher throughput, especially those for template preparation and sequencing reactions. We anticipate, therefore, that many of the specific protocols described above will be replaced over the next 2 years. The overall strategies described in this chapter are likely to persist, however, at least in laboratories performing the shotgun sequencing of cosmids. We encourage the reader not to be intimidated by large-scale sequencing, but rather to see sequencing as an opportunity for refined biological analysis and unexpected discoveries.

REFERENCES

Ausuber, F., Brent, R., Kingston, R., Moore, D., Seidman, J.G. & Struhl, K. (eds) (1991) *Current Protocols in Molecular Biology*. Greene/Wiley, New York.

Sambrook, J., Fritsch, E.F. & Maniatis, T. (eds) (1989) *Molecular Cloning: A Laboratory Manual*, 2nd edn. Cold Spring Harbor Laboratory Press, Cold Spring Harbor, NY.

CHAPTER TWENTY-SIX

Automated Dye-Terminator DNA Sequencing

J.M. KELLEY

The Institute for Genomic Research, 932 Clopper Road, Gaithersburg, MD 20878, USA

26.1 INTRODUCTION

The chemistry of the dye-terminator sequencing reactions differs from that of the dye-primer sequencing reactions in several important ways. The dideoxy sequencing reactions are primed with a custom oligo rather than a universal sequencing primer, the fluorescent dye associated with the products is attached to dideoxynucleotides instead of the primer in the reaction mix, and excess dyes must be removed from the reaction mix before separation and analysis on a sequencer. The reaction mixes contain the template DNA, a custom-picked oligo primer, deoxynucleotides, dye-labeled dideoxynucleotides, a polymerase enzyme, and a buffer system. The reaction is a linear or asymmetric form of the polymerase chain reaction (PCR) utilizing cycle sequencing with the custom primer binding to the template and extending until the reaction is terminated by the incorporation of a dideoxynucleotide. Each nucleotide has a specific fluorescent dye associated with its particular dideoxynucleotide; therefore, when the reaction is terminated by the incorporation of the dye-labeled dideoxynucleotide, the product is labeled with the specific dye corresponding to that specific nucleotide. The excess dye-labeled dideoxynucleotides are removed from the reaction mix, usually by either a precipitation or a spin column purification.

Dye-terminator chemistry can be applied in several situations. In large-scale genomic sequencing projects, dye-terminator chemistry has been used for primer walking. Dye-terminator chemistry has also been used to close the gaps which occur using a shotgun sequencing approach (McCombie *et al.*, 1992a; Martin-Gallardo *et al.*, 1992; Liu *et al.*, 1992; Sulston *et al.*, 1992). By picking primers on both sides of the gap to extend in each direction, often the sequence across the gap can be obtained along with an overlap on each side to confirm the alignments. Confirmation of other difficult areas can be accomplished using custom primers to position the difficult area directly into the best sequence data, usually the first 100–300 bases. Methods have also been developed which use the dye-terminator chemistry to sequence PCR fragments directly (Tracy & Mulcahy, 1991).

26.2 CRITICAL FACTORS FOR DYE-TERMINATOR REACTIONS

In our laboratory, protocols have been developed for using DyeDeoxy terminator reactions with Taq DNA polymerase (ABI) with both single- and double-stranded templates on the ABI Catalyst 800 LabStation. Critical factors in obtaining successful sequencing ladders include the purity of the plasmid template, the primer selection, the parameters of the cycle sequencing temperature profile, and the choice of spin column for purification of the reaction products.

26.2.1 Template purity

We have used M13 preparations (McCombie *et al.*, 1992a; Martin-Gallardo *et al.*, 1992) and plasmid preparations (Kelley *et al.*, 1992; Liu *et al.*, 1992; McCombie *et al.*, 1992b) to give templates of appropriate purity for dye-terminator reactions. The M13 preparations work well but can only extend in one direction. Double-stranded plasmid preparations work well and can extend in both directions with the same template. We have successfully used the plasmid preparation protocol QiaWell-Plus.

26.2.2 Primer selection

We have found several critical factors affecting primer choice, including: location and match in the target sequence, length, concentration, purity, %GC and %GA, and melting temperature (Barnes, 1987; Applied Biosystems, Inc., 1991; Bellis *et al.*, 1992; Kelley *et al.*, 1992).

26.2.3 Location and match

Because the first 30 or so nucleotides of a terminator sequence is less accurate than the next several hundred bases, the location of the primer on the target sequence should be 50 to 100 bases from the area of interest to assure proper alignment and identity with the desired area. There should be no more than three or four of any one base in a row, and the primer should not be conducive to primer–primer or to primer–vector binding. The most critical bases to check are the last 5–10 bases on the 3'-end. These should not have a direct match with other regions of the template or vector. (These problems are more easily recognized with computer programs for primer selection.) Primers can be chosen manually by visually inspecting the sequence or by using available software for primer picking (Rychlik & Rhoads, 1989; Lowe *et al.*, 1990; Hillier & Green, 1991; Applied Biosystems, Inc., 1991). We have used a software package Oligo Selection Program (OSP) (Hillier & Green, 1991) and have also visually selected primers with equal success. OSP has many added advantages and makes it easy to evaluate primer factors relating to annealing temperatures, self-annealing probabilities, primer–vector binding, and specificity.

26.2.4 Length

Primer length is best when it is at least 18–22 nucleotides or greater. Increasing the length will increase specificity (prevents priming at a secondary site) and decrease chances for nonexact hybridization. Increasing the length of the primer is also a method to increase the melting or annealing temperature (T_m) for the reaction (Barnes, 1987; Applied Biosystems, Inc., 1991; Bellis *et al.*, 1992).

26.2.5 Concentration

The concentration of the primer DNA is very important to the stoichiometry of the polymerase reaction. Too much primer will result in very high signal at the beginning of the sequence followed by a rapid decline in the signal, resulting in relatively short usable sequences; too little primer will result in weak signals and high background, also giving short usable sequences. Optimal results in our laboratory are obtained by using a working stock concentration of 3.2 pmol and a final concentration in the reaction mixes of 0.32 pmol.

We have found that the oligos made with the ABI 394 DNA Synthesizer with polystyrene 40–nmol columns do not need to be purified after deblocking and drying under nitrogen (Applied Biosystems, Inc., 1991). If there are problems with nonspecific amplification, oligo purity should be checked.

26.2.6 Purine content, GC content, and T_m

The total content of the purines (G+A) affects the secondary structure and binding properties of the primer. The recommended limit for G+A is about 50%. For the most consistent results, GC content between 40 and 80% is recommended with a GC content of about 50% the best. The melting (annealing) temperature of an oligonucleotide is directly related to its GC content. A T_m greater than 45°C gives better results with the ABI DyeTerminator chemistry. To increase the T_m above 45°C, increase the GC content and/or extend the primer length (Barnes, 1987; Hillier & Green, 1991; Applied Biosystems, Inc., 1991; Bellis *et al.*, 1992).

26.2.7 Temperature profiles

The temperatures used for the denaturing, annealing, and extension steps can be adjusted according to specific primer–template conditions. Using Taq polymerase, a denaturing temperature of 95–96°C can be used with little reduction in enzyme activity. This is especially important when using double-stranded templates. A longer incubation at a slightly higher temperature increases the strand availability for primer binding at the annealing step. A rapid ramp to a lower annealing temperature increases specificity of the reaction. A longer extension time is needed compared with dye-primer sequencing reactions to allow for the incorporation of the dye-labeled dideoxynucleotides.

26.2.8 Removal of excess dye

There are several reliable methods for removal of the excess, unincorporated dyes from the reaction mixes before separation and analysis. These include: ethanol precipitation, phenol/chloroform extraction, and spin column purification. After having tested a number of different methods and different spin columns, we routinely use a modified procedure using Centricep 1.0 ml spin columns (Princeton Separations). A protocol is described below.

26.2.9 Modified procedure for spin column purification of extension products with Centrisep spin columns

To prevent excessive drying of the columns, process only 6–8 samples at a time.

(1) Gently tap the column to cause the gel material to settle to the bottom.
(2) Remove the column stopper and add 0.75 ml of sterile deionized water (dH_2O) to rehydrate.
(3) Stopper the column and invert several times to mix. Allow the gel to hydrate for at least 20 min at room temperature. (Hydrated columns can be stored for a few days at 4°C. Allow columns that have been stored at 4°C to warm to room temperature before use.) When ready to use, remove any air bubbles by inverting or tapping the column and allowing the gel to settle. Remove the upper-end cap *first*, and then remove the lower-end cap. Allow the column to drain by gravity.
(4) Insert the column into the wash tube provided.
(5) Spin in a variable-speed microcentrifuge at 750**g** for 2–5 min to remove the interstitial fluid. (IEC with 8.35 cm rotor, 2800 r.p.m.). Mark the tube to indicate which side is oriented on the outside.
(6) Remove the column from the wash tube and insert it into a sample collection tube.
(7) Carefully remove all of the reaction mixture and load it on the top of the gel material. Carefully drop onto the centre of column, avoiding the edges.
(8) Spin in a variable-speed microcentrifuge at 750**g** for 2–5 min. (IEC with 8.35 cm rotor, 2800 r.p.m.). If using a centrifuge with a fixed angle rotor, place the column in the same orientation that it was in for the first spin—this is important because the surface of the gel will be at an angle in the column after the spin. Spin at the same speed for the same amount of time as for the wash step.
(9) Dry the sample in a vacuum centrifuge for 20–25 min; it may require more time. Do not apply heat. Do not overdry.
(10) Store the dried sample in the dark at −20°C

26.3 METHODS FOR AUTOMATED DYE-TERMINATOR REACTIONS

Sequencing reactions using dye-terminator chemistry can be carried out using DNA thermal cyclers such as the Perkin-Elmer 480 and 9600 or using a robotic LabStation such as the Catalyst 800 from Applied Biosystems, Inc., according to the manufacturers' protocols. Briefly, the following protocols have been used successfully in our laboratory.

26.3.1 Using Perkin-Elmer PE9600

(1) For each sample, add 5 μl sterile dH2O (deionized water), 4 μl DNA template (200–250 ng μl^{-1} for double-stranded or 100 ng μl^{-1} for single-stranded), and 1.5 μl primer DNA (from a 3.2 pmol stock solution) to a thin-walled 0.2-ml reaction tube.
(2) Add 9.5 μl ABI PRISM Ready Reaction Dye-Deoxy Terminator premix (1.58 μM A-DyeDeoxy, 94.74 μM T-DyeDeoxy, 0.42 μM G-DyeDeoxy, 47.37 μM C-DyeDeoxy, 78.95 μM dITP, 15.79 μM dATP, 15.79 μM dCTP, 15.79 μM dTTP, 168.42 mM Tris-HCl (pH 9.0), 4.21 mM $(NH_4)_2SO_4$, 42.10 mM $MgCl_2$, 0.42 U μl^{-1} AmpliTaq DNA polymerase).
(3) Carefully seal the tubes using the tube sealer tool and quick spin to concentrate the reaction mix in the bottom of the tube.
(4) Place the tubes in the thermal cycler that has been preheated to 96°C.
(5) Immediately begin the cycle sequencing program which should be as follows.
    (a) Rapid thermal ramp to 96°C.
    (b) 96°C for 30 s.
    (c) Rapid thermal ramp to 50°C.
    (d) 50°C for 15 s.
    (e) Rapid thermal ramp to 60°C.
    (f) 60°C for 4 min.
    (g) For a total of 25 cycles.
    (h) Rapid thermal ramp to 4°C and hold.
(6) Samples are light sensitive and should be held at 4°C in the dark until further processing.
(7) The excess dyes are removed by spin column purification using CentriCep columns. (See protocol 26.2.9).
(8) After removal of the excess dyes, the samples are dried and stored in the dark at −20°C until loading onto a 373 DNA Sequencer.

26.3.2 Using the ABI Catalyst 800

(1) For each sample, mix a template 'cocktail' by adding 7 μl sterile dH20, 7.5 μl DNA template (200–250 ng μl^{-1} for double-stranded or 100 ng μl^{-1} for single-stranded), and 1.5 μl primer DNA (from a 3.2 pmol stock solution) into a 1.5 ml Sarstedt reaction tube.
(2) Add the proper amount (depending on the number of reactions per run; see User manual for ABI Catalyst 800) of ABI PRISM Ready Reaction DyeDeoxy Terminator mix to a separate tube.
(3) Aliquot the proper amount of sodium acetate 3 M (pH 5.0) to a separate tube.
(4) After a quick spin to ensure that all the liquid is at the bottom of the tube, carefully place the tubes in the proper positions on the Catalyst work surface (according to the work surface diagram for the desired terminator protocol).
(5) Check all fluids and buffers on the Catalyst.
(6) Open the Catalyst software to the DyeTerminator Sequencing program and select the number of templates to be run. Start the program and the software automatically selects the proper temperature profile for the dye-terminator chemistry.
    (a) Rapid thermal ramp to 96°C.
    (b) 96°C for 15 s.
    (c) Rapid thermal ramp to 50°C.
    (d) 50°C for 10 s.
    (e) Rapid thermal ramp to 60°C.
    (f) 60°C for 4 min.
    (g) For a total of 25 cycles.
    (h) Rapid thermal ramp to 4°C and hold.
(7) Samples are light sensitive and should be held at 4°C in the dark until further processing.
(8) The excess dyes are removed by spin column purification using CentriCep columns. (See Section 26.3.3.)
(9) After removal of the excess dyes, the samples are dried and stored in the dark at −20°C until loading onto a 373 DNA Sequencer.

26.3.3 Dye-terminators on the ABI 373 DNA Sequencer

Samples are run on the 373A DNA Sequencer using the same protocol as for dye primers with the following exceptions:

(1) select AnyPrimer (DyeTerminator) for the primer under settings during instrument setup; and
(2) add only 4 μl formamide-EDTA loading buffer (deionized formamide/50 mM EDTA, 5:1 (v/v)).

Following the run, analyze the gel for accurate data. It is a normal pattern for dye terminators to exhibit a weak signal in the first 10–20 bases, followed by a clear strong peak pattern. Because there is no large peak from excess primer for the computer analysis software to recognize and begin the base-calling algorithm, care must be taken to check lane tracking and base-call start with the gel image and raw data for every sample. If these do not agree, reanalyzing with more appropriate Base Call Start data may improve the quality of the data.

Gaps may occur when there is no termination at a nucleotide point (no dye-labeled dideoxynucleotide was incorporated). Secondary structure of the template may cause the polymerase enzyme to 'fall off' before a dideoxynucleotide was incorporated. Some alternatives for this and other difficult areas are given below.

26.4 SUGGESTIONS FOR 'DIFFICULT' AREAS TO SEQUENCE

In genomic and cDNA sequencing there are areas of the template DNA which make it difficult to obtain clean accurate sequencing data. These are usually identified as GC-rich segments, poly(A) or poly(T) stretches, or Alu and other repeat areas. Secondary structure can cause the polymerase to 'fall off' before a dideoxynucleotide is incorporated or can prevent primer binding to the template in this area. Some suggestions for improving the sequencing data from these difficult areas are given below. Several recent publications have addressed this area. (Applied Biosystems, Inc., 1991; Khan *et al.*, 1991; Friedlander *et al.*, 1992; Hawkins *et al.*, 1992; Lee *et al.*, 1992; Bellis *et al.*, 1992).

26.4.1 Resequence with another buffer system

Adjustments in the buffer regarding Mg^{2+} and/or Mn^{2+} concentration have been shown to help. Increasing the Mg^{2+} from 2 mM up to 8.0 mM has resulted in improved sequencing data. For T7 Sequenase, the Mn^{2+} concentration can also be adjusted to give optimal results. If the storage buffer for either the primers or the template DNA contains EDTA, the reaction buffer may need more Mg^{2+} or Mn^{2+} to compensate for the EDTA chelating properties. Some workers have found that adding DMSO to a final concentration of 5–8% will increase the read length of GC-rich areas (Friedlander *et al.*, 1992; Burgett & Rosteck, 1992).

26.4.2 Changes with the enzyme concentration

Changes with the enzyme concentration may improve results: increase the concentration of Taq up to 10–fold (Friedlander *et al.*, 1992).

26.4.3 Change the polymerase enzyme

Taq could be used instead of T7 Sequenase, or vice versa. Taq polymerase seems to sequence better through GC-rich areas (especially with extra Mg^{2+} and/or higher concentration of Taq) while T7 Sequenase yields a more even signal which is helpful with poly A/T stretches and other repeat areas (Friedlander *et al.*, 1992; Hawkins *et al.*, 1992). The T7 Sequenase system can also support the introduction of a new set of fluorescent dyes which improve the accuracy of the sequencing data (Hawkins *et al.*, 1992; Lee *et al.*, 1992).

26.4.4 Try different cycle temperature profiles

Increasing the denaturing temperature and/or time for double-stranded templates can improve results. Some adjustment to the annealing and extension profiles may also help (Barnes, 1987; Applied Biosystems, Inc., 1991; Friedlander *et al.*, 1992).

26.4.5 Check purity of the oligo and of the DNA template

Primer and DNA template purity is critical for the proper binding and extension of the template DNA. Although our experience has found that the oligos made with the ABI 394 DNA Synthesizer do not need to be purified, if there are problems with nonspecific amplification, oligo purity should be checked. The purity of the DNA template is critical, especially when using plasmid preparations and the ABI Catalyst 800 for the sequencing reactions.

26.5 ACCURACY STUDIES FOR AUTOMATED DYE-TERMINATOR SEQUENCING REACTIONS

There has been discussion in the literature regarding the accuracy of sequences from dye-terminator reactions. There is a problem with single-pass dye-terminator sequencing if gaps from nontermination occur. However, these gaps are not significant if dye-terminator reactions are used for gap filling, confirmation, and primer walking. We have conducted some accuracy studies for sequencing reactions using the ABI Catalyst. Accuracy for sequencing double-stranded templates with the Catalyst was determined for dye-primer and dye-terminator reactions by comparing pGEM3Zf+ sequences with the pGEM3Zf+ sequence in GenBank. The percentage accuracy for the dye-primer reaction was calculated and found to be 99% or greater from base 100–450 (with the

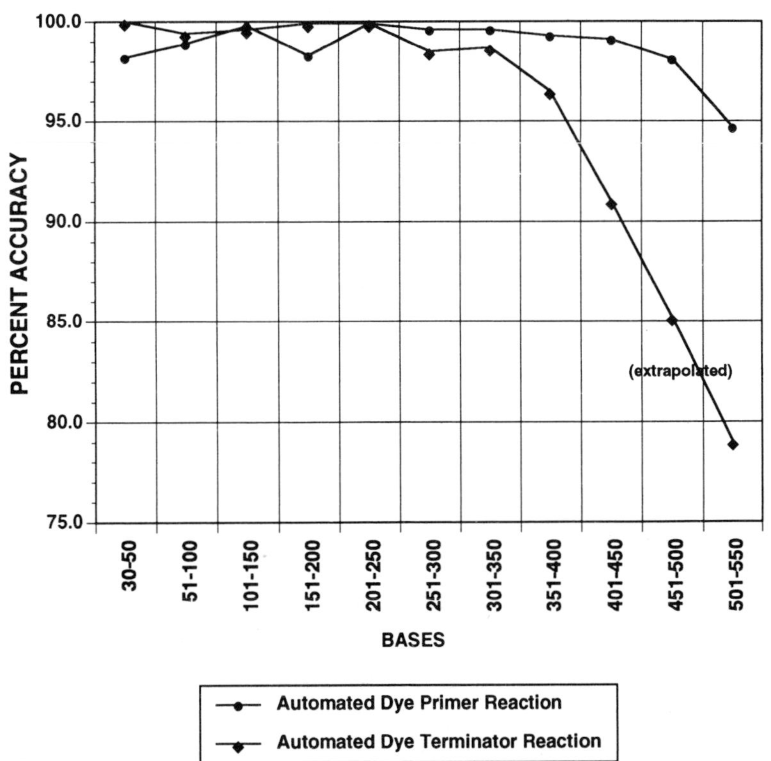

Figure 26.1 Accuracy of automated dye-primer and dye-terminator reactions using the ABI Catalyst.

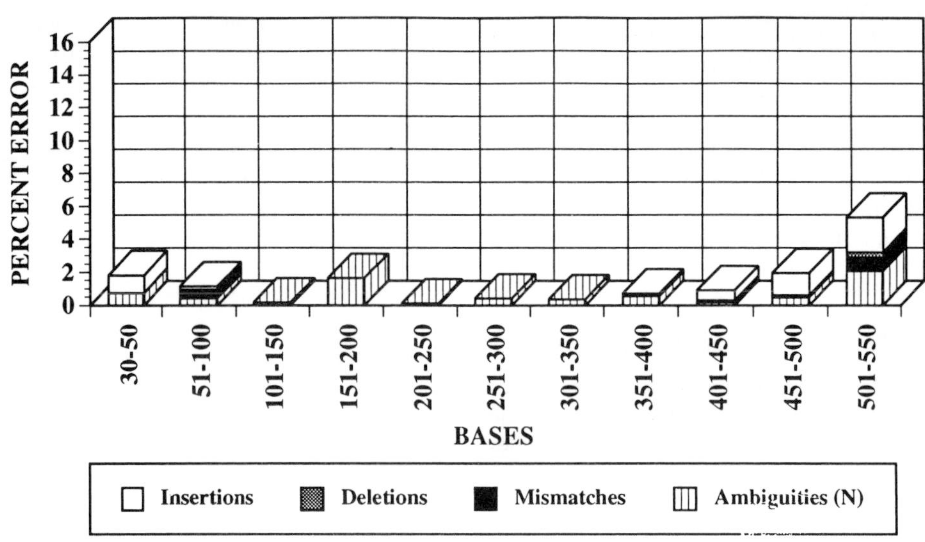

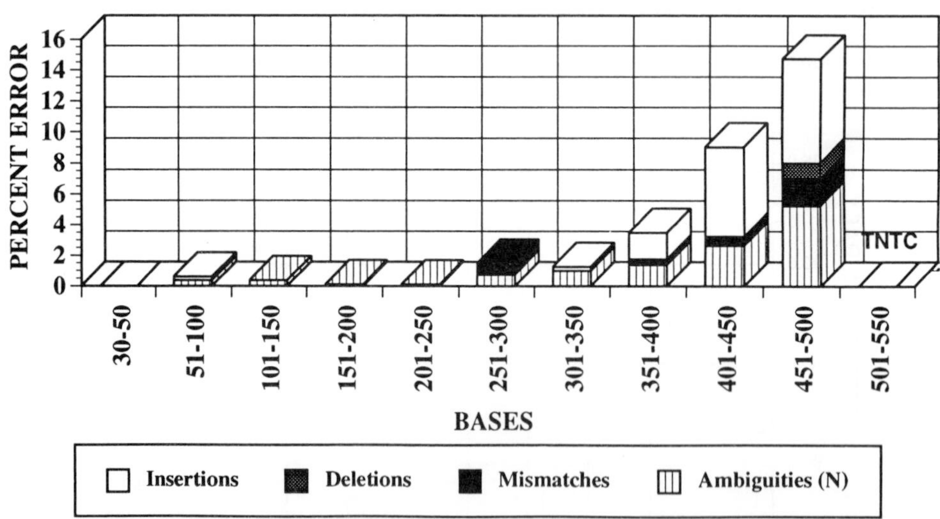

Figure 26.2 Accuracy for sequencing double-stranded templates with the Catalyst was determined for dye-primer (a) and dye-terminator (b) reactions by comparing pGEM3Zf+ sequences with the pGEM3Zf+ sequence in GenBank. The number of errors and ambiguities (Ns) per 50 bases was determined.

exception of one area between bases 151 and 200 where 23/33 dye-primer sequences called an N at the exact same base) and to be about 98% out to base 500. Accuracy rapidly fell off between bases 500 and 550. The percentage accuracy for the dye-terminator reactions was found to be greater than 98% up to about 350 and then fell to about 85% by base 500 (Fig. 26.1). Interestingly, there is no problem at the same base where the dye-primer (m13–21) has the N. We have seen similar results with routine double-stranded templates. Most of the errors were ambiguities or insertions, as with dye-primer sequencing reactions. (Fig. 26.2).

26.6 CONCLUSION

DNA sequencing using dye-terminator chemistry can be carried out by currently available automation. Adjustments in the Taq concentration, plus optimal primer selection and template purity, have improved data obtained with the ABI Catalyst so that sequencing runs of 350–400 bases with 97% or more accuracy are routine with plasmid templates. Although not quite as robust or accurate as dye-primer chemistry, these data are critical for gap-filling and sequence confirmation for difficult areas in large-scale sequencing efforts. Because of the occasional problem with gaps resulting from nontermination, data from dye terminator runs are best used not as single-pass sequences, but in conjunction with multiple sequences for confirmation and gap-filling.

REFERENCES

Applied Biosystems, Inc. (1991) *ABI User Bulletin 19* November.

Barnes, W.M. (1987) *Methods Enzymol.* **152**, 538–556.

Bellis, G.D., Manoni, M., Pergolizzi, R., Redolfi, M.E. & Luzzana, M. (1992) *BioTechniques* **13**, 892–897.

Burgett, S.G. & Rosteck, Jr. P.R. (1992) In *Genome Sequencing and Analysis Conference*, New York, Mary Ann Liebert, IV. p. 30.

Friedlander, E.J., Galvin, M. & Spurgeon, S. (1992) In *Genome Sequencing and Analysis Conference*, New York, Mary Ann Liebert, IV. p. 30.

Hawkins, T., Du, Z., Halloran, N.D. & Wilson, R.K. (1992) *Electrophoresis* **13**, 552–559.

Hillier, L. & Green, P., (1991) *PCR Methods Appl.* **1**, 124–128.

Kelley, J.M., Adams, M.D. & Venter, J.C. (1992) In *Genome Sequencing and Analysis Conference*, New York, Mary Ann Liebert, IV. p. 37.

Khan, A.S., Wilcox, A.S., Hopkins, J.A. & Sikela, J.M. (1991) *Nucleic Acids Res.* **19**, 1715.

Lee, L.G., Connell, C.R., Woo, S.L., Cheng, R.D., McArdle, B.F., Fuller, C.W., Halloran, N.D. & Wilson, R.K., (1992) *Nucleic Acids Res.* **20**, 2471–2483.

Liu, L.-I., Selivanov, N., Massung, R., Utterback, T., Qi, J., Knight, J., Kerlavage, A., Dubnick, M., Esposito, J., Mahy, B.W.J. & Venter, J.C. (1992). In *Genome Sequencing and Analysis Conference*, New York, Mary Ann Liebert, IV. p. 43.

Lowe, T., Sharefkin, Shi Qi Yang, J. & Diffenbach, C.W. (1990) *Nucleic Acids Res.* **18**, 1757–1761.

Martin-Gallardo, A., McCombie, W.R., Gocayne, J.D., Fitz-Gerald, M.G., Wallace, S., Lee, B.M.B., Lamerdin, J., Trapp, S., Kelley, J.M., Liu, L.-I., Dubnick, M., Johnston-Dow, L.A., Kerlavage, A.R., Jong, P., Carrano, A.V., Fields, C. & Venter, J.C. (1992) *Nature Genetics* **1**, 34–38.

McCombie, W.R., Martin-Gallardo, A., Gocayne, J.D., Fitz-Gerald, M., Dubnick, M., Kelley, J.M., Castilla, L., Liu, L.-I., Wallace, S., Trapp, S., Tagle, D., Whaley, W.L., Cheng, S., Gusella, J., Frischauf, A.-M., Poustka, A., Lehrach, H., Collins, F.S., Kerlavage, A.R., Fields, C. & Venter, J.C. (1992a) *Nature Genetics* **1**, 348–353.

McCombie, W.R., Heiner, C., Kelley, J.M., Fitzgerald, M.G. & Gocayne, J.D. (1992b) *DNA Sequence* **2**, 289–296.

Rychlik, W. & Rhoads, R.E. (1989) *Nucleic Acids Res.* **17**, 8543–851.

Sulston, J., Du, Z., Thomas, K., Wilson, R., Hillier, L., Staden, R., Halloran, N., Green, P., Thierry-Mieg, J., Qiu, L., Dear, S., Coulson, A., Craxton, M., Durbin, R. Berks, M., Metastein, M., Hawkins, T., Ainscough, R. & Waterston, R. (1992) *Nature* **356**, 37–41.

Tracy, T.E. & Mulcahy, L.S., (1991) *BioTechniques* **11**, 68–75.

CHAPTER TWENTY-SEVEN

PCR Based Strategies for Gap Closure in Large-scale Sequencing Projects

D.M. MUZNY, S. RICHARDS, Y. SHEN & R.A. GIBBS
Institute for Molecular Genetics, Baylor College of Medicine, 1 Baylor Plaza, Houston, TX 77030, USA

27.1 INTRODUCTION

In general, DNA sequencing strategies fall into the distinct categories of either 'directed' or 'random' approaches. In the directed strategy, individual sequence reads (gels) are ordered and overlapped in relation to the original clone by methods including subcloning, nested deletion construction or primer walking. These approaches reduce the redundancy of sequencing, but require great effort and expense to generate the ordered templates or primers. In contrast, the random or shotgun sequencing strategy requires that fragments from the original clone are sequenced without prior selection. The random method is then dependent upon the capacity of computer assembly (similarity) programs to overlap the gels to generate contiguous DNA sequences.

In applying the random strategy, the challenge has been to minimize the sequencing redundancy while efficiently generating overlaps that close gaps between sets of clones (contigs). To achieve this aim, we have employed the sequence mapped-gap (SMG) method, first described by Edwards & Caskey (1991) (Fig. 27.1.). The advantage of this method is that it considers the physical relationship between sequence information from each end of M13 clones, to order the gels within contigs and to assist in closing gaps between contigs. In a typical project (30–50 kb), random forward sequencing proceeds on single-stranded M13 templates to about 85–95% coverage. Random reverse sequencing of polymerase chain reaction (PCR) generated template also takes place on 10–20% of the M13 clones. After an initial assembly of the accumulated random sequence, contigs are ordered and SMGs established by examining the placement in contigs of the forward and reverse sequencing reads of each M13 clone. If insufficient SMGs are available to span the entire project, further M13 SMGs can then be generated by additional directed reverse sequencing. In this case, the reverse sequence is performed on M13 clones that are placed in the overall assembly to enable further gap closure (see also Chapter 28 in this volume for further discussion). In the final stages of a project, remaining gaps and ambiguities are resolved by designing oligonucleotide primers to sequence through the remaining SMG regions joining the contigs identified by the assembly. A flow chart describing this process is illustrated in Fig. 27.2.

The SMG strategy therefore relies upon reverse sequencing to both build SMGs in the random phase, and to close the majority of gaps between assembled contigs. To accomplish this, we have established PCR

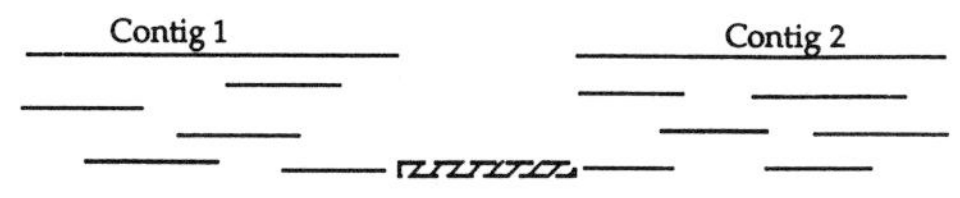

Figure 27.1 Schematic representation of a theoretical view of a partially assembled shotgun sequencing project. Contig 1 and contig 2 are each formed by overlap of sequence reads (solid lines). The knowledge that a gel read (forward sequence) from contig 1 is from the same clone as a gel from contig 2 (reverse sequence) defines the relative position of each contig. The physical link defined by this clone is a sequence mapped gap (SMG).

conditions for generating templates from M13 phage supernatants that reliably produce high-quality reverse sequence data using cycle sequencing protocols. The quality of the PCR generated templates has been compared with alternative methods in 11 projects, totaling more than 200 kb of sequence, and has now become the preferred method for reverse sequencing.

This chapter describes these PCR based protocols and discusses the reliability of the reverse sequencing using data obtained from a large-scale cosmid sequencing project.

27.2 REVERSE SEQUENCING

27.2.1 Oligonucleotide primers and PCR conditions

The PCR methods that we have implemented for reverse sequencing are based upon the asymmetric PCR protocol of Gyllensten & Erlich (1988). Related approaches have been described by Wilson *et al.* (1990); however, we have modified the position of the amplification primers and optimized new PCR conditions. In general, the amplification strategy is to use primers complementary to the M13 vector flanking the polylinker for template amplification, followed by sequencing with a standard M13 sequencing primer (Figs 27.3 and 27.4). In establishing these protocols, shorter oligonucleotides (No. 1720/No. 1721: 20-mers) were found to provide low yields of DNA for sequencing, therefore primer ratios for asymmetric PCR were optimized using 23–24 bp oligonucleotides with 20 pmol of the excess primer, and 0.1 pmol of the limiting primer. To generate the appropriate strand for reverse sequencing, oligonucleotide primers complementary to sequence upstream of the universal sequencing primer binding site were used in excess (e.g. No. 780, 2609, 2399 or 1720). The products were then sequenced with the fluorescent labeled primer M13RP1 (Applied Biosystems Inc.). All primer sets presented in Fig. 27.3 and Table 27.1 have been successfully used to amplify phage template for sequencing.

The asymmetric or symmetric PCRs use a phage supernatant as a convenient starting template (see Section 27.5.1 for details). Optimum asymmetric PCR conditions vary between different model thermocycling instruments. We currently employ the PEC (Perkin–Elmer/Cetus) Gene Amp 9600 thermocycler. Following amplification, each sample is precipitated and resuspended in 25 µl of 1× Sequencing TE and heated to completely dissolve the sample. PCR generated templates can then be sequenced using a variety of sequencing protocols using either Sequenase or Taq DNA polymerase. We currently favour Taq cycle sequencing reactions for easy automation on the ABI Catalyst robot or Beckman Biomeck 1000 workstation (Civitello *et al.*, 1992).

An initial problem associated with the PCR generation of sequencing templates was the occurrence of termination products (stops) and multiple sequences in the data obtained from approximately 10–20% of the samples. Designing amplification primers further from the polylinker (240 bp) as described by Wilson *et al.* (1990), did indeed force termination products farther out in the final sequence, but revealed other problems. Amplified fragments were approximately 500 bp longer than the insert and therefore reduced the effective maximum size of fragment that could be amplified. In addition, samples were still observed to have high backgrounds and multiple sequences up to the point of the stop, followed by clear sequence after

Table 27.1 Oligonucleotide direct sequencing primers for M13.

Primer	*Sequence 5′ to 3′*
545 M13 Reverse Primer	CAGGAAACAGCTATGAC
780 M13 Universal Primer	GTAAAACGACGGCCAGT
1720	GGATAGGTTACGTTGGTGTAG
1721	CCGTCTCGCTGGTGAAAAGA
2608	CTTTATGCTTCCGGCTCGTATGTTG
2609	GCAAGGCGATTAAGTTGGGTAACGC
2398	GGACCGCTTGCTGCAACTCTCTC
2399	GGACGACGACCGTATCGGCCTCAG

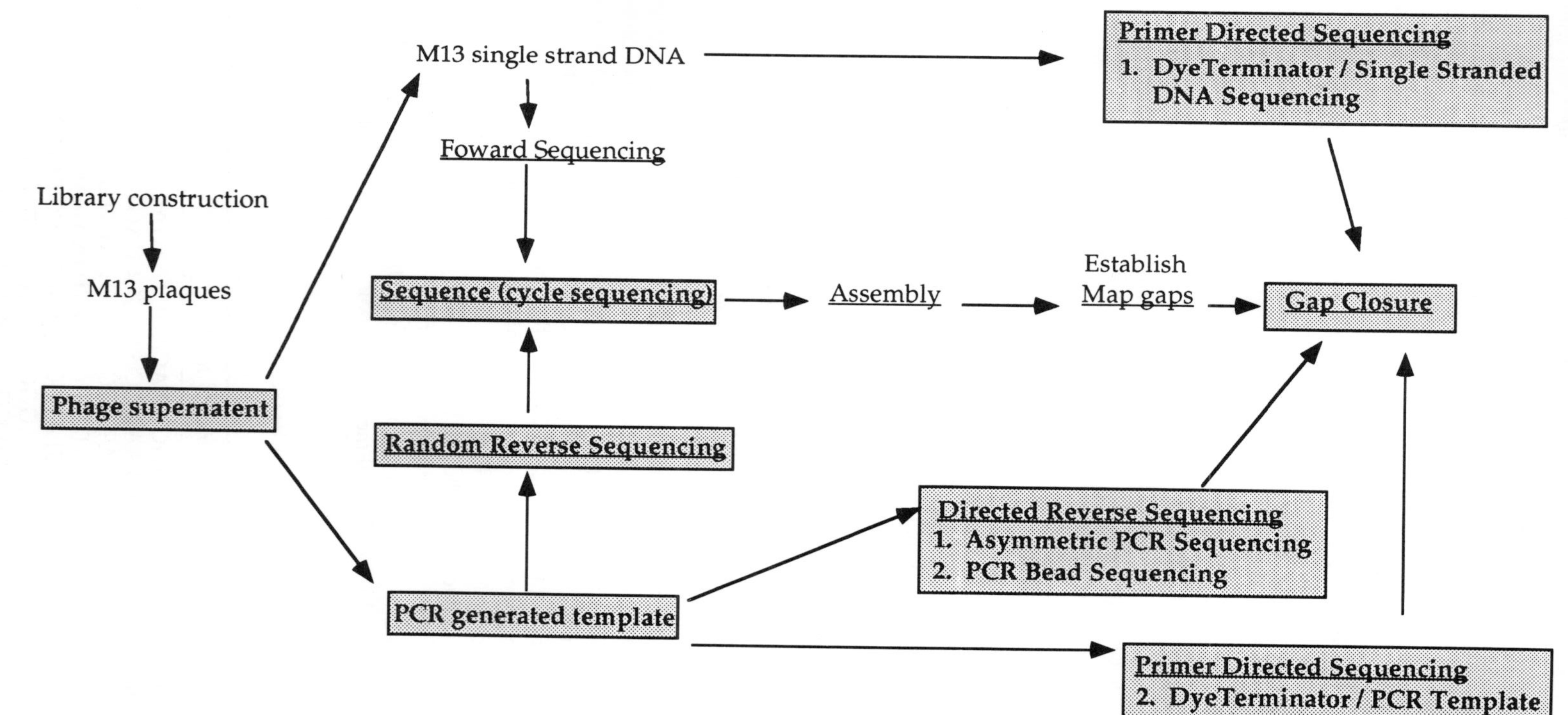

Figure 27.2 This diagram is an overview of the large-scale sequencing process. Shaded areas are directly related to reverse-sequencing or gap-closure activities.

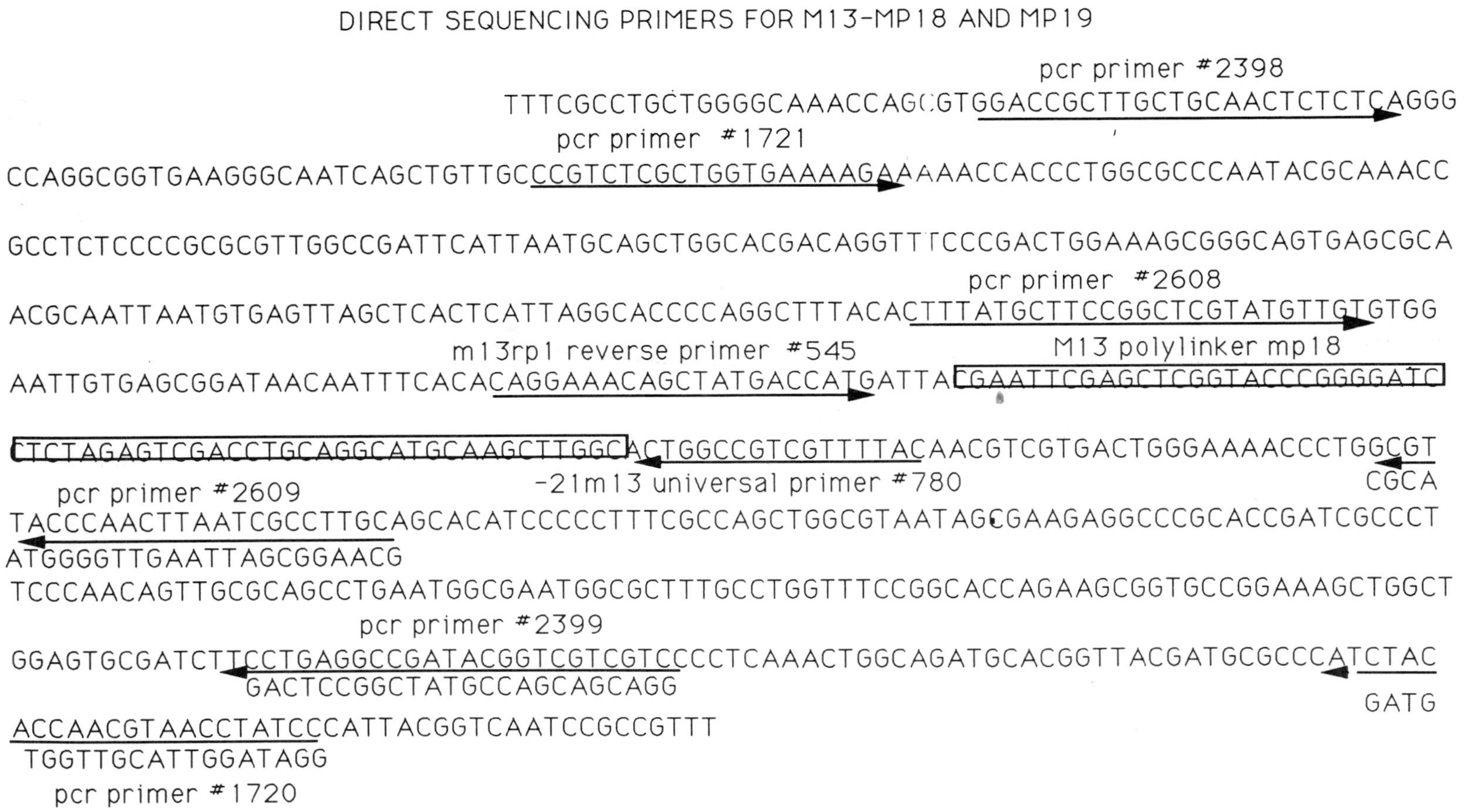

DIRECT SEQUENCING PRIMERS FOR M13-MP18 AND MP19

```
                                                        pcr primer #2398
                          TTTCGCCTGCTGGGGCAAACCAGCGTGGACCGCTTGCTGCAACTCTCTCAGGG
                             pcr primer #1721
CCAGGCGGTGAAGGGCAATCAGCTGTTGCCCGTCTCGCTGGTGAAAAGAAAAACCACCCTGGCGCCCAATACGCAAACC

GCCTCTCCCCGCGCGTTGGCCGATTCATTAATGCAGCTGGCACGACAGGTTTCCCGACTGGAAAGCGGGCAGTGAGCGCA
                                                      pcr primer #2608
ACGCAATTAATGTGAGTTAGCTCACTCATTAGGCACCCCAGGCTTTACACTTTATGCTTCCGGCTCGTATGTTGTGTGG
                        m13rp1 reverse primer #545          M13 polylinker mp18
AATTGTGAGCGGATAACAATTTCACACAGGAAACAGCTATGACCATGATTACGAATTCGAGCTCGGTACCCGGGGATC

CTCTAGAGTCGACCTGCAGGCATGCAAGCTTGGCACTGGCCGTCGTTTTACAACGTCGTGACTGGGAAAACCCTGGCGT
   pcr primer #2609                    -21m13 universal primer #780                CGCA
TACCCAACTTAATCGCCTTGCAGCACATCCCCCTTTCGCCAGCTGGCGTAATAGCGAAGAGGCCCGCACCGATCGCCCT
ATGGGGTTGAATTAGCGGAACG
TCCCAACAGTTGCGCAGCCTGAATGGCGAATGGCGCTTTGCCTGGTTTCCGGCACCAGAAGCGGTGCCGGAAAGCTGGCT
                 pcr primer #2399
GGAGTGCGATCTTCCTGAGGCCGATACGGTCGTCGTCCCCTCAAACTGGCAGATGCACGGTTACGATGCGCCCATCTAC
              GACTCCGGCTATGCCAGCAGCAGG                                            GATG
ACCAACGTAACCTATCCCATTACGGTCAATCCGCCGTTT
 TGGTTGCATTGGATAGG
  pcr primer #1720
```

Figure 27.3 Positions of the M13 amplification primers used to generate template for reverse sequencing reactions in relation to the M13 polylinker.

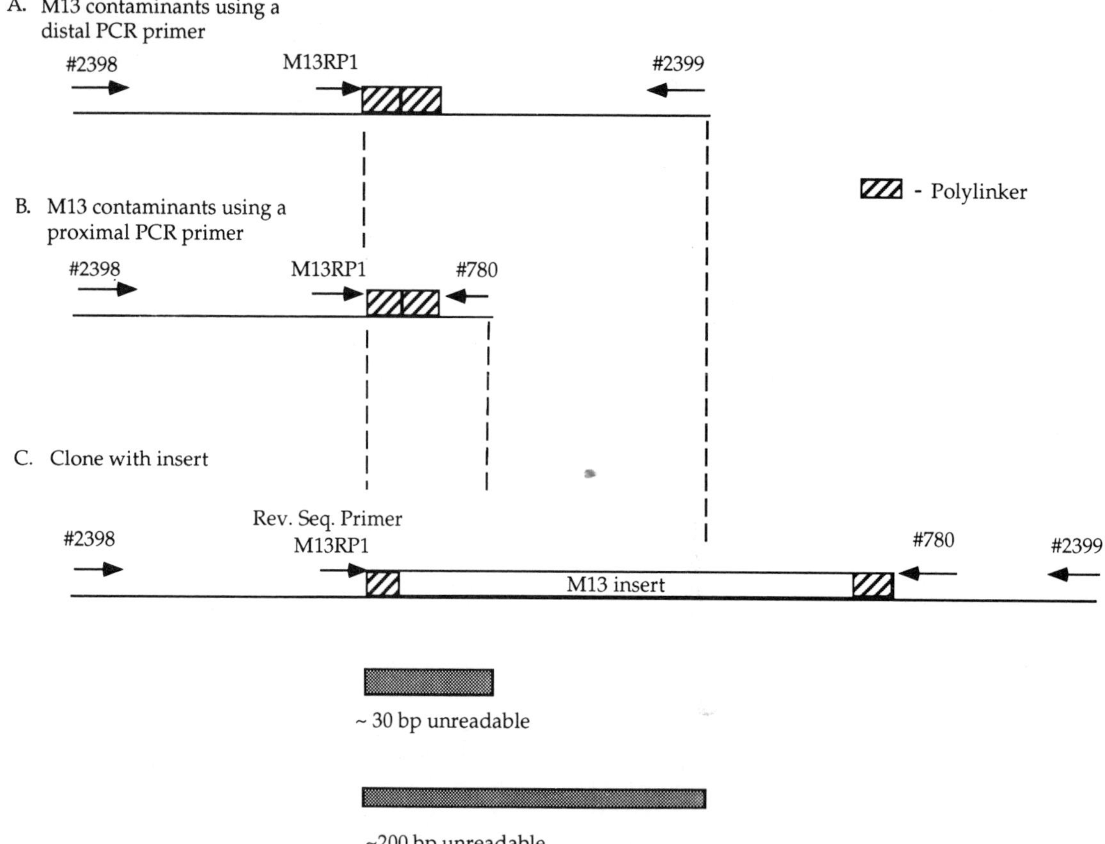

Figure 27.4 Diagram illustrating the effects of PCR primer position in reducing the influence of M13 contaminants on PCR based sequence data. (A) Amplification of M13 contaminants with primers that lie about 240 bp outside the M13 polylinker (Wilson *et al.*, 1990). M13 contaminants can generate up to a 480 bp PCR product that is sequenceable from the M13 reverse primer (M13RP1) priming site to PCR primer No. 2399 resulting in multiple sequences and termination products in the first 200 bp of sequence for the intended clone (C). (B) The results of amplifying with a primer adjacent to the polylinker. The sizes of the M13 contaminant PCR products have been greatly reduced (45–50 bp) and this results in only 25–30 bp of unacceptable sequence in the final sequence chromatogram.

the 'stop' or termination product. The 'stops' occurred at approximately the same distance from the sequence primer site in a number of clones from various sources. These observations, in conjunction with experiments using alternative amplification primers, suggested that the contaminating sequence in most cases is the M13 vector itself. The source of possible contamination with either wild-type or deleted M13 clones is discussed below.

A solution was to perform amplification with one oligonucleotide primer close to the polylinker (e.g. No. 780). The contaminant vector products (amplified polylinker plus vector region) are now shorter, and when sequenced terminate within 20–30 bp from the primer site (Fig. 27.4) Their representation on the final chromatogram is close to the beginning of the read, and can be eliminated using pre-editing computer programs, leaving the majority of the sequence free of termination products and multiple sequences.

In summary, amplifying with vector primers Nos 2398 and 780 provides several advantages including, an increased specificity in amplification using the 24 bp primer No. 2398, an internal site for reverse sequencing, a decreased length of total amplified product and a decreased length of sequenceable vector contaminants.

27.2.2 Phage conditions affecting PCR

We have found several factors concerning phage growth conditions that affect the PCR template generation. An important one concerns the possibility of minor phage contaminations during culture. These possible contaminants include minute amounts of vector phage or other phage from closely spaced plaques, or even contaminated host cells. These subtle phage growth problems can require considerable effort to troubleshoot and the following suggestions in library growth and storage are given.

(1) *Phage plating*. Great care should be taken to use host cell lines for plating that have been properly maintained and proven free of contamination. It is recommended that a fresh overnight culture of host cells is used when plating and picking the library. Additionally, M13 library plaques should be well separated to facilitate selection and minimize contamination from adjacent clones. We routinely plate less than 100 plaques per 100 mm diameter plate.
(2) *Library picking*. M13 plaques should be picked as soon as possible from plated libraries. Leaching of phage from adjacent plaques can introduce contamination at low levels resulting in amplified contamination products and higher background in the reverse sequencing process. A suitable procedure is to pick the phage plaques into 1 ml of 2XYT media in 1.5-ml microfuge tubes or microtiter dishes for immediate growth and phage storage.
(3) *Growth and storage of phage*. Phage clones should be grown for no more than 6 h so that unstable products will remain minimal. Cultures should be spun for 5 min and then stored at a constant 4°C to insure stability of the phage supernatants. We have not found it necessary to supplement the phage stocks for even long-term storage (6 years).
(4) *Bacterial cells in PCRs*. Excess bacterial cells in reverse amplifications have also been a source of contaminant amplification products. When cultures had been accidentally overgrown or the media supplemented to increase the phage yield for single-stranded sequencing, the PCR products contained additional intense multiple contamination bands. PCR amplification of bacterial genomic sequences is a likely source of this contamination, as the problems can be corrected by spinning phage supernatants for an additional 5–10 min in a horizontal microfuge, to remove the bacterial cells. Alternatively, decreasing the amount of phage supernatant added to the PCR from 2–3 to 1–0.5 μl is beneficial. Aliquots of phage supernatant should always be taken from the center of the microfuge tube so as not to disturb the cell pellet or remaining layer of cells at the top of the tube.

In addition to library growth and storage, reaction conditions and the DNA sequence itself may influence the PCR amplifications for reverse sequencing. Many of these difficulties can be overcome with simple adjustments to the PCR protocol.

(1) *Low yield of PCR products*. A low yield of PCR product can be caused by low titer phage supernatants providing insufficient amounts of starting template or by excess phage media inhibiting the PCR reactions. A titering experiment was performed on a stock (1×10^{11} pfu ml^{-1}) M13 phage clone in 100 μl PCRs to establish the amount of phage supernatant necessary for amplification. The results were no amplification at 0.01 μl dilution equivalent of phage supernatant, excellent amplification (0.05–0.1 μg μl^{-1}) using 1–4 μl of phage supernatant, a decrease in amplification at 5 μl and again no visible amplification product from 10 μl of phage supernatant. From these data, we concluded that 1 μl of phage supernatant provided an adequate amount of phage template for amplification and the decrease in PCR generated template at 5–10 μl of supernatant was probably due to inhibitors in the 2XYT media. Thus, when dealing with larger inserts of 2.5 to 3 kb, one can successfully increase the amount of supernatant added to 3–3.5 μl in conjunction with increasing the PCR extension time to 4 min to generate a suitable amount of template.
   Weak PCR products may also reflect the sequence to be amplified. Clones containing tandem arrayed repetitive elements of (15–60 bp) over a 0.5–2 kb region have appeared as smeared PCR products. These yield adequate sequence information to identify the repetitive sequence motif, but often cannot be read for the normal number of bases because of the appearance of mixed signals in the chromatograms. For these difficult clones, symmetric PCR using a biotinylated universal primer followed by magnetic bead sequencing is suggested (see below).
(2) *Weak signal in sequencing reactions*. Weak sequencing reactions associated with normal appearance of PCR products when viewed by agarose gel electrophoresis, followed by a weak signal strength in the fluorescent sequencing, can often be improved by simple adjustments in the ethanol precipitation and resuspension steps. An overnight precipitation at 4°C followed by a 20 min spin in a horizontal microfuge may be necessary to fully recover the amplified template. Finally, we have found it necessary to heat the dried PCR products at 65°C for approximately 5 min to insure complete dissolution of the amplified template.

27.3 APPLICATION OF REVERSE SEQUENCING IN LARGE-SCALE PROJECTS

The PCR based sequencing system has enabled increased automation. Using the Biomeck robot/PEC 9600 thermocycler combination to set up, cycle and then precipitate the PCR reactions as described by

Civitello et al. (1992) has greatly improved throughput. In addition, use of the robotic workstation has standardized the PCRs by routinely generating a set amount of template (about 0.5 μg) for each reaction. Approximately 96 samples can easily be set up, amplified, precipitated and quantitated by gel electrophoresis in 5 hrs by one technician.

To examine the reliability of this reverse sequencing system, the PCRs and related sequence data were examined and categorized from a large-scale project of 41 kb (Table 27.2). This table compares statistics from microbially prepared universal single-strand sequencing data to reverse sequencing data obtained from the PCR products. All sequencing data from the universal and reverse sequencing reactions were processed by a pre-editing computer program, Seqprep (S. Honda, personal communication), which sorts sequence data into three categories according to the number of bases uncalled (Ns) by the ABI proprietors software over a 400 bp region (Table 27.2, section A). A total of 9% of the universal and 15% of the reverse data fell into the categories of conditional and rejected sequence indicating the presence of more than 2% Ns over a 400 bp read. The comparison suggests a slightly lower quality data from PCR generated templates, but further examination reveals this may reflect the sequence itself rather than the template preparation method (Table 27.2). Thirty to fifty percent of the conditional and rejected sequence reads contain repetitive sequences that usually cause poor sequencing reads. When the data were adjusted, to remove these gels, only 5% of the universal sequences and 6.7% of the PCR based reverse sequences were of low quality. Therefore, there does not appear to be a significant difference in the success rate between universal and reverse sequencing using our protocols. While we prefer single-stranded M13 preparations as the superior template in large-scale sequencing, because of the overall low signal-to-noise ratio in single-strand sequencing and because of the difficulties in amplifying large repetitive regions, the PCR methodologies perform quite adequately.

Table 27.2 Comparison of sequencing statistics on a 41 kb large-scale project.

	Universal	*Reverse*
A. Classification of sequencing reads		
Excellent (0–2% Ns)[1]	376	228
Conditional (2–4% Ns)	18	15
Reject (>4% Ns)	19	26
Total reactions sequenced	413	269
B. Breakdown of conditional and rejected sequencing reads		
1. PCR amplification failure	NA	2
2. Sequence extension failure	1	0
3. Poor sequence data		
(a) Repetitive sequence[2]	9	23
(b) Multiple sequences	0	2
(c) Weak sequence data/ short reads[3]	12	11
(d) Noisy reads[4]	9	5
% Failed reads[5]	18/413 = 5	18/269 = 6.7

[1] Sequencing reads categorized as excellent where only 0–2% of the bases were unreadable by the ABI software over a 400 bp region.
[2] Repetitive sequences characterized by tandem sequence arrays 2–3 bp in length covering 50 bp and more, or any polynucleotide sequence over 10 bp.
[3] Sequencing reads where the signal strength is weak and the average read is 350 bp.
[4] Sequencing reads characterized by a high number (>2% Ns) of uncalled bases.
[5] Calculated using subcategories 3(b), 3(c) and 3(d) and the total reactions sequenced for universal and reverse respectively. Category 3(a) was not included because these regions typically cause poor sequence reads unrelated to the sequence methodology used.

27.4 SOLID SUPPORT SEQUENCING APPLICATIONS

A further gap closure technique that we have employed utilizes the solid support strategy of Hultman et al. (1989). Similar protocols have been used extensively in our laboratory for diagnostic sequencing of amplified human DNA fragments (Gibbs et al., 1990), but application in large-scale sequencing has been restricted to the amplification and sequencing of regions resistant to asymmetric amplification for reverse sequencing. Utilizing the oligonucleotide primer sequences previously discussed (Fig. 27.3), symmetric amplification with one biotinylated primer (780) and a normal reverse primer (2398) results in a PCR product that can be captured on an avidin support (Dynabeads M280 Streptavidin), denatured and then sequenced with the fluorescent reverse primer M13RP1. Here, the lower number of amplifications (20 rounds) and increased primer ratios generate sufficient purified template for Taq cycle sequencing reactions using ABI chemistry. Bead sequencing reactions have been automated via a Biomeck 1000 workstation/PEC 9600 using the same protocols and conditions described by Civitello et al. (1992). However, care must be taken to maintain even distribution of the conjugated beads during the sequencing reactions.

Protocols for solid support sequencing have routinely yielded excellent results with low signal-to-noise ratios as in diagnostic sequencing; however, performing bead sequencing does not obviate the problem of contaminating M13 clones or vector products which are still amplified if present in the phage supernatants. One disadvantage of this system is the added cost of the beads at US$1.20 per reaction (Dynal). Another issue is the need for producing biotinylated oligonucleotides that function efficiently in the avidin/biotin binding reactions. For these reasons, and because our reverse PCR sequencing has routinely performed well, as discussed above, we have not implemented this solid phase sequencing protocol on a routine basis.

27.5 PROTOCOLS

27.5.1 Asymmetric PCR amplification for large-scale sequencing

(1) Prepare 1× Cetus buffer (10 mM Tris (pH 8.3), 50 mM KCl, 1.5 mM $MgCl_2$). Pipette buffer (45 μl per reaction) into sterile PCR tubes.
(2) Add 2–3 μl of phage supernatant to each reaction (supernatants should be spun for 5 min in a horizontal microfuge to clarify the supernatant of bacterial cells).
(3) Prepare reagent cocktail: 0.5 μl dNTPs/PCR (25 mM stock), 2.5 U Taq/PCR (5 U μl^{-1} stock -Cetus), 1.0 μl Universal primer No. 780 (20 pmol), 1.0 μl Reverse primer No. 2398 (0.1 pmol). Aliquot 3 μl of the cocktail to each reaction and mix by gentle pipetting. Spin tubes at 100 r.p.m. for 30 s at 4°C. (Cover samples with 30 μl of oil if not using PEC 9600 thermocycler.)
(4) Place into hot thermocycler (94°C), using the following cycling conditions:

	Temp. (°C)	PEC DNA thermal cycler	PEC 9600	
1. Denature	94	7 min	7 min	
2. Anneal	56	1 min	30 s	
Extend	68	4 min	3 min	>35 cycles
Denature	94	30 s	10 s	
3. Extend	68	7 min	7 min	

Spin reactions following amplification to remove condensation.

(5) Examine reactions on a 0.9% agarose gel:
 (a) add 3 μl of reaction to 10 μl of 5× TBE loading dye; and
 (b) electrophorese samples of a 0.9% agarose gel at 100 V for 20 min.
(6) Precipitate reactions by adding 1 volume (50 μl) sterile H_2O, 50 μl 7 M ammonium acetate, and 2.5 volumes ethanol. Reactions are washed with 70% ethanol, dried and then resuspended in 24 μl 1× Sequencing TE buffer (10 mM Tris-HCl (pH 8.0), 0.1 mM EDTA).
(7) Dissolve PCR templates by incubating at 65°C for 10 min, ice quench for 5 min, spin briefly to remove condensation. PCR templates are now ready for cycle sequencing.
(8) **Symmetric PCR reactions** are performed using the above protocol with 5 pM of each primer, 2-min PCR extensions and 18 rounds of amplification.

For a protocol using an automated workstation see Civitello *et al.*, (1992).

27.5.2 Solid phase reverse sequencing protocol

27.5.2.1 PCR amplification

(1) Prepare 1× PCR buffer as stated above. Pipette 1× Cetus buffer (45 μl per reaction) into sterile PCR tubes.
(2) Add 2.5 μl of phage supernatant to each reaction.
(3) Prepare reagent mix: 0.5 μl dNTP/PCR (25 mM stock) 2.5 U Taq/PCR (5 U μl^{-1}, Cetus), 1 μl primer No. 780B (biotinylated) (7 pmol), 1 μl primer No. 2398 (20 pmol). Aliquot 3 μl of reagent mix to each reaction and mix by gentle pipetting. Centrifuge at 1000 r.p.m. for 30 s at 4°C.
(4) Place into hot PEC 9600 thermocycler (94°C) using the following cycling conditions.

1. Denature	94°C	7 min	
2. Anneal	56°C	30 s	
Extend	68°C	90 s	>20 cycles
Denature	95°C	10 s	
3. Final extension	68°C	7 min	

27.5.2.2 Binding to streptavidin beads

Binding and sequencing on the Dynabeads M280 Streptavidin follows the ABI Magnetic Bead Sequencing Protocol as presented in *ABI User Bulletin No. 21* with slight modifications.

(1) Dynabeads M280 Streptavidin are washed twice with 20 μl TTL buffer (6 M LiCl, 300 mM Tris-HCl (pH 8.0), 0.3% Tween-20) at room temperature.
(2) Add 45 μl of biotinylated PCR products to the beads and incubate at 37°C for 30 min.
(3) Immobilize the beads/PCR products and remove supernatant.
(4) To prepare single-strand DNA for sequencing:
 (a) Wash twice with 100 μl of TT buffer (10 mM Tris-HCl (pH 8.0), 0.1% Tween-20).

(b) Denature the strands by resuspending the Dynabeads in 100 μl of 0.15 N NaOH (freshly prepared) and incubating for 5 min at room temperature. Remove the supernatant and resuspend beads again with 100 μl of 0.15 N NaOH and remove without incubation.

(c) To neutralize, wash twice with 100 μl of TT buffer and remove supernatant. Resuspend beads/single-stranded PCR products in 15 μl 1× Sequencing TE (described above).

(5) Sequencing reactions can now proceed using the standard ABI Cycle Sequencing Protocol with an increased bead/template volume from 1 μl for the A and C sequencing reaction to 2 μl, and from 2 μl for the G and T sequencing reaction to 4 μl.

ACKNOWLEDGMENTS

We thank Chris Povinelli and Björn Andersson for valuable discussions and reading the manuscript and Fei Lu for technical help. This work was supported in part by Grants NIH USPHS 1 P30 HG00210 and funds provided by the W.M. Keck Foundation.

REFERENCES

Civitello, A.B., Richards, S. & Gibbs, R.A. (1992) *DNA Sequence* **3**, 17–23.

Edwards, A. & Caskey, C.T. (1991) *Methods: Companion Methods Enzymol.* **3**, 41–47.

Gibbs, R.A., Nguyen, P., Edwards, A., Civitello, A.B. & Caskey, C.T. (1990) *Genomics* **7**, 235–244.

Gyllensten, U. & Erlich, H.A. (1988) *Proc. Natl. Acad. Sci. U.S.A.* **85**, 7652–7656.

Hultman, T., Stahl, S., Hornes, E. & Uhlen, M. (1989) *Nucleic Acids Res.* **17**, 4937–4946.

Wilson, R.K., Chen, C. & Hood, L. (1990) *BioTechniques* **8**, 184–189.

CHAPTER TWENTY-EIGHT

Sequence Map Gaps and Directed Reverse Sequencing for the Completion of Large Sequencing Projects

S. RICHARDS, D.M. MUZNY, A.B. CIVITELLO, F. LU & R.A. GIBBS
Institute for Molecular Genetics, Baylor College of Medicine, 1 Baylor Plaza, Houston, TX 77030, USA

28.1 INTRODUCTION

Most large-scale DNA sequencing projects (10–45 kb) are currently performed using a random sequencing strategy. The DNA to be sequenced is sonicated, size fractionated (1–2 kb) and cloned into a sequencing vector. Clones are then randomly chosen for sequencing and the overlap between individual sequence reads (gels) allows the entire sequence of the original fragment to be reconstructed. This random approach has the advantage of rapid initial accumulation of sequence information; however, as the project proceeds, the probability that additional gels yield new information diminishes. Eventually it is unprofitable to continue further random sequencing and directed sequencing strategies must be employed in order to close the remaining gaps.

In the past, gap closure has been accomplished using primer walking strategies. Disadvantages of these approaches are the requirement of sequence information for primer design and the cost of the oligonucleotide primers themselves. Alternative strategies for gap closure that offer minimal cost in new primer and template production have therefore been sought.

In addition to efficient gap closure, an optimal sequencing strategy should also take advantage of physical information to allow verification of the sequence assembly. This independent confirmation is essential due to the possibility of errors introduced by the computer programs during the assembly. It is especially useful to verify an assembly before the initiation of the gap closure process, so that spurious gap closure can be avoided. Currently restriction maps are the main source of information for the verification of completed edited consensus sequences; however, they are not a reliable or convenient guide to use as a project is progressing.

One approach that has been used to address both gap closure and assembly validation in a sequencing project is reverse sequencing. It is possible to predict where a reverse read of an M13 clone will be located if the length of the clone and the position of the universal read in the assembly are known. Selection of those M13 clones whose reverse reads are positioned within gaps provides a directed closure method. The physical relationship between sequence reads derived from both ends of a M13 clone (a sequence map gap (SMG)) can also be used in the verification of assemblies at any stage in the sequencing process. This chapter further discusses the use of reverse sequencing for gap closure and for verification of sequence assemblies.

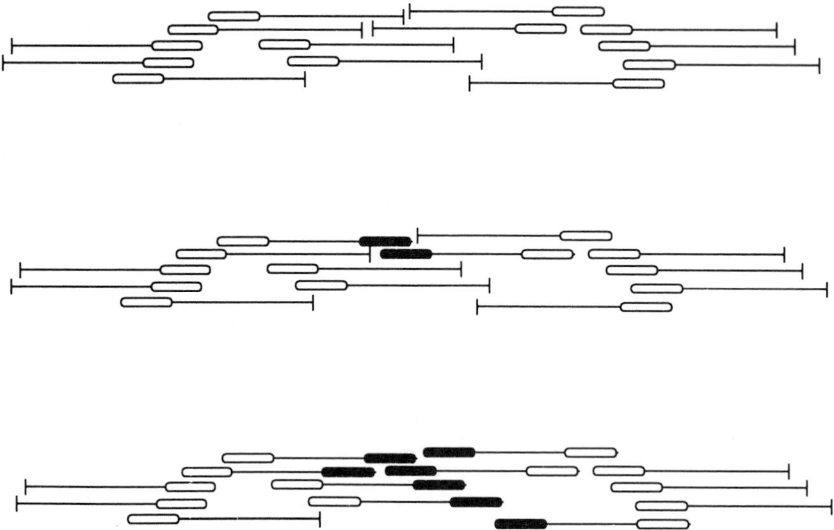

Figure 28.1 Diagram illustrating the use of reverse sequencing in a random sequencing project. The lines represent the M13 clone coverage, whilst the boxed regions represent the actual sequence coverage. Top: An assembly consisting entirely of universal sequencing reads. Note how the length of the M13 clones fills the distance between the two sequence contigs. Middle: Two reverse sequencing reads (shaded boxes) allow ordering of the two contigs. Bottom: Further directed reverse sequencing closes the gap between the contigs. Additional custom oligonucleotides may be required, but the number needed is considerably reduced.

28.2 REVERSE SEQUENCING AND SEQUENCE MAP GAPS

The first large project involving a strategy that included reverse sequencing was the sequencing of the human HPRT locus (Edwards *et al.*, 1991). This primarily involved a strategy where the insert sizes of the M13 shotgun clones were made to be approximately 1.5 kb — significantly larger than the 400 bp sequence read. An important consequence was that most gaps in the consensus sequence were merely gaps in the sequence information, and not gaps in the random clone coverage of the region. Because of this change, gaps in the sequence could be closed using the existing M13 templates.

Additionally, the greater size of the M13 insert allowed both ends to be sequenced without the respective reads overlapping. This led to the development of the SMG concept (Edwards & Caskey, 1991) where the relative physical positions of two sequence reads (universal and reverse) derived from a single shotgun clone allowed confirmation of individual contig assemblies and the ordering of contigs relative to each other (Fig. 28.1). Each SMG in a sequence assembly increases the confidence in the assembly of the region covered by that M13 clone. These SMGs generally overlap so that large regions of the consensus sequence are covered by 'contigs' of SMGs. Problems in the overall assembly are identified where the forward and reverse sequence reads of individual M13 clones are not in the correct position relative to each other.

In around 5% of these cases (a number very dependent upon the accuracy of the assembly algorithm) the problem is due to the presence of a double insert in an M13 clone. Recently, library construction techniques (Povinelli & Gibbs, 1993) involving adaptors to improve the accuracy of the ligation step have significantly reduced the number of double inserts.

In addition to the ordering of contigs, Edwards *et al.* (1991) made limited use of reverse sequencing after the random phase for the closure of gaps, as the strategy had not been optimized with regard to gap closure. Full use of the reverse sequencing was prevented at the time because the reverse sequencing methods used on double-stranded templates did not yield sequence of a consistently high quality.

Subsequently, reliable sequencing with the reverse primer from random library templates has become possible. In particular, the cycle sequencing methodology (Carothers *et al.*, 1989; Murray, 1989) improved the sequencing of double-stranded templates significantly, compared with earlier protocols. The advent of polymerase chain reaction (PCRs) also allowed the production of the opposite strand of an M13 template using asymmetric PCR (Gyllensten & Erlich, 1988). These templates have been successfully used for automated sequencing with the reverse primer (Wilson *et al.*, 1990) and optimization of this method is discussed in this book (Chapter 27). Advances in DNA thermocycler design (Haff *et al.*, 1991) have allowed the simple automation of both cycle sequencing and the PCRs required to obtain the complimentary strand from an M13 clone (Civitello *et al.*, 1992). It is now as simple, rapid and reliable to sequence with the reverse primer from an M13 template as it is with the universal, which has allowed further development of the use of reverse sequencing in random strategies.

28.2.1 How much random sequencing is needed?

A critical decision in a sequencing project is when to cease the random phase, and begin directed approaches. The availability of reverse sequencing techniques complicates this decision because the opportunities for reverse sequencing are related to the total number of the M13 clones used in the project. The number of gaps at the conclusion of the random phase is also related to the total number of clones, and so a proper balance of all of these factors is necessary.

To examine the progression from random to directed reverse sequencing phases, the rate of the accumulation of sequence data was reconstituted from a completed project. The sequence used was a cosmid from DMD intron 44 (Blonden *et al.*, 1989), accession number M86524, insert length approximately 39 kb. Progressively increasing numbers of universal reads each 400 bp long were assembled using the SAM program (Lawrence *et al.*, 1989) in the order that the M13 plaques were picked from the library. The accumulation of consensus sequence data was plotted as a function of the number of universal sequence reads used to create the assemblies (Fig. 28.2). The four different curves shown in Fig. 28.2 are: the total coverage, i.e. the total sum of the lengths of the contigs and the unmerged individual gel lengths; the sum of the lengths of all the contigs; the sum length of all sequence with sufficient coverage to be considered finished, i.e. at least three-fold coverage on one strand, or both strands covered; and the amount of coverage predicted by the equation of Clark & Carbon (1978). All these data are expressed as a percentage of the total length of the cosmid sequence.

The accumulation of total sequence data closely follows the predictions of the Clark & Carbon (1978) expression. The overshoot beginning at 400 gels is due to single sequence reads not being correctly merged into the assembly by the sequence alignment program, thus the same sequence is overrepresented giving a total accumulation of sequence data greater than 100%. Similarly, the percentage of data included in contigs is slightly less than that predicted by the equation because it does not include sequence reads that have failed to merge by the computer into the correct contigs.

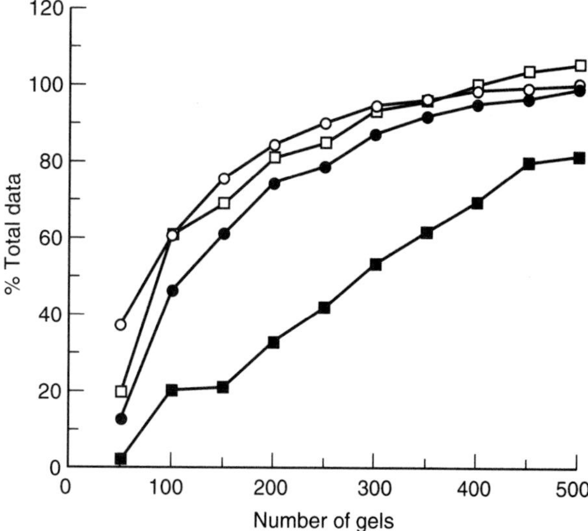

Figure 28.2 Analysis of reconstructed assemblies consisting of progressively increasing numbers of universal reads. (□) The total coverage, i.e. the total sum of the lengths of the contigs and the unmerged individual gel lengths. (●) The sum of the lengths of all the contigs. (■) The sum length of all sequence with sufficient coverage to be considered finished, i.e. at least three-fold coverage on one strand, or both strands covered. (○) The amount of coverage predicted by the equation of Clarke & Carbon (1978). The different types of data are each expressed as a percentage of the total length of the cosmid sequence.

The plateau phase of the curves representing total sequence accumulation has previously led to the conclusion that random strategies involve wasteful redundancy of effort towards the end of a sequencing project. However, the percentage of sequence with sufficient coverage to be declared complete by our criteria shows an approximately linear increase until near the completion of the project (around 80%). Although in the later part of the random phase new sequence information is not being gathered, the additional gels contribute to the accuracy of the sequence by overlapping areas with inadequate coverage. This demonstrates that random sequencing is not truly redundant until the sequencing project is almost complete, and that the simple criterion of total sequence coverage is not an adequate guide for the timing of the transition from the random to directed phase.

Ideally an early change to a directed phase allows rapid completion of a project; however, it is also desirable to close all gaps without resorting to additional cloning and with a minimum of oligonucleotide primer synthesis. It is the additional reverse sequencing opportunities accumulating later in a project that simplifies the gap closure. For example, when 60% of a total sequence is complete, the remaining 40% requires at least one additional read.

It would be difficult to finish the project with the directed reverse possibilities from the 325 randomly sequenced M13 clones available in this example, as not every gap would necessarily be addressed by reverse sequencing. At 80% finished sequence coverage (the point at which the rate of accumulation of complete sequence begins to slow) sequencing approximately 500 M13 clones has, in this example, left 7.8 kb of the same cosmid requiring at least one additional read. In this case gap closure using directed reverse sequencing is facilitated, as the remaining gaps are smaller and there are more positioned M13 clones available. Beyond the 80% complete coverage point, further random sequencing is wasteful by all criteria.

Owing to the successful automation of the random sequencing phase (Wilson *et al.*, 1988; Zimmermann *et al.*, 1988; Mardis & Roe, 1989; Civitello *et al.*, 1992), the consistent buildup of complete sequence coverage, and the reliance of the directed reverse gap closure phase upon the number of positioned M13 clones, this laboratory has routinely continued a random sequencing phase until 80% of the estimated length of the clone has complete sequence coverage. For a 40 kb DNA sequence, this is approximately 500 random sequence reads.

28.3 OPTIMAL USE OF REVERSE SEQUENCING IN RANDOM DNA SEQUENCING STRATEGIES

28.3.1 Reverse sequencing in the random phase: SMG generation

In addition to the value of reverse sequencing in gap closure, the method is important for the construction of the SMGs used for confirmation of the sequence assembly. The advantages of reverse sequencing for SMG generation during the random sequencing phase were explored by a series of reconstructions using data from the same sequencing project discussed above.

The simulated projects containing different fractions of reverse sequences were assembled and compared. The 154 clones sequenced with both the universal and reverse primers in the real project did not provide sufficient data for all reconstructions, and so additional theoretical gels were generated from the final consensus sequence. A total of 250 reverse sequences were selected from the position of the real universal reads and the final completed sequence. For each universal read, a 400 bp theoretical reverse read was generated from a starting position 1 kb from the finish points of the universal read in the opposite orientation. The average length of the reverse primer theoretical sequencing reads was equal to that of the real reverse sequence reads (400 bp). Thus the final length of the theoretical clones generated was 1.4 kb — the average length of the actual clones used in the project. The theoretically generated sequence reads were compared with 154 real reverse sequencing reads and showed good agreement in sequence and read length except for a small amount of variation in the clone length.

Groups of sequencing reads containing different fractions of reverse and universal reads from 0% to 50% reverse (all the M13 clones had been sequenced on both ends) were assembled using the SAM program. The total number of sequence reads was identical in each assembly. This, however, meant that the total number of clones (and thus the total amount of information available) was different in the grouped assemblies. For example, in comparing the 50% reverse assembly to the 0% reverse assembly there were half as many M13 clones used to create an equal number of sequencing reads.

In order to accurately determine the size of any gaps between contigs, the assembly was compared with the known consensus sequence. This also ensured that the sequence was correctly assembled. These assemblies were analyzed for the following characteristics which were expressed as a fraction of the total project length and plotted against the percentage of the reverse sequence reads in the assembly (Fig. 28.3).

As can be seen in Fig. 28.3, the percentage of the total clone length not covered by either any sequence

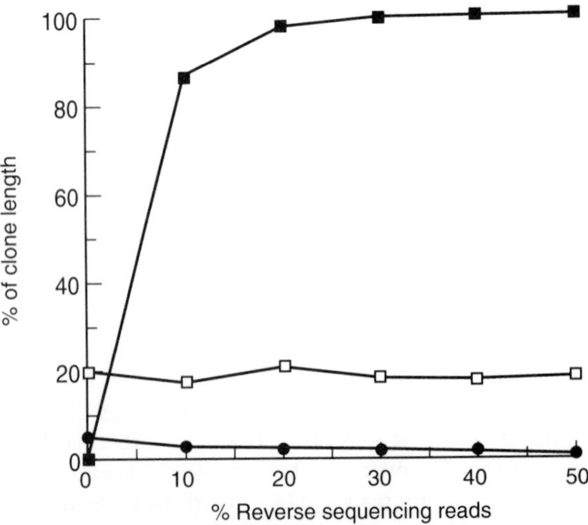

Figure 28.3 Analysis of computer assemblies, at the conclusion of the random phase with different fractions of sequence obtained using the reverse primer. (□) The total length of the gaps in the assembly. (●) The total length of gaps in the finished sequence. (■) The total length of the sequence covered with SMGs. The different types of data are expressed as a percentage of the total sequence length.

(actual gaps) or by sequence of sufficient redundancy (genbank submission gaps) is approximately the same in the different assemblies. This suggests that with an M13 clone length of 1.4 kb, the two ends of the clones can be effectively considered as individual reads for the phase of random accumulation of sequence. In agreement with Edwards *et al.* (1991), sequencing half as many clones from each end does not seem to bias the sequence coverage of the clone; however, the redundancy of the M13 clone coverage (whether sequenced or not) is reduced.

In addition, the data show that the relative number of reverse sequences needed to ensure SMG coverage of an entire project is quite small. At 20% random reverse sequencing 87% of a sequence can be confirmed using the SMGs. This buildup of the SMG information can be modeled with the Clark & Carbon (1978) expression by substituting the average length of the M13 clone insert (in this case 1.4 kb) as the gel length and the number of M13 clones sequenced from both ends as the number of gels. Figure 28.4 shows that in this example, the prediction of SMG coverage closely follows real data. Similar to the accumulation of total real data, the accumulation of SMGs is most rapid in the beginning, but rapidly becomes reduced.

The use of reverse sequencing in the random phase of a project, therefore, does not seem to affect the overall buildup of sequence information as evidenced by the consistent representation of gaps in the different assemblies. The number of random M13 clones is smaller, however, which can limit subsequent opportunities for gap closure without subcloning or oligonucleotide synthesis. Furthermore, even small fractions of reverse sequencing can provide significantly useful SMG information for the verification of assemblies.

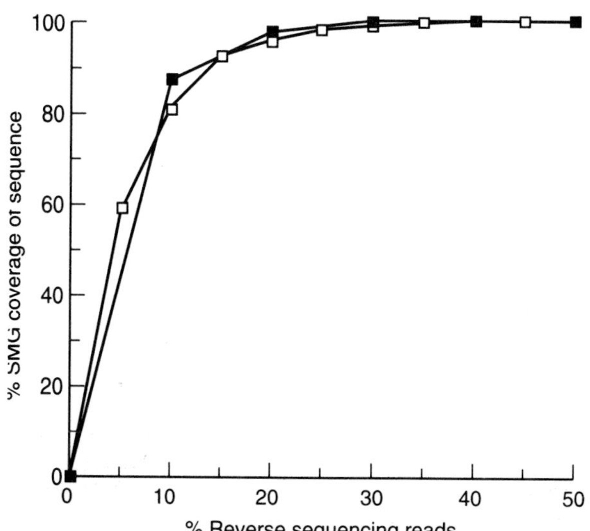

Figure 28.4 Comparison of the predicted (□) and SMG (■) data for the different reconstructed cosmid assemblies. The SMG coverage is expressed as a percentage of the total sequence length.

28.3.2 Directed reverse sequencing in the gap closure phase

At the end of the random phase of a project, gaps in the sequence are defined. At this stage the diminished return of sequencing random M13 clones with either reverse or universal primers forces the transition to a directed strategy. The shift to a directed mode of sequencing involves the definition of sequencing targets, in this case areas of sequence without sufficient coverage, followed by the sequencing of these regions. Ideally, reverse sequencing can be used for gap closure by obtaining reads from selected M13 clones where the corresponding universal sequence gel reads are adjacent to the target region.

When reverse sequencing is applied to every M13 clone in a project, the point at which further random sequencing is unprofitable is reached with the minimum number of M13 clones. However, at the conclusion of the random stage all possible information available from the smaller number of characterized M13 clones by reverse sequencing has already been obtained. This leads to a gap closure scheme relying on custom oligonucleotide sequencing in the middle of the M13 clone to close gaps. At the opposite extreme, sequencing with only the universal primer in the random phase requires twice as many M13 clones as are needed to reach the transition between the random and directed sequencing phases. This increases the number of M13 clones that have their position characterized with a universal read that are available for use as directed reverse templates in the gap closure phase.

To assess the influence of the fraction of total M13 clones that were reverse sequenced in the gap closure phase of a project, the number of possible directed reverse reads that would be positioned in gaps was determined in the assemblies described in Figs 28.3 and 28.4. This estimate assumes 400 bp reads and a 1.4 kb insert size for the M13 shotgun clones. The results were expressed as a function of the total number of sequence reads in the assembly and plotted against the fraction of random reverse sequencing (Fig. 28.5).

It is apparent from Fig. 28.5 that the number of possibilities for directed reverse sequencing decreases with the amount of random reverse sequencing performed before the closure stage of a project. Although complete gap closure by a directed reverse strategy is not likely, theoretically a saving of 85% of the primer reads required for closure is possible. Owing to the redundancy of the directed reverse that

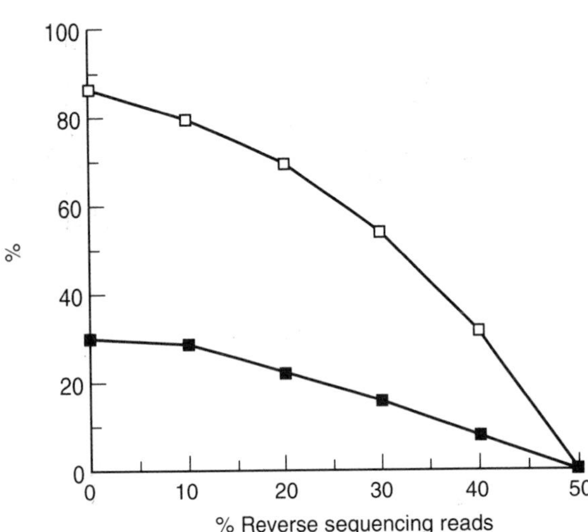

Figure 28.5 The number of directed reverse possibilities (■) expressed as a percentage of the total number of reads in the assembly, and the number of oligonucleotides saved (□) as a consequence of these directed reverse reads, expressed as a percentage of the number of oligonucleotides needed had the reverse sequence not been acquired.

is useful in gap closure, there is a lag in the decline of the curve from the maximum at 0% reverse to zero at 50% reverse where all of the possible reverse sequences have been obtained in the random stages of a project. Accordingly, even if a large fraction of the positioned M13 clones have already had their reverse sequences obtained, there is still room for considerable savings in the number of primer reads required. For example, at 25% reverse sequencing in an assembly it would be possible to save more than half the primer reads with the remaining directed reverse possibilities.

The percentage of oligonucleotides saved by the use of directed reverse sequencing was also estimated (Fig. 28.5). This is derived from the number of primer walks that would be required to complete the sequence, and the number required if all possible directed reverse sequencing was performed. These numbers can be accurately determined when the exact size of the gaps is known. In reality, the savings are slightly lower as variations in the size of the M13 clones makes it difficult to ensure that a gap will always be covered by the reverse sequence of a particular universal read. Secondly, additional primer reads are often required in order to verify sequence in regions where high-quality sequence data are difficult to obtain.

28.3.3 Estimation of the optimal fraction of random reverse sequencing

Balancing each of the factors discussed above suggests that the utility of reverse sequencing in random sequencing projects can be optimized by using a minimal amount of reverse sequencing in the random stage of the project. This fraction should be sufficient to provide the SMG coverage required for contig ordering and verification. If there are still gaps in the SMG coverage, these can easily be closed with directed reverse sequencing.

The greater number of positioned M13 clones in assemblies where less random reverse sequencing has been performed allows more complete use of directed reverse sequencing for gap closure. This facilitates the final stages of a project and considerably reduces the number of oligonucleotides required. Even so, it is unlikely that the entire sequence will be closed without the use of primers. As can be seen from Figs 28.3 and 28.4, the minimal amount of random reverse sequencing needed to obtain full SMG coverage is approximately 10–15%. This corresponds to around 50 random reverse sequencing reads for an average-sized cosmid project (39 kb).

28.3.4 Real examples of savings made with directed reverse sequencing

For the random phase of sequencing projects in our laboratory, about 10% of the M13 clones are routinely sequenced with both universal and reverse primers. This ensures that most of the sequence is covered with SMGs, which provide the ability to order contigs, estimate the size of the gaps between them and increase confidence in an assembly. This fraction of reverse sequencing at the random phase of the project also allows for ample directed reverse sequencing, making it possible to reduce the number of synthesized primer reads required for closure by at least half. Directed reverse sequencing is sometimes required to complete the SMG coverage and validate the assembly. However, owing to the large size of the M13 insert only a small number of such reads are required.

The directed closure of four cosmid sequences where different proportions of reverse reads had been used was compared with the theoretical oligonucleotide savings described above (Fig. 28.6). The savings are less than the predicted maximums, but in three of the cases they are only 20% lower. On average, the synthesis of 45 specific primers was obviated following the directed reverse closure stage, equating to an average saving of approximately US $2400 per cosmid sequenced. In addition, significant savings in the manual time and effort required in the closure process were made.

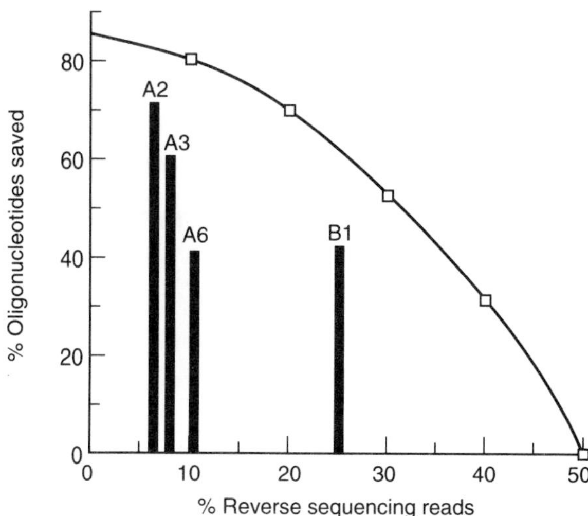

Figure 28.6 A comparison of the savings made (bars) with directed reverse sequencing in four cosmid projects performed in this laboratory with the theoretical savings (curve) that can be made. The number of oligonucleotides saved is again expressed as a percentage of the number of oligonucleotides required had the directed reverse sequences not been acquired.

28.4 PCR MAP GAPS

Complete coverage of the clone with SMGs only ensures correlation between the final sequence and the clone used in the construction of the library. A different approach is needed in order to detect any rearrangements that may have occurred during the cloning of the original genomic fragment or the library production. In many cases, exon order, determined from cDNA sequence, will validate the assembly of the sequenced clone, but with non-coding pieces of DNA this information is not available. In an effort to address the issue of whether or not the consensus sequence truly represents the sequence in the genome we have turned to PCR map gaps (PMGs). Here primers are designed from the consensus sequence and a fragment of the predicted length is obtained by PCR from genomic DNA. As a positive control, the original clone is used. The PMGs can then be joined with contigs of SMG coverage. Thus it is now a goal to provide complete coverage of the consensus sequence with SMGs, and then to relate this single SMG contig to the genomic sequence with three or four (depending on the fragment length) PMGs in different parts of the clone.

28.5 CONCLUSIONS

Since the original use of the longer M13 insert length in the sequencing of the human HPRT gene to ensure clone coverage of any gaps in the sequence contigs, we have expanded our use of the reverse sequencing to take full advantage of the larger insert size. This has led to two main advantages: greater confidence in the final consensus sequences due to complete SMG coverage of the clone and facilitated closure of sequencing projects. The investigator does not have to do any additional work to obtain SMG data as, although the M13 clone pool is smaller, the sequence obtained with the reverse sequencing is as useful in the random build up of sequence as a universal sequencing read. The advantage of being able to target a large number of reverse sequence reads into contig gaps devoid of sequence information and generate these sequence reads in an automated manner increases the speed of the closure process. It also reduces the number of oligonucleotides required for gap closure. The technique can also ensure a supply of SMG data almost anywhere to resolve questions about the integrity of the sequence assembly. The facile automation of the PCRs and cycle sequencing reactions required to obtain the reverse sequence have made this an appropriate strategy for the completion of large-scale sequencing projects.

ACKNOWLEDGMENTS

We thank Christine Povinelli, Björn Andersson and Mike Metkzer for valuable discussions and reading the manuscript. This work was supported by an NIH grant No. USPHS 1 P30 HG00210, and from funds provided by the W.M. Keck Foundation.

REFERENCES

Blonden, L.A.J., den Dunnen, J.T., van Passes, H.M.B., Wapenaar, M.C., Grootscholten, P.M., Ginjaar, H.B., Bakker, E., Pearson, P.L. & van Ommen, G.J.B. (1989) *Nucleic Acids Res.* **17**, 5611–5621.

Carothers, A.M., Urlaub, G., Mucha, J., Grunberger, D. & Chasin, L.A. (1989) *BioTechniques* **7**, 494–496, 498–499.

Civitello, A.B., Richards, S. & Gibbs, R.A. (1992) *DNA Sequence* **3**, 17–23.

Clarke, L. & Carbon, J. (1978) *Cell* **9**, 91–99.

Edwards, A. & Caskey, C.T. (1991) *Methods: Companion Methods Enzymol.* **3**, 41–47.

Edwards, A., Voss, H., Rice, P., Civitello, A.B., Stegemann, J., Schwager, C., Zimmermann, J., Edfle, H., Caskey, C.T. & Ansorge, W. (1991) *Genomics* **6**, 593–608.

Gyllensten, U. & Erlich, H.A. (1988) *Proc. Natl. Acad. Sci. U.S.A.* **85**, 7652–7656.

Haff, L., Atwood, J.G., DiCesare, J., Katz, E., Picozza, E., Williams, J.F. & Woudenberg, T. (1991) *BioTechniques* **10**, 102–112.

Lawrence, C.B., Shalom, T.Y. & Honda, S. (1989) *SAM: A Software Package for Sequence Assembly Management for UNIX Systems*. Molecular Biology Information Resource, Department of Cell Biology, Baylor College of Medicine, Houston, TX.

Mardis, E.R. & Roe, B.A. (1989) *BioTechniques* **7**, 840–850.

Muzny, D.M., Richards, S., Shen, Y., Gibbs, R.A. (1994) This volume, Ch. 27.

Murray, V. (1989) *Nucleic Acids Res.* **17**, 840–850.

Povinelli, C.M. & Gibbs, R.A. (1993) *Analytical Biochemistry* **210**, 16–26.

Wilson, R.K., Yuen, A.S., Clark, S.M., Spence, C., Arakelian, P. & Hood, L.E. (1988) *BioTechniques* **6**, 776–787.

Wilson, R.K., Chen, C. & Hood, L. (1990) *BioTechniques* **8**, 184–189.

Zimmermann, J., Voss, H., Schwager, C., Stegemann, J. & Ansorge, W. (1988) *FEBS Lett.* **233**, 432–436.

CHAPTER TWENTY-NINE

Optimized Methods for Large-scale Shotgun DNA Sequencing in Alu-rich Genomic Regions

F.J.M. IRIS

CEPH, 27 rue Juliette Dodu, 75010 Paris, France

29.1 INTRODUCTION

The most commonly used approach to large-scale DNA sequencing programs is based on the shotgun approach (Deininger, 1983). The genomic region to be sequenced is randomly broken into fragments by physical means, such as sonication or pressure shearing (Schriefer *et al.*, 1990), and all fragments in a size range varying between 500 bp and 3.0 kb are cloned into a single-stranded phage vector, such as M13 (Yanish-Perron *et al.*, 1985). Since the genomic region of interest is entirely disorganized by the fragmentation method used, the M13 recombinant clones cannot be directly ordered into a linear array and must be sequenced blindly. As a result, large-scale sequencing projects (over 50 kb) require a considerable number of clones in order to accumulate sufficient sequence data from which the region of interest could theoretically be reconstituted into a single linear consensus sequence. For a region of 50 kb, about 1200 clones must be considered.

The conventional shotgun approach, however, leads to considerable contigation problems when applied to large DNA fragments (40 kb or more) characterized by a high density of repeated sequences such as the various members of the Alu families. Thought to have evolved from ancestor genes similar to BC200 (Labuda & Striker, 1989; Jurka & Miloslavljevic, 1991), the members of the Alu group are characterized by strong homologies between their 5′- and 3′-ends over 130 or so bases. The total number of full-length Alu repeats, averaging 285 bases, is estimated at 5×10^5 per genome (Britten & Kohne, 1968; Schmid & Jelinek, 1982). Taking into consideration the numerous blocks of partial Alu sequences also present, the various members of the Alu family represent over 10% of the human genome (Sanger, 1982). When present at high density within a genomic region, these repeats considerably impede and may sometimes entirely block the process of contig construction from random genomic sequences. These difficulties can be largely alleviated by adopting particular cloning and sequencing strategies coupled to extremely stringent computer-based analytical methods.

I describe here, a series of simple and highly effective methodological and analytical strategies that have been successfully used to sequence and assemble, by the shotgun technique, a 90 kb HLA class III genomic region containing at least 92 Alu repeats in three dense clusters, 41 of which were spread over a 24.5 kb DNA segment.

29.2 METHODOLOGY

Raw sequencing data acquisition is based on fluorescence-labeled sequencing using the automated ABI 373A system.

29.2.1 Cosmid clones and shotgun sequencing libraries

Three cosmids were used in the construction of M13 sequencing libraries. Cosmid 4.2 contained an insert of approximately 43 kb mapping into the TNF region. Cosmid 1.11 mapped into the Bat2 (HLA-B associated transcript) region, some 45 kb downstream from the TNF locus (Fig. 29.1). The two cosmid inserts did not overlap. Cosmid 6.1 (B144 region) contained the 10 kb gap separating the Bat2 and TNF region cosmids. This last cosmid was voluntarily partly sequenced.

29.2.2 Clone DNA purification

Over two-thirds of the TNF region clones (980 templates) were treated by the classical phenol/chloroform method. DNA from the remaining TNF clones and all the Bat2 and B144 region clones were prepared in microtiter plates by preferential precipitation (see Section 29.3).

29.2.3 Template screening

Sequencing templates were selected by dot–blot hybridization to the relevant cosmid insert radiolabeled (α^{32}P]dCTP) by the random priming method. Hybridizations were carried out in 50% formamide, 5×SSPE, 1% SDS, 0.5% nonfat milk powder, 5% dextran sulfate, and 250 μg ml^{-1} sonicated salmon sperm DNA at 43°C for 4 to 6 h. Final stringency washes were in 0.1× SSC, 0.2% SDS at 65°C for 10 min twice. Autoradiography was at −80°C for 4–8 h using either Amersham Hyperfilm MP or Dupont cronex X-ray films in the presence of intensifying screens. Only clones strongly hybridizing to the insert (generally over 75% of the clones screened) were selected for sequencing. All other clones were ignored.

29.2.4 Physical contig mapping

The relevant cosmid is digested with a restriction enzymes pannel (4–5 enzymes, 5 μg cosmid per digest) in order to produce cosmid fragments spread between 12 and 0.5 kb. The appropriate digests (at least four) are then individually loaded into the wells of 1% agarose minigels, electrophoresed under standard conditions and the separated cosmid fragments then transferred to nylon membranes. The membranes are serially probed with the clones located at each end of the unplaced contigs. Only the clones located at the adjacent extremities of two contigs on either side of a gap will systematically hybridize to the same cosmid fragments in each digest under high stringency conditions. This allows a direct gap-filling PCR approach. The most useful cosmid restriction fragment containing the gap region of interest is now identified. The complete sequences and the relative orientations of the two contigs on either side of the gap are now known. Identifying unique PCR 20–mer primers localized on each side of the gap and amplifying the gap region from the isolated cosmid fragment becomes perfectly feasible, if not a formality.

29.2.5 PCR reaction purification procedure

The filtration medium (P60 Bio-Gel) is equilibrated in TE (pH 8, 4–5 g P60/100 ml TE) for at least 24 h. Delivery into the wells of the filtration plate (Millipore Millititer GV) is carried out with an Eppendorf multipipette fitted with a combitip dispensing 250 μl per delivery. Following loading, the filtration plate fitted over a 96–well collecting plate is centrifuged at 350**g** for 3 min. The wells are then reloaded once more and the plate centrifuged as described above. The collecting plate is cleaned and carefully fitted to the now-ready filtration plate. The PCR reactions (50 μl) are then loaded over the gel-filled wells and the filtration unit centrifuged once more. The collecting plate is recovered, and the filtered PCR reaction products, free of unincorporated dNTP and excess primer, are now ready for sequencing without further manipulation.

29.2.6 Purifications of cycle sequencing reactions

The unincorporated fluorescence-labeled dideoxynucleotides must be removed from the sequencing reactions. This is accomplished by filtration over Biogel P30 equilibrated in TE. The procedure is identical to that followed for purification of PCR reaction products. The sequencing reactions (30 μl) are loaded over P30–filled wells. To each well is then added 20 μl of TE and filtration is carried out by centrifugation at 350**g**. The reaction products recovered in the wells of the collecting plate are lyophylized and resuspended in 5.0 μl deionized formamide.

29.2.7 DNA sequence assembly and analysis

Data from the ABI 373A DNA sequencers were collected on Macintosh llcx computers and subsequently transferred to a Sun server 4/470. In-house software was used to remove vector and low-quality read lengths. After several months of trying to read

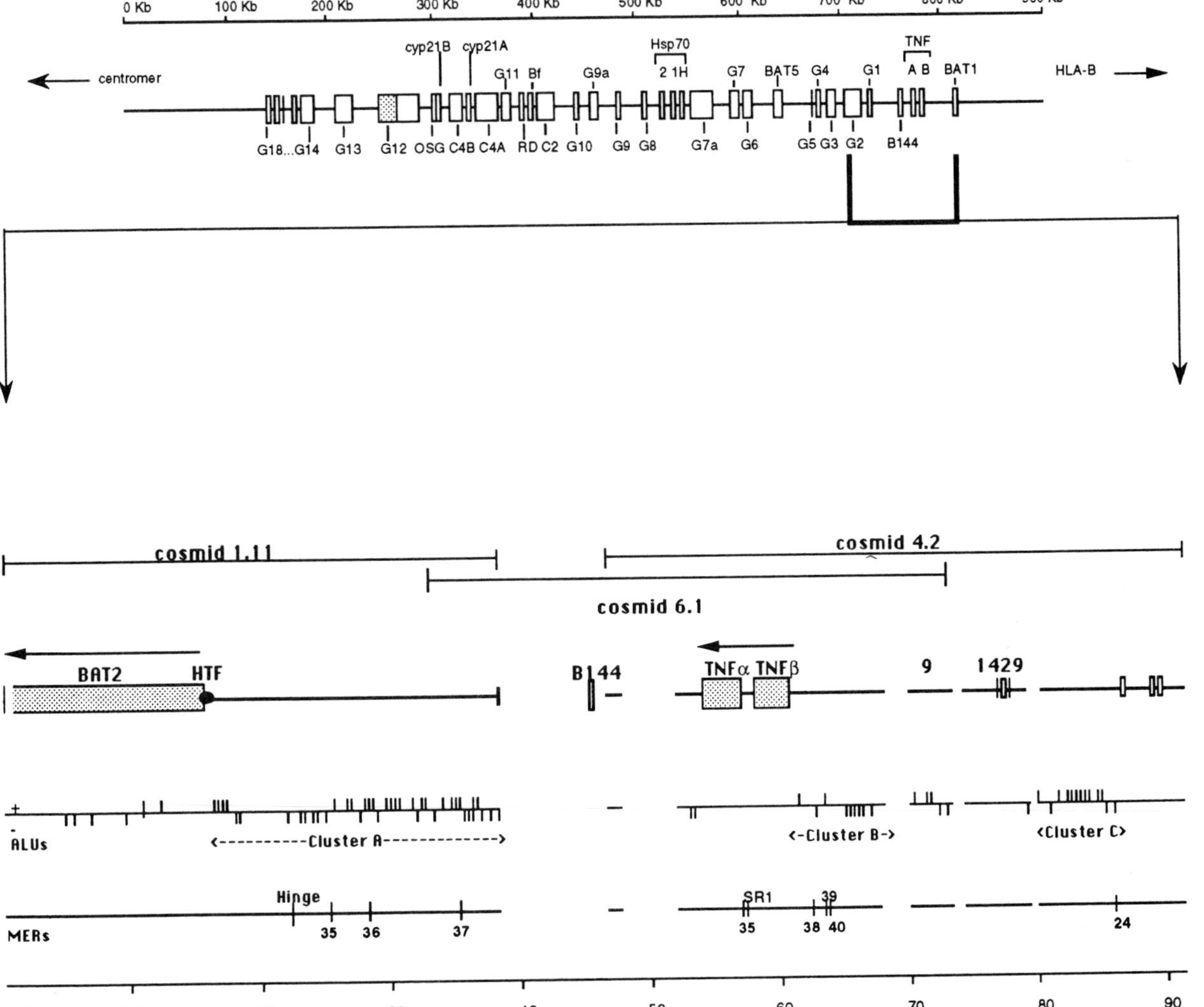

Figure 29.1 Linear representation of the class III region sequenced and analyzed. A detailed map of the whole HLA class III region is presented at the top of the diagram together with a linear scale (in kb). The segment analyzed here is defined by the area in brackets. The relative position of the cosmid clones utilized is represented under extended form immediately below the class III map. The regions entirely sequenced are represented by heavy lines. The blank areas correspond to incompletely sequenced regions. The numbers identify the contigs not yet definitely orientated. The direction of transcription of *Bat2* and the *TNF* genes is indicated by arrows. The potential exons identified at the telomeric end of the *TNF* region (cos 4.2) are represented by open boxes. The positions of Alu repeats are indicated by vertical bars. The orientation of the bars (upwards or downwards) corresponds to the respective orientation of the Alu sequences in standard DNA strand notation. The position and classification of each MER element identified in this region is indicated just above the linear length scale (in kb) at the bottom of the diagram.

as far as possible, we decided the maximum authorized length for a gel reading would be 400 bases. After that position the sequence became rapidly unreliable, mainly because of random insertions. To assemble the sequence, we used the shotgun package of Dear & Staden (1991) (including Dap/Xdap and sap). While the portions of 'unique' sequences could be assembled almost automatically, the *Alu* clusters made it necessary to check the assembly process manually. For the analysis of the assembled sequence we made extensive use of the Fasta (Pearson & Lipman, 1988) and Blast softwares (Karlin & Altschul, 1990). Detection of potential new genes was undertaken using three different methods: a new version of Predictor (Claverie & Bougueleret, 1986); NCBI programs XNu, Xblast & Exon; and CRM software on the GRAIL server (Uberbacher & Mural, 1991). The *Alu* sequence classification was carried out using the Pythia server according to the latest Alu classification criteria (Jurka *et al.*, 1992)

29.3 PROTOCOLS

29.3.1 Sonication

(1) Place a 1.5 ml Eppendorf tube containing 5 μg cosmid DNA in 50 μl sterile water into the cold bath of the sonication apparatus.
(2) Adjust the bottom of the tube to about 1 mm above the sonication probe.
(3) Fill the bath with ice-cooled water to about half the tube height. Remove all traces of ice fragments.
(4) Give a 30-s continuous sonication pulse at full power.
(5) Change the cold bath and give another 30-s pulse.
(6) Repeat.
(7) Check the fragmentation range by loading 1 μg sonicated DNA into a 1% agarose minigel, using ϕX174 DNA cut with *Hae*lll as size marker (size range 1.35–0.2 kb).

29.3.2 End repair

(1) In a 500 μl Eppendorf tube, place: 20 μl sonicated DNA (2 μg); 6 μl 5× T4/Klenow buffer (250 mM Tris-HCl (pH 7.5), 50 mM $MgSO_4$, 0.5 mM DTT); and 2 μl (1 unit) T4 DNA polymerase diluted in 1× buffer.
(2) Place at 37°C for 4 min. Transfer to an ice/water bath and add: 1 μl dNTP mix (10 mM of each dATP/dCTP/dTTP/dGTP), and 1 μl (2.5 units) of Klenow diluted in 1× buffer.
(3) Place at 37°C for 60 min.
(4) Stop reaction by addition of 50 mM EDTA and store at −80°C if required.

29.3.3 Fragment recovery

29.3.3.1 Electrophoresis

(1) Prepare an 8-well 1% minigel using low-melting-point (LMP) agarose in TBE electrophoretic buffer (*do not use TAE buffer*).
(2) Load the two outer wells with a 150 ng size marker (*Hae*III ϕX174).
(3) Load wells 3 and 6 with the end-repaired sonicated DNA.
(4) Run for 2–3 h at 55 V (3 V cm^{-1}).
(5) Thoroughly clean an ultraviolet (UV) transparent plastic sheet (Maylard type).
(6) Place the sheet over the UV box and slide the gel onto the plastic sheet.
(7) Cut out from each lane the sonicated size-range window of interest. *Do not use a wide window; use two different but overlapping windows from different electrophoretic lanes.*

29.3.3.2 Agarase treatment

(1) Soak the agarose block in a large volume (1 ml) of agarase buffer. Leave 30 min at room temperature.
(2) Remove the buffer and repeat for 15 min twice.
(3) Remove the buffer and melt the agarose block at 55°C.
(4) Add 1 volume of buffer and 1.5 units of β-agarase (NEB: 10 mM BisTris (pH 6.5), 1 mM EDTA). Leave at 40°C for 90 min.
(5) Extract by the phenol/chloroform method and precipitate the DNA.
(6) Wash the pellet twice in a large volume (0.5 ml) of 70% ethanol and resuspend in 5 μl sterile water or 1× ligase buffer.
(7) Use 1 μl to quantify final DNA recovery by electrophoresis (1% agarose).

29.3.4 Ligation

This is carried out in a final volume of 15 μl.

(1) In a 500 μl Eppendorf tube, place: 4 μl of resuspended sonicated DNA fragments (25–40 ng); enough *Sma*I-cut, phosphorylated M13 vector to ensure a vector ends (v) to insert ends (I) molar ratio of 3:1 in a final volume of up to 7 μl (i.e. for fragments with an average size of 1 kb, if I=40 ng, V=840 ng); Add 3 μl of 5×ligase buffer; and 1 μl (2 units) of T4 ligase diluted in 1×ligase buffer.
(2) Leave overnight at 14°C.
(3) The ligation can be used immediately or stored at −80°C.

Chimeric clones must imperatively be absent from a sequencing library. A vector:ends molar ratio of 3:1 is important in this respect.

29.3.5 Transformation

This step is carried out using ⅓ to ¼ of the ligation.

(1) Competent host cells are prepared by using the method of Hanahan (1983) or purchased from commercial suppliers.
(2) 5 μl of ligation reaction is diluted to 50 μl with TFB (3 mM hexamine cobalt chloride ($Co(NH_3)_6Cl_3$), 45 mM $MnCl_2$;100 mM RbCl; 10 mM$CaCl_2$; 10 mM k-MES (pH 6.3).
(3) 200 μl of competent host cells is added (JM 101 or DH5−α).
(4) The cells are incubated 30 min on ice and, immediately before plating out, are subjected to a 90-s heat shock at 42°C. Plating out is carried out on 2YT agar plates according to the standard soft-agar method, containing 200 μl of indicator bacteria (noncompetent host cells, $A_{600} = 0.03$ to 0.05), 25 μl of 2.5% IPTG and 25 μl of 50 mg ml^{-1} X-gal.

It is highly inadvisable to incubate the transformed bacterial cell in 2YT prior to plating out. The resulting higher apparent transformation efficiency is not only entirely artificial but may also lead to the loss of infrequently cloned fragments by dilution effects.

29.3.6 Template preparation

29.3.6.1 Primary amplification

The procedure is carried out under sterile conditions in duplicate round-bottom 96-well microtiter plates.

(1) 175 μl of (2YT RichL (1:1) 2YT: 8.0 g bactotryptone, 5.0 g yeast extract, 2.5 g NaCl/500 ml. RichL: 5.0 g bactotryptone; 2.5 g yeast extract; 2.5 g NaCl/500 ml, 1% glucose) is dispensed into the wells of each plate.
(2) A well-defined white plaque is picked with a toothpick.
(3) The toothpick is plunged once into a duplicate well on each plate.
(4) The plates are closed with their lids (*not* an acetate sheet sealer), placed on a plate shaker at medium speed and incubated at 37°C for 6–7 h.
(5) Using multichannel pipettes, the contents of one growth plate are transferred to a filtration plate adapted over a flat-bottom collecting plate. The other duplicate plate is utilized for the production of glycerolated master stocks stored at −80°C (equal volumes of bacterial suspension and 2× freezing medium (100 ml: 1.26g K_2HPO_4, 0.09 g sodium citrate, 0.018 g $MgSO_4.7H_20$, 0.18 g $(NH_4)_2SO_4$, 0.36 g KH_2PO_4, 8.8 g glycerol, H_2O to 100 ml).
(6) The filtration block is centrifuged at 1200**g** for 15 min at 4°C.
(7) The collecting plate is recovered, sealed with an acetate sheet and stored at 4°C. This supernatant retains a good infective activity for at least 6 months.

29.3.6.2 Secondary amplification

The first amplification allows production of a good infective supernatant but not enough phage progeny for template preparation. A second round of amplification is therefore required.

(1) 175 μl of a JM 101 preculture ($A_{600} = 0.3$ to 0.5) in 2YT/RichL is loaded into the wells of duplicate round-bottom 96-well microtiter plates.
(2) Each duplicate well is then seeded with 10–15 μl of infective supernatant from the corresponding well of the 'infective' plate.
(3) Growth and filtration are carried out as above.

The resulting supernatant can now be treated for template preparation.

29.3.6.3 Template purification

All operations are carried out on duplicate microtiter plates using multichannel pipettes. Do *not* treat more than four plates concurrently.

(1) To 150 μl of supernatant/well in a flat-bottom plate,
add 25 μl 20% PEG and 2.5 M NaCl. Using a multichannel pipette, transfer the supernatant-PEG to a round-bottom plate.
(2) Place on ice for 20 min and then centrifuge at 3200**g** for 15 min.
(3) Aspirate the supernatant and invert the plate on tissue paper. Leave for 1 min.
(4) Add not more than 45 μl TE (pH7.5)/5% Triton X100/40 mM EDTA. Place on a shaker for 5 min.
(5) To the Triton solution, add not more than 20 μl 5.0 M NH_4OAc. Mix by pipetting up and down.
(6) Leave at room temperature for 5 min and spin at 3200**g**, 4°C for 15 min to precipitate the proteins.
(7) Carefully transfer the supernatant to a new plate, avoiding the protein precipitate by tilting the plate to aspirate the supernatant.
(8) Add 2 volumes of isopropanol to the supernatant. Mix by pipetting up and down a few times and leave at room temperature for 5 min.
(9) Centrifuge at 3200**g**, 4°C for 30 min. Aspirate the isopropanol and replace with not less than 250 μl 70% ethanol. centrifuge at 2000**g** for 10 min and aspirate the ethanol as thoroughly as possible.
(10) Invert the plate on tissue paper and leave for 5 min, *not more*.

(11) Add 25 μl sterile H_2O, seal the plate and place on a shaker for 30 min.

The templates are now ready for sequencing. Store the plate at −20°C.

29.3.7 Direct single-stranded reverse sequencing

This PCR-based method uses the single-stranded templates already prepared and partly sequenced by the standard Sanger (1982) method. The aim is to obtain the sequence of the unknown extremity.

29.3.7.1 Primary PCR amplification (6–10 cycles)

(1) In a 500 μl Eppendorf tube, place: 10 μl diluted single-stranded template (200–300 ng), 1 μl standard dNTP mix (10 mM each), 1 μl standard M13 forward sequencing primer (4–6 pM), 6 μl 5× sequencing buffer (see above) and sterile water to a final reaction volume of 25 μl.
(2) Place at 98°C for 3 min, cool in an ice-water bath and add 5 μl of Taq polymerase (1 unit) diluted in 1× sequencing buffer.
(3) Mix well and add a drop of paraffin oil (if required) and start the reaction.

Thermocycling profiles
Tubes/mlcrotiter-plate thermocycler:
97°C for 1 s; 96°C for 15 s; 53°C for 1 s; 55°C for 15 s; 72°C for 1 s; 70°C for 90 s. The short thermal overshoots are necessary to increase the ramping rates.

Capillary thermocycler:
95°C for 5 s; 55°C for 5 s; 70°C for 60 s; with the cooling rate at the maximum setting.

Remove excess primer by filtrating the PCR reaction over P60–Biogel equilibrated in water (see Section 29.2).

29.3.8 Direct double-stranded gap bridging

This method directly utilizes very small amounts of a well-defined restriction fragment easily detectable after electrophoresis in LMP agarose run in TBE buffer. The fragment selected must contain a sequencing gap defined on either side by well-characterized primers.

(1) Fit a P20 Gilson pipette (or equivalent) with a disposable cone.
(2) Cut the tip of the cone to a diameter of about 1.5 mm.
(3) Place the gel over the UV box and stab the band of interest with the pipette.
(4) Expel the small plug of agarose into a 500 μl Eppendorf tube.
(5) Add 16 μl of sterile water, 2 μl (10–20 ng) of nonamplifiable (yeast) DNA and place at 60°C for a few minutes to melt the agarose plug.
(6) Add 1 μl (4–8 pM) of each PCR primer, 5 μl 5× sequencing buffer and mix well. Then add 5 μl (1 unit) of Taq polymerase diluted in 1× sequencing buffer.
(7) Place on the thermocycler set at 25 amplification cycles.

The thermal parameters are identical to those described in Section 29.3.7.1.

The PCR product is recovered and sequenced.

29.4 PRACTICAL ASSESSMENT

29.4.1 Standard shotgun sequencing

The sequencing library produced from the *TNF* region (cos 4.2) was constructed from a single sonicated DNA size-range window (0.4–1.0 kb). Over 97% of the sequence raw data (446.2 kb) was generated from about 1300 M13 clones. The average read length obtained was 382 bases (±53) per clone sequenced. Over two-thirds of the templates were purified by phenol/chloroform extraction. Prior to gap bridging, these data were organized into 25 separate contigs and 458 lone sequences corresponding to 175 kb of raw data. The process of contig construction by random sequence overlap was entirely blocked at this level by the presence of high densities of very similar Alu repeats, thus presenting analytical difficulties that could not be resolved by standard sequencing strategies. At completion, the final average exploitable read length was 292 bases per sequence for a mean redundancy of 4.46 readings per base. Only partial reconstruction of cosmid 4.2 into three large contigs could be achieved (Fig. 29.1), leaving sequencing gaps of approximately 2.5, 2.2 and 1.7 kb, respectively.

29.4.2 Mixed library shotgun sequencing

The *Bat2* region random M13 sequencing library was constructed from two independent size ranges of sonicated cosmid fragments. This considerably improved the ease of initial contig assembly from random sequences in spite of an Alu sequences density even greater than that encountered in the TNF region. Over 98% of the total raw data (356 kb) were obtained from approximately 750 M13 clones. The average read length obtained from each clone was 457±72 bases. All the templates were prepared according to the microtiter-plate method (see above). Prior to gap-bridging, these data were organized into 14 contigs covering 52.6 kb of linear sequence and 51 lone sequences corresponding to 20.4 kb of uncontigated data. At completion, a single contig of 49.058 kb was produced with a mean redundancy of 7.3 readings per base. The final average exploitable read-length was 380 bases per sequence.

29.4.3 Gap-bridging strategies

Reverse sequencing of contig-end clones (Fig. 29.2) allowed direct bridging of 16 sequencing gaps in the TNF region and 10 sequencing gaps in the Bat2 region. Furthermore, this approach led to the unambiguous identification of eight misalignments within dense clusters of repeated sequences. The overall sequencing success rates (inserts fully sequenced) were 63.7% and 73.6% in the TNF and Bat2 regions, respectively. The overall mean average read length was 473±71 bases per sequence.

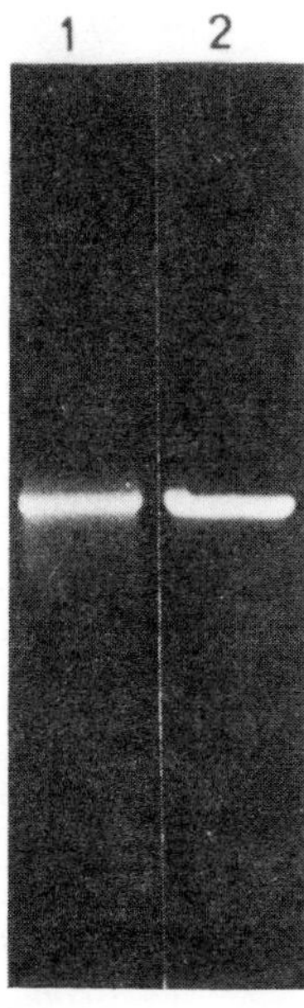

Figure 29.2 Nondenaturing agarose gel electrophoretic analysis of PCR products before (lane 1) and after (lane 2) filtration on P60 Bio-Gel. A unidirectional PCR reaction was carried out on an M13 clone containing a human DNA insert of approximately 900 bases, using the M13 universal sequencing primer as PCR primer. The reaction conditions were as detailed in the text. Electrophoretic analysis was carried out on a 1.5% agarose minigel at 95 V for 15 min in TAE buffer with 5 μl (1/10) of PCR reaction loaded.

Direct double-stranded gap bridging (Fig. 29.3) allowed closure of only one of the four gaps remaining in the TNF region after reverse sequencing. Several smaller contigs could be placed into the gaps by contig mapping, but these could not be joined to the main sequence or positioned and oriented with respect to one another. Furthermore, the rather large gap regions systematically corresponded to particularly dense Alu clusters and PCR amplifications on the corresponding cosmid fragments produced incoherent results (data not shown).

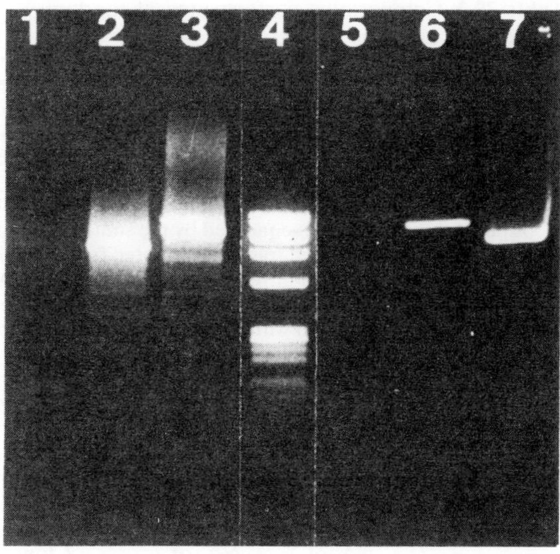

Figure 29.3 Nondenaturing agarose gel electrophoretic analysis of PCR amplifications carried out on a cosmid clone containing 45 kb of human DNA characterized by high density of repeated sequence (over 60% of the cloned fragment). Standard PCR amplifications, carried out on 5–10 ng of total clone DNA using 20-mer primers derived from two noncoding regions of 1.3 kb and 0.9 kb, were consistantly marred by the presence of several amplification products and high background levels (lanes 2 and 3). Reducing the concentration of template DNA to less than 50 pg led to PCR amplification failure (lane 1). Adding 10–20 ng of nonamplifiable DNA (yeast DNA) to the diluted template (10 pg) restored specific amplification of the targets (lanes 6 and 7), producing material which could be directly sequenced. Amplifications carried out with the same primers on yeast DNA alone (20 ng) did not yield a detectable product (lane 5). The molecular weight marker used was 150 ng ϕX 174 DNA cut with *Hae*III.

All three gaps remaining in the Bat2 region after reverse sequencing were successfully and rapidly bridged by this approach. These data are summarized in Tables 29.1 to 29.3.

29.4.4 Cost efficiency of the strategies utilized

There really is only one incompressible cost associated with large-scale automated sequencing—that of machine maintenance. The cost per base of the final consensus sequence depends principally upon (i) the efficiency of template production, (ii) the total number of random sequencing reactions carried out, (iii) the computing time required for contigation of the random sequences produced, and (iv) the level of expenditure associated with gap bridging. Each of these parameters is open to compression, bearing in mind that two

Table 29.1 Comparative performance of random sequence acquisition.[a]

	No. of sequences	*Bases sequenced (kb)*	*Average read length*	*No. of contigs*	*Redundancy readings*	*% Locus contigated*	*No. of free sequences*
Cos 4.2	1135	433.7	382 ± 53	25	4.2	38	458
Cos 1.11	708	349.9	457 ± 72	14	4.7	64	51

[a] Cosmids 4.2 and 1.11 were sequenced by the standard shotgun method. In the case of cosmid 4.2, the sequencing templates were derived from sonicated fragments isolated from a single size-range window whereas those of cosmid 1.11 were derived from two independent and different windows (see text). A direct comparison of the random sequencing results obtained from these two cosmids of comparable lengths and structures demonstrates the effects of the mixed cloning approach. These are four-fold. As compared with the standard approach (cos 4.2), the mixed approach (cos 1.11) resulted in a much better contigation of random sequences (64% versus 38%), with considerably fewer gaps (54% less), obtained from a smaller number of random sequences (37.6% less), and considerably fewer unattached random sequences remaining in the database (a nearly nine-fold reduction). The average redundancy of readings per base remained comparable in both sets of data.

Table 29.2 Comparative performance of directed and reverse sequencing.[a]

	No. of primers	*No. of clones*	*Clones sequenced*		*Average read length*	*No. of gap bridges*	*Clones fully sequenced*	
			%	*No.*			*%*	*No.*
Directed sequencing								
Cos 4.2	51	154	73.4	113	424 ± 61	16	63.7	72
Cos 1.11	35	123	76.4	94	409 ± 47	3	19.1	18
Reverse sequencing								
Cos 1.11		93	97.8	91	483 ± 85	10	73.6	67
Cos 4.2		41	95	39	464 ± 77	0	100	39

[a] Bridging of the sequencing gaps at the conclusion of random sequence accumulation and contigation was undertaken on clones positioned at the extremities of each contig. Two sequencing approaches were utilized, based on directed (internal priming) and reverse sequencing. Directed sequencing was considerably more effective as a gap-bridging method on a database derived from a uniformly sized sequencing bank and characterized by the presence of numerous small contigs and unattached sequences. On a considerably more compact database derived from a mixed sequencing bank, reverse sequencing was the gap-bridging method of choice.

Table 29.3 Comparative overall performances of cloning and sequencing strategies.[a]

Cosmid	*Cloning window size (kb)*	*Total no. of sequences*	*Total bases sequenced (kb)*	*Redundancy reading*	*Final contig no.*	*Unattached sequences or contigs*	*% Locus sequenced*
Cos 4.2	0.7–1.0	1522	446.2	4.46	9	137	97.2
Cos 1.11	0.7–1.0 0.4–0.8	935	356	7.3	1	3	100

[a] A direct comparison of the final results demonstrates the considerable superiority of the mixed sequencing bank over a standard, uniformly sized bank. The mixed bank allowed complete contigation of the locus sequenced, a final result impossible to achieve from the standard bank, using 38.5% fewer sequences, for a greater fiability of sequence determination (a 1.6-fold increase in redundancy readings per base) resulting in a final database presenting orphan sequences.

readings represent the absolute minimum redundancy per base in a trustworthy consensus database. Optimization of the system requires efficient cloning procedures. Ideally, all the templates required should be obtained from one transformation and chimerism should be absent. Random sequencing of a cosmid insert requires about 1200 clones. The methods described here allowed the production of over 1600 clones in one transformation. Cosmids isolated on CsCl gradients are never entirely free of bacterial genome contamination. The sonicated material utilized will thus necessarily contain some bacterial host genomic fragments. In our case, these clones represented about 5% of the sonicated libraries. The selection of sequencing templates cannot therefore be based on color selection alone, the more so since any

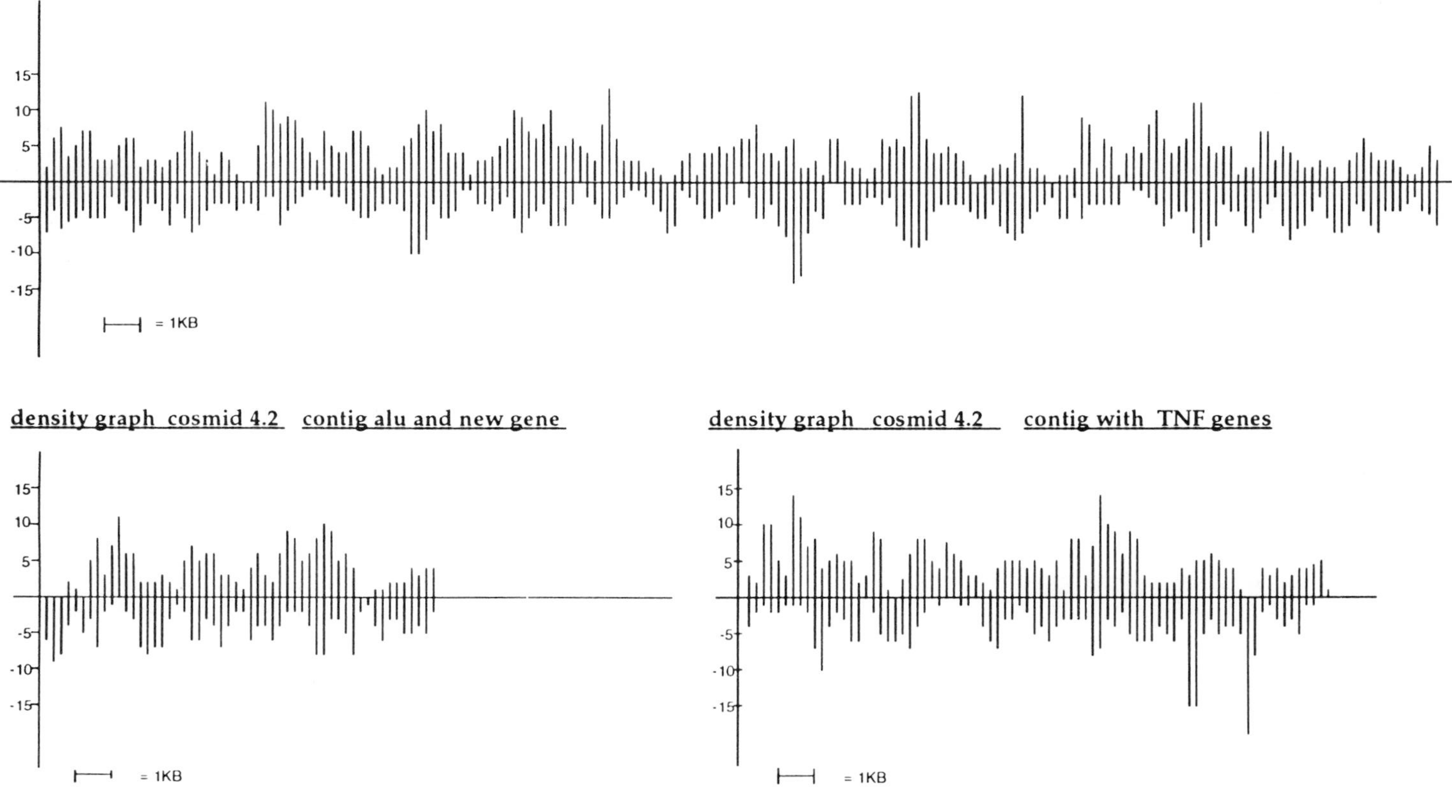

Figure 29.4 Density profile of random sequence accumulation over the length of the largest contigs obtained from cosmid 4.2 (TNF region) and the whole of the Bat2 region. The scale at the extreme left of the diagrams indicate the number of sequences accumulated in the + (upward) and − (downward) orientation. The vertical bars along the central line represent the number and orientation of random sequences accumulated every 200 bases. The diagrams reveal three peculiarities of the sequencing libraries. Regions presenting preferential cloning orientation tend to follow one another in opposite directions, producing characteristic sinusoidal density profiles. This phenomenon is not necessarily synonymous with under-representation of the regions concerned. Only a few regions present either absolute directional cloning preferences or complete absence of preferences, but most sequences presented a marked preference of cloning orientation. These effects were apparently independent of factors such as coding sequences as opposed to repetitive sequences. The average size of cloned fragments could, however, play a role in this context. As compared with the uniformly sized TNF region library (850 bp average insert size), the mixed library (Bat2 region) presented fewer heavily under-represented regions (three versus at least six) and an evident decrease in the density of regions cloned in one orientation only.

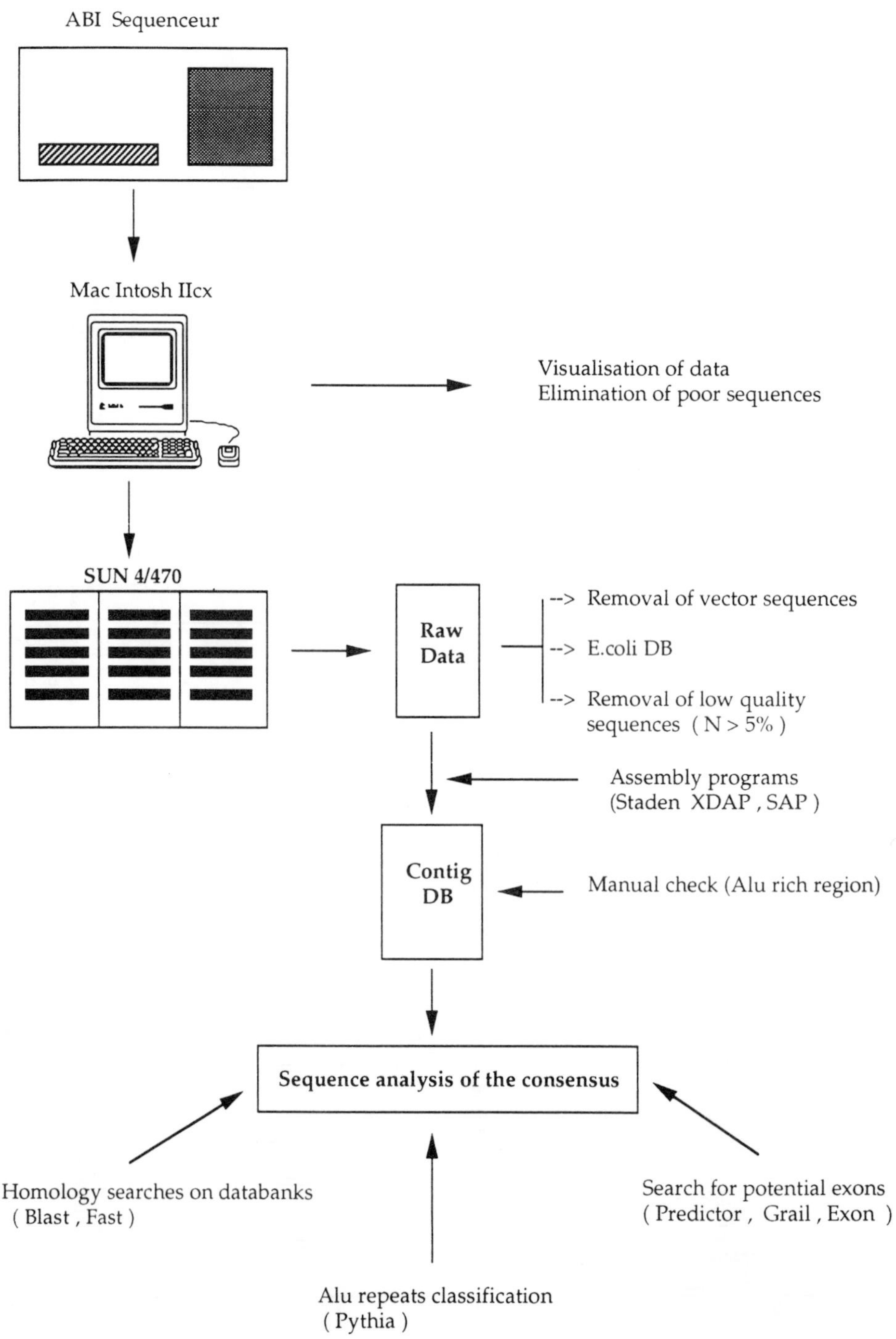

Figure 29.5 Schematic representation of the overall approach utilized for data treatment, from raw sequence production to final consensus sequence analysis. The types of computer program and the data-handling procedures used are described in the text.

frameshift in the *lacZ* gene of the vector will lead to expression of an apparent recombinant phenotype (white plaques). In our experience, false recombinants account for nearly 10% of the clones produced. It is therefore essential to utilize effective screening strategies in order to obtain an efficient sequencing library. The method of choice consists of dot–blot screening of the prospective templates using the isolated cosmid insert as a probe. Isolating the insert reduces to a minimum the effects of genome coli contamination of the probe as well as the effects due to homologies between cosmid and sequencing vectors. This form of screening usually eliminated one-quarter of all the clones initially selected, leading to a sequencing library containing very few, if any, noncosmid-derived templates.

The second parameter that needs compression is the cost of template production. Efficient screening requires purified DNA in order to reduce background signal. Furthermore, the quality of the sequencing results is also largely dependent on the quality of the purified templates. A fast and highly effective template purification method is therefore essential. The method decribed above allows a single worker to produce over 400 high-quality sequencing templates per day.

The next process requiring compression is the number of sequencing reactions needed to obtain over 90% of the region analyzed under the form of random sequences. Ideally, each sequencing reaction should produce directly exploitable results. In other words, each template should be sequenced only once. This aspect is governed by three unrelated parameters: (i) the reproducibility of high-quality template preparation, (ii) the optimization of sequencing reaction chemistry, and (iii) the reproducibility of high-resolution sequencing gels. Under well-optimized conditions, every sequence produced should present less than 1% overall indetermination (less than 5% over the last 100 bases) and the average raw sequence read length should exceed the maximum exploitable read length. As an example, the maximum exploitable read length in the context of our work was determined to be 400 bases. For maximum efficiency, any of our raw sequencing read length had to be in the range 420–500 bases. Using the protocols described above, we achieved this aim in over 80% of all sequences obtained (Tables 29.1 and 29.2).

The final process that needs to be compressed is the cost of gap bridging. The most costly component is represented by the acquisition of highly specific internal primers, which, in most cases, will be used only once. It its essential to reduce, as much as possible, the need for this highly cost-inefficient means of gap bridging. To do so, one needs first to produce a sequencing library made from inserts with an average size no greater than two average sequencing read lengths and containing very few cloning gaps (Fig. 29.4). Given that these requirements can be easily fulfilled (see Section 29.2), highly effective gap bridging can be carried out at no extra cost over that of standard sequencing, using only the standard M13 forward and reverse sequencing primers. Within this context, the need for synthetic oligonucleotide primers is restricted to those few regions severely under-represented within the sequencing library.

If obeyed, this multilevel optimization has a direct and drastic effect upon computer-assisted contig construction, the last parameter that needs cost compression. Under the scheme detailed above, automatic contigation is highly facilitated if extremely stringent computer-based analytical methods are used (Fig. 29.5) and very few misalignments will occur. As a result, a trustworthy consensus sequence is rapidly obtained. The only factor that will induce time-consuming computer analyses will be the density of repetitive sequences encountered over the region analyzed and this, by definition, is neither operator- nor strategy-dependent. Cost compression at this level can only be achieved through the use of task-adapted computer programs.

The optimization process described here reduced the overall sequencing cost from $5 per base, under the standard shotgun sequencing approach, down to a little over $2 per base, while leading to both greater sequence accuracy and complete contigation of an extremely Alu-rich genomic region. It must, however, be borne in mind that these strategies have their limitations. It is our considered opinion that, while perfectly satisfactory on Alu-rich cosmid clones of 40–50 kb, structurally similar genomic regions in the range 60–70 kb constitute the upper performance limits of these techniques.

REFERENCES

Britten, R.J. & Kohne, D.E. (1968) *Science* **161**, 529–540.

Claverie, J-M. & Bougueleret, L. (1986) *Nucleic Acids Res.* **14**, 179–196.

Dear, S. & Staden, R.A. (1991) *Nucleic Acids Res.* **14**, 3907–3911.

Deininger, P.L. (1983) *Anal. Biochem.* **129**, 216–223.

Hannahan, D. (1983) *J. Mol. Biol.* **166**, 157–162.

Jurka, J. & Miloslavljevic, A. (1991) *J. Mol. Evol.* **32**, 105–121.

Jurka, J. Walichiewicz, J. & Milosavljevic, A. (1992) *J. Mol. Evol.* **35**, in Press.

Karlin, S. & Altschul, S.T. (1990) *Proc. Natl. Acad. Sci. U.S.A.* **87**, 2264–2268.

Labuda, D. & Striker, G. (1989) *Nucleic Acids Res.* **17**, 2477–2491.

Pearson, R.W. & Lipman, D.J. (1988) *Proc. Natl. Acad. Sci. U.S.A.* **85**, 2444–2448.

Sanger, M.F. (1982) *Cell* **28**, 433–434.

Schriefer, L.A., Gebauer, B.K., Qiu, L.Q.Q., Waterston, R.H. & Wilson, R.K. (1990) *Nucleic Acids Res.* **18**, 7455–7456.

Schmid, C.W. & Jelinek, W.R. (1982) *Science* **216**, 1065–1070.

Uberbacher, E.C. & Mural, R.J. (1991) *Proc. Natl. Acad. Sci. U.S.A.* **88**, 11261–11265.

Yanisch-Perron, C., Vieira, J. & Messing, J. (1985) *Gene* **33**, 103–119.

CHAPTER THIRTY

Use of Dimethyl Sulfoxide to Improve Fluorescent, Taq Cycle Sequencing

S.G. BURGETT & P.R. ROSTECK, JR
Lilly Research Laboratories, Eli Lilly and Company, Indianapolic, IN 46285, USA

30.1 INTRODUCTION

Recent advances in the automation of DNA sequencing have greatly increased the throughput and consistency of the sequencing process (Smith *et al.*, 1986). Methods utilizing base-specific, fluorescent-dye-labeled dideoxynucleotide triphosphate terminators improve the process of preparing nucleotide sequencing reactions for single-lane analysis on instruments such as the DuPont Genesis and Applied Biosystems (ABI) model 373A sequencers (Prober *et al.*, 1987). Asymmetric polymerase chain reaction (PCR), or cycle sequencing, methods utilizing the thermostable DNA polymerase from *Thermus aquaticus* (Taq polymerase) (Carothers *et al.*, 1989) further simplify preparation of sequencing reactions by eliminating the time-consuming steps required for template denaturation and purification, and primer annealing.

We have successfully implemented cycle sequencing with Taq polymerase to analyze most template types including: single-strand DNA; double-strand, supercoiled plasmids; large, linear, double-strand DNA fragments; and purified PCR products. However, double-strand templates containing a high mol% guanine plus cytosine (G+C) generally produce shorter readable sequences than other templates when analyzed with fluorescent cycle sequencing. We have developed an improved reaction buffer for use with (G+C)-rich and other types of template.

30.2 RESULTS

Results of a typical fluorescent, cycle sequencing reaction with a (G+C)-rich, double strand template are presented in Fig. 30.1. Low signal strength and a low signal-to-noise ratio are frequently observed when sequencing templates of this type. Computerized base-calling of the data is only accurate for approximately the first 120 nucleotides following the primer annealing site. Altering the thermal cycling conditions was initially attempted to extend the amount of accurate sequence data obtained with high (G+C) templates. Higher initial denaturation temperatures and elevated denaturation temperatures at each amplification cycle failed to produce data consistently superior to that shown in Fig. 30.1. These results suggest that strong, local secondary structures or rapid reannealing of template DNA strands during the annealing or extension phases of each PCR cycle is responsible for the inefficient sequencing through difficult templates.

Rapid or snap cooling of chemically or thermally denatured (G+C)-rich or other difficult double-strand DNA templates has been attempted to prevent

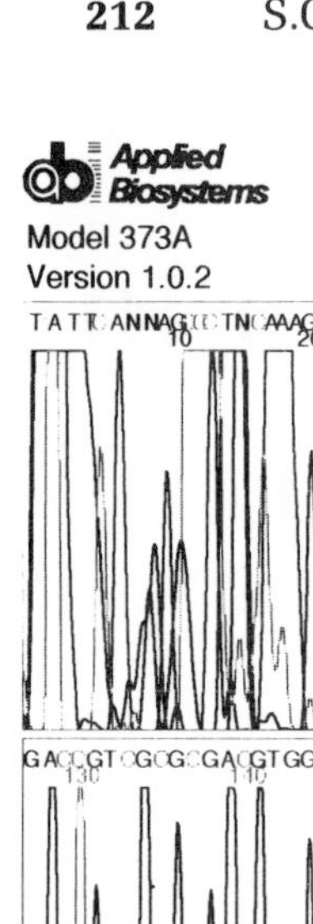

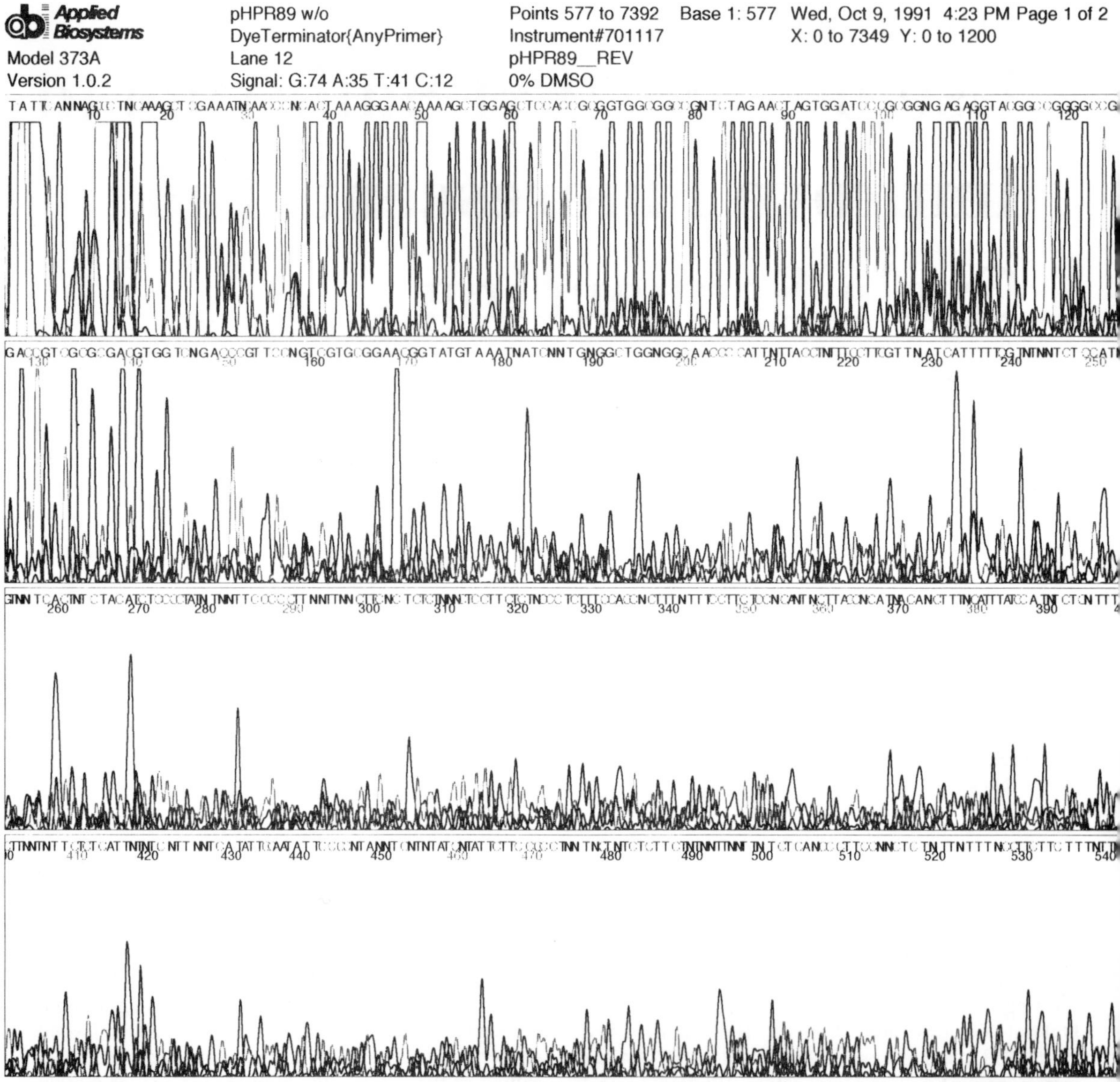

Figure 30.1 Automated, fluorescent cycle sequence analysis of a (G+C)-rich, double-strand DNA template. The plasmid pHPR89 is a derivative of pBluescript SK+ (Stratagene, La Jolla, CA) containing a 2.7 kb *Bam*HI insert from *Streptomyces fradiae* with a (G+C) content of 74.3 mol% (Rosteck *et al.*, 1991). Supercoiled plasmid DNAwas purified from *Escherichia coli* MM294 (Bachmann, 1987) by cesium chloride gradient centrifugation according to Maniatis *et al.* (1982), except that propidium diiodide was substituted for ethidium bromide, desalted by chromatography on Sephadex G50 (PD10 columns, Pharmacia, Piscataway, NJ) and ethanol precipitated by the addition of 0.1 volume 3 M sodium acetate (pH 8.0) and 2 volumes of ethanol. Plasmid DNA (1 μg) and 3.2 pmol of M13 reverse primer (Boehringer Mannheim, Indianapolis, IN) were reacted in a 20 μl cycle sequencing reaction using the proportions of 5× TACS buffer (400 mM Tris-HCl, 10 mM $MgCl_2$, 100 mM $(NH_4)_2SO_4$, pH 9.0), deoxynucleotide triphosphate mix (750 μM dITP, 150 μM dATP, 150 μM dTTP, 150 μM dCTP), G, A, T, and C DyeDeoxy Terminators, and Taq polymerase recommended by the supplier (ABI, Foster City, CA). Reactions were overlayed with mineral oil and reacted in a model 480 DNA thermal cycler (Perkin-Elmer Cetus, Norwalk, CT) for 25 cycles of 1 min at 95°C, 15 s at 50°C, and 4 min at 60°C (note that a denaturation temperature of 95°C is used rather than the 96°C used in standard Taq cycle sequencing reactions). Excess, unincorporated dideoxynucleotide terminators were removed by chromatography on Sephadex G50 spin columns (Quick Spin columns, Boehringer Mannheim, Indianapolis, IN) and the reaction was concentrated by ethanol precipitation. The resulting precipitate was dissolved in 4 μl of 80% deionized formamide, 10 mM EDTA, heated at 95°C for 2 min and cooled on ice. Sequence reaction products were analyzed on an 8.3 M urea, 6% polyacrylamide gel electrophoresed for 14 h at 30 W on an ABI model 373A DNA sequencer. Automatic basecalling was performed by the ABI version 1.0.2 analysis program.

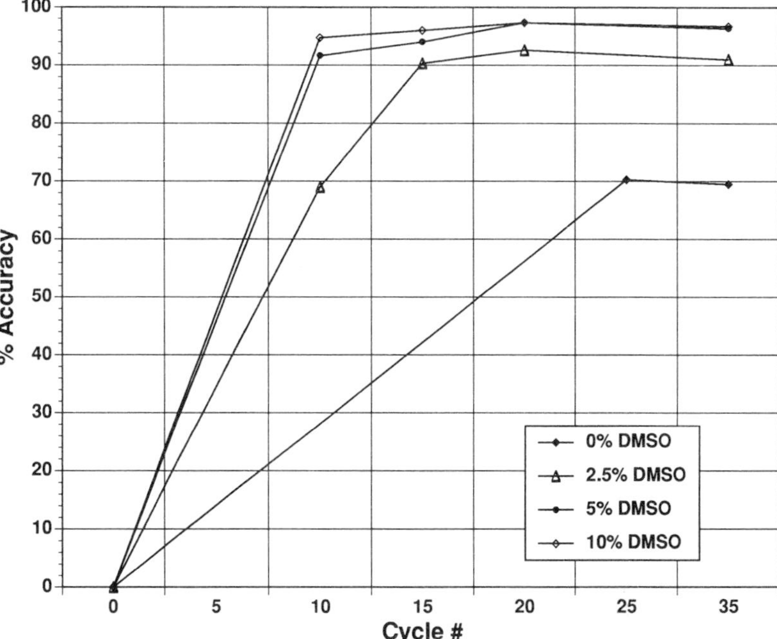

Figure 30.2 Sequence accuracy with a (G+C)-rich template and varying concentrations of DMSO. Cycle sequencing reactions containing a 74 mol% (G+C) template DNA and either no DMSO or 2.5%, 5%, or 10% DMSO (EM Science, Gibbstown, NJ) were prepared and analyzed as described in the legend to Fig. 30.1. Accuracy of the data for the first 300 nucleotides was measured after 10, 15, 20, or 35 thermal cycles by comparing the computerized basecalls to the known sequence of the template.

renaturation of template DNA strands for PCR or nucleotide sequencing (Cassanova *et al.*, 1990; Cusi *et al.*, 1992). High-throughput sequencing strategies require elimination of most template preparation steps making manual denaturation and subsequent purification of the template DNA undesirable. Incorporation of the nucleotide analogs inosine and 7–deaza-2′-deoxyguanosine to improve the efficiency of the PCR or sequence analysis of difficult DNA templates with Taq polymerase has been suggested previously (McConologue *et al.*, 1988; Innis *et al.*, 1988; Gyllensten 1989). Since inosine is routinely included in the fluorescent, cycle sequencing reaction mixtures used in this work, it was felt that an alternative to nucleotide analogs should be investigated to maximize the length of readable sequence data with high GC DNA templates.

The effects of adding the organic solvents formamide, glycerol, or dimethyl sulfoxide (DMSO) to cycle sequencing reactions with Taq polymerase were compared. Fluorescent sequence data obtained with a (G+C)-rich template in reactions containing glycerol were of poorer quality than seen in Fig. 30.1, while reactions containing 1–10% formamide showed little or no improvement over the quality seen in Fig. 30.1. However, data generated from a reaction containing 2.5% (v/v) DMSO suggested that difficult templates could be more efficiently sequenced in the presence of this reagent. Varying concentrations of DMSO up to 10% (v/v) were tested in cycle sequencing reactions with a high GC template and the accuracy of automatic basecalling of the first 300 nucleotides was measured as a function of the number of thermal cycles. Figure 30.2 illustrates the results of this time-course study. Maximum sequence accuracy was obtained in reactions containing 5% or 10% DMSO. Accuracy of automated basecalling increased through the first 20 thermal cycles at all concentrations of DMSO tested and then remained essentially constant through 35 cycles.

An example of fluorescent, cycle sequencing data obtained with a (G+C)-rich template in the presence of 5% DMSO is presented in Fig. 30.3. Both signal strength and signal-to-noise ratios are significantly improved relative to the reaction without DMS0 shown in Fig. 30.1. The computerized basecalls are over 97% accurate through at least the first 450 nucleotides and the data contains fewer ambiguous basecalls (Ns) than is typically obtained with high GC templates. The majority of ambiguous basecalls represent a common artifact that is frequently observed with Taq DyeDeoxy Terminator cycle sequencing chemistry and most template types in which C residues are highly suppressed when they follow a G in the sequence. Additional comparisons of 5% and 10% DMSO with a variety of double-strand templates demonstrated minimal differences between reactions containing these levels of solvent, thus 5% DMSO was chosen for routine use.

To assure that cycling sequencing of nonproblematic templates is not adversely affected by the addition of DMSO to the sequencing reactions, a time-course

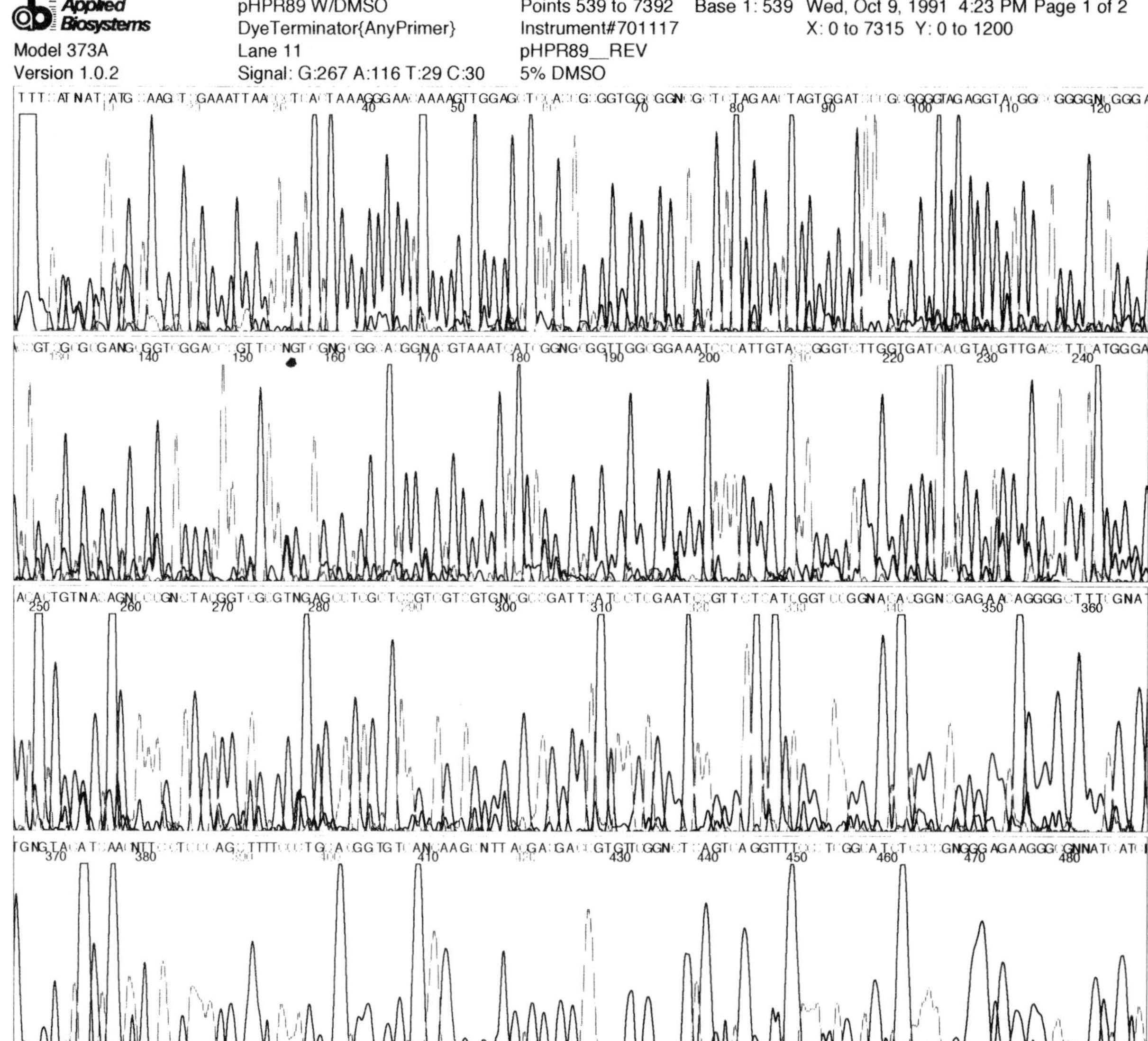

Figure 30.3 Automated, fluorescent cycle sequencing of a (G+C)-rich, double-strand template with 5% DMSO. pHPR89 plasmid DNA (1 μg) was reacted in a Taq, cycle sequencing reaction containing 5% DMSO processed and analyzed as described in the legend to Fig. 30.1.

study of the accuracy of sequence data generated from a control template reacted with varying concentrations of DMSO was performed. The plasmid vector pBluescript SK+ typically produces sequence data of up to 99% accuracy at 500 nucleotides when analyzed in standard Taq DyeDeoxynucleotide Terminator cycle sequencing reactions and was chosen for this study. The results of fluorescent, cycle sequencing of SK+ in the presence of 2.5%, 5%, or 10% DMSO are compared to results obtained in the absence of DMSO in Fig. 30.4. No significant differences are seen in the accuracy of automatic basecalling at 500 nucleotides under any of the conditions tested even when reactions were extended to 35 thermal cycles. The proportions of ambiguous basecalls were also similar for each of the four reaction conditions evaluated (data not shown).

DMSO can be conveniently and routinely incorporated into all Taq-based cycle sequencing work. DNA templates containing inverted repeat sequences with the potential to form significant intrastrand secondary structures often are difficult to sequence with Taq polymerase. However, cycle sequencing reactions containing DMSO frequently produce highly

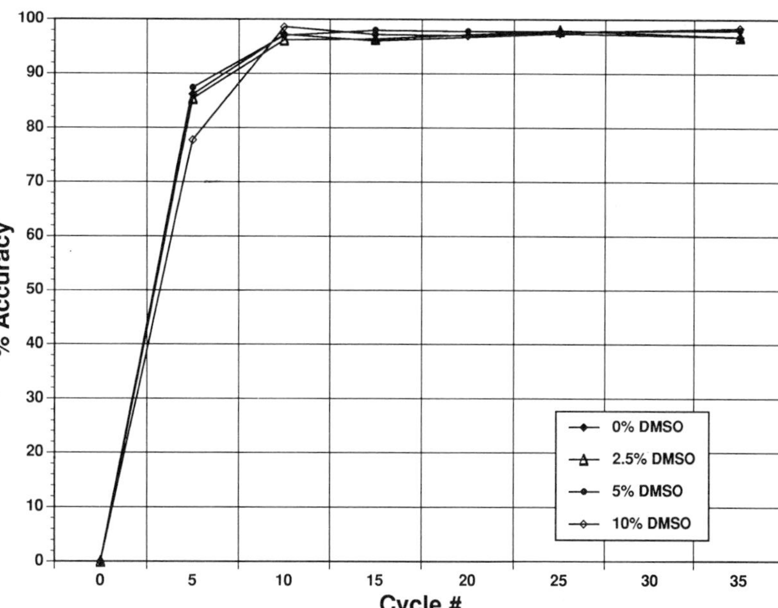

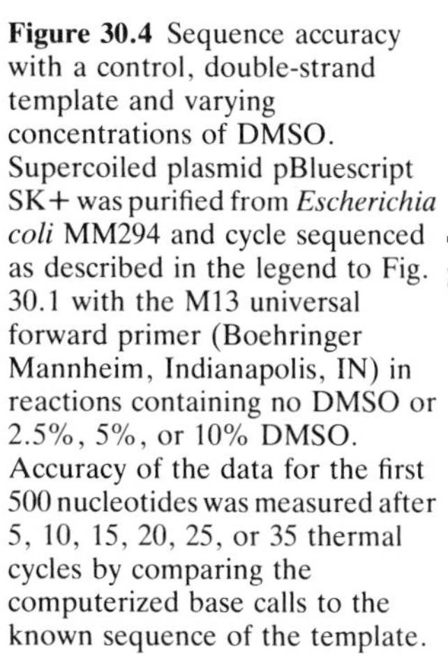

Figure 30.4 Sequence accuracy with a control, double-strand template and varying concentrations of DMSO. Supercoiled plasmid pBluescript SK+ was purified from *Escherichia coli* MM294 and cycle sequenced as described in the legend to Fig. 30.1 with the M13 universal forward primer (Boehringer Mannheim, Indianapolis, IN) in reactions containing no DMSO or 2.5%, 5%, or 10% DMSO. Accuracy of the data for the first 500 nucleotides was measured after 5, 10, 15, 20, 25, or 35 thermal cycles by comparing the computerized base calls to the known sequence of the template.

accurate sequence data through these difficult regions. Additionally, many templates that prove difficult to sequence with standard cycle sequencing procedures for more obscure reasons produce more accurate sequence data in the presence of 5% DMS0. The addition of DMS0 has proved to be a useful modification to increase the efficiency and robustness of fluorescent, cycle sequencing reactions with Taq polymerase. Finally, this modification can be readily incorporated into schemes to automate the preparation of sequencing reactions to further improve the process of generating nucleotide sequence data.

ACKNOWLEDGMENTS

We especially thank our collegues T. Bennett, B. DeHoff, B. Glover, M. Greaney, I. Jenkins, and K. Sutton in the DNA Sequencing and Synthesis Core Laboratories of Lilly Research Laboratories for their encouragement and assistance, and the management of Lilly Research Laboratories for supporting this work.

REFERENCES

Bachmann, B.J. (1987) In *Escherichia coli and Salmonella typhimurium Cellu lar and Molecular Biology*, F.C. Neidhardt, J.L. Ingraham, B.K. Low, B. Magasanik, M. Schaechter, & H.E. Umbarger (eds). American Society for Microbiology, Washington, DC, pp. 1190–1230.

Carothers, A.M., Urlaub, G., Mucha, J., Grunberger, D. & Chasin, L.A. (1989) *BioTechniques* **7**, 494–499.

Cassanova, J.-L., Pannetier, C., Jaulin, C. & Kourilsky, P. (1990) *Nucleic Acids Res.* **18**, 4028.

Cusi, M.G., Cioé, L. & Rovera, G. (1992) *BioTechniques* **12**, 502–503.

Gyllensten, U. (1989) *BioTechniques* **7**, 700–708.

Innis, M.A., Myambo, K.B., Gelfand, D.H. & Brow, M.A.D. (1988) *Proc. Natl. Acad. Sci. U.S.A.* **85**, 9463–9440.

Maniatis, T., Fritsch, E.F. & Sambrook, J. (1982) *Molecular Cloning: A Laboratory Manual.* Cold Spring Harbor Laboratory Press, Cold Spring Harbor, NY, pp. 90–91.

McConologue, L., Brow, M.A.D. & Innis, M.A. (1988) *Nucleic Acids Res.* **16**, 9869.

Prober, J.M., Trainor, G., Dam, R., Hobbs, F.W., Robertson, C.W., Zagursky, R.J., Cocuzza, A.J., Jensen, M.A. & Baumeister, K. (1987) *Science* **238**, 336–341.

Rosteck, Jr, P.R., Reynolds, P.A. & Hershberger, C.L. (1991) *Gene* **102**, 27–32.

Smith, L.M., Sanders, J., Kaiser, R.J., Hughes, P., Dodd, C., Connel, C.R., Kent, S.B.H. & Hood, L.E. (1986) *Nature* **321**, 674–679.

PART III

Informatics

A. SEQUENCE ASSEMBLY THEORY AND ALGORITHMS

CHAPTER THIRTY-ONE

Neural Networks for Automated Base-calling of Gel-based DNA Sequencing Ladders

C. TIBBETTS, J.M. BOWLING & J.B. GOLDEN III

Department of Microbiology and Immunology, Department of Mechanical Engineering, Vanderbilt University Schools of Medicine and Engineering, Nashville, TN 37232–2363, USA

31.1 INTRODUCTION

Contemporary automated DNA sequencing instruments gather data describing distributions of labeled oligonucleotides in electrophoretically separated sequencing ladders. Automated sequencers represent two classes of instrumental design and rationale. Densitometric film scanners digitize images of sequencing ladders from films or autoradiograms, exposed and developed after fixed periods of electrophoresis. Other systems have fixed or scanning detectors which monitor electrophoretic transport of labeled oligomers through the gel, generating digital images of sequencing ladders in real time. Both classes of instrument use computer software to translate their raw data, the digitized images of the sequencing ladders, to specific DNA sequences.

The base-calling software supporting an automated DNA sequencer represents a recursive process of: (1) scanning the ladder image to locate the trace of the next oligomer in the sequence, and (2) evaluating the particular attributes of the oligomer's image which specify its terminal nucleotide. Evaluation of a primary determinant, related to individual system design, specifies the 3′ terminal nucleotide of each oligomer. Some real-time systems and film scanners use single-labeled oligomers, in familiar arrays of four parallel ladders to discriminate spatially among the four possible terminal nucleotides. Other real-time scanning instruments employ selective bandpass filters for spectroscopic discrimination of four base-specific, fluorescent labels on the terminal nucleotides.

Effective separation and resolution of oligomers by gel electrophoresis and the sensitivity and resolution of the detector system together determine practical limits of sequence accuracy and length which may be obtained by an automated DNA sequencing system. Under ideal conditions, the determination of successive terminal nucleotides represented in a sequencing ladder image is straightforward. However, when separations or signals are suboptimal, ambiguities or errors in the final sequence are more likely. These appear as miscalled bases, extra or missing bases, or unidentified bases in DNA sequence files. A significant performance goal for automated sequencers is accuracy which matches or exceeds the interpretation of the raw data by human experts. Based on experience from manual sequencing methods, practised in the field over the last decade, we should expect at least 97–99% correct calls of the first 300–400 bases of a ladder.

Unfortunately, the sequences generated by primary base-calling software using the Du Pont Genesis 2000 system often reach only 90–95% correct calls before oligomer signals abate to levels of noise or before gel and detector resolution preclude further ordering of oligomers. Investigators often attempt to improve accuracy through systematic review of the raw-data images, seeking and editing ambiguous and erroneous calls in the DNA sequence files. We are interested in minimizing such time-consuming and discouraging review of primary data. One option may be automation of the review and editing process, with new software emulating the base-calling skills of human experts. Alternatively, base-calling software can be designed to incorporate these skills during the primary analysis of raw data images.

In this chapter, we describe an approach to base-calling which treats the translation of ladder images to DNA sequences as a problem of pattern recognition. DNA sequencing ladders are described in terms of four informative parameters: (1) the size order of the oligonucleotides; (2) the lane position or fluorescence of each oligomer's terminal nucleotide; (3) the signal intensity or yield of each oligomer; and (4) the separation of each oligomer from the preceding oligomer. Contextual, local arrays of these informative parameters define image patterns which correspond to specific sequences of nucleotides. We have developed simple neural networks which operate as base-calling programs for automated DNA sequencers. These networks exploit the multiple informative parameters of sequencing ladder image patterns to improve both accuracy and length of primary DNA sequence determinations. Neural networks have also been developed for raw-data processing in real time, expediting the extraction of multiple informative parameter arrays from raw-data images. Finally, a neural network editor is under development to automatically review the raw data and correct many of the ambiguous or erroneous calls in primary DNA sequence files.

31.2 SEQUENCING LADDER IMAGES

31.2.1 Multiple informative parameters

Automated DNA sequencing systems most often employ variations of the biochemical sequencing strategy described by Sanger *et al.* (1977). The alternative, base-specific chemical cleavage strategy of Maxam & Gilbert (1977) can also be adapted to various automated sequencing systems. Both methods rely upon the physical separation of homologous series of oligonucleotides on the basis of length, through denaturing polyacrylamide gel electrophoresis (Maniatis *et al.*, 1975). The determination of sequence by automated DNA sequencing instruments is based on analysis of the label associated, directly or indirectly, with the four species of 3′ terminal nucleotides on sequencing ladder oligomers. This may be achieved by the location of single-labeled oligomers in an array of four parallel, base-specific ladders. Alternatively, four base-specific fluorescent labels may be distinguished, spectroscopically, within single lane ladders comprising all the oligomers of the sequence. Ansorge *et al.* (1990) and Tabor & Richardson (1989, 1990) proposed analysis of singly labeled oligomer ladders in single lanes by exploiting unique properties of the T7 DNA polymerase in reactions with divalent manganese ions. In their approach, four different ratios of the concentrations of dideoxy- to deoxy-nucleotides, present in the DNA synthesis reaction, specify four base-specific ranges of oligomer yields through the ladder.

In general, the products of Sanger sequencing reactions show significant variances of oligomer yields and electrophoretic separations. Figure 31.1 illustrates a portion of a sequencing ladder monitored through only one photometer of a Du Pont Genesis 2000 fluorescence-based sequencer (Prober *et al.*, 1987). The variation in oligomer yields is evident from the quantitative time profile of fluorescence. Consistent, sequence-associated variation of oligomer yields in sequencing ladders reflects events during the *in vitro* synthesis of the oligomers (Hindley, 1983; Jensen *et al.*, 1991; Smith *et al.*, 1986; Ansorge *et al.*, 1987; Connell *et al.*, 1987; Tabor & Richardson, 1987, 1989, 1990; Kristensen *et al.*, 1988; Toneguzzo *et al.*, 1989). The DNA polymerase and template DNA are intimately associated in the locale of the 3′ terminus hydroxyl group of the new DNA strand. Pools of the next complementary dNTP and ddNTP analog compete for net incorporation. Chain termination with the ddNTP increments the yield of oligomers of the particular length associated with the site of addition. Elongation with the dNTP commits the new chain to represent one of the longer oligomers in the ladder.

We have frequently observed sequence-associated variation of oligomer yields in parallel ladders representing sequences that differ only by single point mutations. Our analysis of such ladders also led to the discovery of sequence-associated variation of oligomer separations in sequencing ladders (Bowling *et al.*, 1991). Reptation, or the end-on migration of the single-stranded DNA through the polyacrylamide matrix (Lerman & Frisch, 1982; Lumpkin & Zimm, 1982; Bean & Hervet, 1983a, b; Edmonson & Gray, 1984; Hervet & Bean, 1987), allows subtle conformational differences of the oligomers to measurably affect their separation from one another. An automated DNA sequencer or film reader facilitates quantitative analysis of the mobility differences of the successive oligomers. These differences are determined by

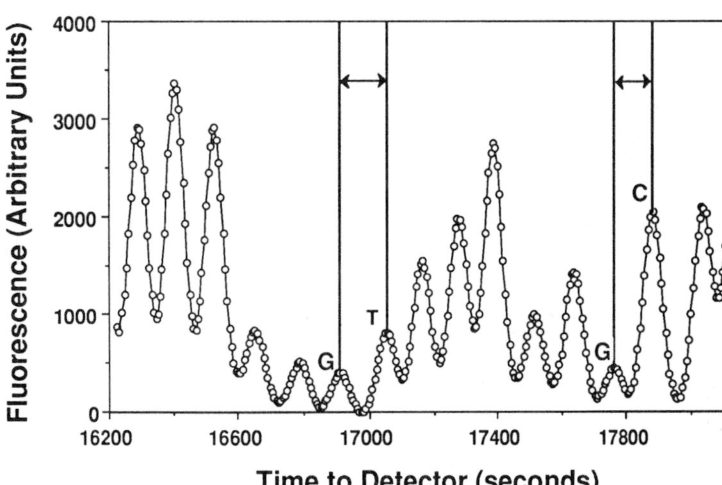

Figure 31.1 Segment of DNA sequencing ladder illustrating variation in the yield (signal intensity) and separation of the oligonucleotides. The sequence is from M13mp18 phage DNA, using T7 DNA polymerase and fluorescent dideoxy terminators (Bowling *et al.*, 1991). The data stream is derived from one of the two parallel filter-photometers of a Du Pont Genesis 2000 sequencer. The unprocessed raw data streams are accessed through our own software interface to the instrument, developed as a collaboration with the Du Pont Co. Peak finding, and the evaluations of oligomer separations and intensities are accomplished using a second-derivative type signal processing algorithm.

nearest neighbor interactions among the 3′ terminal two to three nucleotides of the oligomers (Bowling *et al.*, 1991). The ladder shown in Fig. 31.1 highlights the separations of two oligomer pairs, with the 3′ sequences –dGddC and –dGddT. The separations of these two oligomers from the next shorter oligomers differ by ±17% from the average separation for this length of oligomer.

Oligomer yields represent biochemically determined informative parameters of DNA sequencing ladders. Oligomer separations represent biophysically determined informative parameters. The separations of successive oligomers and their relative intensities are easily measured during the image processing of the ladder data. Our approach to improve accuracy of base-calling (Tibbetts & Bowling, 1991; Bowling, 1992) includes analysis of both the oligomer separations and relative yields, in contextual arrays, together with the primary sequence determinant of a particular system (lane position or fluorescence filter ratio).

31.2.2 Contextual arrays

Relative intensities and separations of a set of oligomers were evaluated from Du Pont Genesis 2000 sequencing data. The upper panels in Fig. 31.2 illustrate strong correlations ($R^2 = 0.97$) of associated data pairs obtained from identical, aligned DNA sequences but from different ladders and different gels. If data from another ladder of unrelated DNA sequence is compared, there is no significant correlation (Fig. 31.2, lower panels). Furthermore, if the arrays of relative intensities and separations are taken from the same sequence, ladder, and gel, but offset from alignment by a single nucleotide, then no correlations are observed.

Extended arrays of the informative parameters are reproducible and attributable to specific sequences. Contextual arrays of relative intensities and separations, together with a model for peak shape (such as an error function), can be used to closely approximate the original ladder image (normalized). The informative parameters are readily evaluated in the normal course of image processing and analysis of an automated system's primary determinants of sequence. Contextual arrays of informative parameters should positively contribute to accurate DNA sequence determination.

31.3 NEURAL NETWORKS FOR PATTERN RECOGNITION

Multiple informative parameters, in contextual arrays, characterize the image of a DNA sequencing ladder. The relationships of this new information to specific DNA sequences appear to be very complicated. Some effort has been directed towards a systematic statistical survey of correlations between the informative parameters and specific DNA sequences. If this were a simple problem of multivariate analysis, then the experience of human experts would have already led to a set of interpretive rules for base-calling using these parameters. Few such rules are familiar.

An alternative to development of an expert system or rule-based system for analysis of sequencing ladders is treatment of the problem in terms of pattern recognition. This approach has been fruitful in recent studies of DNA sequences. A variety of pattern recognition-based programs have focused on searches for biological landmarks represented as short domains

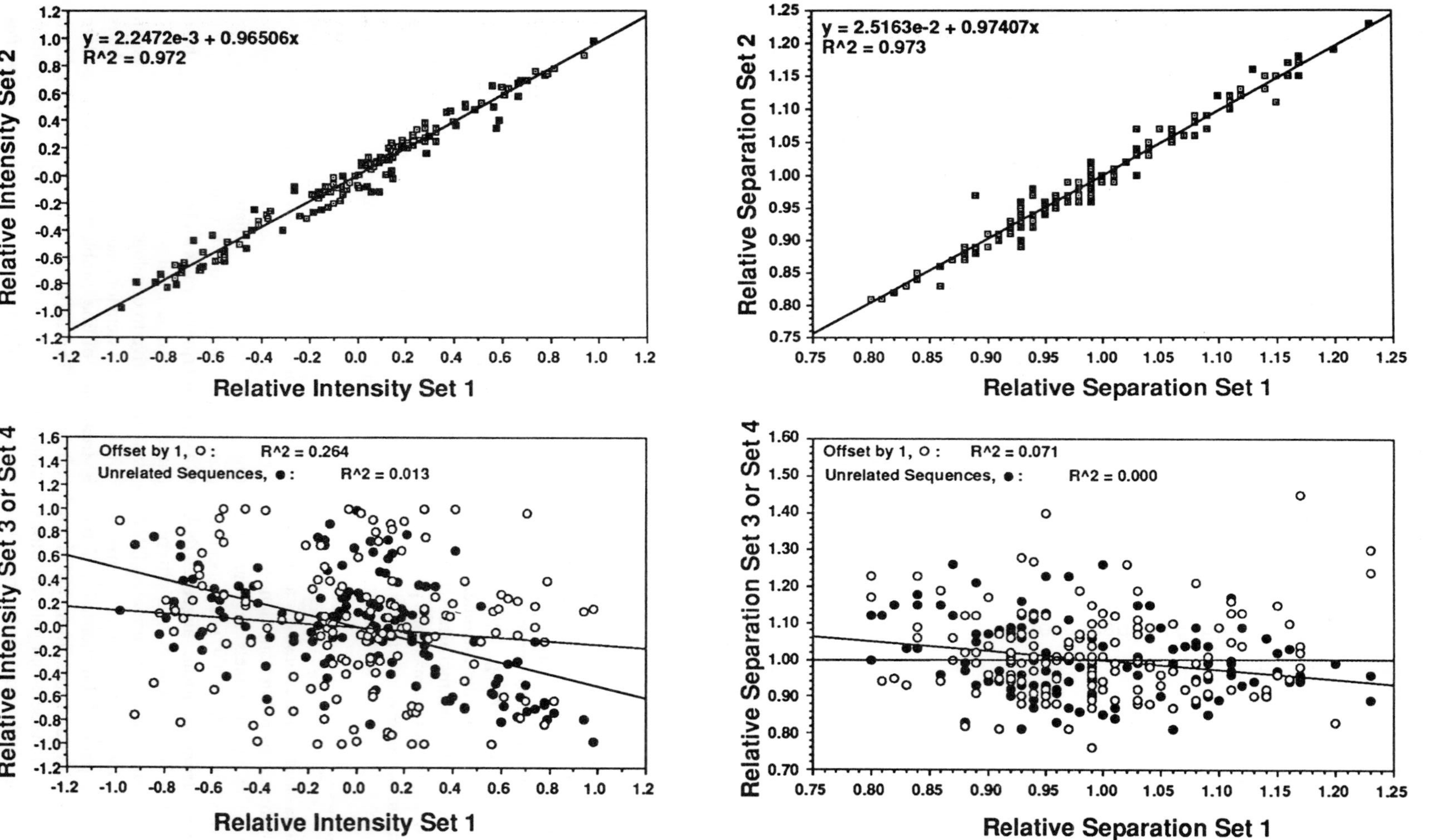

Figure 31.2 Correlations of contextual arrays of relative intensities and separations of oligomers in DNA sequencing ladders. Data similar to that shown in Fig. 31.1 were analyzed for several ladders of 250–350 nucleotides. Relative intensity of an oligomer is evaluated as the ratio of the integrated signal of the oligomer band to that of the preceding (shorter) oligomer in the ladder. Relative separation is calculated as the time of separation of an oligomer and the preceding oligomer, divided by the average separation expected based on oligomer length (Bowling *et al.*, 1991). The upper panels illustrate strong correlation of contextual arrays of relative intensity (left) and separation (right) for identical, aligned DNA sequences analyzed in different ladders of different gels. The lower panels show no correlation of arrays representing different DNA sequences, or identical DNA sequence data with offset alignment.

within long DNA sequences (Stormo *et al.*, 1982; Petersen *et al.*, 1990; Brunak *et al.*, 1990a, b; Hunter, 1991; Uberbacher & Mural, 1991). We have utilized simple back-propagation neural networks as a pattern recognition tool (Jones & Hoskins, 1987; Kosko, 1992a, b; Pao, 1989; Rumelhart, 1986; Simpson, 1990; Wasserman, 1989) for the lower level problem of analyzing raw sequencer data for translation to finished DNA sequences.

An artificial neural network is a paradigm of computer learning. It is represented by layered arrays of processing elements as illustrated in Fig. 31.3(A) and 31.3(B). Each processing element, or node, within the network performs simple computations based on the values of data presented as input. The arrays of processors representing different layers of the network are connected by variable weights. The sum of products of input values and variable connection weights (Σ) is translated to output values in the range of 0 to 1, based on a nonlinear, sigmoid transfer function (∫). If the sum of input products, Σ, is large and negative, output from the node, ∫, approaches limit zero; if large and positive, output approaches limit one.

As each example of an input data array is fed forward through the network, the output nodes present an array of values based on the present state of the network's connection weights (memory). Initially the connection weights are all set to a distribution of small, random values, typically in the range of −0.1 to +0.1. At this stage, the network has no intrinsic capacity to perform its pattern-matching functions.

Training takes place on comparison of output values with those expected for particular examples of data input. Differences between expected and actual outputs of the network are used to make small adjustments of the weights in order to reduce the magnitude of the output error. Through a reiterative process of feeding forward input vectors and back-propagating errors through the weight matrix, the network eventually is trained for optimum performance in matching data to results.

The next three sections of this chapter describe our applications of simple artificial neural networks in automated base-calling programs.

31.4 NEURAL NETWORK BASE-CALLING

31.4.1 Simple analysis of multiple parameters

Sequencing ladders run through the Du Pont Genesis 2000 platform were used to generate archives of input data for training a base-calling neural network (Figs 31.3(B) and 31.3(C)). Values of fluorescence ratio (two photometer channels, primary determinant of system), relative intensity (one channel) and relative separation were recorded for several ladders with 250–350 oligomers. The network was then trained to map the arrays of three variables to one of four binary output node arrays.

Input		*Output*					
Fluorescence ratio		1	0	0	0	=	ddG
separation	———	0	1	0	0	=	ddA
intensity		0	0	1	0	=	ddT
		0	0	0	1	=	ddC

Assignments of terminal dideoxynucleotide codes to each input vector were based on matching the data to known DNA sequences of the samples used for construction of the training data sets.

Figure 31.3(C) illustrates learning curves of the network with different training data sets. Network base-calling accuracy was evaluated throughout the training episodes, as the average percent correct base-calls for six separate training cycles with each set of training data. Note that the data used for the evaluation of network base-calling are from different sequencing ladders than the data used for training of the network.

One training set had all input data values replaced by the constant value 1. Under these conditions, lacking any informative input data, the network fails to learn at all, calling only 25% of bases correctly by random guesswork (not shown). When fluorescence ratio data were included in the training data set, base-calling performance reached a plateau of 95% accuracy after 16 000 iterations (Fig. 31.3(C)). When the training set included the data values for oligomer separations and relative intensities, as additional informative parameters, the performance increased to 99% correct calls (Fig. 31.3(C)). This five-fold increase in accuracy was often maintained over a 5–15% longer sequence than base-calling of the same ladders with fluorescence data only. It is also of interest to note that the most recent and upgraded Du Pont BaseCaller software, versions 4.0 and 5.0, made 95% correct calls on analysis of the same ladders used for training the network. That software is based on signal processing and analysis of the fluorescence ratio as the primary determinant of sequence.

31.4.2 Extended pattern recognition

The relative intensity and separation parameters included in the neural network analysis of Fig. 31.3 make reference only to the immediately shorter oligomers of the ladder image data. The simple network (14 nodes, one hidden layer, 49 connections)

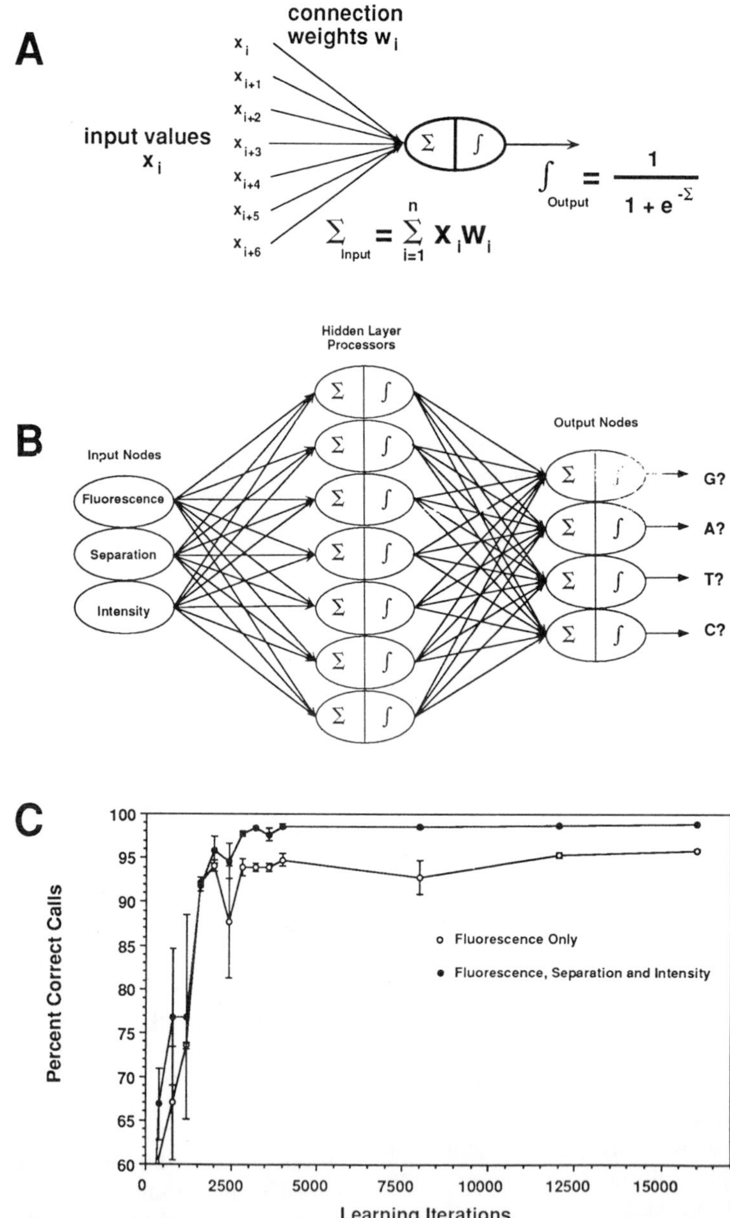

Figure 31.3 (A) A model processing node as an element of an artificial neural network. Values of input (data or output values from higher processors) are multiplied by variable connection weights. The sum of these products is transformed by a nonlinear sigmoid function to determine the response (output) of the node for a given state of inputs. (B) A simple back-propagation neural network for base-calling using the multiple informative parameters of fluorescence ratio (primary determinant), relative separation and relative intensity. The outputs calculated for each input data vector are compared; the node with the largest output (range 0–1) determines the base-call. Output values are clustered about 0.5 early in training, then approach the limits of 0 or 1 as the network learns to recognize patterns of input variables and map these to specific bases. (C) Learning curves for the network with different training sets. (○) 95% correct calls at plateau; this curve was trained with fluorescence ratio data only. (●) 99% correct calls at plateau; this curve was trained with the fluorescence ratio data together with relative separations and intensities. Each point represents the average performance from six independent training sessions at the particular stages of training. Standard error bars are shown, although these are too small to show in the plateau region (the average SE in the plateau region is ±0.2%).

does not take into account longer range correlations of ladder image patterns with sequences.

It is a simple matter to construct larger networks supporting a larger array of input parameters. We have analyzed larger networks with informative parameters representing up to three successive oligomers as short ladder image patterns. Such networks typically have 40–50 nodes, one or two hidden layers, and 500 to 600 connections. Nine input nodes (three parameters, three oligomers) present the somewhat larger patterns of ladder image data. Four output nodes are trained to specify the nucleotide in the centre of the trinucleotide input data array. With limited data sets, we have observed modest improvements in accuracy with such networks, at the expense of greater time for training and for preparation of larger training data sets.

In the long run, we envision training networks on patterns of up to seven oligomers (21 input nodes), mapping their arrays of ladder image parameters to the 64 trinucleotides (12 output nodes for three bases) that may appear in the middle of the input arrays. Such a network would likely have two hidden layers, 100–200 nodes and about 2000 connections, Input parameter arrays will be sequentially evaluated, as oligomers [1–7], [2–8], [3–9], [4–10], etc., mapping to the base triplets [3,4,5], [4,5,6], [5,6,7], [6,7,8], etc. This approach will provide a redundant evaluation of each oligomer at each of three positions within patterns of the heptanucleotide data input.

This scaling up of a base-calling neural network raises an important issue regarding the number of examples to be used in training. Experience and common sense suggest that effective training sets should represent multiple examples of each of the patterns likely to be analyzed by the trained network. In our limited studies thus far, using data generated with a Du Pont Genesis 2000 sequencer, several ladders of 300–400 bases have been sufficient to represent experimentally redundant examples of the 64 varieties of trinucleotides. Our proposed larger network requires discrimination among 4^7 = 16 384 examples of different heptanucleotide sequence patterns. A raw-data archive approaching 100 000 bases of sequence may be necessary to experimentally address pattern-recognition-based sequence determination at this level.

Several large-scale sequencing projects are now in progress under the support of the human genome initiative. The Stanford University Yeast Genome Group has made available their extensive archive of finished and confirmed, cosmid-cloned yeast DNA sequences, about 360 kb of raw sequencing data. These data represents almost 900 ladders of Taq DNA polymerase/thermal cycle sequencing reactions, analyzed with the Applied Biosystems ABI 373 DNA sequencer. Access to even more ABI 373 raw data has been offered from another archive of human cDNA sequences. We have developed new software tools (see Section 31.5) to provide access to the raw data within these large ABI 373 data archives. Relative oligomer yields and separations, in contextual arrays, will be analyzed separately and together with the four-channel fluorescence ratio data that is the primary determinant of the ABI 373 sequencing system.

31.5 REAL-TIME SIGNAL CONDITIONING AND FEATURE EXTRACTION

The Applied Biosystems ABI 373 DNA sequencer employs four base-specific fluorescent dye labels in single lane ladders (Smith *et al.*, 1986; Connell *et al.*, 1987). Four data records of the instrument's single photometer are generated for each cycle of scanning the sequencing gel. On each scan, fluorescence from the illuminated position of the gel is sampled as it passes through each of four bandpass filters.

In general, the base-calling accuracy and lengths of sequences obtained from the ABI 373 system have been quite satisfactory. However, the raw data streams recorded by the instrument are transformed by the system's image processing and base-calling software. Their algorithms for this data transformation impede a user's access to the actual separations and relative intensities of the oligomers in the ladder. We are, of course, most interested in the recovery and analysis of these sequence-informative parameters from the raw, unprocessed data streams.

This situation led us to dissect the structure of ABI 373 sample data files. Near the end of ABI 373 data files we found an array of tagged, fixed-length data records which describe the detailed structure of the entire file. The records tagged as 'DATA 0001' through 'DATA 0004', and 'DATA 0009' through 'DATA 0012', give the locations and sizes of data blocks corresponding to the sample's raw and processed data streams. The raw and transformed photometer data are stored as tandem arrays of short integers (2 byte).

We have recently developed a neural network (about 400 nodes and 1600 connections) which operates as a signal conditioner for the raw data streams generated by the ABI 373 sequencer. Sets of the four-channel raw-data values, representing several successive scans, are mapped through the neural network to four output nodes. These are encoded to indicate if one of the four bases is represented at the centre of the array of image data.

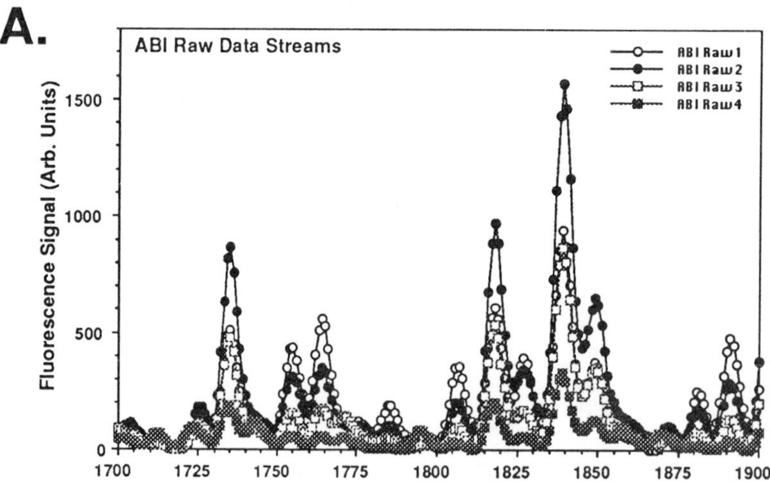

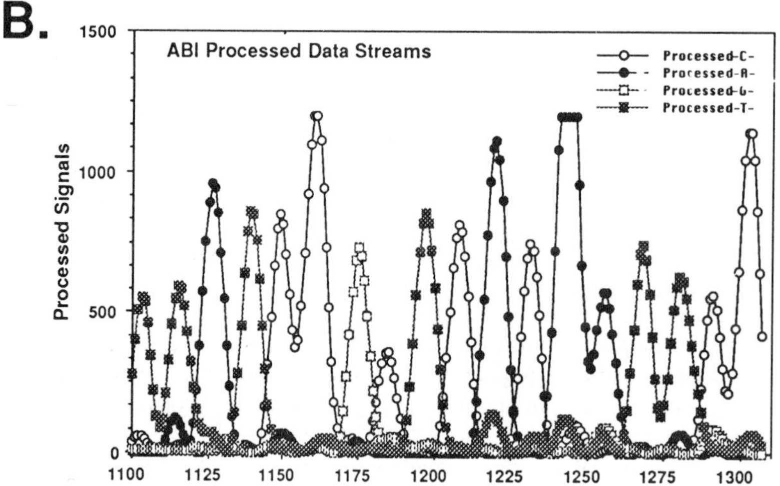

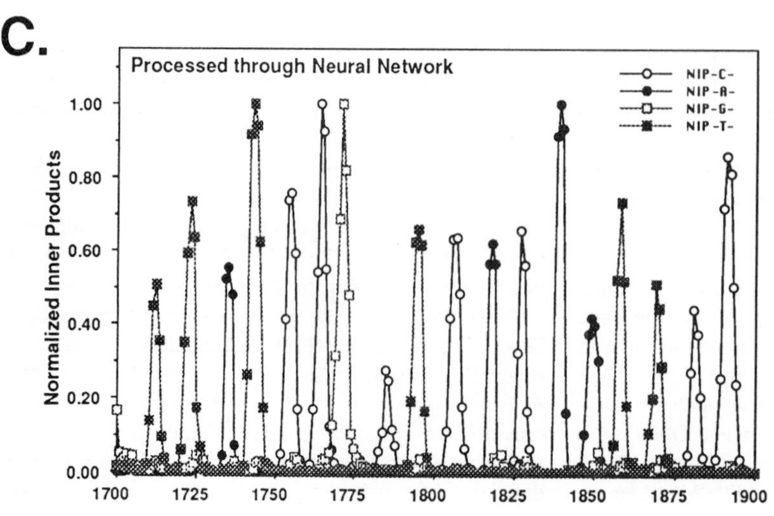

Figure 31.4 Unprocessed and processed data streams from the ABI 373 sequencing system. (A) Raw data streams representing 17 nucleotides of an M13mp18 sequencing ladder. Channels 1, 2, 3, and 4 correspond to the filters which preferentially transmit the fluorescence from the labeled primers of the ddC, ddA, ddG and ddT reactions, respectively. The average intensity of oligomers in the ddT channel is typically low. (B) ABI-transformed data streams corresponding to the interval shown in the upper panel. Coefficients from an inverted matrix of filter-dye-fluorescence crossover are used to transform the data streams to profiles of base-specific signals. Oligomers are displayed with relatively uniform spacing; signal intensities are normalized across the four channels, and the peaks of strong bands are occasionally truncated as the A peak at 1250. Limited channel crossover of base-specific signals is observed. (C) The same raw data in panel (A), transformed with a floating neural network signal conditioner. The network is trained to recognize the peaks of oligomers bearing the four different labels. The data from the four output nodes of the network are multiplied by the corresponding signals in the four channels of the raw-data streams (inner product), then normalized within each channel across the display window. The normalization is only for the display, and is not used in the evaluations of relative intensities of the oligomers. The transform effected by the neural network leads to sharp, well deconvolved bands with very little cross-channel signal. The separations of the oligomers are unchanged with respect to the original raw-data streams.

Input (20 nodes)	*Output (4 nodes)*	
scan i, channels 1, 2, 3, 4	0 0 0 0	= No call
scan $i+1$, channels 1, 2, 3, 4	1 0 0 0	= G
scan $i+2$, channels 1, 2, 3, 4 ←	0 1 0 0	= A
scan $i+3$, channels 1, 2, 3, 4	0 0 1 0	= T
scan $i+4$, channels 1, 2, 3, 4	0 0 0 1	= C

Figure 31.4 illustrates segments from a sequencing ladder as the raw-data streams (A) and processed data streams (B) of the ABI 373 sequencer. The DNA sequence represented is

–TTATCCGCTCACAATTCC–

as is apparent from the processed data streams. The bottom panel of Fig. 31.4(C) shows the profile which results from analysis of the scan-by-scan outputs from the four output nodes of our signal conditioning network. Inner products of the output node values and the corresponding raw fluorescence data provide a transformed data stream which preserves the informative image parameters of the oligomers (relative intensities and separations).

Several advantages are apparent from this application of a neural network for signal-image processing, compared with the processing of the data streams by ABI 373 software system. The traces of the individual oligomers are narrower and better separated within each data channel. There is less crossover of signals between the data channels. The neural network operates without reference to an external file of coefficients, used in the ABI program's four dye multicomponent analysis. These features of the neural network transformation of the raw data facilitate peak finding and evaluation of the informative parameters of oligomer separations and intensities.

Another advantage of this approach to signal processing is that it could be readily implemented for on-the-fly processing during the prolonged process of data acquisition. A very small buffer of recent scan data is sufficient to process each incoming scan line of data through the neural signal conditioner. The instrument's scanning control software would be minimally encumbered while generating arrays of informative parameters for the oligomers as they pass the detector. We estimate that this approach could reduce the post-run processing and base-calling time per DNA sample by a factor of 10 to 100.

31.6 AUTOMATED SEQUENCE PROOFREADING

Considerable time and experience have already been invested in the development of base-calling programs for various automated DNA sequencing systems. Each system has idiosyncrasies of instrument design which have been exploited and managed to finely tune base-calling performance. This situation led us to a feasible and expeditious alternative to a completely restructured, pattern recognition-based base-calling system. A neural network can review the raw data and base-calls generated by a conventional base-calling pro- gram, to identify and specify correction of erroneous or ambiguous calls. This neural network would operate as an auto-editor or proofreader of base-calls. The accuracy of called DNA sequences would improve, because the auto-editor network would exploit more information from the sequencing ladder images than is used by the conventional base-calling programs.

Table 31.1 Encoding of editing actions.

	Output node									
Editing of base No. 2	*1*	*2*	*3*	*4*	*5*	*6*	*7*	*8*	*9*	*10*
No change	1	0	0	0	0	0	0	0	0	0
Delete call	0	1	0	0	0	0	0	0	0	0
Add G	0	0	1	0	0	0	0	0	0	0
Add A	0	0	0	1	0	0	0	0	0	0
Add C	0	0	0	0	1	0	0	0	0	0
Add T	0	0	0	0	0	1	0	0	0	0
Change to G	0	0	0	0	0	0	1	0	0	0
Change to A	0	0	0	0	0	0	0	1	0	0
Change to C	0	0	0	0	0	0	0	0	1	0
Change to T	0	0	0	0	0	0	0	0	0	1

A simple back-propagation network was designed (about 400 nodes and 4000 weighted connections) to demonstrate the feasibility of this software approach to improving base-calling accuracy. Intensities and separations of oligomers are routinely evaluated in the course of image processing and base-calling of sequencing ladders. The input data vectors for the network are these informative parameters together with corresponding base-calls, in arrays of four successive calls. The network is trained to map input vectors to 10 output nodes, corresponding to 10 possible editing actions on the second position of the input array. These editing actions are encoded among the output nodes as indicated in Table 31.1. The desired output vectors for the training data sets are determined by which, if any, editing action may be required to establish consensus of the called sequence with the previously determined sequence of the DNA samples.

Networks have been trained using sets of experimental data derived from an automated DNA sequencer (Millipore BaseStation). Figure 31.5 shows an example of early results of training an auto-editor network. Network output values, representing no action or editing, are plotted for sorted arrays of erroneous and correct base-calls. Among the incorrect base-calls,

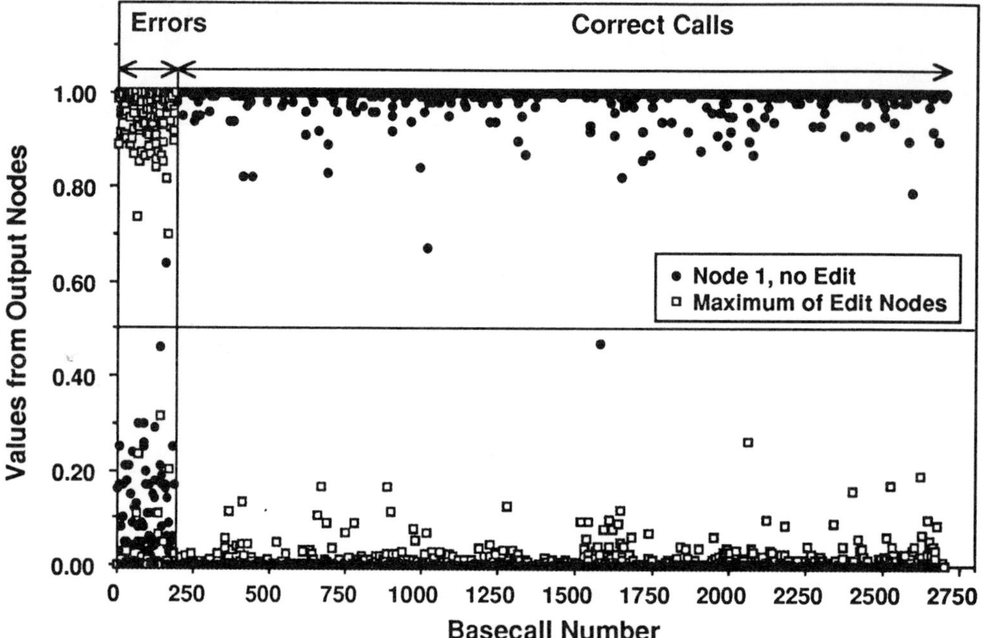

Figure 31.5 Analysis of base-calls with a neural network sequence auto-editor. A neural network was trained to review base-calling together with the raw data of an automated DNA sequencer (Millipore BaseStation). Training sets of raw data and sequences were generated from experimental files representing 2000–3000 base-calls, from 8 to 12 sequencing ladders. The sequences used in these experiments had approximately 93–95% correct calls. Thus each network was trained with approximately 200 erroneous calls. The *x*-axis represents a sorted array of training set base-calls, clustering incorrect calls to the left (not all of the correct calls are shown). The outputs of node 1 (●, no-edit) frequently fall below a 0.5 threshold in the region of the error calls. (○) Maximum output from nodes 2–10 (see text) on review of each base-call, indicating the need for specific editing action. Many of the error calls are flagged with high values among these nine output nodes. Approximately 50–75% of the errors in the training sets are recognized for correct editing. Among the data shown, only one correct call is erroneously flagged by a low value of the no-edit node.

clustered together at the left of each *x*-axis array, between 50 and 80% of the errors are flagged by low values (<0.1) of output node 1 (no change), or high values (>0.9) of one of the output nodes 2 through 10 (specific editing action). Through the remaining input vectors, corresponding to correct base-calls in the original sequence file, there are few, if any, outputs from the network that flag an error or prescribe a corrective editing action. If such networks can reliably trap and correct 50–90% of erroneous base-calls, then the error frequency of base-calling would decrease by a factor of 2–10.

These early experiments have been performed with diverse sets of data representing a different gel and sequencing reaction conditions. We are evaluating relationships between the selection of data for the training sets and the performance of the networks. This will likely involve decisions which balance the benefits of relative accuracy and breadth of application. We believe that optimum performance will require data derived from specific conditions of DNA sequencing chemistry and gel electrophoresis. This could be managed by training individual networks to accommodate varieties of sequencing strategies now in common practice. This is not a serious problem for practical application of this type of program. An identical feed-forward network architecture, the engine of the neural auto-editor, can refer to one of several external files (about 16 kbyte) representing an appropriately trained set of connection weights.

31.7 CONCLUSIONS

The relative intensities and separations of oligomers, represented in images of DNA sequencing ladders, are informative parameters. Analysis of these parameters, together with the primary determinant of an automated sequencer, can lead to DNA sequences with significantly increased accuracy over a modestly increased length.

Artificial neural networks are very useful for rapid development of programs which implement a pattern-recognition-based approach to DNA sequence determination.

(1) Neural networks can be trained to rapidly translate multiple informative parameter arrays from raw data to accurate DNA sequences.
(2) A neural network signal conditioner facilitates peak finding and expedites extraction of multiple informative parameter arrays used for determination of DNA sequences.
(3) A neural network sequence auto-editor may improve the accuracy of DNA sequence files, while taking advantage of the specialized signal and image processing software developed for individual automated DNA sequencing systems.

ACKNOWLEDGMENTS

This work was initiated, and supported in part, as a Collaborative Research Program of E.I. Du Pont de Nemours & Co. and Vanderbilt University. Additional and ongoing support has been provided through a grant from the United States Department of Health, National Center for Human Genome Research (HG00562). Access to a large archive of raw sequencing data (ABI 373) was made available through collaboration with the Stanford University Yeast Genome Project. We appreciate the enthusiastic support and cooperation of the members of that group, including David Botstein, Ron Davis, John Mulligan, Kevin Hennessy and George Hartzell. The Institute for Genomic Research (TIGR) (Craig Venter, Tony Kerlavage, and Jenny Kelley) has provided access to similar data from the archives of their human cDNA sequencing program.

We gratefully acknowledge the interest and support of the Millipore Corporation and its BioImage group: Wally Welch, Todd Dahlberg, Andrew Creasey and Tish Dunn-Creasey. They provided feature tables from DNA sequencing data, generated using the Millipore BaseStation automated DNA sequencing instrument.

REFERENCES

Ansorge, W., Sproat, B., Stegemann, J., Schwager, C. & Zenke, M. (1987) *Nucleic Acids Res.* **15**, 4593–4602.

Ansorge, W., Zimmermann, J., Schwager, C., Stegemann, J., Erfle, H. & Voss, H. (1990) *Nucleic Acids Res.* **18**, 3419–3420.

Bean, C.P. & Hervet, H. (1983a) *Biophys. J.* **41**, A289.

Bean, C.P. & Hervet, H. (1983b) *Bull. Am. Phys. Soc.* **28**, 444.

Bowling, J.M. (1992) Contextual interpretation of oligodeoxynucleotide arrays for determination of DNA sequences. Ph.D. Dissertation, Vanderbilt University, Nashville, TN, March 1992.

Bowling, J.M., Bruner, K.L., Cmarik, J.L. & Tibbetts, C. (1991) *Nucleic Acids Res.* **19**, 3089–3097.

Brunak, S., Engelbrecht, J. & Knudsen, S. (1990a) *Nature* **343**, 123.

Brunak, S., Engelbrecht, J. & Knudsen, S. (1990b) *Nucleic Acids Res.* **18**, 4797–4801.

Connell, C., Fung, S., Heiner, C., Bridgham, J., Chakerian, V., Heron, E., Jones, B., Menchen, S., Mordan, W., Raff, M., Recknor, M., Smith, L., Springer, J., Woo, S. & Hunkapiller, M. (1987) *BioTechniques* **5**, 342–348.

Edmondson, S.P. & Gray, D.M. (1984) *Biopolymers* **23**, 2725–2742.

Hervet, H. & Bean, C.P. (1987) *Biopolymers* **26**, 727–742.

Hindley, J. (1983) In *Laboratory Techniques in Biochemistry and Molecular Biology*, T.S. Work & R.H. Burdon (eds). Elsevier Biomedical Press, Amsterdam.

Hunter, L. (1991) *AI Mag.* **11**, 27–37.

Jensen, M.A., Zagursky, R.J., Trainor, G.L., Cocuzza, A.J., Lee, A. & Chen, E. (1991) *DNA Sequence* **1**, 233–239.

Jones, W.P. & Hoskins, (1987) *BYTE Mag.* **Oct**.

Kosko, B. (1992a) *Neural Networks and Fuzzy Systems*. Prentice Hall, Englewood Cliffs, NJ.

Kosko, B. (1992b) *Neural Networks for Signal Processing*. Prentice Hall, Englewood Cliffs, NJ.

Kristensen, T., Voss, H., Schwager, C., Stegeman, J., Sproat, B. & Ansorge, W. (1988) *Nucleic Acids Res.* **16**, 3487–3496.

Lerman, L.S. & Frish, H.L. (1982) *Biopolymers* **21**, 995–997.

Lumpkin, O.J. & Zimm, B.H. (1982) *Biopolymers* **21**, 2315–2316.

Maniatis, T., Jeffrey, A. & van deSande (1975) *Biochemistry* **14**, 3787–3794.

Maxam, A.M. & Gilbert, W. (1977) *Proc. Natl. Acad. Sci. U.S.A.* **74**, 560–564.

Pao, Y.-H (1989) *Adaptive Pattern Recognition and Neural Networks.* Addison-Wesley, New York.

Petersen, S.B., Bohr, H., Bohr, J., Brunak, S., Cotterill, R.M.J., Fredholm, H. & Lautrup, B. (1990) *TIBTECH* **8**, 304–308.

Prober, J.M., Trainor, G.L., Dam, R.J., Hobbs, F.W., Robertson, C.W., Zagurski, R.J. Cocuzza, A.J., Jensen, M.A. & Baumeister K. (1987) *Science* **238**, 336–341.

Rumelhart, D.E., Hinton, G.E. & Williams, R.J. (1987) In *Parallel Distributed Processing: Exploration in the Microstructures of Cognition.* Vol. 1: *Foundations*, (D.E. Rumelhart & J.L. McClelland (eds). MIT Press, Cambridge, MA, pp. 318–362.

Sanger, F., Nicklen, S. & Coulson, A.R. (1977) *Proc. Natl. Acad. Sci. U.S.A.* **74**, 5463–5467.

Simpson, P.K. (1990) *Artificial Neural Systems: Foundations, Paradigms, Applications, and Implementations*. Pergamon Press, New York.

Smith, L.M., Sanders, J.Z., Kaiser, R.J., Hughes, P., Dodd, C., Connell, C.R., Heiner, C., Kent, S.B.H. & Hood, L.E. (1986) *Nature* **321**, 674–679.

Stormo, G.D., Schneider, T.D., Gold, L. & Ehrenfeucht, A. (1982) Nucleic Acids Res. **10**, 2997–3011.

Tabor, S. & Richardson, C.C. (1987) *Proc. Natl. Acad. Sci. U.S.A.* **84**, 4767–4771.

Tabor, S. & Richardson, C.C. (1989) *Proc. Natl. Acad. Sci. U.S.A.* **86**, 4076–4080.

Tabor, S. & Richardson, C.C. (1990) *United States Patent No. 4,962,020*, 9 October 1990.

Tibbetts, C. & Bowling, J.M. (1991) *United States Patent Application*, (pending).

Toneguzzo, F., Beck, J., Cahill, P., Ciarkowski, M., Page, G., Glynn, S., Hungerman, E., Levi, E., Ikeda, R., McKenney, K., Schmidt, P. & Danby, P. (1989) *BioTechniques* **7**, 866–877.

Uberbacher, E.C. & Mural, R.J. (1991) *Proc. Natl. Acad. Sci. U.S.A.* **88**, 11261–11265.

Wasserman, P.D. (1989) *Neural Computing: Theory and Practice*. Van Rostrand Reinhold, New York.

CHAPTER THIRTY-TWO

Advances in Sequence Assembly

E.W. MYERS

Department of Computer Science, University of Arizona, Tucson, AZ 85721, USA

32.1 INTRODUCTION

The scale of DNA sequencing projects has grown to the point that data analysis is becoming a significant aspect of the process. Shortly after the development of the sequencing techniques of Sanger *et al.* (1977) and Maxam & Gilbert (1977) just 15 years ago, investigators were sequencing stretches of 1000 to 5000 nucleotides and were quite content with assembling the fragment data from their gel runs manually. Even as projects became more ambitious, the assembly step itself was not very time-consuming, and simple software that assisted manual assembly by pointing out the overlaps between fragments was all that was required. But today, projects to sequence entire cosmids or small yeast artificial chromosomes (YACs) of 40 000 to 200 000 nucleotides are routinely being undertaken, propelled by goals such as the Human Genome Initiative (US Department of Energy, 1992).

A variety of schemes have been proposed for sequencing such large stretches. The traditional shotgun approach still has a great deal of appeal because it is economical, parallelizable, and automatable (Sanger *et al.*, 1982). However, for the large stretches anticipated, coverage is a significant problem. Statistical consideration alone shows that if one sequences six genome equivalents of 1 kbp fragments of a 50 kb clone, then one can expect two to six gaps in the resulting assemblages of the data (Lander & Waterman, 1988). Of course, biological and experimental considerations tend to further increase this number. It is clear that in order to achieve closure other techniques such as directed sequencing or polymerase chain reaction (PCR) must be employed near the end of a large-scale shotgun project. However, investigators aware of the problem have concocted the following variations that can reduce the number of contig gaps that need to be covered by more costly alternatives.

(1) *Dual end*: sequence both ends of inserts that are designed to be larger than twice the average gel run length.
(2) *Sequencing 'noise'*: run each gel run well past the point where accurate data are being gathered. The 'noisy' tail still contains enough information for computing overlaps (but is not used for determining sequence content).
(3) *Sampling without replacement*: sequence only those clones that do not hybridize with any sequenced so far. This accelerates one to the 'knee' of the coverage probability curve.

It is thus clear that in any large-scale 'shotgun' style project, there will actually be additional information about the relative position and quality of sequence data that a good software solution should take into account.

Besides shotgun strategies, there are many other distinctive approaches that are actively being pursued in an attempt to produce more economical, accurate, and faster methods. For example, there is much excitement about sequencing by hybridization (Drmanac *et al.*, 1989). However, it should be noted that most of these proposals really amount to very rapid ways to produce fragment data and still beg the problem of assembling the fragments. In another direction, improvements in the cost of producing primer oligonucleotides may make it quite economical to directly sequence large stretches via primer walking (Studier, 1989). PCR using frequently occurring subsequences or transposable elements as primers can be used to obtain multiple start points for such directed walks, so that a higher level of parallelism can be achieved. More exotic technologies, such as directly reading the sequence of a single strand with a scanning tunneling microscope (Allison *et al.*, 1990), are contemplated but well into the future. While these methods produce data which are easier to 'assemble', it seems that the most robust approach from the view of developing software is to build a system that handles the combinatorially most complex problem, i.e. 'pure' shotgun, but which can gracefully handle problems for which more is known about how the fragments assemble. For example, even for directed sequencing the computational aspect still presents the common problems of aligning walks in opposite directions and producing a consensus over all walks.

For the remainder of the paper, we consider the computational problem posed by a large shotgun sequencing project where the strategy is not 'pure' in the sense that there is additional information about how fragments overlap. We divide our treatment into aspects that affect the underlying combinatorial algorithms, and those that are basically engineering issues essential for a high capability and user-useful system. Well-designed and practical algorithms for the fragment assembly problem should be capable of handling all of the following:

(a) *Errors.* Sequencing error generally runs at about 1% but can be as high as 5%. The ability to tolerate high error rates is desirable: all data is useful according to 'noisy' sequencing advocates. Errors impact the determination of fragment overlap and require multialignment in high-coverage regions in order to determine an overall consensus reconstruction.

(b) *Unknown orientation.* For many experimental protocols one does not know if a fragment is from the 5′ or 3′ strand.

(c) *Incomplete coverage.* Needless to say, one cannot expect a given data set to assemble into a single contig. One should be able to assemble the fragments available at any point during a project in order to get some sense of their progress in terms of coverage and the desirability of continuing to shotgun.

(d) *Vector removal and multiple inserts.* Invariably, some initial portion of a fragment sequence belongs to the vector, and occasionally, for very short inserts, some of the final portion as well. Also, for some experimental protocols, two or more inserts may be spliced into a single vector. These artifacts must be *detected* by the algorithm.

(e) *Overlap and orientation constraints.* As mentioned several times, a project may employ protocols for which there is additional information about where a fragment comes from or how it overlaps with others. For example, when sequencing without replacement one knows the fragments *do not* overlap, and when dual-end sequencing one knows that the ends are in opposite orientations, non-overlapping, and at some approximate distance apart. A method that accommodates such information and, even better, is sped up by such information, is desirable.

(f) *Alternative solutions.* While a given computer-generated solution may 'look' very reasonable, how do you know it is clearly the best or that there are no other assemblages that are also consistent with the data? An algorithm that has the built-in ability to generate alternative solutions in some rank order of 'goodness' is very desirable. One can, with confidence, accept a given solution if the next best is clearly inferior.

Similarly, the following list enumerates a set of capabilities that should be available to the investigator using a software system that has at its heart a suite of algorithms meeting the specifications listed above.

(1) *Fragment database.* When fragments are entered into the system, one should be able to enter additional information such as experimental conditions, responsible investigator, date of entry and, if at all possible, a digitized image of the raw data. Over time, one should be able to annotate the removal of vector sequence, the discovery of multiple inserts, and corrections to the raw data. The raw data must never be removed. All this requires a modest 'database' of the fragment data.

(2) *Assembly browser.* For a large project, the assemblages will be very large, so the interface must include an easy-to-navigate, window-based, browser for examining assemblages. Essential to this end is autocorrelation amongst views of the data.

(3) *Multialignment editor.* The multialignments produced in high-coverage regions will reveal sequencing errors. An editor for correcting these is a must. Such an editor should be able to automatically bring the user to a region that needs examination, and should be able to display the original raw data, e.g. an ABI waveform or digitized gel image, on screen and synchronized with the current editing focus.
(4) *Revision control.* A simple revision control mechanism is needed to manage and keep track of refinements to the fragment data from (a) the removal of vector sequence, (b) the discovery of multi-inserts, and (c) manual correction with the sequence editor. Such revision control should be transparent to the user except when needed.
(5) *A posteriori constraints.* Any automated system should permit a user to override the decisions it makes. The user should be able to manipulate and constrain the assemblages produced *a posteriori*, based on human insight.
(6) *A priori constraints.* Mixed strategies must be employed for closure near the end of a sequencing project, and strategies such as sampling without replacement and dual-end sequencing may be used to further alleviate the problem. In these cases there is additional *a priori* information that constrains the nature of the assembly and one should be able to signal these to the system.
(7) *Seamless integration.* The software system should permit one to easily apply other sequence analysis tools to the reconstructions it produces and to export said to other software, such as similarity searches against national databases.

There is no current system and/or algorithm suite that meets all the requirements in the two lists above, and most meet very few. However, the level of sophistication is rising rapidly and current systems under development should come very close to meeting all the capabilities above. The sections that follow review the progress to date on algorithms and software systems; We conclude with future developments.

32.2 ALGORITHMS FOR FRAGMENT ASSEMBLY

Given short fragment sequences randomly sampled from a long unknown DNA sequence, the shotgun fragment assembly problem is to determine the most likely reconstruction of the original sequence. The earliest work was so *ad hoc* in nature that a formal statement of the problem the software was attempting to solve was not given and the underlying algorithms did little more than assist in 'melding' fragments together (Staden, 1979; Gingeras *et al.*, 1979). As work progressed it became clear (Peltola *et al.*, 1984; Turner, 1986; Kececioglu, 1991) that the formal problem for a shotgun problem addressing items (a), (b), and (c) in Section 32.1 was in essence a 'noisy' shortest common superstring problem:

> *The fragment assembly problem.* Given a collection of fragments $\mathcal{F}$ and a small error tolerance $\varepsilon \in [0,1]$, find the *shortest* string R such that for every fragment $F \in \mathcal{F}$, F or its Watson–Crick complement aligns with a substring of R with $\varepsilon|F|$-or-less differences (insertions, deletions, and substitutions).

R is a common superstring in that every fragment F or its complement is (approximately) a substring of R.

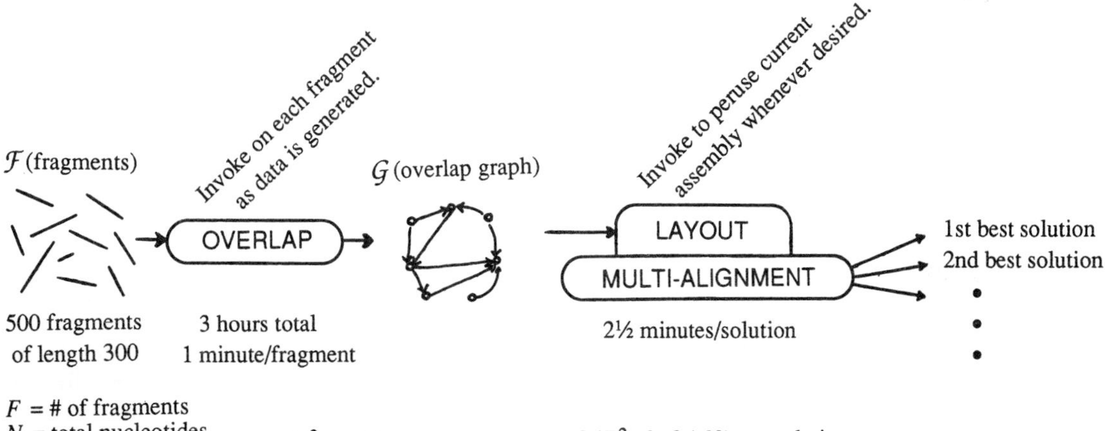

Figure 32.1 A three-phase sequence assembly algorithm.

The 'noise' is modeled by ε which is an upper bound on the sequencing error rate, e.g. $\varepsilon = 0.05$ asserts that no gel runs have more than 5% of their base-calls in error. The criterion that R be as short as possible is an appeal to the principle of parsimony: R is the shortest possible explanation of all the data $\mathcal{F}$.

Theoretically, the problem is in the class of NP-hard problems (Maier, 1978), implying that a procedure that is guaranteed to be computationally efficient for all problem instances probably does not exist. However, empirically the problems that arise in practice can be solved quite well by sufficiently powerful algorithms. Such algorithms generally decompose the problem into the three phases depicted in Fig. 32.1. We discuss each phase and, in particular, our approaches to them (Kececioglu & Myers, 1992).

32.2.1 Overlap phase

This first phase compares all pairs of fragments to determine significant approximate overlaps. We advocate a full-sensitivity approach, as opposed to heuristics which occasionally miss significant overlaps. Our own algorithms score each overlap based on its statistical significance. We are the only ones to do so and can do it because of an incremental alignments algorithm we developed specifically for this problem (Myers, 1986). Efficiency, without losing sensitivity, is afforded by the further use of a Four-Russians prescreener (Wu *et al.*, 1992). For a problem with 500 fragments of average length 300, we can perform this computationally intensive step in 3 h on a 20 MHz workstation. Moreover, this computation can be amortized over the period of data entry which for the 500 fragment project above was 6 months. The results of the overlap comparisons are encoded as edges in a directed, weighted *overlap graph*: each fragment is modeled as a vertex, an edge from A to B models an overlap between A and B, and its weight is the score of the overlap. In our model, a fragment is 'entered' into the overlap graph once by comparing it against those already in the graph. The computation need not be repeated and, as an example, it took less than 2 min to enter the last fragment into the graph for the project above.

32.2.2 Layout phase

This phase takes the overlap graph, call it $\mathcal{G}$, as input and generates a series of alternate assemblies or layouts of the fragments based on the pairwise overlaps therein. A layout specifies the relative locations and orientations of the fragments with respect to each other and is typically visualized as an arrangement of overlapping, directed lines, one for each fragment. One desires a layout that is as short as possible. In terms of the overlap graph, this problem reduces to one of finding a maximum weight *Hamiltonian path* through (Turner, 1986; Kececioglu, 1991). Whereas previous investigators have used a simple greedy heuristic (Staden, 1982; Peltola *et al.*, 1984; Huang, 1992) to find a Hamiltonian path whose weight is near the optimal, we are the first to use a relaxation approach. We took an efficient algorithm for the computationally tractable problem of generating *directed spanning trees* in order of score, and ran this generator until (1) either a tree that was a path was generated, or (2) too much time had elapsed. If one terminates with case (1) then one has the optimal solution in hand. In case (2), a postprocess converts the generated spanning trees into paths by locally optimal edge exchanges, and then reports the highest scoring 'repaired' tree as the solution. It is always the case that these greedily repaired spanning trees produce layouts whose scores are at least as good as those produced by the simple greedy heuristic. Moreover, in practice we find that our algorithm usually terminates via case (1) with an optimal solution. In those cases where it has to repair, we are at least able to report how far from optimal our reported solution might be. Even more important, our approach is fundamentally generative due to the use of the generative spanning tree algorithm and so our software suite meets criterion (f) in Section 32.1. By simply continuing to run the generator, the approach can deliver a second-best solution, then a third-best solution, and so on.

32.2.3 Multialignment phase

The final phase simultaneously aligns the sequences of the fragments in a layout produced by the layout phase giving a final consensus sequence as the desired reconstruction of the original strand. This phase, like the previous one, represents another intractable problem requiring time exponential in the overlap depth to solve optimally. We proceed by producing an initial alignment consistent with all the pairwise alignments of the edges in a branching of the previous phase. This is always possible, computationally efficient and, since the error rate is typically less than 10%, produces a very good first approximation (unlike the related work on protein sequences that are 70% different (Feng & Doolittle, 1987)). In a second step, a 'window' is swept over this initial alignment to optimize the alignment in subregions where the use of global overlap alignments produced locally non-optimal subalignments. With this window sweep we empirically find the resulting multialignment to be almost-everywhere optimal, especially when the error rate is less than 5%.

In closing this section, we note several interesting

points to ponder in considering algorithm designs for this problem. First, most complaints about the quality of solutions produced are attributable to weak algorithms in the overlap and multialignment phases. Some have asserted that it is not worth the seemingly prohibitive amount of time required to do a full-sensitivity dynamic programming comparison of fragments, or a second sweep in the multialignment phase. But it is exactly failures here – poor or undetected alignments, or obviously improvable multialignments – that most annoy the end-user. Moreover, the argument against the amount of time required diminishes with each year as the power and speed of machines continues to rise. The second point concerns the layout phase. While the simple greedy heuristic works on most occasions, it does occasionally fail. The question to ask is whether one can afford these failures and what one is willing to pay to reduce them. We think that if the process is to be automated, then the automaton must be as reliable as possible. So we again will choose more sophisticated approaches in an attempt to increase reliability. Given that the time to gather the data is measured in units of days and months, why should one be disturbed that the accompanying computation takes a few minutes?

32.3 SOFTWARE SYSTEMS FOR FRAGMENT ASSEMBLY

Early software systems did little more than assist in melding fragments together (Staden, 1979; Gingeras *et al.*, 1979). The first software to be based on a reasonably firm analytic foundation was the SEQAID tool (Peltola *et al.*, 1984). These systems were not highly interactive and designed for a simple ASCII terminal interface. While the most recent systems have significantly improved window-based, menu-driven interfaces (e.g. Dear & Staden, 1991), the capabilities of the current workstation technology have not been fully utilized. Moreover, in many cases the underlying algorithm suite is still quite naive and facilities such as revision control ((4) in Section 32.1), constraints C(5) and (6) in Section 32.1) and seamless integration ((7) in Section 32.1) to other tools are not realized. Megabase sequencing projects can benefit, and arguably require, the robust software solutions and sophisticated user interfaces specified in the introduction. We are currently in the process of building such a system (our second), and we discuss some of our design decisions here.

The first and most important design decision was to separate issues of interface and environment from the suite of algorithms solving the underlying combinatorial problem. The suite of algorithms have been packaged into an object-oriented 'kernel' of routines that manipulate overlap graph, sequence, and multialignment objects. This internal interface divides the software into two layers that may be modified or upgraded independently of each other.

As a preliminary design experiment we built a simple browsing system (Miller & Myers, 1991) we call 'fragment assembly browser' (FAB) that realizes capabilities (1) through (4) (Section 32.1). FAB is built upon the X-11 windowing system and Fig. 32.2 shows a sample screen image. Once data have been entered and system parameters set through a number of other menu options, one can elect to browse the current set of fragment assembly solutions. An initial window lists the alternate assemblages for a given sequencing project. Selection of one of the lines summarizing an assemblage opens a window containing a listing of the contigs or islands for that assemblage. Selection of one of these contig lines opens a window graphically portraying the corresponding layout as a series of lines. Finally, selection of a layout opens a window on a multialignment showing the individual bases in the layout. Thus, the overall organization allows one to start at the highest conceptual view of a solution and navigate down towards the most detailed level. All windows are scrollable, for example, multialignments typically do not fit on a screen and so the user can scroll it either left or right to reveal different portions within the window. X-11 permits one to move windows about, change their size, reduce them to icons, and change their priority.

One may open as many windows on each object – contig lists, layouts, and multialignments – as they wish, including several windows on the same object if desired. A window opened as the result of a selection on another is termed the child of the originating or parent window. Among layouts and multialignments a box in the parent shows the region of the child currently being displayed in the child's window. Scrolling the child moves the box in the parent, and either dragging the box or scrolling the parent updates the view in the child. This *autocorrelation* feature greatly facilities navigation and prevents one from 'getting lost'. One can zoom in or out on layouts, compare two layouts, and there is a limited editing capability for multialignments. All the interaction is mouse/button driven with menus much in the 'Macintosh' style.

Conceptually FAB maintains a project database consisting of a collection of fragments and an overlap graph modeling the approximate overlap relationships between them. As new fragment data becomes available it is entered in this database, given a name, and compared against all other fragments in order to keep the overlap graph up to date. Additional information, such as the date of entry and sequencing

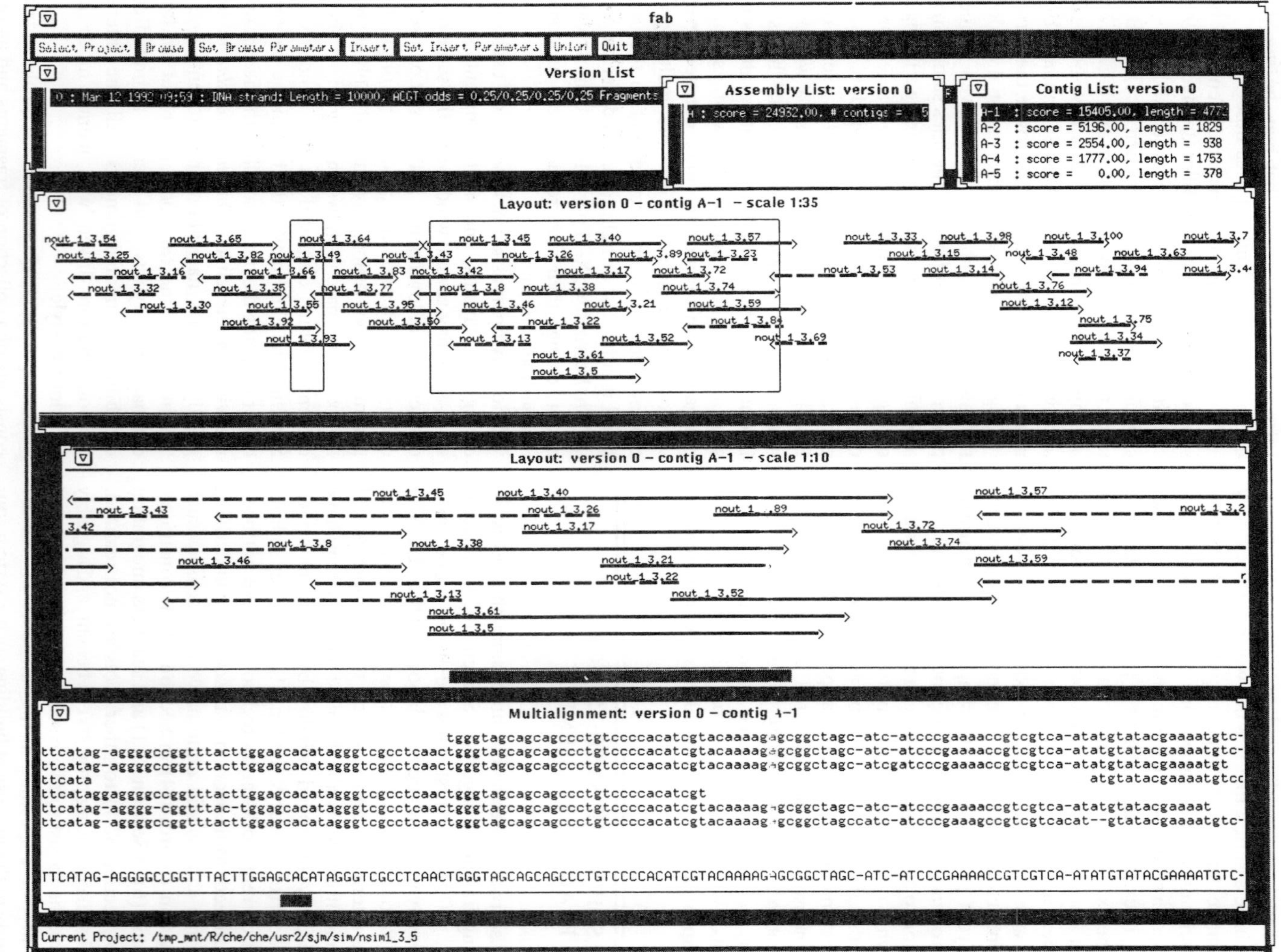

Figure 32.2 Sample browser screen.

conditions, is also associated with each fragment. (Ultimately FAB will also be capable of maintaining a digitized gel image or other representation of the raw data.) At any time during the course of the project users can invoke the assembler upon the fragments or a subset of the fragments currently in the database. Users are immediately placed in the browser with the results of the assemblage as described above. While browsing the results, they can edit multialignments (ultimately in consultation with the stored digitized gel images). Furthermore, they can explore alternate layouts of the fragments by asking where a fragment or contig fragments might otherwise be incorporated, and then constraining the assembler to utilize such an overlap relationship. When such a session is over, users are asked if they wish to save the changes to a fragment's sequence implied by a multi-edit, and if they wish to save a set of assembly constraints used in editing a layout. These changes, if saved, must be reflected back into the project database, thus closing a feedback cycle.

This feedback cycle is potentially dangerous. What if a user later decides that the multialignment changes were not correct, or that a layout rearrangement was not advisable based on new sequence data? If the project database is updated destructively, then this previous state is lost. Thus, we argue that revision control is necessary in an assembly environment. Our context presents a much simpler problem than that classically required for software maintenance. Revision control for fragment assembly need only support the model that a series of progressive refinements lead ultimately to correct sequence. In this model, the *history* of a fragment is a *linear* series of revisions leading from the raw data to the current revision. Each revision, except the current one, has exactly one successor. Using standard technology, such revision control typically requires only an additional 20–30% space overhead over holding just the raw data.

All raw fragment data entered into the system are placed in project revision 0. Each set of edits performed and saved in a session with the browser creates a new project revision that supersedes the previous one. Revisions of a project are numbered sequentially and time-stamped by default. They can be given explicit names, and comments can be associated with them. By default, a user is always working with the most recent revision, so that revision control is effectively *invisible* until a user explicitly needs to utilize the capability, say to undo some changes or re-examine a series of earlier corrective decisions. The most recent set of revisions may be removed, changes to an individual fragment can be undone, and histories of fragments or groups of fragments may be 'rolled back'.

Finally, we turn attention to constraints (5) and (6) (Section 32.1). Such capabilities must be supported by the underlying algorithms for assembly ((e) in Section 32.1). To date, no system has such capabilities and we are the first to address them at the algorithmic level (Kececioglu & Myers, 1992). Our relaxation approach to the layout phase naturally supports constraints: the spanning tree generator algorithm already produces alternate trees by placing constraints on the *edges* (i.e. overlaps) that must be in or not in the next spanning tree. In a completely analogous manner one can modify such generators to produce solutions with the fragments in given orientations. This algorithmic capability will translate at the system level into users being able to assert that they wish to see all solutions in which certain overlaps are 'IN', certain overlaps are 'OUT', and certain pairs of fragments are in either the 'SAME' or 'OPPOSITE' orientation. Users may use such a feature to override decisions automatically made by the assembly kernel whose parsimony-based optimization criterion may produce erroneous results in, for example, highly repetitive regions. We qualify such constraints as *a posteriori*, because they are imposed after investigators have viewed the results of an assembly and applied their human expertise.

A priori constraints are those that are intrinsic to the data collection strategy, such as dual-end sequencing or the use of directed sequencing or nested deletions near the end of a project when investigators move to other techniques to span gaps between contigs. All such protocol constraints can be mapped internally to a set of overlap and orientation constraints, with the exception of dual-end sequencing which requires one to be able to assert that a given pair of fragments occur at a certain distance range from one another. With this additional internal constraint mechanism a rich set of *a priori* constraint scenarios can be handled. At the interface level we plan to support a menu of protocol types, such as 'NESTED', 'DIRECTED', 'DUAL', etc., that can be applied to subsets of the fragments and are automatically translated into the appropriate set of overlap, orientation, and distance constraints.

32.4 FUTURE DEVELOPMENTS

Software development for handling the computational aspects of an experimental protocol typically lags behind the innovation of the methods. In the case where techniques are not well established, no one is willing to invest the manpower to produce a production quality software system. However, shotgun assembly and its variations have become so commonplace that a significant number of computational scientists have studied the problem and a number of commercial concerns have perceived enough of a market to produce products. None currently achieves all the specifications of the list in the introduction, but there are no technical barriers to such an end. As the nature of the needed computational support becomes better delineated, such products will appear on the scene.

On the purely algorithmic and analytic front, refinements to the accuracy and speed of the underlying computation continue to be made. For example, our next suite of kernel algorithms will improve on the speed of the overlap computation by at least a factor of 30 by virtue of further algorithmic advances. By moving from a relaxation approach using spanning trees to one involving the graph-theoretic concept of a *matching*, we expect to greatly increase the reliability of the layout phase. Finally, analytic advances in the analysis of multialignments will improve the error correcting/detecting accuracy of the multialignment phase. All these advances are currently in development and we anticipate their arrival over the course of the next few years.

ACKNOWLEDGMENTS

This work was supported in part by the National Laboratory of Medicine under grant No. R01 LM04960 and the Aspen Center for Physics.

REFERENCES

Allison, D.P., Thompson, J.R., Jacobson, K.B., Warmack, R.J. & Ferrell, T.L. (1990) *Scanning Microsc.* **4**, 517–522.

Dear, S. & Staden, R. (1991) *Nucleic Acids Res.* **19**, 3907–3911.

Drmanac, R., Labat, I., Brukner, I. & Crkvenjakov, R. (1989) *Genomics* **4**, 114–128.

Feng, D. & Doolittle, R. (1987) *J. Mol. Evol.* **25**, 351–360.

Gingeras, T.R., Milazzo, J.P., Sciaky, D. & Roberts, R.J. (1979) *Nucleic Acids Res.* **7**, 529–545.

Huang, X. (1992) *Genomics* **14**, 18–25.

Kececioglu, J.D. (1991) *Exact and Approximate Algorithms for DNA Sequence Reconstruction* (*Technical Report 91–26*). Department of Computer Science, University of Arizona, Tucson, AZ 85721.

Kececioglu, J.D. & Myers, E.W. (1992) *Algorithmica* in press.

Lander, E.S. & Waterman, M.S. (1988) *Genomics* **2**, 231–239.

Maier, D. (1978). *J. ACM* **25**, 322–336.

Maxam, A.M. & Gilbert, W. (1977) *Proc. Natl. Acad. Sci. U.S.A.* **74**, 560–564.

Miller, S. & Myers, E.W. (1991) *A Fragment Assembly Project Environment* (*Technical Report 91–17*). Department of Computer Science, University of Arizona, Tucson, AZ 85721.

Myers, E.W. (1986) *Incremental Alignment Algorithms and Their Applications* (*Technical Report 86–2*). Department of Computer Science, University of Arizona, Tucson, AZ 85721.

Peltola, H., Söderlund, H. & Ukkonen, E. (1984) *Nucleic Acids Res.* **12**, 307–321.

Sanger, F., Nicklen, S. & Coulson, A.R. (1977) *Proc. Natl. Acad. Sci. U.S.A.* **74**, 5463–5467.

Sanger, F., Coulson, A.R., Hong, G.F., Hill, D.F. & Petersen, G.B. (1982) *J. Mol. Biol.* **162**, 729–773.

Staden, R. (1979) *Nucleic Acids Res.* **7**, 2601–2610.

Staden, R. (1982) *Nucleic Acids Res.* **10**, 4731–4751.

Studier, F.W. (1989) *Proc. Natl. Acad. Sci. U.S.A.* **86**, 6917–6921.

Turner, J. (1986) *Approximation Algorithms for the Shortest Common Superstring Problem* (*Technical Report WUCS–86–16*). Department of Computer Science, Washington University, St Louis, MO 63130.

US Department of Energy (1992) *Human Genome: 1991–92 Program Report*. National Technical Information Service, US Department of Commerce, 5285 Port Royal Road, Springfield, VA 22161.

Wu, S., Manber, U. & Myers, E.W. (1992) *A Sub-quadratic Algorithm for Approximate Limited Expression Matching* (*Technical Report 92–36*). Department of Computer Science, University of Arizona, Tucson, AZ 85721.

CHAPTER THIRTY-THREE

Computer-aided Sequence Reconstruction: Software Support for Multiple Large-scale Sequencing Strategies

S. HONDA, N.W. PARROTT & C.B. LAWRENCE

Computational Molecular Biology Group, Department of Cell Biology, Baylor College of Medicine, 1 Baylor Plaza, Houston, TX 77096, USA

33.1 INTRODUCTION

The Genome Reconstruction Manager (GRM) is a long-term software engineering project to develop a system to support large-scale sequencing efforts. The system provides a comprehensive environment for completing all tasks related to sequence reconstruction from data generated by automated sequencing machines. GRM supports a range of sequencing strategies from random to fully directed, as well as hybrid strategies that use combinations of both the random and directed approaches. This chapter provides a user's perspective of the GRM environment and begins with an interpretation of the sequence reconstruction problem. The figures included in this chapter are screendumps from a working prototype of the system demonstrated at the Genome Sequencing and Analysis Conference IV.

Biologists play a pivotal role in successful design of molecular biology software. Not only are they the end users of a system, but they follow a certain logic in going about their daily work. The methodology and logic they employ is an integral part of what must be incorporated into a useful sequence reconstruction environment. GRM is being developed using a principle known as 'user-centered design' (Norman & Draper, 1986). In short, the design is based on what users see as their goals, what tasks they want to accomplish and how they want to go about doing them in terms of the methods that are employed as well as the physical interaction they have with the software. This is achieved through collecting information from the users and targeting the system towards goals resulting from a lengthy process of analysis, design and evaluation. The process consists of two phases: requirements analysis followed by a design/implementation cycle. The process is iterative, gathering feedback from users with each iteration.

The software engineering team consists of three types of individual: (1) a 'super-user' or domain expert, who has knowledge of the scientific process that is being rendered; (2) a team of software engineers, who collectively work to design and implement the system; and (3) a group of key scientific collaborators (usually computer scientists) who excel at solving the computational problems that the system needs to address. The domain expert is responsible for outlining the requirements that the system must meet. Hence, it is important that this individual have the necessary scientific background to fully understand the process and the implications of all activities. Information is obtained by interviewing users and observing them during the normal course of their everyday work. Note is taken of the available software tools currently being used, to what extent human intervention is required and the logic of the process. Requirements are passed down from domain expert to algorithm developer and software engineer; they are evaluated for feasibility and cohesiveness as a design evolves. Implementation specifications are written and the work of coding and testing the system begins.

33.2 THE RECONSTRUCTION PROCESS

33.2.1 Sequencing strategies and their effect on the process

Current technology allows accurate sequencing runs of approximately 450 bases. Consequently, generating cosmid-sized sequences necessarily involves deriving them from smaller overlapping fragments. These overlapping components most often come from subclones of a longer sequence. The sequence reconstruction process is defined as the combination of methods required to resynthesize or 'reconstruct' the original target sequence from its disparate parts.

The algorithmic solutions to the reconstruction problem are in a large part affected by the sequencing strategies themselves. Our collaborations with laboratories using the random or 'shotgun' approach and a partially directed transposon-mediated approach have allowed us to do a formal requirements analysis of the problem. We present the random strategy as commonly used by many sequencing laboratories (Edwards & Caskey, 1991) and the partially directed transposon-mediated approach (Palazzolo *et al.*, 1991) as two examples to illustrate how this review of the various sequencing strategies has given way to the specification of a more automated approach to sequence reconstruction.

In the random or shotgun strategy, a clone (usually around 50 kb in length) is fragmented by physical methods and fragments are size-selected from gels. A single-stranded phage subclone library (for this example) is made from these fragments (usually 1 kb) and, depending on the expected size of the final consensus, a calculated number of these clones are randomly selected and sequenced using one primer. A small number are also sequenced using a primer originating from the opposite end of the insert. The fragments are then assembled using an assembly algorithm. Algorithms currently in use are not able to use the fact that two nonoverlapping fragments originating from different ends of the same clone and having opposite orientations should fall into the contig some defined distance apart based on the size of the clone insert. (This is a type of 'map'.) To compensate for this shortcoming, the user evaluates the output from the assembler and is forced to readjust parameters and fragment order to achieve the correctly 'mapped' results. The methods employed in GRM take into account the 'maps' *prior* to assembly and use that information to generate the final output. The user commits much less time to achieving the desired solution.

For the partially directed strategy described in Fig. 33.1, ordered subpools of fully ordered transposon-containing clones are generated. The ordering information can be specified to the reconstructor in the form of 'constraints' upon the output of the assembler and the final solution presented to the user. Again, no user intervention is required to 'fix' aberrant assemblies.

The key finding in our review of sequencing strategies is that constraints, or prior knowledge about fragment order and orientation, should play an integral part in the reconstruction process. The user should not be forced to 'fix' aberrant assemblies because the system is not capable of considering all of the facts. By contrast, GRM will possess this capability as an essential component of automating the process of sequence reconstruction.

A practical look into what constraints are is best done by example. Some common constraints for medium-scale (cosmid-sized) sequence reconstruction projects are listed below.

(1) *Mapping gaps* – fragment A and fragment A′ should be in opposite orientations at a distance of approximately 500 bases apart.
(2) *Fixing an end* – fragment J should map to the extreme right end of the final contig; fixes orientation of final contig.
(3) *Overlaying a restriction map* – final contig should conform to known restriction map for this clone using knowledge that small fragments and multimers sometimes go undetected.
(4) *Fragment order known* – directed strategies.

33.2.2 Understanding the data flow of reconstruction

The flow of data from sequence fragment to finished consensus is somewhat strategy dependent. Each strategy makes use of the same algorithmic components: assembly and constraint propagation. The difference lies in the order and the extent to which any single component is utilized. Constraints upon a rigorous assembly solution play a critical and essential part in the final solution. They are key elements in all strategies thus far examined. As examples, Fig. 33.1 can be reviewed again in the context of data flow for the random strategy with directed closure of gaps and

Random or Shotgun Strategy Reconstruction Cycle

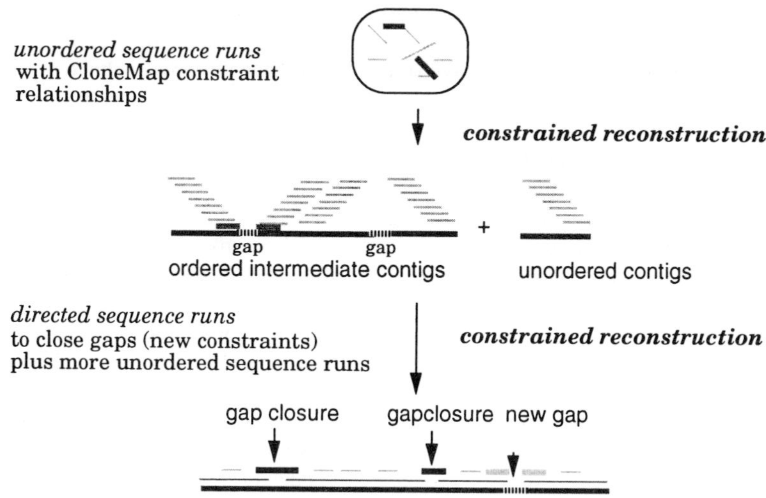

LBL Transposon-Mediated Strategy Reconstruction Cycle

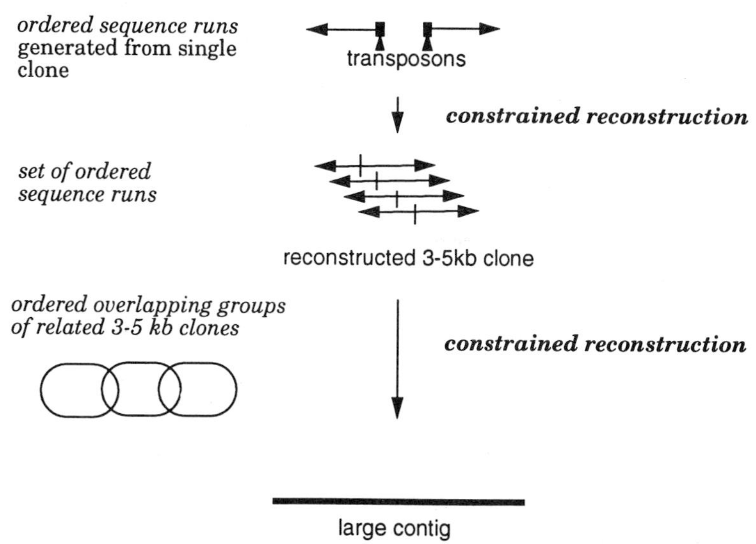

Figure 33.1 Sequence reconstruction cycles for the random or shotgun sequencing strategy (upper) and the Lawrence Berkeley Laboratory (LBL) directed transposon-mediated strategy (lower).

a transposon-mediated (directed strategy originally described by the Lawrence Berkeley Laboratory (Palazzolo *et al.*, 1991).

The system must artfully manage the data as discrete projects. The strategy defines the useful size and boundaries of a project. For the random strategy, the project is generally the final contig. For the directed strategy described here, the useful size is the small clone containing one transposon even though the goal of the process is to construct a very large P1-sized contig. Projects are merged at different stages of the process depending on the strategy. A project can be thought of as the user's primary unit of work.

Once the data are delegated to a single project, data that are not useful in building the final contig must be removed. This includes regions containing vector, sequences with a high degree of ambiguity, library contaminants and so forth. We call this screening process the 'preprocessing step', since it occurs prior to submission to the assembly kernel. Constraints are then specified indicating the known relationships of fragments to each other and to the final contig map. The constraints and fragments are submitted to the assembly algorithm, a computationally intensive step. The overlap information from the assembler and the constraint-reasoning system produces a solution (or a set of solutions) that best matches the user's expectations (as indicated by the constraint set). The final contigs are presented to the user for examination and editing. Original data should be available during this process. This process occurs many times before the final contig is produced. With each iteration, new fragments are usually added, some problematic ones deleted and constraints modified. The final finished consensus can be explored for similarities to known sequences and subjected to analysis in the context of the reconstruction.

33.2.3 Evolution of a new reconstruction paradigm

Our goal is to provide a total environment to support the reconstruction of DNA sequences from overlapping fragments and to produce a sequence suitable for annotation and submission to databases. The process of sequence annotation, in fact, involves a complex set of tools for sequence analysis or exploration and is one which we have also begun to address in our work.

The key finding of our requirements analysis is a redefinition of the reconstruction paradigm. All available packages provide a means for finding overlaps between fragments. They view the transformation of data from a 'primary sequence' state to a 'finished' state as an 'assembly problem'. User-centered analysis has led us to view the process as one of 'reconstruction' of a target sequence. This subtle difference in interpretation has resulted in a design more in harmony with the goal of the user.

Existing systems require and expect a high degree of user intervention. Overlaps are edited manually and fragment order manipulated when necessary by circumventing the idiosyncracies of the assembly algorithm. Our view (and the biologist's view) of the process is to use prior knowledge about order and orientation of fragments as part of the whole process. This use of auxillary information is a more complex process requiring a system which can reason about the assembly algorithm's solution. In effect, GRM will permit the biologist to actively direct the reconstruction process as it is being performed, rather than it being a passive process which the biologist acts to 'fix' once the deed is done. Hence, we choose the phrase, 'reconstruction process' to supersede the more commonly accepted 'assembly' to give a more realistic sense to the global scope of tasks the user faces in generating a finished consensus sequence. We view 'assembly' as one small component of this complex process.

No existing software package has been designed to efficiently meet this goal. Two major areas not supported by any package are project management beyond the file system level and support of the full reconstruction paradigm. Another important issue is scalability. In order for any system to maintain its usefulness for the lifetime of the Genome Project, there must be vast improvements in the automation of the reconstruction process. Here we are not talking about technological advances in the physical laboratory work that is being performed, but, rather, the computational process of reconstruction once the data have been generated. As stated previously, major efforts must be put toward removing human intervention from the reconstruction process as much as is deemed reasonable. GRM is attempting to meet these goals.

Packages being used by large-scale sequencing laboratories include the highly popular xdap (Dear & Staden, 1992), Seq-Man (DNAStar, Madison, WI), GCG (Genetics Computer Group, Madison, WI), SAM (C. Lawrence, Baylor College of Medicine, Houston, TX), and the IntelliGenetics (IntelliGenetics, Palo Alto, CA) GEL packages. Others supporting strategy-specific needs are also being used by a small number of laboratories.

33.3 THE GRM ENVIRONMENT – A USER PERSPECTIVE

33.3.1 Definitions and concept

Here we provide definitions and descriptions of concepts essential to the workings of GRM.

SeqRun. The readable sequence generated by a sequencing reaction or set of reactions.

Usable region. The region or regions of a SeqRun useful in the reconstruction process.

Contig. SeqRuns are merged by an assembly algorithm to produce Contigs.

Constraint. The statement of the spatial relationships or orientational relationships between any set of SeqRuns.

Metacontig. Each contig may have a relationship to one or more contigs by a constraint. These 'complex' contigs are called 'metacontigs' ('meta' meaning 'later in development') since their formation occurs 'postassembly' or after the assembly engine has done its work.

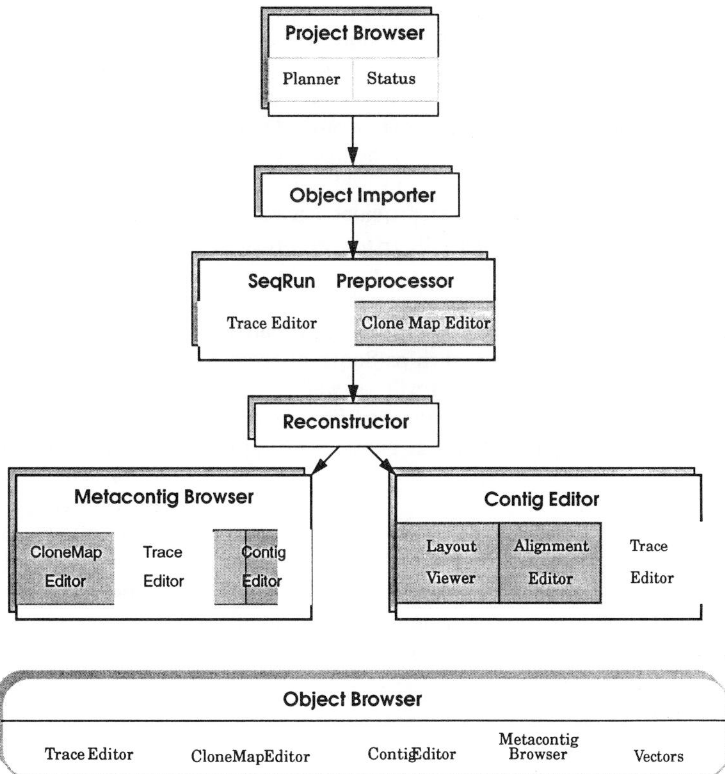

Figure 33.2 Data flow through GRM task modules as it currently appears in the working prototype.

33.3.2 Overview of the system

Figure 33.2 presents the data flow through the various task modules of the system. Primary hardware-specific sequence data are brought into a project by the object importer. The project is an organizational unit containing all of the data necessary to produce the final contig. (We will defer our discussion of the bounds of a project to Section 33.3.3.1) They are then sent on to the SeqRun preprocessing unit, where sequence features affecting assembly are identified. Once preprocessing is complete, sequences are sent to the reconstructor, which generates contigs and metacontigs. The constraints editors can be used to place orientation, distance and order relationships on fragments in the reconstruction. For instance, in the map-gap closure strategy (Edwards & Caskey, 1991), forward and reverse primers are used to generate sequence runs off of the same clone. Each sequence run has an opposite orientation and relative distance that can be 'mapped' and used to assign primers for closing a critical gap. Specifying this information to the reconstructor through one of the constraints editors (in this case, the clone map editor) (or importing the constraints into the project via the constraints interpreter) tells the reconstructor which assembly outcome should be selected to best meet the user's expectation.

The result of a completed reconstruction event can be viewed in one of two ways: in the context of the traditional contig editor, where each contig is presented for viewing and editing, or via a higher order metacontig viewer, which shows how the traditionally created contigs might be related to one another by the constraints information. Objects can be viewed independently of tasks (such as reconstruction, preprocessing or editing) via the object browser.

33.3.3 Description of the GRM components

33.3.3.1 Project management

The goal of any sequencing project is to obtain with a high degree of certainty the nucleotide sequence of some defined segment of DNA. For those using the shotgun strategy and closing small gaps by directed methods, each project may contain a cosmid-sized clone, while for the transposon-mediated strategies, a project may be quite small and merging of many small projects may actually constitute what the biologist sees as the final product. Each project must, therefore, have the potential to form a relationship with other projects. The relationship may be an overlapping one, or simply an ordered interval with gaps in between. The project manager is responsible for maintaining this information, for sorting it out, and for viewing the hierarchical relationships among projects.

The project browser is a viewport into an individual project. The project browser lists the projects available for selection. Once a project is selected, information regarding its status and access to a project planner is allowed. The project browser is the main screen for the GRM environment. A row of pushbuttons on the right-hand side of the screen provides access to all options. The options are: SeqRun preprocessor, reconstructor, constraints editors, metacontig viewer, contig editor, trace editor, object browser, object importer, autoassembler, and multiproject manager.

All user-visible objects, such as SeqRuns, contigs, metacontigs, vectors, and so forth, as well as objects essential to the internal function and flow within the system, are maintained in a commercial object-oriented database.

33.3.3.2 SeqRun preprocessor

The goal of this processing step is to present to the assembler with a set of sequencing fragments of sufficient quality to be accurately assembled. The SeqRun preprocessing unit is responsible for defining the region or regions (i.e. usable region) of a SeqRun which should be submitted to the sequence reconstructor.

Preprocessing is a multistep process. The system, however, has the flexibility to allow switching on or off any of the options. The options provided are:

(1) vector (or other sequence) removal from the ends of SeqRuns;
(2) screening out of low-quality SeqRuns and internal regions of SeqRuns of poor quality;
(3) removal of library contaminants (e.g. λ) and dimerized vector clones; and
(4) prefiltering of sequences which might potentially cause problems in the reconstruction phase (e.g. Alu).

Matches to vector (or other sequences which might occur at the 5′- or 3′-ends of a SeqRun) at the 5′ (all strategies) and 3′ (some strategies) ends are detected by a local similarity (SIM) match algorithm (Huang & Miller, 1991) and are excluded from the assembly process. Low-quality regions are sieved out by a simple method which evaluates the spacing and overall frequency of Ns in a SeqRun. SeqRuns containing repetitive sequences, i.e. Alu, and not containing unique sequence at each end are excluded from the reconstruction process.

The results of the SeqRun preprocessor can be viewed using the trace editor or passed directly on to the reconstructor. The full-blown trace editor fully describes the data resulting from a preprocessing run. Positions of Ns, viewing and manual modification of the excluded regions, and editing capabilities are all part of the normal functioning of the trace editor. A text report displaying SeqRun properties for all processed sequences can be printed to a file as part of the project's permanent history.

33.3.3.3 Sequence reconstructor

The reconstructor panel controls the reconstruction of target SeqRuns from primary sequence data and constraints. Two alternative algorithmic solutions are embedded in GRM to support reconstruction from SeqRuns generated by the random sequencing strategy: CAP (contig assembly program) (Huang, 1991) and FAK (fragment assembly kernel) (Kececioglu, 1991). The Reconstructor in GRM ensures that reconstructions satisfy the constraints provided by the directed component of a strategy.

A unique capability of the GRM reconstructor is its ability to distribute compute intensive tasks on other networked hosts, such as idle workstations or a compute server. This occurs transparently to the user.

33.3.3.4 Constraint editors

Constraints are used to represent higher order knowledge that a biologist demands of the reconstruction results. Existing packages do not support constraints.

If one were to think of the project's final finished contig as a map with all of the SeqRuns associated and the final contig as being placed positionally along that map, then the constraints represent the approximate location of some of those SeqRuns on that final map (Honda *et al.*, 1993). This is one form of constraint. The 'map relationship' may also be between SeqRuns rather than to the final map. This is the case in the pairing of forward and reverse primer runs from the same clone. In this case, a spatial and orientation relationship can be defined for this pair of SeqRuns. We currently have implemented a 'clone map' editor (Fig. 33.3) to specify this special type of prevalent constraint for labs using the mapped gap closure strategy (Edwards & Caskey, 1991) in combination with the random method.

A second type of ordering is used in the transposon-directed strategies. With these methods, orders of subpools of SeqRuns can be defined as part of the reconstruction process. This can be achieved through a general constraints editor which we call the 'map palette', allowing the graphical manipulation of map relationships between multiple SeqRuns.

Another important class of constraints is exemplified by the placement of the final contig on a high-level restriction map. Frequently, a restriction map (or map of other such 'landmark' sequences) is used to assist

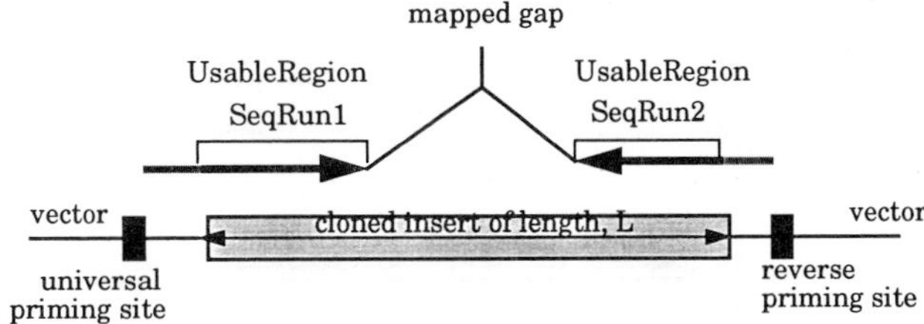

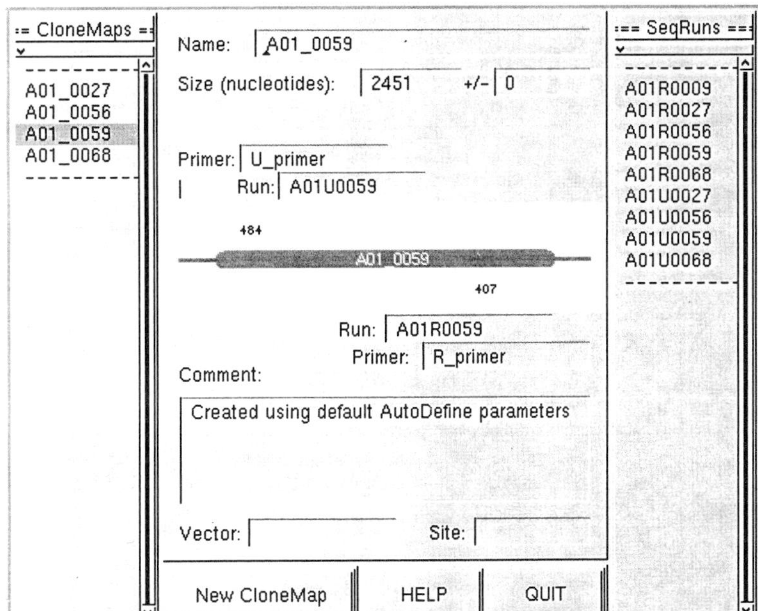

Figure 33.3 Screendump of the GRM clone map constraint editor. Upper panel shows diagrammatic representation of the clone-map-type constraint. The lower panel shows the GRMs clone map editor.

in the ordering of contigs. Our goal is to support such activities in GRM. Our general interpretation of the constraints problem should provide a solution to this class of constraints. A future version of the system will provide access to this capability.

Constraints and both the analytical and visual tools to view and use them are a large focus of our work in building a system to fully support multiple sequencing strategies. We expect that these tools will evolve rapidly and extensively over the next 2 years.

33.3.3.5 Metacontig viewer

The metacontig is a representation of the traditional contigs with the constraint information imposed on it. The metacontig viewer (Fig. 33.4) assists the user in identifying regions potentially useful for the closure of gaps. A graphical representation of the metacontig shows the position of the traditional contigs, the constraints and the sources of the constraints. Double-clicking on any of the graphical objects invokes the editor for the object selected.

A text version of the metacontig layout can be obtained through a 'print to a file' utility.

33.3.3.6 Contig editor

The contig editor (Fig. 33.5) is used for viewing and editing contigs that result from passing data through the assembly algorithm. The order of the SeqRuns in a contig is affected by the constraints as defined by the user through one of the constraints editors. The location and effect of constraints on the contigs can only be viewed via the metacontig viewer.

The contig editor consists of three panels: the layout panel, the alignment editor and the features panel. The three panels are tiled into one window.

The top panel is the layout panel. It maintains a list

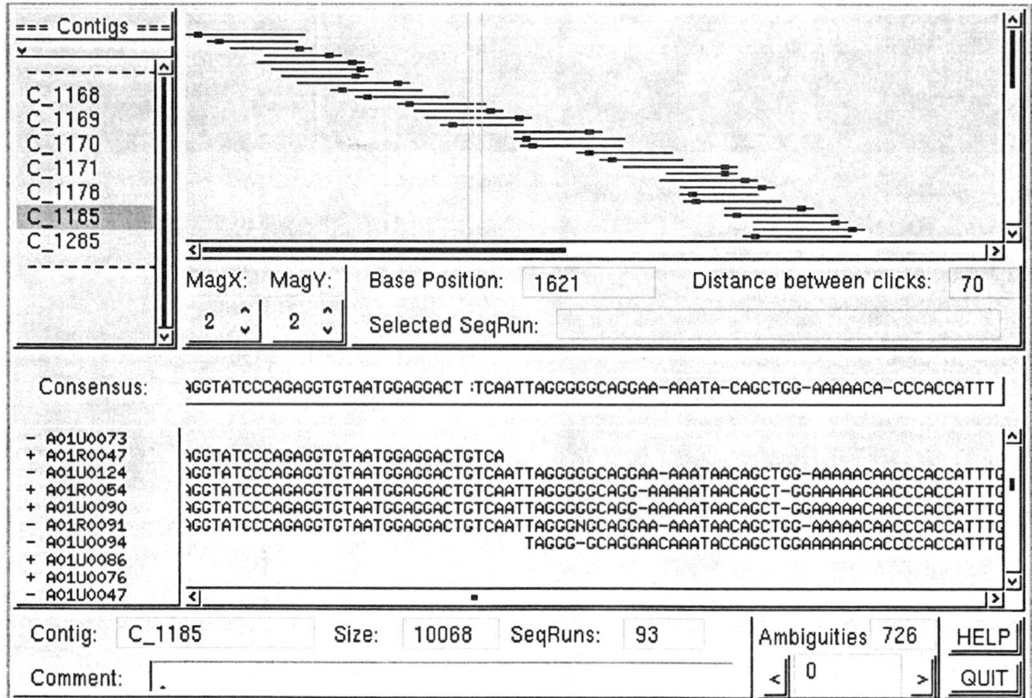

Figure 33.4 Screendump of the GRM contig editor. A set of trace files can be spawned from the alignment window of the contig editor. The trace files can be edited in the context of the contig editor.

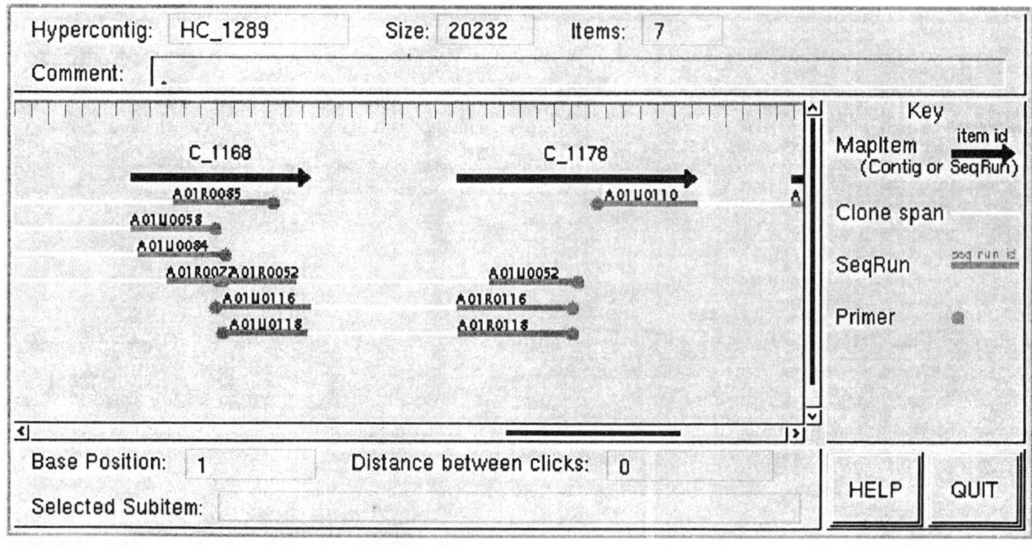

Figure 33.5 Screendump of the GRM metacontig editor. Two 'assembled' contigs (heavy arrows) are linked together in a larger metacontig by multiple SeqRun pair clone map constraint relationships. SeqRuns are shown in grey, clone map relationships in white.

for selecting the contig to view. To the right of the selection list, is the panoramic schematic overview of the selected contig. Each SeqRun in the contig is represented by an arrow which designates orientation. The magnification of this view can be adjusted easily to view as much or as little of the panorama as desired. A base scale displays contig coordinates and SeqRun identities are clearly shown on each arrow.

The layout panel has two cursors, both of which relate to the alignment window. One corresponds to the entire contents of the alignment window and one, internal to the first, corresponds to the column (individual base or column of bases) cursor in the alignment window.

The second panel serves as the multiple sequence alignment editor, where base-calls can be edited in the context of the overall alignment. Individual base-calls and entire columns (base-calls plus consensus) can be edited. The orientation and SeqRun identity are clearly displayed for each SeqRun.

Several capabilities make this editor unique. First, the policy for determining the consensus sequence can be selected by the user. Three policies now exist: majority rule, unanimous decision and plurality rule. Majority rule states that the base-call exceeding 50% of the total base-calls in any given position will be used as the consensus call, whereas for plurality rule the most prevalent base will be called in that position. Unanimous decision defines the consensus call only if total agreement exists for the entire column. Apart from the policies that can be used to determine the consensus, the alphabet can also be selected. Currently, [AGCTN] and IUPAC are supported.

The alignment editor contains another useful feature relevant to consensus determination. In determining the consensus sequence (and performing the reconstruction), only the usable regions of a SeqRun are used. It is often useful to view the entire SeqRun in the context of the alignment, if a region with much conflict or poor coverage is involved. We have implemented a method which selectively 'unmasks' the excluded regions of a SeqRun or group of SeqRuns within the alignment editor.

SeqRun editors can be spawned directly from the contig editor. The contig editor is unique as compared to other packages in that the SeqRun editors that are invoked, the multiple alignment editor and the panoramic layout screen are all synchronized for viewing such that the cursors in each of the windows track with each other. The alignment window has a column cursor to edit individual bases in the column or the entire column, including the consensus. Upon invoking the SeqRun editors, a stack of editors appears with the SeqRun editor cursor also in the same position. This position is also represented by a single base cursor (within a window cursor representing the part of the alignment being viewed) in the layout window. Repositioning of any one of these three cursors repositions the cursor in the other two windows synchronously. Edits in the alignment are reflected in the SeqRun editor's base-call edit line. Original base-calls are also displayed in uneditable form. The SeqRun editors called from the contig editor are minimal editors. They do not contain the full-blown capabilities of the full editor when invoked via the SeqRun editor button on the main project panel or via the sequence preprocessing unit.

The third panel (not shown) consists of a diagrammatic representation of features identified along the contig. Column conflicts and depth of coverage are two examples that can be used to assist in the editing process. The location of Alu sequences is often helpful in reconstructing human sequences since these repeats often cause misplacement of SeqRuns into the contig. We are currently implementing a short list of features that can be used to assist in verifying the correctness of order and for identifying potential problem areas or areas of conflict in the reconstruction process.

Layouts and alignments of each of the contigs can be saved to a file and subsequently printed for permanent book-keeping.

33.3.3.7 Trace editor

The trace editor provides viewing and editing capabilities for ABI-generated data. We will be extending the system to also support other file types, including *Pharmacia*, in the near future.

The trace editor displays the trace file with original base-calls and an editable line in a scrolling window. Magnification can be adjusted from a 1:1 pixel ratio to 1:10. Bases are color-coded to match the ABI color scheme. Features of the trace editor include regular-pattern string matching, a listing of all ambiguous base-calls (N), and an indicator showing the excluded regions of the SeqRun (excluded regions are in grey) with a hand cursor to move the limits for the excluded regions. Sequence quality as determined by the sequence preprocessor is also displayed.

The list of N positions within a SeqRun is a selectable set of positions. Clicking on any of the numerical positions shifts the trace to that position and the 'N' is highlighted for easy reference.

The trace editor exists in two forms: in the complete form described here and in an abbreviated form. The abbreviated form is invoked by the contig editor and has been designed as such to minimize use of screen space. This minimalistic trace editor displays the file and is editable, but does not contain the more complicated quality and N assessment information. Traces are stacked and tiled when invoked from the contig editor.

33.3.3.8 Object browser

The object browser is a centralized location for all user-visible objects. Objects and groups of objects can be created or imported from the file system. Objects include SeqRuns, contigs, metacontigs, constraints, vectors, special sequences (these are sequences used by the sequence preprocessing unit and the explorer to identify special features on SeqRuns and contigs). A short summary report for each type of object or group of objects can be printed to the screen, to a file or to a printer for permanent record keeping.

33.3.3.9 Automated processing in GRM

GRM possesses automated processing capabilities. By selecting this option, a group of SeqRuns can be subjected to preprocessing and reconstruction without user intervention. Default values for preprocessing and reconstruction are used unless redefined by the user through a simple menu selection.

33.4 EXTENDING THE ENVIRONMENT FOR SEQUENCE EXPLORATION

A natural consequence of having sequences is delving into the biology of the sequence at hand. This involves locating 'landmarks' or features on the sequence and identifying the relationship between any one or group of landmarks with physical and genetic markers.

Many of these exploratory capabilities are also important in verifying the correctness of the order of SeqRuns in the final contig. We have begun implementing several key exploratory features and intend to expand our work to a more sophisticated set of tools for general sequence exploration as relevant to genomic sequencing activities. The object model that we have developed (Honda *et al.*, 1993) should prove extremely valuable in creating a seamless reconstruction/exploration environment for genome researchers.

ACKNOWLEDGMENTS

This work is funded by the Department of Energy (CL) and National Library of Medicine (CL).

REFERENCES

Dear, S. & Staden, R. (1992) *Nucleic Acids Res.* **19**, 3907–3911.

Edwards, A. & Caskey, C.T. (1991) *Methods* **3**, 41–47.

Honda, S., Parrott, N.W. & Lawrence, C.B. (1993) In *Proceedings of the 26th Hawaii International Conference on Systems Sciences Vol. 1.* T. Mudge, V. Milutinovic & L. Hunter (eds) IEEE Computer Society Press, Los Alamitos.

Huang, X. (1992) *Genomics* **14**, 18–25.

Huang, X. & Miller, W. (1991) *Adv. Appl. Math.* **12**, 337–357.

Kececiouglu, J.D. (1991) *Technical Report TR 91–26.* Department of Computer Science, University of Arizona, Tucson, AZ.

Norman, D.A. & Draper, S.W. (eds) (1986) *User-Centered System-Design*. Lawrence Erlbaum, Hillsdale, NJ.

Palazzolo, M.J., Sawyer, S.A., Martin, C.H., Smoller, D.A. & Hartl, D.L. (1991) *Proc. Natl. Acad. Sci. U.S.A.* **88**, 8034–8038.

CHAPTER THIRTY-FOUR

Stochastic Optimization Tools for Genomic Sequence Assembly

C. BURKS,[1,2] M.L. ENGLE,[1] S. FORREST,[3] R.J. PARSONS,[4] C.A. SODERLUND[1] & P.E. STOLORZ[5]

[1] Theoretical Biology and Biophysics Group, T-10, MS K710, Los Alamos National Laboratory, Los Alamos, NM 87545, USA

[2] Center for Human Genome Studies, Los Alamos National Laboratory, Los Alamos, NM 87545, USA

[3] Department of Computer Science, University of New Mexico, Albuquerque, NM, USA

[4] Computer Research and Applications Group, Los Alamos National Laboratory, Los Alamos, NM 87545, USA

[5] Santa Fe Institute, Santa Fe, NM, USA

34.1 INTRODUCTION

A number of groups have recently published the results of cosmid insert-sized and larger sequencing projects. These recent efforts are noteworthy because of their conception and implementation as short-term, globally comprehensive, sequencing projects; this contrasts with previously published sequences of comparable size, which have been the result of piecing together many smaller projects (e.g., based on λ clones) generated over a much longer period, often by several independent research groups.

The limitation of experimental methods to several hundred contiguous bases for direct determination of DNA sequences has meant that longer regions of DNA have to be determined as overlapping fragments. Based on their determined sequence overlap (and often relying on any available ancillary information), one assembles the sequence fragments into a layout reflecting their relationship to each other and the parent sequence from which they were derived.

Large-scale sequencing of DNA can be done with directed methods, random methods, or (most often) a combination of the two (Howe & Ward, 1989; Hunkapiller *et al.*, 1991a,b). The random approach makes the fragment assembly calculation more challenging: the high coverage typically required to cover the parent increases the combinatorial complexity of exploring possible layouts tremendously, and the lack (in most cases) of overlap information independent of the fragment sequences makes the layout determination particularly vulnerable to complications arising from genomic repeats, experimental artifacts, and base-calling errors. Though the frustration with the inefficiencies of the random approach has led to an increased interest in developing more efficient and less expensive directed strategies for large-scale sequencing, random strategies are – and, for the immediate future, will be – a major component of large-scale sequencing projects.

The assembly of overlapping sets of fragments into large segments of contiguous sequence, described as *contigs* (Staden, 1980), requires the pairwise alignment of the fragments among themselves, followed by the synthesis of these pairwise results into an optimal global alignment expressed as a consensus sequence. In practice (Staden, 1987; Messing & Bankier, 1989) one decides that enough fragments have been generated and sequenced by monitoring the depth of coverage (to minimize the impact of experimental errors arising during sequencing of individual fragments) and the size and minimal multiplicity of the contigs being generated. When gaps persist independently of the depth of coverage, one can bridge those gaps using directed methods. (When a random strategy is used, the progress of a sequencing project can be predicted using formulae given by Lander & Waterman (1988).)

Gallant *et al.* (1980) demonstrated that the determination of the shortest common superstring – in which the fragment assembly problem can be recast (see discussion below) – with a four-letter alphabet to be NP-complete (i.e. computationally impractical for large data sets). Even for the simplified approach of modeling fragment assembly as determining the order of beads (fragments) on a string of the order of N-factorial ($O(N!)$ operations would be required for N fragments. The determination of overlap strengths is also daunting, requiring $O(N^2)$ operations.

Other factors further complicate assembly. The input fragment sequences usually include experimental ambiguities or errors – including rearrangements, substitutions, insertions, or deletions of bases – that affect assessment of pairwise overlap strength and subsequent detailed alignments. Though sequencing both strands contributes to minimizing errors arising in base-calling, it also leads to the necessity for assigning and tracking strand sense through the assembly and alignment calculations. Naturally occurring DNA sequences tend to be repetitive on many different scales. Repetitive sequences longer than individual fragments may cause ambiguities in the sequence assembly which cannot be resolved without additional information; this problem is exacerbated by higher rates of conservation among and larger repeat units in a repeat family. Finally, there are potential sources of ancillary information that could contribute to the overall solution (e.g. pairwise proximity among fragments in an input set of data may be known by independent physical mapping).

Computers are playing an increasing role – approaching a continuum – in the information flow associated with large-scale sequencing projects (Burks, 1989; Cinkosky *et al.*, 1991). As with other components of the generation, management, and analysis of nucleotide sequence data, the increase in size of data sets associated with these projects makes impractical the manual assembly of the resulting sequence fragments into a coherent consensus sequence corresponding to the clone insert (parent sequence). Furthermore, the inherent computational complexity of sequence assembly makes software tools intended to facilitate the assembly by emulating (rather than improving upon) the manual approach increasingly impractical. In particular, processing of raw sequence data becomes a significant bottleneck when trying to ramp up sequence throughput to the levels (10^5–10^7 bp day^{-1}) that have been proposed in the context of the human genome project (Hunkapiller *et al.*, 1991b). There are several stages in this process that require the development of better algorithms and/or tools; one example is the assembly of DNA fragment sequences generated during random sequencing strategies.

Rawlings' (1986) directory lists 15 software packages with at least some sequence assembly capabilities; unfortunately, the algorithmic basis of these and more recently developed tools has not been extensively reviewed (although brief surveys have been compiled (Kececioglu, 1991; Churchill *et al.*, 1993)). In the following discussion, we focus primarily on the issues surrounding layout generation, which is more computationally intensive than calculation of overlap strength; it should be noted, however, that several of the papers cited (e.g. Kececioglu, 1991; Huang, 1992) also focus on improved algorithms for overlap strength determination.

Probably the best known sequence assembly package, and the inspiration for many of the others that have been developed, is that of Staden (1979, 1980). This approach was originally created for application (i) to sets of sequence fragments corresponding to parent sequences that were short compared to those being determined now, and (ii) in a highly interactive, user-directed mode (see, for example: Staden, 1987; Rice *et al.*, 1991) that is adequate for occasional use when sequence data are accumulating relatively slowly, but which quickly becomes impossible in a steady stream of sequence data. In addition, most of these approaches are based on 'greedy' procedures that extend a contig a fragment at a time from a given starting point. This approach has the advantage of efficiency gained by calculating only a single solution from the set of all possible layouts, but the distinct drawbacks of (i) inexactness due to the high degree of manual intervention required to achieve the end result, and (ii) potentially ending up at a less-than-optimal final solution.

The first drawback was addressed by Peltola *et al.* (1983, 1984), who used interval graph theory (briefly sketched by Sedgewick (1990) for developing a formal representation of the greedy approach, extending a given contig by the strongest overlapping fragment among the remaining fragments not already in a contig. This allowed one to automate the assembly process; this approach was used most recently by Huang (1992). Over the past few years, several groups (Gallant, 1983; Tarhio & Ukkonen, 1988; Turner, 1989; Kececioglu & Myers, 1989; Li, 1990; Blum *et al.*, 1991; Kececioglu, 1991) explored extending this approach to solving the shortest common superstring problem, and, conversely, casting the sequence assembly problem in terms of finding the shortest common superstring. They found that greedy solutions, in the worst case, were not guaranteed to have length less than within a factor of 2 (or more) of the global solution. Kececioglu & Myers (1989) were the first to allow for errors in the input fragment sequences, and to provide theoretical analysis of the effect of this constraint on finding shortest common superstring solutions; they present several alternative strategies for sequence assembly taking this into account. Cull & Holloway (1992) presented an approach to sequence assembly based on suffix array methods that takes mismatch errors in the input strings into account.

Descriptions thus far of sequence assembly implementations have not presented extensive, systematic, independent exploration of the effects of individual parent sequence, project size, coverage, error rate, or repetitive density on the computational run time and/ or approach to a globally optimal solution. More typically, only one or two 'successful' examples of assembly is given. This makes it very difficult to compare one algorithm with another, or to select an otpimal algorithm for a particular input data set. In addition, most of the implemented tools provide little capability for introducing ancillary input on fragment integrity or overlaps, forcing one to implement these constraints with manual editing that can take as much as several orders of magnitude more time than was required for the initial assembly. Finally, every current widely available tool presents only one assembly algorithm to the user; experimenting with (let alone production use of) different assembly strategies involves the very cumbersome passing of data sets back and forth from different data formats, software systems, and hardware platforms.

Faster, more flexible, and better defined assembly tools should provide several important benefits for large-scale, high-throughput projects: (i) projects need to be monitored in real time, allowing one to quickly adjust and optimize the choice of experimental strategies and corresponding consumption of reagents and experimentalists' time as the project progresses; (ii) the final sequence should include measures of the reproducibility and robustness of the final result; and (iii) users should be able to experiment with modular recombination of the various computational stages in sequence assembly, facilitating the development of customized solutions to match new experimental strategies and/or new input data characteristics.

34.2 ALGORITHMS

We focus here on the specific challenge, given the pairwise overlap strength among a set of fragments, of generating an optimal layout describing the offsets and strandedness of the input fragments. We have described in detail elsewhere (Churchill *et al.*, 1993) the other necessary modules of the overall assembly process (determination of overlap strengths among input fragment pairs, conversion of fragment layout to a detailed multiple sequence alignment, and conversion of the multiple sequence alignment to a consensus sequence) and the implementations that we use.

Sequence assembly software packages traditionally include a single algorithm for generating fragment layouts (to date, most of these algorithms have been greedy). We are interested in comparing the performance of a number of algorithms, both with respect to their rate of convergence on an optimal value and the closeness of that value to the global optimum. This comparison may be useful both for governing the selection of particular algorithms for particular input data sets and for building hybrid algorithms based on one or more of the traditional 'pure' algorithms. The particular layout determination algorithms we have implemented reflect our immediate interest in determining whether or not stochastic combinatorial optimization strategies will be useful in this context. Thus, we are currently experimenting with greedy algorithms and several stochastic search algorithms: shuffling with relaxation, shuffling with simulated annealing (linear and adaptive), and genetic algorithms.

34.2.1 Greedy approaches

For each step of a greedy algorithm a decision is made (in this case whether or not to add a particular fragment to the current contig being constructed), after which the chosen course of action is never reconsidered. Our implementation selects the best pair of fragments from the available set and then groups all remaining connected (overlap strength > threshold) fragments into a contig by recursively looking in the remaining fragments for the first overlap to the rightmost fragment in the contig; when there are no more rightward-extending fragments present, the contig is extended leftwards. When the contig cannot be extended any further in either direction, a new contig is started; this is repeated until all the fragments have been placed in a contig. If the overlap strengths contain neither false positives nor false negatives, this algorithm will construct internally correct contigs (though they may represent subcontigs of the input, parent contigs). However, when the overlap strengths reflect the presence of repeats and/or experimental sequencing errors, the greedy approach often produces incorrect solutions. Note that one can identify a number of heuristic modifications that could be used to improve this algorithm; we plan to experiment with these (and use more sophisticated algorithms developed by other groups) in the future.

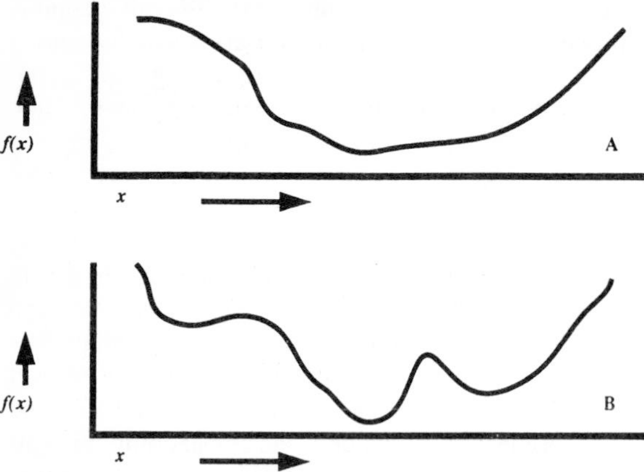

Figure 34.1 *Objective functions*: (a) single, global minimum; (b) multiple, local minima.

34.2.2 Combinatorial optimization

Consider, in Fig. 34.1(a), a parameter, x, the function, $f(x)$, and the objective of minimizing $f(x)$ within the range $i<x<j$. Optimization of $f(x)$ in this case corresponds to minimizing $f(x)$. If the objective function, $f(x)$, can be expressed as a smooth and differentiable equation, the methods of calculus yield the minimum as $f'(x)=0$. Alternatively, one can find the minimum by selecting an initial value for $f(x)$ and by moving down the steepest local gradient in solution space. For details, see, for example, the book by Gill *et al.* (1981).

However, these approaches are less practical if $f(x)$ has multiple minima and/or is not continuously differentiable. In such a case (see Fig. 34.1(b)), one can still find the global minimum by enumerating $f(x)$ over all values x of interest and empirically selecting the minimum (or come arbitirarily close if x ranges over nonintegers).

This approach, though yielding an exact solution, is very inefficient, and is thus impractical for an $f(x)$ being characterized over any significantly large range and number of parameters x_i. Purely stochastic sampling decreases the cost of enumerating a particular volume of solution space, but is ultimately and similarly limited by its lack of efficiency. Unfortunately, many optimization problems are computationally complex (Johnson & Papdimitriou, 1985), yielding large solution spaces that are both multimodal and noisy, requiring alternative, approximative strategies. Combinatorial optimization problems (Lawler *et al.*, 1985), where one is optimizing not over a continuum but over a large number of discrete states, and including sequence assembly, fall into this category. For this class of problems, one redefines optimization to mean achieving an objective function value within an acceptable range of the global minimum, and within an acceptable amount of time.

34.2.3 Shuffling

We are experimenting with several stochastic search algorithms for determining optimal fragment layouts during sequence assembly. These algorithms (with the exception of genetic algorithms) are implemented for a model of layout optimization developed by Churchill *et al.* (1993). Layouts are optimized for an ordinal, or 'beads on a string', model of fragment clustering by permuting, or 'shuffling', the fragment order in the model. This is done in the context of an $[N \times N]$ matrix for N fragments, where the i,jth element contains the overlap strength between fragments i and j. In this context, 'better' layouts are achieved by clustering higher values near the matrix diagonal, analytically expressed as minimizing the objective function:

$$f = \mathrm{T}_{i,j} |i - J| s_{ij} \qquad [34.1]$$

where s_{ij} is the overlap strength for potentially overlapping pairs of fragments and is zero otherwise.

34.2.4 Layout determination

When the shuffling is completed, contigs are selected and output using the following mechanism for extending and defining the termini of a contig. Starting from a fragment i, if $s_{i,i}{+}_1 > 0$, we say the fragments are connected and move on to i+1. We continue until fragments i and i+1 are not connected ($s_{i,i}{=}_{1\,+}\,0$) and assign i as the right end of the contig. The left end of the contig is defined similarly, by checking for connectedness to the left ($s_{i,i}{-}_1 > 0$). The leftmost fragment, i, is placed in its input orientation; the fragment i+1 is placed relative to fragment i. The relative orientation and offset of fragments i and i+1 (determined during and brought forward from the pairwise overlap strength calculations) are used to compute the placement; this is continued iteratively until the right end fragment of the contig has been output. The result is a layout containing fragments placed relative to a global frame of reference.

34.2.5 Relaxation

One class of approaches to combinatorial optimization is based on stochastic searches (distinct from the purely stochastic sampling described above). Given a starting point in solution space, one selects by a purely or partially stochastic process from a set of 'neighboring' positions in solution space, and applies some rule to determine whether or not to 'move' to the new position in solution space. Relaxation is a simple example: if, for successive iterations, k:

$$\Delta f = f_{k+1} - f_k \qquad [34.2]$$

and $\Delta f<0$, the new solution is an improvement and we move to it. Thus, one tests stochastically selected neighboring positions x, and moves downhill when the corresponding $f(x)$ is less. Relaxation can lead to the same problem mentioned above for gradient methods: one can end up in a local minimum.

34.2.6 Simulated annealing

A second example of stochastic search is simulated annealing (Kirkpatrick *et al.*, 1983; Davis, 1987; van Laarhoven & Aarts, 1987; Aarts & Korst, 1989), also described as stochastic relaxation. Simulated annealing addresses the problem of local minima by providing some probability of moving uphill towards higher values of the objective function. As with relaxation, if $\Delta f<0$, one adopts the new solution. However, if $\Delta f>0$, we move to the new solution with probability

$$P = \exp(-\Delta f/T) \qquad [34.3]$$

where T is decreased by a constant value at each iteration. Based on the analogy with annealing, the probability, P, of moving uphill ('higher energy') is greater early on in the system iteration when the 'temperature', T, is high, and decreases with iterations as T drops, with the expectation that the uphill moves will allow one to escape from shallow, nonglobal minima early on in the iterative process. Our implementation of annealing has been described in detail elsewhere (Churchil *et al.*, 1993); our choice of starting temperature, decrement of temperature at each iteration, and number of iterations is arrived at by experimentation with a given data set.

We have begun to explore increasing efficiency with adaptive annealing schedules (as distinguished from the linear annealing described above). Extensive studies of many optimization problems have indicated that predefined annealing schedules are not optimal in general. In particular, it has been found that there are certain temperature ranges (their location depending upon the specific problem at hand) in which annealing schedules must be very slow and careful to ensure a good final solution, and others in which the rate of annealing is not at all crucial. Therefore, the efficiency of annealing algorithms can often be substantially improved by the systematic application of adaptive annealing schedules, in which the response of the algorithm at any given temperature is used to modify the rate of alteration of the temperature itself. We are experimenting with the following three schedules (Aarts & Korst, 1989):

$$T(k+1) = T(k)\left[1 - \frac{T(k)^2}{c_1 t(k)^2}\right] \qquad [34.4a]$$

$$T(k+1) = T(k)1/\left[1 + \frac{T(k)\log(1+c_2)}{3t(k)}\right] \qquad [34.4b]$$

$$T(k+1) = T(k)\exp[-c_3 T(k)/t(k)] \qquad [34.4c]$$

In each of these expressions, $T(k)$ denotes the temperature at annealing step k. The parameter c_i represents a constant set once at the beginning of an annealing run. The main quantity driving the computation, $t(k)$, represents the variance of the objective function averaged over all the solutions selected during the annealing procedure (see equation [34.3]) up until annealing step k.

Why can the adaptive schedules described here be expected to work? Briefly, the reason is that they all monitor the variance, $t(k)$, and then decrement the temperature roughly inversely as the size of this quantity. It has been shown that temperature ranges in which the variance is larger than average are precisely those in which delicacy of annealing schedule is critical; the schedules above have each been shown to be optimal in different contexts characterized by different qualitative behaviors of the objective function in solution space for the specific problem under consideration (Aarts & Korst, 1989). In general, it is extremely difficult to characterize the objective function surface in question with an accuracy which justifies the choice of one of these three forms above any of the others. The question is best addressed empirically.

34.2.7 Genetic algorithms

A final example of stochastic search is genetic algorithms (Holland, 1975; Davis, 1987; Goldberg, 1989), which model the identification of new positions in solution space after the common modes of mutation (e.g. base substitution and chromosomal crossover) in genomic DNA, and model the rules for moving to these new solution space positions after the process of natural selection. Individual positions in solution space are modeled as bit strings (in the canonical form of genetic algorithms), or 'chromosomes', and one maintains a population of these bit strings. Given a starting population of strings, the next iteration ('generation') of the population is derived by (i) applying substitutions and recombinations to and among the current population to generate candidate new strings, and (ii) populating the next generation with numbers of these candidates in proportion to the fitness of their individual objective function values.

We have made an initial implementation of genetic algorithms for the sequence assembly problem (Parsons *et al.*, 1993). We allow substitution and recombination

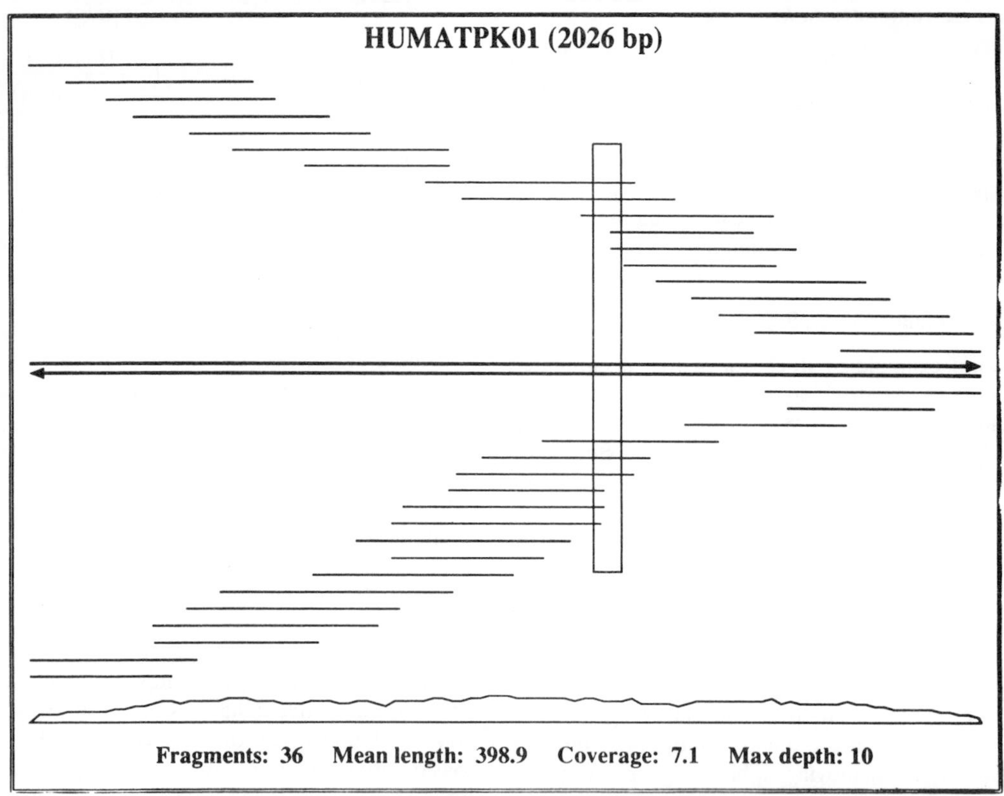


```
gcagtgtagtgagaccgccatttctagaaaaaaattaaaaattagccgggcgtggtggcac
gcagtgtagtgagacAgccatttctagaaaaaaattaaaaattagccgggcgtggtggcac
gcagtgtagtgagaccgccatttctagaaaaaaattaGaaattagccgggcgtAgtggcac
                                    aaaaattagccgggcgtggtggcac
                                     aaaattagccgggcgtggtggcac

GCAGTGTAGTGAGACCGCCATTTCTAGAAAAAAATTAAAAATTAGCCGGGCGTGGTGGCAC
CGTCACATCACTCTGGCGGTAAAGATCTTTTTTTAATTTTTAATCGGCCCGCACCACCGTG

cgtcacatcactctggcgTtaaagatctttttttaatttttaatcggcccgcaccaccgtg
cgtcacatcactctggcggtaaagatctttttttaatttttaatcTgcccgcaccaccgtg
cgtcacatcacCctggcggtaaagatctttttttaatttttaatcggcccgcaccaccgtg
Tgtcacatcactctggcggtaaag
cgtcacatcactctggcggtaaag
cgtcacatcactctggG

|          |   || |                  |       |       |
```

Figure 34.2 *Sample data set used as input for layout assembly.* A sample fragment set generated by applying *genfrag* (Engle & Burks, 1993) to the 2029 bp GenBank (Rel. 72.0) entry HUMATPK01 (Acc. No. M55090 (Sverdlov *et al.*, 1987). This is a human brain DNA sequence containing two Alu repetitive elements. (a) *Genfrag* produced 36 fragments with a mean length of 399 nucleotides for an average depth of coverage across the parent of 7.1. Random errors (in this case, single nucleotide substitutions) were introduced at an effective rate of 2.01%. (b) An enlarged section of the boxed region in (a) is shown to demonstrate the complexity of errors introduced. This corresponds to a 61 bp section (spanning 1200–1260 on the parent) which includes the central poly(T) repeat of an Alu element. Vertical bars denote alignment columns with at least one error present.

operations on the individual solutions, and are experimenting with the 'random keys' representation of Bean (1992) to address the problem of fragments being duplicated during recombination (which is not a problem in canonical genetic algorithm applications). We are also exploring extending the mutational operator set and other representations for the layout optimization problem.

Stochastic search combinatorial optimization methods are disadvantaged by uncertainty about exactly how close one is to the global optimum; however, they provide the great advantage, for some applications, of reliably finding a solution reasonably near the global optimum in reasonable search time.

34.3 SYSTEM AND DATA

The software was developed on a Sun Microsystems (Mountain View, CA) workstation running SunOS UNIX. Programs were written in the C programming language, and the interface was implemented in the OpenWindow (X-window compatible) windowing environment. The genetic algorithms implementation was done with the GENESIS development environment (Grefenstette, 1984). All fragment sets used here were generated by extracting known nucleotide sequences from GenBank (Burks *et al.*, 1992) and fragmenting them computationally to conform to a range of desired values for coverage, fragment length, repeat density, and error rate (Engle & Burks, 1993).

34.4 COMPARISON OF ALGORITHMS

34.4.1 Single algorithm, single seed

Using the fragments shown in Fig. 34.2, we did a single run (based on a particular random seed) with our implementation of equation [34.4b]. Figure 34.3 presents the optimization as a function of number of iterations (i.e. the number of times new solutions were generated and a decision made on whether or not to move to that solution).

34.4.2 Single algorithm, multiple seeds

Before making claims about comparisons of different algorithms, it is pertinent to compare the effects of different random seeds on individual algorithms. In Fig. 34.4, the effects of varying the random seed are shown for linear annealing (see equation [34.3]). Though the initial descent is similar among the five runs, they diverge at a high number of iterations, with two runs ending at local minima that are suboptimal (and which resulted in multiple contigs) compared to the global minimum reached by the other three runs.

34.4.3 Multiple algorithms, single seed

Figure 34.5 compares a single run for each of seven alternative algorithms, on a single data set and with the same random seed for initiating optimization (this does not apply to the greedy algorithm). All of the stochastic search algorithms converge on a much better minimum than the greedy algorithm. Relaxation and

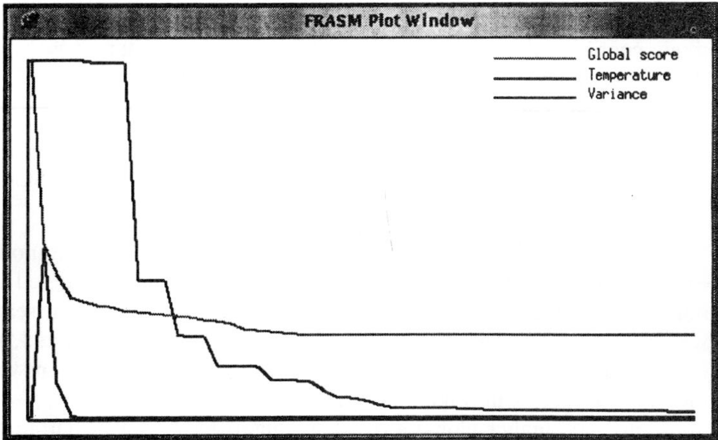

Figure 34.3 *Single algorithm, single random seed: adaptive annealing.* Objective function (see equation [34.1]) is plotted as 'global score' against number of optimization iterations. Input data were those shown in Figure 34.2. Starting temperature, T, was 500 with c_2=2.0 (see equations [34.3] and [34.4b]. 10 000 iterations took 19 s of CPU time and generated a single contig with a consensus sequence 99.9% similar to the parent sequence. This figure is taken from the software interface and includes overlay plots of the global score (ranging from 9.0×10^5 to 2.2×10^5), the descent of temperature (ranging from 500.0 to 7.8), and variance of the objective function score (ranging from 1.5×10^9 to 0).

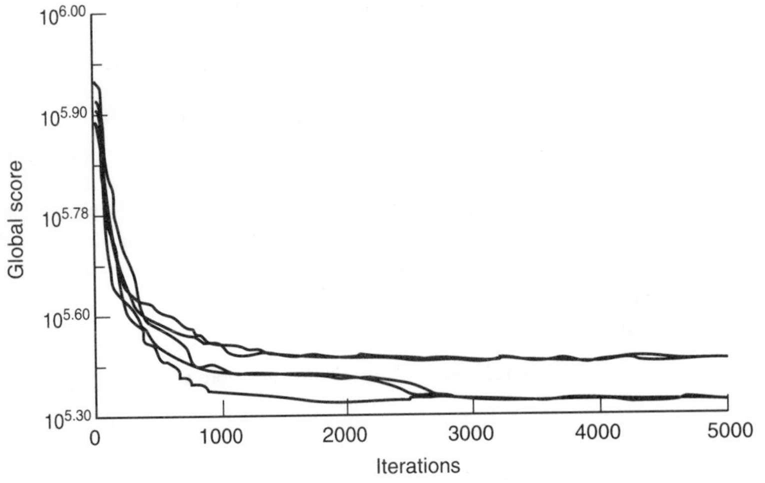

Figure 34.4 *Single algorithm, multiple random seeds: linear annealing.* Objective function ('global score'; see equation [34.1]) is plotted against number of optimization iterations. Input data were those shown in Figure 34.2. Initial T was 500 and $\triangle T$=0.055556 (see equation [34.3]).

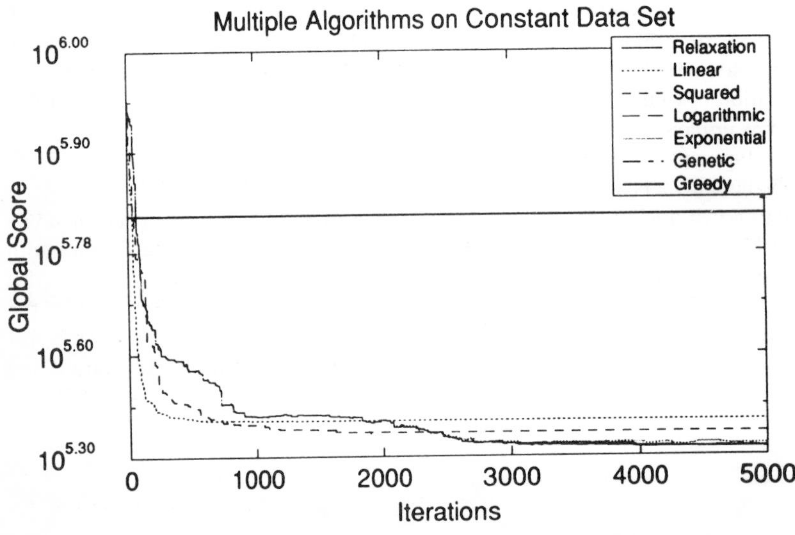

Figure 34.5 *Multiple algorithms, single random seeds.* Objective function ('global score'; see equation [34.1]) is plotted against number of optimization iterations. Input data were those shown in Figure 34.2. Annealing parameters were set as in Figures 34.3 and 34.4, with the addition of c_1=0.2 and c_3=0.5. The genetic algorithm run was done with a population of 200 strings of length 222; crossover rate was 0.8 and mutation rate was 0.005. For comparison to the other strategies, the best global score among the 200 strings was plotted version generation number (where, for these purposes, each new generation constitutes a new 'iteration').

all the simulated annealing algorithms resulted in a single contig; both the genetic and the greedy algorithms ended up with multiple contigs. Among the stochastic search algorithms, there are variations both with respect to rate of descent and with respect to the final minimum on which they converge.

34.4.4 Multiple algorithms, multiple seeds

Because the variation arising with changing the random seed (Fig. 34.4) appears comparable to that seen with changing algorithms (Fig. 34.5), we have experimented with plotting both changes (by sampling the descent path) in Fig. 34.6. Though at early iterations (and higher global scores) the algorithms' performances are relatively indistinguishable, at higher iterations they perform differently, even when averaged over several seed variations.

34.5 DISCUSSION

We have shown that assembly layout optimization can be modeled and run with a number of alternative stochastic search algorithms. In the limited number of cases tested, these algorithms are able to produce the correct answer (in this case, all fragments placed in one contig, with correct offsets relative to one another) in a matter of seconds. This is consistent with our

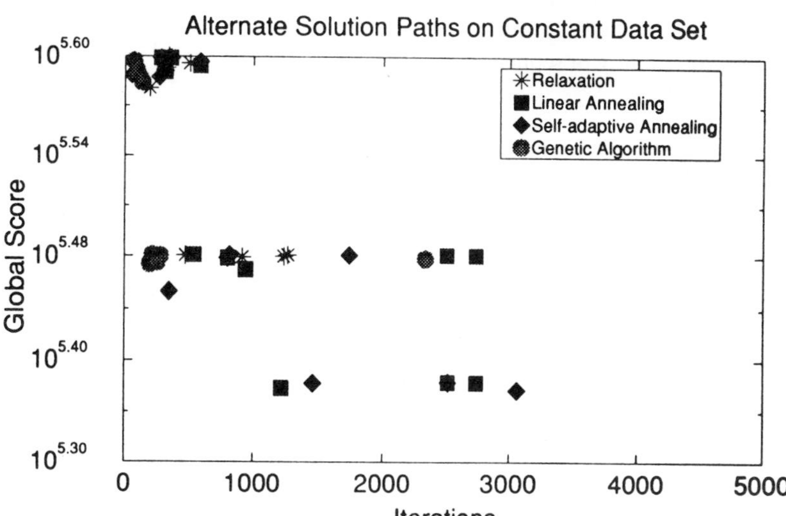

Figure 34.6 *Multiple algorithms, multiple seeds.* Objective function ('global score'; see equation [34.1]), monitored for four alternative stochastic search algorithms run successively with five different seeds, is sampled at three objective function values and plotted against number of optimization iterations. Input data were those shown in Figure 34.2. Run parameters were set as for Figures 34.3 to 34.5.

general goal of identifying algorithms that can very rapidly assemble a fragment set and arrive at a solution reasonably close to the global optimum.

However, we have not reached the point of asserting that these approaches are 'better' than available tools or recent alternative algorithms that have not yet been implemented as software tools. We believe that testing such as assertion will require: (i) extensive fine-tuning of our algorithms; and (ii) testing our and other algorithms over a much broader range of data sets than has yet been done. Furthermore, we predict that the choice of better algorithms, or combinations of algorithms, will depend on the kind of data sets being assembled.

The particular problem of sequence fragment assembly we have addressed here is one of several examples of the more general problem of assembling pieces of DNA based on experimental information about their degree of overlap (Kececioglu, 1991; Alizadeh *et al.*, 1992; Churchill *et al.*, 1993). In our case, overlap strength is determined by comparing sequences; in other cases, the complete sequence of each fragment is not known and overlap is determined using less information.

One example of a related problem is the generation of a restriction map from the patterns of restriction enzyme susceptibility in double digest strategies (Lander, 1989). Stochastic relaxation approaches have been applied to this problem (Goldstein & Waterman, 1987; Grigorjev & Mironov, 1990). Another such example, used in physical map assembly, is the comparison of patterns of susceptibility to restriction enzymes and/or hybridization to one or more DNA probes to determine fragment overlaps and layouts (Stallings *et al.*, 1990; Balding & Torney, 1991; Fickett & Cinkosky, 1993; Pratt & Dix, 1993; Soderlund *et al.*, 1993).

In both cases, given a set of overlap strengths for a set of fingerprinted clones (Balding & Torney, 1991), or an analogous determination of overlap strengths in the context of restriction digests, the shuffling and simulated annealing strategy could be used to generate the fragment layout for the clones. For example, we have extended the shuffling optimization model used here to the context of building contigs based on single digest restriction analysis of individual clones (Soderlund *et al.*, 1993). Similarly, Cuticchia *et al.* (1992) have applied simulated annealing to physical map assembly based on oligo-based fingerprints of the relevant clones.

Multiple sequence alignment is another related problem, if one considers individual nucleotides as the sequence fragments. Though using a different optimization model than that we have presented in our context, Lukashin *et al.* (1992) have applied simulated annealing to this context.

Finally, it is worth noting that solving and/or increasing the efficiency of the problem of layout determination does not in itself solve the overall problem of sequence assembly. In addition to the need for separate solutions for other recognized algorithmic modules of the assembly process discussed above, it is currently the case that much time (weeks, for cosmid-sized projects) currently goes into manually editing the output from sequence assembly packages to compensate for the presence of various kinds of error in the input data (Dear & Staden, 1991); the translation of these manual editing protocols onto algorithmic and computational platforms is a significant challenge that will greatly increase the overall efficiency and reproducibility of sequence assembly.

ACKNOWLEDGMENTS

This work was funded in part by the DOE Human Genome Program (ERW-F137, R. Moyzis, P.I.; ERW-F159, R. Keller, P.I.), under the auspices of the US Department of Energy. C.S. was supported by a DOE Human Genome Postdoctoral Fellowship, and R.P. by a LANL Director's Postdoctoral Fellowship.

REFERENCES

Aarts, E. & Korst, J. (1989) *Simulated Annealing and Boltzmann Machines: A Stochastic Approach to Combinatorial Optimization and Neural Computing*. Wiley, New York.

Alizadeh, F., Karp, R.M., Newberg, L.A. & Weisser, D.K. (1992) *Physical Mapping of Chromosomes: A Combinatorial Problem in Molecular Biology* (*Technical Report No. TR-92-066*). International Computer Science Institute, Berkeley, CA.

Balding, D.J. & Torney, D.C. (1991) *Bull. Math. Biol.* **53**, 853–879.

Bean, J.C. (1992) *Genetics and Random Keys for Sequencing and Optimization* (*Technical Report 92–43*). Department of Industrial & Operations Engineering, University of Michigan, Ann Arbor, MI.

Blum, A., Jiang, T., Li, M., Tromp, J. & Yannakakis, M. (1991) In *Proceedings of the 23rd ACM Symposium on Theory of Computing*. pp. 328–336.

Burks, C. (1989) In *Computers and DNA*, G.I. Bell & T. Marr (eds). Addison-Wesley, Reading, MA, pp. 35–45.

Burks, C., Cinkosky, M.J., Gilna, P., Hayden, J.E.-H., Keen, G.M., Kelly, M., Kristofferson, D., Fischer, W.M. & Lawrence, J. (1992) *Nucleic Acids Res.* **20**, 2065–2069.

Churchill, G.A. & Waterman, M.S. (1992) *Genomics,* **89**, 89–98.

Churchill, G.A., Burks, C., Eggert, M., Engle, M.L. & Waterman, M.S. (1993) Assembling DNA sequence fragments by shuffling and simulated annealing (in preparation).

Cinkosky, M.J., Fickett, J.W., Gilna, P. & Burks, C. (1991) *Science* **252**, 1273–1277.

Cull, P. & Holloway, J.L. (1992) In *Proceedings of the Twenty-Fifth Hawaii International Conference on System Sciences. Vol. I: Architecture and Emerging Technologies*, V. Milutinovic & B.D. Shriver (eds). IEEE Computer Society Press, Los Alamitos, CA, pp. 620–629.

Cuticchia, A.J., Arnold, J. & Timberlake, W.E. (1992) *Genetics* **132**, 591–601.

Davis, L. (1987) *Genetic Algorithms and Simulated Annealing*. Pitman, London.

Dear, S. & Staden, R. (1991) *Nucleic Acids Res.* **19**, 3907–3911.

Engle, M.L. & Burks, C. (1993) *Genomics* **16**, 286–288.

Fickett, J.W. & Cinkosky, M.J. (1993) In *Proceedings of the Second International Conference on Bioinformatics, Supercomputing, and Complex Genome Analysis*, H.A. Lim, J.W. Fickett, C.R. Cantor & R.J. Robbins (eds). World Scientific, Singapore, pp. 273–286.

Gallant, J.K. (1983) *J. Theor. Biol.* **101**, 1–17.

Gallant, J.K., Maier, D. & Storer, J. (1980) *J. Comput. Systems Sci.* **20**, 50–58.

Gill, P.E., Murray, W. & Wright, M.H. (1981) *Practical Optimization*. Academic Press, New York.

Goldberg, D.E. (1989) *Genetic Algorithms in Search, Optimization and Machine Learning*. Addison-Wesley, Reading, MA.

Goldstein, L. & Waterman, M.S. (1987) *Adv. Appl. Math.* **8**, 194–207.

Grefenstette, J.J. (1984) In *Proceedings of a Conference on Intelligent Systems and Machines*. Rochester, MI, pp. 160–168.

Grigorjev, A.V. & Mironov, A.A. (1990) *Comput. Applic. Biosci.,* **6**, 107–111.

Holland, J.H. (1975) *Adaptation in Natural and Artificial Systems*. University of Michigan Press, Ann Arbor, MI.

Howe, C.M. & Ward, E.S. (eds) (1989) *Nucleic Acids Sequencing: A Practical Approach*. IRL Press, Oxford.

Huang, X. (1992) *Genomics* **14**, 18–25.

Hunkapiller, T., Kaiser, R.J., Koop, B.F. & Hood, L. (1991a) *Curr. Opinions Biotechnol.*, **2**, 92–101.

Hunkapiller, T., Kaiser, R.J., Koop, B.F. & Hood, L. (1991b) *Science,* **254**, 59–67.

Johnson, D.S. & Papdimitriou, C.H. (1985) In *The Traveling Salesman Problem*, E.L. Lawler, J.K. Lenstra, A.H.G. Rinnooy Kan & D.B. Shmoys (eds). Wiley, New York, pp. 37–85.

Kececioglu, J. (1991) *Exact and approximate algorithms for sequence recognition problems in molecular biology*. Ph.D. Thesis, Department of Computer Science, University of Arizona, Tucson, AZ.

Kececioglu, J. & Myers, E. (1989) *A Procedural Interface for a Fragment Assembly Tool* (*Publication TR-89-5*). Department of Computer Science, University of Arizona, Tucson, AZ.

Kirkpatrick, S., Gellat, Jr., C.D. & Vecchi, M.P. (1983) *Science* **220**, 671–680.

Lander, E.S. (1989) In *Mathematical Methods for DNA Sequences*, M.S. Waterman (ed.). CRC Press, Boca Raton, FL, pp. 35–51.

Lander, E.S. & Waterman, M.S. (1988) *Genomics* **2**, 231–239.

Lawler, E.L., Lenstra, J.K., Rinnooy Kan, A.H.G. & Shmoys, D.B. (eds) (1985) *The Traveling Salesman Problem*. Wiley, New York.

Li, M. (1990) In *Proceedings of the 31st IEEE Symposium on Foundations of Computer Science*. pp. 125–134.

Lukashin, A.V., Engelbrecht, J. & Brunak, S. (1992) *Nucleic Acids Res.* **20**, 2511–2516.

Messing, J. & Bankier, A.T. (1989) In *Nucleic Acids Sequencing: A Practical Approach*, C.M. Howe & E.S. Ward (eds). IRL Press, Oxford, pp. 1–36.

Parsons, R., Forrest, S. & Burks, C. (1993) In *Proceedings of the First International Conference on Intelligent Systems for Molecular Biology*, T. Hunter, D. Searls & J. Shavlik (eds). AAAI/MIT Press, Menlo Park, CA, in press.

Peltola, H., Söderlund, H., Tarhio, J. & Ukkonen, E. (1983) *Information Processing* **83**, 59–64.

Peltola, H., Söderlund, H. & Ukkonen, E. (1984) *Nucleic Acids Res.* **12**, 307–321.

Pratt, D.M. & Dix, T.I. (1993) In *Proceedings of the Twenty-Sixth Hawaii International Conference on System Sciences. Vol. I: Systems Architecture and Biotechnology*, T.N. Mudge, V. Milutinovic & L. Hunter, (eds). IEEE Computer Society Press, Los Alamitos, CA, pp. 756–762.

Rawlings, C.J. (1986) *Software Directory for Molecular Biologists*. Macmillan, London.

Rice, P.M., Elliston, K. & Gribskov, M. (1991) In *Sequence Analysis Primer*, M. Gribskov & J. Devereux (eds). Stockton Press, New York, pp. 1–59.

Sedgewick, R. (1990) *Algorithms in C.* Addison-Wesley, Reading, MA.

Soderlund, C., Torney, D. & Burks, C. (1993) In *Proceedings of the Twenty-Sixth Hawaii International Conference on System Sciences. Vol. I: Systems Architecture and Biotechnology*, T.N. Mudge, V. Milutinovic & L. Hunter (eds). IEEE Computer Society Press, Los Alamitos, CA, pp. 620–629.

Staden, R. (1979) *Nucleic Acids Res.* **6**, 2601–2610.

Staden, R. (1980) *Nucleic Acids Res.* **8**, 3673–3694.

Staden, R. (1987) In *Nucleic Acid and Protein Sequence Analysis: A Practical Approach*, M.J. Bishop & C.J. Rawlings (eds). IRL Press, Oxford, pp. 173–217.

Stallings, R.L., Torney, D.C., Hildebrand, C.E., Longmire, J.L., Deaven, L.L., Jett, J.H., Doggett, N.A. & Moyzis, R.K. (1986) *Proc. Natl. Acad. Sci. U.S.A.* **87**, 6218–6222.

Sverdlov, E.D., Monastyrskaya, G.S., Broude, N.E., Ushkarev, Y.A., Melkov, A.M., Smirnov, Y.V., Malyshev, I.V., Allikmets, R.L., Kostina, M.B., Dulubova, I.E., Kiyatkin, N.I., Grishin, A.V., Modyanov, N.N. & Ovchinnikov, Y.A. (1987) *Dokl. Biochem.* **297**, 426–431.

Tarhio, J. & Ukkonen, E. (1988) *Theor. Comput. Sci.* **57**, 131–145.

Turner, J. (1989) *Inform. Comput.* **83**, 1–20.

van Laarhoven, P.J.M. & Aarts, E. (1987) *Simulated Annealing: Theory and Applications*. D. Reidel, Boston, MA.

PART III

Informatics

B. DATA ANALYSIS TOOLS

CHAPTER THIRTY-FIVE

Software Tools for Protein Similarity Searching

G. SUTTON & A.R. KERLAVAGE
The Institute for Genomic Research, 932 Clopper Road, Gaithersburg, MD 20878, USA

35.1 INTRODUCTION

Protein similarity searching seeks to find meaningful similarities between proteins based on their amino acid sequences. Current technology allows for large-scale sequencing of DNA. Several large-scale projects are underway to sequence mRNA. Sequencing, although a goal in itself, is secondary to understanding the structure and function of the proteins encoded by the mRNA. The easiest method of obtaining indications of an unknown protein's structure and function is by comparison of the unknown protein to previously characterized proteins. Amino acid sequence similarity between proteins often indicates a corresponding similarity in structure and function.

The similarity of two amino acid sequences is based on an alignment of the two sequences. The alignment is optimized so that identical and similar amino acids have a one-to-one correspondence between the two sequences. Biochemical or physical similarity between amino acids, e.g. the similarity of arginine and lysine or leucine and isoleucine, is typically represented by a numerical score in a 'substitution matrix' such as one of the PAM matrices (Dayhoff *et al.*, 1978; Schwartz & Dayhoff, 1978). Several reviews on sequence alignment are available (Sankoff & Kruskal, 1983; Waterman, 1989). The optimal alignment depends on the similarity or substitution matrix which defines the similarity measure for amino acids, and gap costs which define the penalty for introducing gaps of various sizes into the alignment. One of the difficulties in comparing protein similarity algorithms is that the choice of similarity matrix and gap costs has a significant impact on the performance of the algorithm.

Many computer software tools exist for identifying similarity between an unknown protein and databases of known proteins. Most of these tools are based on the same underlying definition of sequence similarity (these tools can usually be used for DNA as well as amino acid sequences). There are two driving forces which differentiate these tools: computer resource requirements and parameter tuning for optimization of domain-dependent goals. There have been limited comparisons of these tools for some applications (Argos *et al.*, 1991; Pearson, 1991; Gish & States, 1993). This chapter does not attempt to provide a thorough evaluation or comparison of these tools for protein similarity searches. The intent of this chapter is to make researchers aware of some of these tools.

This chapter gives a brief explanation of six protein similarity database search tools: FASTA, BLAST, BLAZE, BLITZ, PLSEARCH, and BLOCKS/ PATMAT. References which provide a more thorough treatment of the algorithms discussed in this chapter are also provided. A brief explanation of the classic dynamic programming approach to sequence alignment is presented first. Each software tool is covered in a separate section. A brief explanation of the tool's algorithm along with how to obtain and use the tool is given.

The dynamic programming approach to sequence alignment provides a basis for discussing most protein similarity database search tools. BLAZE and BLITZ use a variation of the dynamic programming algorithm. These tools are only practical because they have been implemented on MasPar parallel processor computers which reduces the time necessary for a search. FASTA and BLAST use heuristic techniques to reduce the complexity of the search algorithm making execution on a workstation feasible. PLSEARCH uses covering patterns for protein families to reduce the size of the database and focus the search on features significant to protein families. BLOCKS constructs compact representations of highly conserved regions within protein families. PATMAT searches for instances of these highly conserved regions in unknown protein sequences.

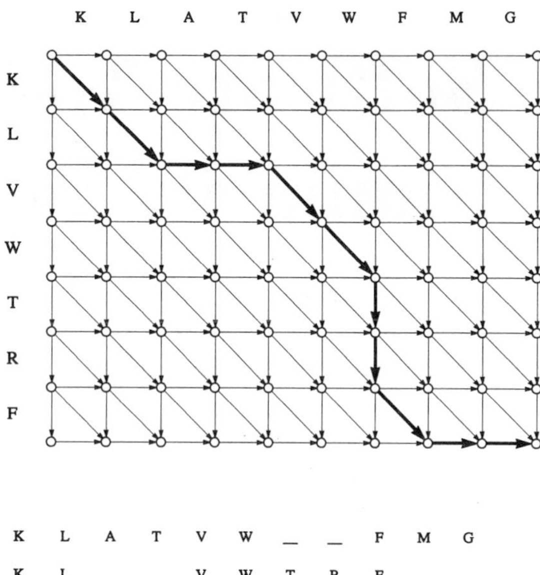

Figure 35.1 A dynamic programming search grid for aligning two amino acid sequences. All edges of the search grid are shown. Edges shown in bold indicate a path through the search grid corresponding to one alignment. This alignment may be optimal depending on the specific gap costs and similarity matrix used. This alignment is also shown in standard notation below the search grid.

35.2 SEQUENCE ALIGNMENT USING DYNAMIC PROGRAMMING

The dynamic programming algorithm constructs a two-dimensional grid with the two amino acid sequences to be aligned laid out along the horizontal and vertical axes. If the sequences are of length M and N amino acids, then the grid consists of $(M+1)\times(N+1)$ points laid out in a Cartesian plane. Every point is connected to its nearest neighbors by horizontal and vertical edges, resulting in M rows of vertical edges and N columns of horizontal edges. A way to visualize the algorithm is to label the M rows of edges, in order, with the M amino acids of the query sequence putting the first amino acid at the top, and the N columns of edges, in order, with the N amino acids of a database sequence putting the first amino acid at the left. A complete alignment is represented by a path of edges from the upper left point to the lower right point (Fig. 35.1). Horizontal edges can only be traversed from left to right and vertical edges from top to bottom. In addition, diagonal edges connecting neighboring points from top left to bottom right are added and can only be traversed in that direction. A horizontal edge in the ith column between the jth and $j+1$th rows represents a gap being introduced into the query sequence between the jth and $j+1$th amino acids which is aligned with the ith amino acid in the database sequence. A vertical edge represents an analogous gap in the database sequence. A diagonal edge which crosses the ith column and the jth row represents an alignment of the jth query sequence's amino acid with the ith database sequence's amino acid. A positive cost (penalty) or a negative cost (reward) is associated with each edge. Gap costs are assigned to the horizontal and vertical edges and similarity or substitution costs are assigned to the diagonal edges. The cost of a path (and hence the alignment the path represents) is the sum of the costs of the edges which make up the path. An optimal alignment is defined to be the alignment represented by the minimum cost path. The dynamic programming algorithm efficiently searches all possible paths to find the optimal path. This is achieved in order $M\times N$ time by filling in an $M\times N$ array of path costs.

For a good introduction to sequence alignment including the dynamic programming algorithm, see Sankoff & Kruskal (1983). Application of the dynamic programming algorithm to DNA and protein sequences is shown in Needleman & Wunsch (1970). Generalization from complete alignment to maximal partial alignment is shown in Smith & Waterman (1981). Reduction of the number of paths searched to improve efficiency is shown in Gotoh (1982).

35.3 BLAZE AND BLITZ

BLAZE and BLITZ basically provide database searching using the classic dynamic programming approach described above. The optimal alignment of the query sequence against every database sequence is determined and high-scoring alignments are returned. The performance of these tools is subject to the choice of similarity matrix and gap costs which are specifiable to some extent. No other factors are considered by the dynamic programming algorithm. The dynamic programming algorithm produces good results and has gained wide acceptance. Its biggest drawback is computational resource requirements. BLAZE and BLITZ address this issue by parallelizing the algorithm to run on the MasPar parallel processor. BLAZE is available from Intelligenetics, Inc. (Palo Alto, CA) as either a pay-per-use E-mail server or as a commercial product for the MasPar. BLITZ is currently available as an E-mail server (for information send E-mail with HELP as the first line to Blitz@EMBL-Heidelberg.DE). There are other implementations of the dynamic programming algorithm for other parallel processors (Jones, 1992). Implementations also exist for workstations and personal computers but are too slow for large-scale database searching.

35.4 FASTA

FASTA speeds up the search for an optimal alignment by reducing the search space of the dynamic programming algorithm using an efficient initial estimate of the best region of the dynamic programming grid to search. FASTA first marks all diagonal edges which correspond to exact matches of amino acid subsequences of length *ktup* (a user-specified parameter – default value of 2 for amino acid searches). Diagonals with the largest number of marked edges are then reevaluated using a similarity matrix. The highest scoring diagonal is then connected to other nearby high-scoring diagonals using the dynamic programming approach. FASTA is much faster than the full dynamic programming approach because it limits the application of the dynamic programming approach to a very narrow diagonal band of the search grid. The drawback to this approach is that optimal alignments which contain large gaps may be missed. FASTA has been shown to produce very good results in general (Pearson & Lipman, 1988; Pearson, 1991). A modification to the FASTA algorithm which marks diagonal edges corresponding to very similar amino acid matches as well as exact matches is called FASTDB (Brutlag *et al.*, 1990). The FASTA code is available via anonymous ftp from uvaarpa.virginia.edu in directory/pub/fasta. The FASTDB tool is available as a commercial product from Intelligenetics.

35.5 BLAST

BLAST is similar to FASTA in that it speeds up the search for an optimal alignment by limiting the search space using an efficient estimate of high-scoring regions of the search space. BLAST (Altschul *et al.*, 1990) uses an efficient mechanism to find all high-scoring diagonal segments (based on a similarity matrix) in the dynamic programming search grid of length w (a user-specified parameter – default value of 4 for amino acid searches). These high-scoring segments are combined when adjacent and extended further diagonally based on a similarity matrix. These extended segments are called maximal segment pairs (MSP). MSPs do not contain any gaps because they contain only diagonal edges – vertical and horizontal edges represent gaps. BLAST reports all high-scoring MSPs between the query sequence and a database sequence. The advantages of BLAST are speed (faster than FASTA) and not having to specify gap penalties – high-scoring MSPs are reported regardless of where they appear in the search grid. The disadvantages of BLAST are that distantly related proteins may not contain high-scoring MSPs but still have a significant optimal alignment using the dynamic programming approach, and that alignments are not generated to combine nonoverlapping MSPs which must instead be done by postprocessing. A nice feature of BLAST is that a DNA sequence can be used as the query sequence so that all three or six frames can be searched at the same time. The output of this search is combined so that some frame-shift errors are easy to identify. The BLAST code is available via anonymous ftp from ncbi.nlm.nih.gov in directory/pub/blast. BLAST is also available as an E-mail server (for information send E-mail with HELP as the first line to blast@ncbi.nlm.nih.gov.).

35.6 BLOCKS/PATMAT

BLOCKS constructs a database of highly conserved protein domains from multiple alignments of protein families. PATMAT searches the BLOCKS database for matches of these domains to the query sequence. High-scoring matches are reported. By identifying these matches, BLOCKS/PATMAT finds protein family markers that are independent of gap costs and may not seem significant in alignments with individual protein family members. BLOCKS/PATMAT is discussed much more fully in Chapter 42 in this volume.

35.7 PLSEARCH

PLSEARCH uses a database of covering patterns for each known protein family. The covering pattern is a string of symbols which includes gaps, amino acids and classes of amino acids. The covering pattern abstracts important features of the protein family. PLSEARCH uses the dynamic programming approach to search a query sequence against its database of covering patterns. The similarity matrix is generalized to include gaps and amino acid classes. There are indications that PLSEARCH can find proteins which are members of a covering pattern's family which cannot be found by comparison to individual members of the family (Smith & Smith, 1990). The PLSEARCH code is available via anonymous ftp from mbcrr.harvard.edu in directory MBCRR-Package.

35.8 OTHER SOFTWARE TOOLS

There are other tools for protein similarity searching, such as the sensitive dot-matrix algorithm (Argos, 1987; Rechid *et al.*, 1989). This algorithm overlays several different estimates of sequence similarity on the same diagonal edges of the dynamic programming search grid. These estimates can include similarity matrices and physical characteristics such as hydrophobicity. Another approach is to use multiple alignments to improve detection of members of protein families in a fashion similar to PLSEARCH and BLOCKS/PATMAT (Gribskov *et al.*, 1987; Vingron & Argos, 1989, 1991; Depiereux & Feytmans, 1992; Higgins *et al.*, 1992). A variant on this is to build a statistical model of a protein family and see if query sequences fit the model (Baldi *et al.*, 1993; Haussler *et al.*, 1993). A good introduction to a wide range of protein similarity searching approaches is available in Doolittle (1990).

REFERENCES

Argos, P. (1987) *J. Mol. Biol.* **193**, 385–396.

Argos, P., Vingron, M. & Vogt, G. (1991) *Protein Eng.* **4**, 375–383.

Altschul, S.F., Gish, W., Miller, W., Myers, E.W. & Lipman, D.J. (1990) *J. Mol. Biol.* **215**, 403–410.

Baldi, P., Hunkapiller, T., Chauvin, Y. & McClure, M. (1993) In *Neural Information Processing Systems 5* (S. Hanson, J. Cowan & C.L. Giles, eds) pp. 747–754. Morgan Kaufmann, San Mateo, CA.

Brutlag, D.L., D'Autricourt, J., Maulik, S. & Relph, J. (1990) *CABIOS* **6**, 237–245.

Dayhoff, M., Schwartz, R.M. & Orcutt, B.C. (1978) In *Atlas of Protein Sequence and Structure*, Vol. 5, Suppl. 3, M. Dayhoff (ed.). National Biomedical Research Foundation, Silver Spring, MD, pp. 345–352.

Depiereux, E. & Feytmans, E. (1992) *CABIOS* **8**, 501–509.

Doolittle, R.F. (ed.) (1990) *Methods Enzymol.* **183**.

Gish, W. & States, D.J. (1993) *Nature Genet.* **3**, 266–272.

Gotoh, O. (1982) *J. Mol. Biol.* **162**, 705–708.

Gribskov, M., McLachlan, A.D. & Eisenberg, D. (1987) *Proc. Natl. Acad. Sci. U.S.A.* **84**, 4355–4358.

Haussler, D., Krogh, A., Mian, I.S. & Sjolander, K. (1993) In *Proceedings of the Hawaii International Conference on System Sciences* (T. Mudge, V. Milutinovic & L. Hunter, eds) Vol. 1, pp. 792–802, IEEE Computer Society Press, Los Alamitos.

Higgins, D.G., Bleasby, A.J. & Fuchs, R. (1992) *CABIOS* **8**, 189–191.

Jones, R. (1992) *Int. J. Supercomput. Appl.* **6**, 138–146.

Needleman, S. & Wunsch, C. (1970) *J. Mol. Biol.* **48**, 444–453.

Pearson, W.R. (1991) *Genomics* **11**, 635–650.

Pearson, W.R. & Lipman, D.J. (1988) *Proc. Natl. Acad. Sci. U.S.A.* **85**, 2444–2448.

Rechid, R., Vingron, M. & Argos, P. (1989) *CABIOS* **5**, 107–113.

Sankoff, D. & Kruskal, J.B. (1983) *Time Warps, String Edits, and Macromolecules: The Theory and Practice of Sequence Comparison*. Addison-Wesley, Reading, MA.

Smith, R.F. & Smith, T.F. (1990) *Proc. Natl. Acad. Sci. U.S.A.* **87**, 118–122.

Smith, T.F. & Waterman, M.S. (1981) *J. Mol. Biol.* **147**, 195–197.

Schwartz, R.M. & Dayhoff, M. (1978) In *Atlas of Protein Sequence and Structure*, Vol. 5, Suppl. 3 (M. Dayhoff, (ed.). National Biomedical Research Foundation, Silver Spring, MD, pp. 353–358.

Waterman, M.S. (1989) In *Mathematical Methods for DNA Sequences*, M.S. Waterman (ed.). CRC Press, Boca Raton, FL, Ch. 3, pp. 53–92.

Vingron, M. & Argos, P. (1989) *CABIOS* **5**, 115–121.

Vingron, M. & Argos, P. (1991) *J. Mol. Biol.* **218**, 33–43.

CHAPTER THIRTY-SIX

Large-scale Sequence Analysis

J.-M. CLAVERIE

National Center for Biotechnology Information, National Library of Medicine, National Institutes of Health, Bethesda, MD 20894, USA

36.1 INTRODUCTION

Recent progress in DNA sequencing, mostly with the introduction of automated sequencers and biochemistry robots, has changed the pace by which new sequences are determined. As a consequence, these technological advances have also modified the scientific context of sequence determination. Formerly the final step along the way of well-worked-out biochemical or genetic study, DNA sequencing is now being turned into an early discovery tool, at the same time providing both the primary structure and the first evidence of the existence of a gene and/or a protein.

The concept of the expressed sequence tag (EST) (Adams *et al.*, 1992a,b; Waterston *et al.*, 1992; McCombie *et al.*, 1992) is at the forefront of this emerging strategy to gene identification. Here, partial cDNA sequences are determined at a high rate and stored in searchable computer databases, where they await eventual identification and further characterization.

For organisms with compact genomes (e.g. *Escherichia coli* and yeast), systematic genome sequencing is also becoming an efficient way to identify new genes and to precisely locate known ones. These efforts are producing increasingly larger contigs of 'anonymous' (i.e. with unknown or incompletely known gene content) sequences (e.g. Oliver *et al.*, 1992; Daniels *et al.*, 1992; Martin-Gallardo *et al.*, 1992) to be analyzed by computer techniques.

Finally, human genetics is also relying on large-scale DNA sequencing as a key step in the strategy of positional cloning of finely mapped disease genes. Once the chromosomal location of a gene is sufficiently narrowed, sequence contigs of 50–100 kb are routinely determined. These long anonymous DNA sequences must then be analyzed by computer in order to delineate putative coding exons, to be used as probes for the isolation of relevant cDNAs (Legouis *et al.*, 1991; Iris *et al.*, 1993).

Up to now, most sequence analysis programs and commercial packages have been implemented with the processing of individual cDNA or protein sequences in mind. The current algorithms (see reviews in: Doolittle, 1986, 1990; Boguski *et al.*, 1992) for database similarity search, multiple sequence alignment, or protein structure analysis are quite satisfactory. The quality of their user interface has improved steadily with time, from previously cumbersome command lines and listings to the fully interactive displays and color windows of today. However, these programs are designed for the detailed study of a single (or a few) sequences, and are poorly suited to the quick processing of either very large data sets (e.g.

thousands of partial cDNA sequences) or very large sequences. In the context of large-scale sequencing projects, interactive color displays become a nuisance to the scientist if he or she must spend hours before the screen reviewing the results. Paradoxically, one then needs to revert to 'batch' processing with all analyses being run overnight and the significant findings recorded into a database without human intervention. The scientists will then interact with the resulting database according to their specific interest, and only be presented with the *a priori* most interesting pre-interpreted data. For instance, one could register a set of 'standing' sequence queries and be automatically informed by the computer (e.g. by electronic mail) each time new sequences related to them have been encountered as the project progresses.

However, suppressing human intervention is not trivial. Programs produce cold results, while human researchers are mostly interested in hot 'findings'. The difference between results and findings is often what separates *statistical* from *biological* significance. While this discrimination is easily (and often subconsciously) performed by the scientist interacting with the results, a new layer of software has to be added to the classical algorithms when the raw data are to be processed without human intervention.

The main part of the analysis of large sequence sets concerns the identification of their putative functions, using similarity searches against databases of known genes and proteins. This chapter presents a collection of programs specially designed for this task in the context of large-scale analysis. As an example, I show how the preliminary identification of anonymous sequences can be fully automated by combining the execution of those programs with *ad hoc* 'shell scripts' (UNIX system command files). A particular emphasis is made on XBLAST and XNU, two second-layer tools. Those two modules implement a general masking strategy, and allow biologically significant results to emerge from otherwise lengthy and 'noisy' similarity search outputs.

The various software tools and the philosophy behind their utilization will be presented in a practical setting, consisting of three typical test cases: (1) the annotation of a large partial cDNA sequence data set; (2) the search for a gene in a 67 kb contig of human genomic DNA; and (3) the analysis of a large sequence from the *E. coli* chromosome.

36.2 ANALYZING LARGE DATA SETS: THREE TEST CASES

36.2.1 An EST project

An EST project might consist of the following challenge: every week, the project is going to produce 1000 anonymous partial cDNA sequences. The task is to analyze them and return the result without sophisticated or costly database management software. The most informative analysis we can do is to run all the new sequences against established databases like GenBank (Burks *et al.*, 1992) or SwissProt (Bairoch & Boeckmann, 1992) and generate a report on these searches in a manageable way.

Two main classes of program are used to perform database similarity searches. Programs like FASTA (Pearson & Lipman, 1988) or FASTDB (Brutlag *et al.*, 1990) produce *global* alignments, by matching query and target sequences over their entire lengths. This usually results in pairwise alignments containing gaps and deletions. The second type, like BLAST (Altschul *et al.*, 1990) only finds the best *local* alignments, that is optimal matching segments constituted of contiguous (ungapped) residues. Despite a small loss in sensitivity, BLAST has many distinctive advantages for exhaustive database scans. First, its simpler algorithm lends itself to a much faster implementation. It is also mathematically tractable. This allows each alignment to be associated with a straightforward measure of its statistical significance (Karlin & Altschul, 1990). Finally, local alignments are *a priori* best suited to the analysis of experimental data. For instance, both in the cases of EST projects or genomic sequencing (exons), we expect most matches with known proteins to be partial and to correspond to short contiguous segments.

Before performing the search, the EST data need to be rewritten into a suitable format. BLAST (and FASTA) requires the so-called 'FASTA format'. This is a simple way of recording a nucleotide or amino acid sequence and some accessory information in the same text file. The first line, the 'definition line', begins with a special '>' character followed by a one-word sequence identifier. The following lines specify the sequence in the standard one-letter code. A typical FASTA-formatted EST sequence looks like this:

```
>EST1234 small human brain cDNAlibrary#234
ATATATGGAGGAGATGAGC
GCAGATCAGCNNCATGCGT
GCAGXCTAGACAGATCNTA
...
```

The convention used by all programs using this format is that the first word after the > is a unique identifier for the sequence (here EST1234), but the rest of the line (as long as we wish) may be used to store any additional information (Fig. 36.1). I will take advantage of this useful feature throughout this chapter.

The program SIZEDB (Table 36.1) will help us to prepare the collection of EST sequences in the correct format. This program is a general tool with multiple options to reformat, clean up and verify sets of

A ESTs

```
>1590 | HS9B_HUMAN HEAT SHOCK PROTEIN HSP 90-BETA  +1 480 4.2e-64
GCATTGCTATTTATTCCTCGTCGGGCTCCCTTTGACCTTTTTGAGAACAAGAAGAAAAAGAACAACATCA
AACTCTATGTCCGCCGTGTGTTCATCATGGACAGCTGTGATGAGTTGATACCAGAGTATCTCAATTTTAT
CCGTGGTGTGGTTGACTCTGAGGATCTGCCCCTGAACATCTCCCGAGAAATGCTCCAGCAGAGCAAAATC
TTGAAAGTCATTCGNAAAAACATTGTTAAGAAGTGCCTTGAGCTCTTCTCTGAGCT
>1591 | ALUA_HUMAN: ALU CLASS A WARNING ENTRY  +1 166 2.7e-39
TTTTTTTTGAAACGAAGTCTCAGTCTGTCACCCAGGCTGGAGTGCAGTGGCACGATCCCGGCTCACTGCA
ACCTCTGTNTCCCAGGCTCAAGCTAGTCTCCTGCCTCAGCTGCCCGAGCAGACGGGACTACAGGCACCCC
CACCACGCCCGGCCAATCTCCAAATGGTTCTTTTTTTCCGGAGTAGTAAGTTACAATATGGGAGATTATT
>1594 Hippocampus, Stratagene (cat. #936205) Homo sapiens
ATGGGTCACTGAGGCTTTTNATTTTGAGCACAAAACCACCGGGGATCTAGCCTNTGGCCACCCCGGNGAT
GACACGAGGCTCACATGACTCTAGACACTTGGTGGAAAGTGAGGCGAGAAAAACAATGACTTGGGCCAAT
TACACGACTGCAAAGCTAGAGCTGCCAACAGGGCTCCAGGGAGCTTNGCTTCT
```

B Kallmann's syndrome gene candidate exons

```
>Kallmann_strdl_3389_3498     size: 110   index: 56.6286
gsgncseartllcdfsagrtsvnhgriwrqvrnsrlv
>Kallmann_strdl_16019_16282   size: 264   index: 55.5483
rcgcfttvqiatslylhlcsftgavlsfgqmsafsskaqsqnisfiqllttapmpseed
teeshrewilspakvlgisqrpkatplc
>Kallmann_strdl_35254_35397   size: 144   index: 55.1511
rfssgkaqsiqdmpgdnasyiglqqwlkgvgdqqlqlplvlgmrmlrc
```

C E. coli ORFs

```
>coli_54597_55628 | CORA_ECOLI MG / CO TRANSPORT PROTEIN  1740 2.3e-245
MLSAFQLENNRLTRLEVEESQPLVNAVWIDLVEPDDDERLRVQSELGQRPATRPELEDIEASARFFEDD
DGLHIHSxxxxEDAEDHAGNSTVAFTIRDGRLFTLRERELPAFRLYRMRARSQSMVDGNAYELLLDLFE
TKIEQLADEIENIYSDLEQLSRVIMEGHQGDEYDEALSTLAELEDIGWKVRLCLMDTQRALNFLVRKAR
LPGGQLEQAREILRDIESLLPHNESLFQKVNFLMQAAMGFINIEQNRIIKIFSVVSVVFLPPTLVASSY
GMNFEFMPELKWSFGYPGAIIFMISRAWHRICTLSGRTGCKKRERWLSLAKPPVxxxxxxxxxxxxxx
>coli_81796_82806 | No similarity [SwissProt 22 score: 90]
MSGEIFCTHVTKSTALRTAHRGANCIDNDNLFHCDSLSRFQVANRTGWLNRRIVRFVMTRFITQTFRHF
IQRAEVLILLCHIEVFCATEGVQPAERAAVERREAQAVDQRHIRFRRSGDNAFLQTAHYFVDHRDHHAG
DNLFFAEIALRLAHFCQQVFNGGVFFFLRLTFAVFFITPETEAVLLTKAVGIKQRVDGIAVIFLHPLRE
ACCHDSLSVMRGINAHNVQQICRAHRPAKLFFHHFVDLAEIRAVAQQLAETGEIREQHAVNKEAGAVVN
HNRRLAHLARPGxxxxxxxxxxxxxxNNFYQRHSVDRVKEVHPAEVFRAFQRVGQFADRNG
```

Figure 36.1 Samples of FASTA-formatted sequences.
(A) EST sequences after automatic annotation by DFBLAST. Three sequences are shown. The definition line begins with the '>' immediately followed by a unique identifier (here a number). The rest of the line is used to store additional information. DFBLAST recorded the similarity of EST 1590 with HSP 90-β heat shock protein together with the reading frame, the similarity score and its probability. EST 1591 is most similar to an Alu element. EST 1594 is not similar to a known coding region, and its original definition line has been preserved by DFBLAST. (B) Three candidate exons from the Kallmann's syndrome gene region. EXON generated the unique identifier from the location of the exon in the original 67 kb DNA sequence. The definition line is also used to store the size of the sequence together with an index characterizing its 'exonic' propensity. (C) Three *E. coli* ORFS. The sequence identifier was generated by ORFDB from the location of this ORF within the original 91.4 kb DNA sequence (Daniels *et al.*, 1992). The rest of the definition line was generated by DFBLAST from a search against SwissProt release 22 (for score ⩾90).

Table 36.1 Some programs for large-scale sequence analysis.

Name	*Usage*	*In*	*Out*
orfdb	orfdb [option] [start_codons] <seq> [min [max]]	DNA seq	DB of aa seq
exondb	exon <seq> [min][max]	DNA seq	DB of aa seq
sizedb	sizedb [options] <DB of seq> [min[max]]	seq db	seq db
dfblast	dfblast [options] <blast.output> <seq> [no_txt]	seq	annotated seq
xblast	xblast [options] <blast.output> <seq> [char]	seq	masked seq
xnu	xnu <seq> [options]	aa seq db	masked seq db

FASTA-formatted sequences. For instance, we use SIZEDB to filter nonstandard characters from the sequence (e.g. 'X'), to adjust the sequence line length, to modify the definition line in various ways (for instance appending the sequence size), or to select a subset of sequences according to a [min., max.] size range. Once the EST sequences are in a suitable format, we are ready to run a similarity search for every one of them.

Because we are mostly interested in potential coding regions, the first similarity search will be run against a protein sequence database. For this, we use the BLASTX program (Gish & States, 1993) which automatically generates the six translations (reading frames 1, 2, and 3 on both strands) of the nucleotide query (each individual EST) prior to comparing them with every protein database entry.

The command line will look like:

blastx prot.db EST1234 E=0.01 > EST1234.out

where the '*E*=0.01' option instructs the program to report only highly significant matches (this option will be implied throughout the rest of this chapter). '*E*=0.01' will cause the program to report matches which are expected to occur by chance alone with a frequency of 0.01 (or less) for the entire database search.

Obviously, we do not wish to run the same command 1000 times, and neither do we wish to browse through the 1000-page output corresponding to those searches. A more practical solution is to have an additional program read this output for us and annotate the EST query accordingly. This module, DFBLAST ('def' blast), is a filter that will generate a new definition line from the best match (if any) reported in the search output. We must now replace the previous command line by:

blastx prot.db EST1234 E=0.01 > EST1234.out
dfblast EST1234.out EST1234 > EST1234.ann

However, DFBLAST is a filter (in the sense given to that term by the UNIX system) and we can directly feed it with the search output without creating an intermediary file. For this we will invoke the 'piping' feature of the UNIX system:

blastx protein.db EST1234 E=0.01 |
dfblast + EST1234 > EST1234.ann

where the '+' is a place holder. In its standard mode, DFBLAST simply appends the original query definition line with a summary of the best similarity found. Alternative options allow for an explicit annotation of the nonmatching queries (e.g. 'No similarity within SwissProt 25.0') as well as replacing (instead of appending) the original definition line right after the unique sequence identifier (Fig. 36.1).

Finally, in order to process the entire EST set, we need to write a simple command file (e.g. UNIX 'shell script') to process each query one after another thus performing:

For (all ESTs) do
{*blastx prot.db EST | dfblast + EST > EST.ann*}

After running this search, we may want to search GenBank with the nucleotide sequences with the hope of detecting a few more matches in noncoding regions. For this we will use the set of annotated EST sequences we just created as queries for a new round of search. Now we use BLASTN for nucleotide versus nucleotide comparisons (Altschul *et al.*, 1990). The complete processing of the data will then look like:

For (all ESTs) do
{*blastx protein.db EST |*
dfblast + EST >EST.ann}

For (all EST.ann) do
{*blastn GenBank EST.ann |*
dfblast + EST.ann > EST.end}

The collection of annotated ESTs (EST.end) is the result. The processing has been performed with minimal human intervention. It provides most of the scientific information pertinent at this stage, in a compact and searchable form.

36.2.2 Finding a human disease gene

Our second test case is the analysis of a 67 kb human genomic sequence in the region of the Kallmann's syndrome gene (Legouis *et al.*, 1991). Here, the challenge was to pinpoint candidate exons to be used for probing cDNA libraries. Because we are looking for protein coding exons (the only ones with recognizable features), we are interested in sequence segments of the form:

[donor splice site] [ORF] [acceptor splice site]

Vertebrate splice sites do not have strict sequence consensus (except for the GT/AG rule) and we have to rely on a score/position matrix (Senapathy *et al.*, 1990) method to detect them. We will then retain every putative splice donor/acceptor pair separated by an open-reading frame (ORF). However, vertebrate exons are of virtually any size (on the small side), and those criteria are not discriminating enough. For instance, there are approximately 1000 candidates (overlapping) putative exons of length ⩾60 nucleotides for each strand of the Kallmann's syndrome gene 67 kb region. This means that the simple [splice site]/ORF detection method has to be supplemented with an assessment of the *a priori* likelihood for every one of those sequences segments to be coding exons.

This assessment will take advantage of: (1) a differential usage of nucleotide hexamers observed in exons versus introns (Claverie *et al.*, 1990; see also the review by Fickett & Tung, 1992); and (2) the eventual similarity of the putative exons with known proteins or translated ESTs.

A first program, EXON, takes each separate strand of the DNA sequence as input, together with two matrix/position scoring matrices representing our best knowledge about human splice sites (Senapathy *et al.*, 1990). Other input parameters are score thresholds for accepting candidate splice sites, as well as the minimal exon size we are willing to consider. Then, a command line like:

exon 67_kb_strand1 > strand1.candidates

is used to generate a collection of FASTA-formatted candidate coding exons translated into amino acid sequences. For each sequence in this file, the definition line indicates the position of the ORF in the nucleotide sequence, its size, and the value of an index describing (from its hexamer composition) how likely this segment is to be a coding exon (Fig. 36.1). In the standard option, EXON ranks the candidate exons according to this index value, best candidates first.

The output of EXON is a collection of *bona fide* FASTA-formatted amino acid sequences, and can immediately be used as a set of queries for database similarity search. This time we will use the BLASTP program (Altschul *et al.*, 1990) because we want to compare amino acid with amino acids. The corresponding command line will look like:

For (all candidates) do
{*blastp prot.db candidate > candidate.out*}

At this point, we may want to look at the output ourselves, provided there are not too many matches. More likely, we may want to proceed with further tests. In this case, we simply record the similarity information *in situ* on the definition line of each matching candidate exon. The command line will then become:

For (all candidates) do
{*blastp prot.db candidate |
dfblast + candidate > candidate.ann*}

We can now use those sequences as input for other programs. For instance, we might want to scan an EST database to see if any of those candidates resemble an expressed cDNA. Because we want to compare amino acids with nucleotides, we will use another version of BLAST called TBLASTN (Gish & States, 1993):

For (all candidate.ann)
{*tblastn EST.db candidate.ann |
dfblast + candidate.ann > candidate.end*}

At this point, we have to manually browse through the first ranking elements of the resulting candidate end file and start thinking about synthesizing a few probes.

36.2.3 Analyzing an *E. coli* chromosomal region

Our final test case is to locate the putative genes in a large *E. coli* genomic contig (e.g. the 91.4 kb sequence recently determined by Daniels *et al.* (1992)). In addition, we want to know which ones were previously known. For this, we first generate a collection of all ORFs above a certain size, in FASTA format. This is accomplished with the ORFDB program with a command line like:

orfdb ecoli.contig [min][max] > contig.orf.

This will generate a collection of FASTA-formatted amino acid sequences with a unique identifier built from the position of the ORF in the original nucleotide sequence (Fig. 36.1).

To determine which of the ORFS are related to known proteins or published ESTs, we will use:

For (all orfs) do
{*blastp prot.db orf | dfblast + orf > orf.ann*}

For (all orf.ann) do
{*tblasn EST.db orf.ann |
dfblast + orf.ann > orf.end*}

We now have an annotated collection of all potential genes in that region, to be used in further analyses.

36.3 METHODOLOGICAL ARTIFACTS AND SOLUTIONS

The simple data processing outlined in Section 36.2 is, in fact, not sufficient. For pedagogical reasons, we ignored a number of methodological artifacts. These are the problems (and solutions) we want to review in this section. A closer look at the annotations just generated in Sections 36.2.1 to 36.2.3 shows that they are not of the best quality. A large proportion of them (over 20% for both the ESTs and the candidate exons) are derived from misleading or biologically insignificant database matches. Here, 'biologically significant' means a result that either: (i) suggests a genuine common ancestry for the matched sequences; or (ii) is unexpected and provides information about a putative function. The analysis of many database search results indicates three main reasons for which our previous straightforward approach needs to be corrected: (1) protein databases contain erroneous, misleading entries; (2) database entries and natural sequences contain ubiquitous repeat elements; and (3) database entries and natural sequences exhibit segments of atypical residue composition.

The problem induced by the ubiquity of Alu repeats in human DNA will serve to illustrate points (1) and (2). Analyzing the results obtained in Section 36.2.1 shows that 10% of the EST sequences with a match in the protein database, in fact, contain various types of Alu elements. In turn, these elements match erroneous entries (as well as *warning* entries) in the database. This problem has been discussed in detail elsewhere (Claverie, 1992). Alu elements are also a major problem in the analysis of primate genomic sequence. For instance, they are responsible for most of the apparent protein database matches in the Kallmann's syndrome gene 67 kb contig.

Other ubiquitous elements account for more misleading matches both in the EST and genomic sequences: Line-1 repeats (matching reverse transcriptase and virus components), and MER sequences. Once recognized, the problem can be readily solved by a first example of 'second layer' analysis tool: XBLAST.

36.3.1 Xblast: a masking tool for *a priori* unworthy matches

The purpose of XBLAST is to clean from the search output the voluminous report of *a priori* unworthy results: matches that we can expect to occur, but that we do not wish to see. For this, XBLAST will mask in the original query sequence (e.g. EST, or candidate exon or ORF sequences) the precise segments bearing the troublesome similarity. Because 'worthy' matches can be obscured by 'unworthy' ones (Fig. 36.2), masking the sequences with problems has to be preferred over simply discarding it. The information used to direct the masking will be supplied by a BLAST search against a database of preselected 'junk' sequences (Fig. 36.2).

For instance, let us create a database of various Alu elements, Line-, MER, etc. To be safe, we may want to add other potential sources of artifacts, like sequences from the cloning or sequencing vectors. To purge our collection of ESTs from any trivial match with these sequences, we will use a command line like:

For (all ESTs) do
{*blastn junk.db EST* | *xblast* + *EST* > *EST.mask*}

Here, we are using a junk database of nucleotide sequences and thus invoked the BLASTN program. Alternatively, the junk.db might consist of a six-frame translation of the same troublesome entries:

For (all ESTs) do
{*blastx junk.db EST* | *xblast* + *EST* > *EST.mask*}

The latter works better for processing segments as polymorphic as Alu repeats. It also has the advantage of greater consistency when the subsequent analyses are going to involve the translations of the ESTs (e.g. Section 36.2.1). When the same similarity threshold is used for both the masking and the following database search, we can be assured that not a single residual 'junk' match will contaminate the final result.

For ESTs, XBLAST masking may also serve to automatically eliminate redundant clones such as those coding for tubulin, actin or mitochondrial and ribosomal proteins. In this case, the 'junk' database will consist of a number of these 'unwanted' sequences. Note that DFBLAST and XBLAST have a similar syntax. With XBLAST, the information from the BLAST search output is used to modify the sequence part of the query instead of its definition line.

Similarly, a masking step against an 'Alu/Line-1/MER' junk database is definitely required for the proper processing of the 67 kb Kallmann's syndrome gene contig. It will be performed automatically by:

For (all exons) do
{*blastp junk.db exon* | *xblast* + *exon* > *exon.mask*}

For (all exon.mask) do
{*blastp protein.db exon.mask* |
dfblast + *exon.mask* > *exon.ann*}

Repeated elements and junk sequences are less of a problem for organisms with more compact genomes like yeast or *E. coli*. Masking by XBLAST may nevertheless be useful to filter out cloning or sequencing

A Query: EST1591 (210 nucleotides)

```
                                                      Reading      Probability
Sequences producing High-scoring Segment Pairs:        Frame  Score    P(N)

ALUA_HUMAN !!!! ALU CLASS A WARNING ENTRY !!!!          +1     166  2.7e-39
ALUC_HUMAN !!!! ALU CLASS C WARNING ENTRY !!!!          +2     153  1.9e-34
COX1_MOUSE CYTOCHROME C OXIDASE POLYPEPTIDE I.          -2      92  4.8e-07
COX1_RAT   CYTOCHROME C OXIDASE POLYPEPTIDE I.          -2      92  4.8e-07
COX1_XENLA CYTOCHROME C OXIDASE POLYPEPTIDE I.          -2      90  9.8e-07
COX1_BALPH CYTOCHROME C OXIDASE POLYPEPTIDE I.          -2      90  9.8e-07
COX1_BOVIN CYTOCHROME C OXIDASE POLYPEPTIDE I.          -2      90  9.8e-07
COX1_HUMAN CYTOCHROME C OXIDASE POLYPEPTIDE I.          -2      90  9.8e-07
```

B Query: *Alu* masked EST1591

```
>1591 - Alu masked
nnnnnnnnnnnnnnnnnnnnnnnnnnnnnnnnnnnnnnnnnnnnnnnnnnnnnnnnnnnnnnnnnnnnnnn
nnnnnnnnnnnnnnnnnnnnnnnnnnnnnnnnnnnnnnnnnnnnnnnnnnnnnnnnnnnnnnnnnnnnnn
nnnnnnnnnnnnnnnnnnnnnnnnnnnnnnnnnnnnnnnnnnnnnnnnnnnCCAAATGGTTCTTTTTTCCG
GAGTAGTAAGTTACAATATGGAGATTATT
```

```
                                                      Reading      Probability
Sequences producing High-scoring Segment Pairs:        Frame  Score    P(N)

COX1_MOUSE CYTOCHROME C OXIDASE POLYPEPTIDE I.          -2      92  4.8e-07
COX1_RAT   CYTOCHROME C OXIDASE POLYPEPTIDE I.          -2      92  4.8e-07
COX1_XENLA CYTOCHROME C OXIDASE POLYPEPTIDE I.          -2      90  9.8e-07
COX1_BALPH CYTOCHROME C OXIDASE POLYPEPTIDE I.          -2      90  9.8e-07
COX1_BOVIN CYTOCHROME C OXIDASE POLYPEPTIDE I.          -2      90  9.8e-07
COX1_HUMAN CYTOCHROME C OXIDASE POLYPEPTIDE I.          -2      90  9.8e-07
```

C Best biologically significant alignment

```
COX1_MOUSE:  CYTOCHROME C OXIDASE POLYPEPTIDE I. (Length = 514)

     1591:   209  IISHIVTYYSGKKEPF 159          reading frame -2
                  ::::+:::::::::::
     COX1:   253  IISHVVTYYSGKKEPF 268
```

Figure 36.2 Masking Alu segments with XBLAST. (A) Summary of the BLASTX output for EST1591. The search was run against SwissProt (for score ≥90). More relevant matches are obscured by the presence of a high scoring (S=166, S=153) Alu-related segment. (B) Alu-masked EST1591 query and resulting BLASTX output. The Alu-related segment [1-208] was masked (n) by searching a subset of representative translated Alu elements (Claverie, 1992) and piping the output into XBLAST. The few remaining nucleotides are sufficient to identify this sequence as coding for cytochrome *c* oxidase polypeptide I. (C) Best local alignment between Alu-masked EST1991 and cytochrome *c* oxidase.

vector sequences prior to the final assembly of the sequence. XBLAST can also be used in more creative ways. For instance, if our interest is only in the new genes within our contig, we may prepare a database of all known *E. coli* protein and mask out all the previously known sequences (in this case using a very high similarity score threshold). The command lines will then look like:

For (all orfs) do
{*blastp E.coli.db orf* | *xblast + orf* > *orf.new*}

For (all orf.new) do
{*blastp protein.db orf. new* |
dfblast + orf.new > orf.new.ann}

which will automatically generate an annotation of *E. coli* putative new genes based on their significant similarity to proteins of other organisms.

XBLAST recognizes various options, including '-r' which causes the output to be reversed, i.e. nonmatching segments to be masked, and '-d' which constitutes a collection (in FASTA format) of the matched segments contained in every query. There are numerous creative ways these options can be used to: (1) enhance the *a priori* biological significance of database searches; and (2) adapt them to a specific interest.

36.3.2 XNU: a tool for delineating and masking repeats within proteins

Masking our test queries with XBLAST dramatically improves the biological relevance of our automated annotation procedure. However, a more subtle problem is still causing statistically significant but misleading matches to be reported. Many ESTs, candidate exons or *E. coli* putative coding regions appear to be similar to a limited set of database entries including transcription factors, homeotic proteins, surface glycoproteins and keratins. A closer look at these matches reveals that they involve segments constituted of a smaller than usual number of distinct amino acids. Because of their sequence simplicity, these segments are called 'low complexity' or 'low entropy' repeats. When analyzing anonymous DNA sequences, these segments often arise from the translation of homopolymeric regions frequently found outside coding regions. Unfortunately, a number of real proteins exhibit low complexity segments as well. They correspond to the ones we cited before. For instance, many transcription factors exhibit a poly(glutamine) tail, while keratin and a number of cell wall proteins contain poly(proline) and poly(glycine) segments. Other types of troublesome motifs consist of the repetition of a small motif such as is found in antifreeze proteins (poly[TAAT]) or collagens (poly[GXX] or poly[GXP]).

Because of their biased (nonrandom) composition, these regions greatly increase the probability of spurious matches between unrelated proteins (Karlin & Altschul, 1990). This point can be illustrated by a simple computer experiment. We first build a theoretical query made of 20 successive homopolymeric segments such as:

80[L]–64[A]–56[G]–56[V]–56[S]–48[K]–48[T]–
48[E]–40[R]– . . . –16[H]–16[C]–10[W]

with each numerical coefficient approximating the natural abundance of each amino acid. Used as a query, this low complexity sequence matches 690 SwissProt (release 21) entries with highly significant ($p<10^{-6}$) scores (using BLASTP, a PAM120 matrix and score $\geqslant$90). If we now randomly shuffle this 802-residue sequence and use it as a query for the same database scan, not a single match is observed. This is the expected result, since no protein in the database can possibly be related to our theoretical query.

To correct the problem induced by low complexity segments in large-scale sequence analysis, we have developed XNU, a program to locate and mask internal repeats in amino acid sequences (Claverie & States, 1993). Figure 36.3 illustrates the basic algorithm behind this program. Briefly, the sequence is compared to itself; the local alignments are scored with a PAM matrix and their significance is estimated according to Karlin & Altschul (1990). In a dot-matrix representation, the matched internal repeats appear as off-diagonal segments (Fig. 36.3). Low complexity characterizes those segments found very close to the main diagonal (often as a smear) while 'normal' repeats appear at a distance at least equal to their length. This property immediately suggests an algorithm to discriminate between low-complexity regions and regular repeats which does not involve any *ad hoc* measure of the local information content. Low-complexity segments will simply be recognized from the fact that they repeat themselves with a periodicity of a few positions, while regular internal repeats are associated with much larger periodicity (e.g. 100 residues for an immunoglobulin or a fibronectin domain).

The elements repeated in the sequence are mapped by projecting (vertically and horizontally) the off-diagonal elements onto the main diagonal. A selective mapping (and eventual masking) of the low-complexity segments is accomplished by only taking into account the off-diagonal elements close to the main diagonal (Fig. 36.3). With a short periodicity threshold (the default value is 10), XNU operates on low-complexity repeats. With a higher threshold (up to half the sequence length), XNU operates on both regular and low-complexity repeats.

In the current implementation, the XNU command

initial self-comparison

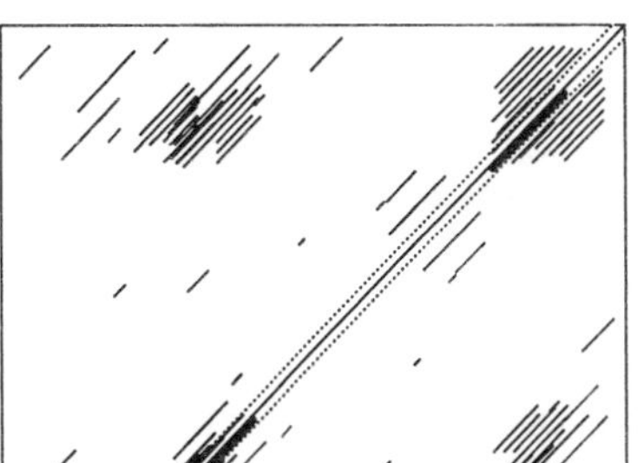

after masking with xnu

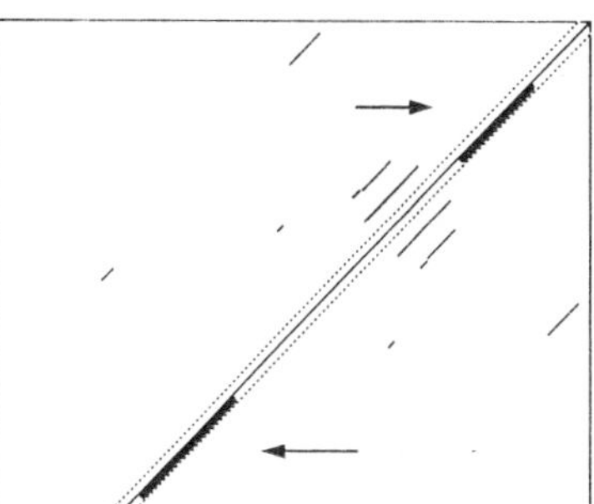

Figure 36.3 Masking internal and/or intrinsic repeats with XNU. (Left) Dot-matrix representation of the self-comparison of the sequence of mouse epithelial growth factor. Pairing segments are scored with the PAM120 matrix, and only alignments with $p<0.01$ are shown. Low-complexity segments induce a characteristic pattern of off-diagonal elements, very close to the main diagonal. (Right) Self-comparison of the same sequence after processing by XNU with default parameters. Low-complexity segments (indicated by arrows) are located by considering off-diagonal elements at a distance of 10 residues (indicated by dotted lines) from the main diagonal (short period repeats).

line recognizes a variety of parameters controlling the repeat periodicity, the score threshold (or probability) for considering local alignments, and the PAM matrix to be used. Various options allow the output to take alternative forms. In the standard mode (no parameters or options specified) XNU detects and masks the low-complexity repeats, thus ensuring that each reported match will reveal a biologically significant similarity (e.g. common ancestry).

Figure 36.4 shows the performance of XNU on one of the worst cases, i.e. collagen. Processing a collagen sequence with XNU allows a typical BLASTP search output to be reduced from 130 reported matches (most of them without any biological significance) to only five, all of them with *bona fide* collagen-related entries. Figure 36.5 illustrates the combined usage of XBLAST and XNU on the analysis of the transcription factor NF-κB p49 sequence. The initial database search output is very confusing, featuring matches with Alu-related entries, DNA-binding proteins, cell wall structural protein, surface receptor, keratin and myosin! In contrast, the processed entry is only found to match truly biologically relevant entries: transcription factors and DNA-binding proteins (Fig. 36.5).

Because XBLAST and XNU involve the same basic matching algorithm and rely on the same statistical model of alignments (Karlin & Altschul, 1990) they can be used in combination with each other and, with subsequent BLAST searches, in a consistent way. Equivalent score/significance thresholds can be used through the entire process of: (1) masking the query for unworthy matches (XBLAST); (2) masking its low complexity segments (XNU); and (3) running the final database search. For instance, using a significance threshold of $p \leq 0.01$ for each masking step will warrant that no biologically relevant match of equal or greater significance is at risk of disappearance from the final output.

The optimal processing of our three test cases should now involve a supplementary step to prevent low complexity segments to induce artifactual matches. For the best possible result, XNU masking should be applied on both the query sequences and the protein database being scanned. For instance, the script for a fully automatic annotation of the candidate exons in the Kallmann's syndrome gene region now becomes:

For (all exons) do
{*xnu exon* > *exon.xnu*}

For (all exon.xnu) do
{*blastx junk.db exon.xnu*|
xblast + *EST* > *exon.mask*}

For (all exon.mask) do
{*blastp prot.db exon.mask* |
dfblast + *exon.mask* > *exon.ann*}

This final procedure reduces the search output from over 500 to only three putative candidates with a significant match in the database. One of them was successfully used as a probe for the ADML-X cDNA, the Kallmann's syndrome gene (Legouis *et al.*, 1991).

36.4 DISCUSSION

The various modules introduced here have been used in a variety of real large-scale sequence analysis projects, including the annotation of a recently established comprehensive database of all published partial cDNAs (DbEST) (Boguski & Tolstoshev, 1992), the identification of the gene responsible for

A Original query : Collagen ALPHA 1(IX)

```
>CA19_HUMAN COLLAGEN ALPHA 1(IX) CHAIN PRECURSOR.
MKTCWKIPVFFFVCSFLEPWASAAVKRRPRFPVNSNSNGGNELCPKIRIGQDDLPGFDLISQFQVDKAASRRAIQRVVGS
ATLQVAYKLGNNVDFRIPTRNLYPSGLPEEYSFLTTFRMTGSTLKKNWNIWQIQDSSGKEQVGIKINGQTQSVVFSYKGL
DGSLQTAAFSNLSSLFDSQWHKIMIGVERSSATLFVDCNRIESLPIKPRGPIDIDGFAVLGKLADNPQVSVPFELQWMLI
HCDPLRPRRETCHELPARITPSQTTDERGPPGEQGPPGASGPPGVPGIDGIDGDRGPKGPPGPPGPAGEPGKPGAPGKPG
TPGADGLTGPDGSPGSIGSKGQKGEPGVPGSRGFPGRGIPGPPGPPGTAGLPGELGRVGPVGDPGRRGPPGPPGPPGPRG
TIGFHDGDPLCPNACPPGRSGYPGLPGMRGHKGAKGEIGEPGRQGHKGEEGDQGELGEVGAQGPPGAQGLRGITGLVGDK
GEKGARGLDGEPGPQGLPGAPGDQGQRGPPGEAGPKGDRGAEGARGIPGLPGPKGDTGLPGVDGRDGIPGMPGTKGEPGK
PGPPGDAGLQGLPGVPGIPGAKGVAGEKGSTGAPGKPGQMGNSGKPGQQGPPGEVGPRGPQGLPGSRGELGPVGSPGLPG
KLGSLGSPGLPGLPGPPGLPGMKGDRGVVGEPGPKGEQGASGEEGEAGERGELGDIGLPGPKGSAGNPGEPGLRGPEGSR
GLPGVEGPRGPPGPRGVQGEQGATGLPGVQGPPGRAPTDQHIKQVCMRVIQEHFAEMAASLKRPDSGATGLPGRPGPPGP
PGPPGENGFPGQMGIRGLPGIKGPPGALGLRGPKGDLGEKGERGPPGRGPNGLPGAIGLPGDPGPASYGKNGRDGERGPP
GLAGIPGVPGPPGPPGLPGFCEPASCTMQLVSEHLTKGLTLERLTAAWLSA
```

B Original database search output

```
                                                                      Probability
Sequences producing High-scoring Segment Pairs:                 Score  P(N)       N

CA19_HUMAN COLLAGEN ALPHA 1(IX) CHAIN PRECURSOR.                 4922  0.0        1
CA25_HUMAN PROCOLLAGEN ALPHA 2(V) CHAIN PRECURSOR.                730  0.0        5
CA12_BOVIN COLLAGEN ALPHA 1(II) CHAIN (FRAGMENTS).                696  0.0        4
CA2Y_HUMAN COLLAGEN ALPHA 2(XI) CHAIN (FRAGMENT).                 659  1.2e-318   4
CA11_MOUSE COLLAGEN ALPHA 1(I) CHAIN (FRAGMENT).                  726
..................................................
120 matches !!!
.....................................................
MABA_RAT   MANNOSE-BINDING PROTEIN A PRECURSOR (MBP-A) (M...       79  1.3e-07    3
ANX7_HUMAN ANNEXIN VII (SYNEXIN).                                  76  1.5e-07    2
EXON_HSV2  ALKALINE EXONUCLEASE (EC 3.1.11.-).                     74  1.7e-07    4
GRP1_PETHY GLYCINE-RICH CELL WALL STRUCTURAL PROTEIN PREC...       88  1.0e-06    2
APEG_XENLA APEG PROTEIN PRECURSOR (FRAGMENT).                      82  5.6e-06    5
MSAP_PLAFC MEROZOITE SURFACE ANTIGENS PRECURSOR (PMMSA).           95  5.9e-06    1
KSGZ_DROME PROTEIN KINASE SHAGGY, ZYGOTIC (EC 2.7.1.-) (P...       75  0.00018    2
SRP1_YEAST SERINE-RICH PROTEIN.                                    71  0.00035    2
```

C xblast (*Alu*) and xnu-masked query

```
>NF-kappa B p49  [xblast and xnu-masked]
MESCYNPGLDGIIEYDDFKLNSSIVEPKEPAPETADGPYLVIVEQPKQRGFRFRYGCEGPSHGGLPGASSEKGRKTYPTV
KICNYEGPAKIEVDLVTHSDPPRAHAHSLVGKQCSELGICAVSVGPKDMTAQFNNLGVLHVTKKNMMGTMIQKLQRQRLR
SRPQGLTEAEQRELEQEAKELKKVMDLSIVRLRFSAFLRASDGSFSLPLKPVTSQPIHDSKSPGASNLKISRMDKTAGSV
RGGDEVYLLCDKVQKDDIEVRFYEDDENGWQAFGDFSPTDVHKQYAIVFRTPPYHKMKIERPVTVFLQLKRKRGGDVSDS
KQFTYYPLVEDKEEVQRKRRKALPTFSQPXXXXXXXXXXXXXXXXXXXXXXXXXXVLMEGGVKVREAVEEKNLGEAGRxx
xxxxxxxxxxxxxxxxxxxxxxxxxxxxxxxxxxxxxxxxxxxxRRL
```

D new database search output

```
                                                                      Probability
Sequences producing High-scoring Segment Pairs:                 Score  P(N)       N

KBF1_HUMAN DNA-BINDING FACTOR KBF1 (NUCLEAR FACTOR NF-KAP...      336  2.1e-111   3
TREL_MELGA C-REL PROTO-ONCOGENE PROTEIN (C-REL PROTEIN) (...      257  1.2e-50    2
TREL_CHICK C-REL PROTO-ONCOGENE PROTEIN (C-REL PROTEIN) (...      253  1.4e-50    2
TREL_AVIRE REL TRANSFORMING PROTEIN (P58 V-REL).                  257  3.6e-49    2
DORS_DROME EMBRYONIC POLARITY DORSAL PROTEIN.                     240  3.0e-47    2
TREL_MOUSE C-REL PROTO-ONCOGENE PROTEIN (C-REL PROTEIN).          231  1.1e-38    4
```

Kallmann's syndrome (Legouis *et al.*, 1991) and the analysis of a large Alu-plagued human genomic contig in the HLA class III region (Iris *et al.*, 1993). They are also indispensable in making sense out of statistical studies involving the self-comparison of the entire protein sequence database (Green *et al.*, 1993; Claverie, 1993).

The modules are simple tools, coded in the C programming language and adapted to the UNIX environment. Their purpose is the scientific analysis of large sequence data sets by a single scientist, not the daily management of a large-scale sequencing project in a big laboratory. The latter has to take advantage of the more secure and powerful environment of object-oriented (Sulston *et al.*, 1992) or relational database systems (Kerlavage *et al.*, 1993). The programs listed in Table 36.1 are available upon request. However, they are research tools and are thus not maintained with the rigor which characterizes 'professional' software. Rather than to advertise the use of specific programs, the purpose of this chapter was more to expose a few key concepts. For instance, we showed that large-scale sequence analysis can be performed with minimal human intervention, using a handful of modular programs, the execution of which is chained via simple command files (e.g. UNIX 'shell' script). When the correct methods are used, computer performance is not a limiting factor and very large sequence analyses can be run on typical UNIX workstations.

ACKNOWLEDGMENT

I thank Dr Warren Gish for helpful discussions concerning the use of BLAST and his careful review of this manuscript.

REFERENCES

Adams, M.D., Dubnick, M., Kerlavage, A.R., Moreno, R.F., Kelley, J.M., Utterback, T.R., Nagle, J.W., Fields, C. & Venter, J. (1992a) *Nature* **355**, 632–634.

Adams, M.D., Kelley, J.M., Gocayne, J.D., Dubnick, M., Polymeropoulos, M.H., Xiao, H., Merril, C.R., Wu, A., Olde, B., Moreno, R.F., Kerlavage, A.R., McCombie, W.R. & Venter, J. (1992b) *Science* **252**, 1651–1656.

Altschul, S.F. (1991) *J. Mol. Biol.* **219**, 555–565.

Altschul, S.F., Gish, W., Miller, W., Myers, E.W. & Lipman, D. (1990) *J. Mol. Biol.* **215**, 403–410.

Bairoch, A. & Boeckmann, B. (1992) *Nucleic Acids Res.* **20** (Suppl.), 2019–2022.

Brutlag, D.L., Dautricourt, J.-P., Maulik, S. & Relph, J. (1990) *Comput. Appl. Biosci.* **6**, 237–245.

Burks, C., Cinkosky, M.J., Fischer, W.M., Gilna, P., Hayden, J.E., Keen, G.M., Kelly, M., Kristofferson, D. & Lawrence, J. (1992) *Nucleic Acids Res.* **20** (Suppl.), 2065–2069.

Boguski, M.S. & Tolstoshev, C. (1992) *NCBI News* **1.3**, 6.

Boguski, M.S., Ostell, J. & States, D.J. (1992) In *Protein Engineering: A Practical Approach*, A.R. Rees, R. Wetzel & M.J.E. Sternberg (eds). IRL Press, Oxford, Ch. 5, pp. 57–88.

Claverie, J.-M. (1992) *Genomics* **12**, 838–841.

Claverie, J.-M. (1993) *Nature* **364**, 19–20.

Claverie, J.-M., Sauvaget, I. & Bougueleret, L. (1990) *Methods Enzymol.* **183**, 237–252.

Claverie, J.-M. & States, D. (1993) *Comput. Chem.* **17**, 191–201.

Daniels, D.L., Plunkett, G. & Blattner, F.R. (1992) *Science* **257**, 771–778.

Doolittle, R.F. (1986) *Of Urfs and Orfs. A Primer on How to Analyze Derived Amino-acid Sequences.* University Science Books, Mill Valley, CA.

Doolittle, R.F. (ed.) (1990) *Methods Enzymol.* **183**.

Fickett, J.W. & Tung, C.-S. (1992) *Nucleic Acids Res.* **20**, 6441–6450.

Gish, W. & States, D. (1993) *Nature Genet.* **3**, 266–272.

Figure 36.4 XNU masking of low-complexity repeats in collagen. (A) Human collagen sequence used as initial query. (B) Summary of the BLASTP search output on SwissProt (PAM120, score with $p<0.01$). Collagens are dreadful because of their long [GXX]/[GXP] intrinsic repeats. A standard search produces thousands of highly statistically significant matches (not shown) with 135 different proteins. The complete BLASTP output (reporting every significant local alignment) is about 200 000 lines long! (C) The same sequence after processing by XNU (default parameters). The low-complexity segments have been replaced by 'x'. (D) Summary of the BLASTP search output with the XNU-masked query (same parameters as in A). Only the closest relatives of this collagen are now retained as significant matches with the remaining 'unique' segments of the query.

Figure 36.5 Combination of XBLAST and XNU masking. (A) Sequence of a p49 form of NF-κB as published by Schmid *et al.* (1991). (B) Summary of a BLASTP search (PAM120, for scores with $p<0.01$) with this sequence. The confusing output suggests similarities with transcription factors, Alu elements, and unrelated enzymes and structural proteins. (C) New NF-κB query obtained after masking with XBLAST for Alu-like segments (x) and with XNU for low complexity (X). The search output only retains the biologically significant relationship with the transcription factors. From Schmid *et al.*, 1991.

A Original query : NF-kappa B p49

```
>NF-kappa B p49
MESCYNPGLDGIIEYDDFKLNSSIVEPKEPAPETADGPYLVIVEQPKQRGFRFRYGCEGPSHGGLPGASSEKGRKT
YPTVKICNYEGPAKIEVDLVTHSDPPRAHAHSLVGKQCSELGICAVSVGPKDMTAQFNNLGVLHVTKKNMMGTMIQ
KLQRQRLRSRPQGLTEAEQRELEQEAKELKKVMDLSIVRLRFSAFLRASDGSFSLPLKPVTSQPIHDSKSPGASNL
KISRMDKTAGSVRGGDEVYLLCDKVQKDDIEVRFYEDDENGWQAFGDFSPTDVHKQYAIVFRTPPYHKMKIERPVT
VFLQLKRKRGGDVSDSKQFTYYPLVEDKEEVQRKRRKALPTFSQPFGGGSHMGGGSGGAAGGYGGAGGGEGVLMEG
GVKVREAVEEKNLGEAGRGLHACNPAFGRPRQADYLRSGVQDQLGQQRETSSLLKIQTLAGHGGRRL
```

B Original database search output

Sequences producing High-scoring Segment Pairs:		Score	Probability P(N)	N
KBF1_HUMAN	DNA-BINDING FACTOR KBF1 (NUCLEAR FACTOR NF-KAP...	345	7.3e-132	3
TREL_MELGA	C-REL PROTO-ONCOGENE PROTEIN (C-REL PROTEIN) (...	257	5.6e-51	2
TREL_CHICK	C-REL PROTO-ONCOGENE PROTEIN (C-REL PROTEIN) (...	253	6.3e-51	2
TREL_AVIRE	REL TRANSFORMING PROTEIN (P58 V-REL).	257	1.7e-49	2
DORS_DROME	EMBRYONIC POLARITY DORSAL PROTEIN.	240	1.4e-47	2
TREL_MOUSE	C-REL PROTO-ONCOGENE PROTEIN (C-REL PROTEIN).	231	5.1e-39	4
GRP1_PETHY	GLYCINE-RICH CELL WALL STRUCTURAL PROTEIN PREC...	80	7.5e-14	4
K1CJ_HUMAN	KERATIN, TYPE I CYTOSKELETAL 10 (CYTOKERATIN 1...	70	1.7e-11	4
GRP2_PHAVU	GLYCINE-RICH CELL WALL STRUCTURAL PROTEIN GRP ...	84	1.3e-08	3
HM27_HUMAN	HOMEOBOX PROTEIN HOX-2.7 (HOX2G).	71	4.4e-07	2
GRP1_PHAVU	GLYCINE-RICH CELL WALL STRUCTURAL PROTEIN GRP ...	70	1.0e-06	2
ALUF_HUMAN	!!!! ALU CLASS F WARNING ENTRY !!!!	99	1.0e-06	1
ANDR_HUMAN	ANDROGEN RECEPTOR.	72	3.7e-06	2
KG3A_RAT	GLYCOGEN SYNTHASE KINASE-3 ALPHA (EC 2.7.1.37)...	79	5.4e-05	3
ALUB_HUMAN	!!!! ALU CLASS B WARNING ENTRY !!!!	85	0.00013	1
ALUE_HUMAN	!!!! ALU CLASS E WARNING ENTRY !!!!	78	0.0014	1
K1C6_BOVIN	KERATIN, TYPE I CYTOSKELETAL VIB (CYTOKERATIN ...	73	0.0039	3
MYSB_ACACA	MYOSIN HEAVY CHAIN IB.	71	0.016	1

C xnu-masked query : Collagen ALPHA 1(IX)

```
>collagen.xnu
MKTCWKIPVFFFVCSFLEPWASAAVKRRPRFPVNSNSNGGNELCPKIRIGQDDLPGFDLISQFQVDKAASRRAIQRVVGS
ATLQVAYKLGNNVDFRIPTRNLYPSGLPEEYSFLTTFRMTGSTLKKNWNIWQIQDSSGKEQVGIKINGQTQSVVFSYKGL
DGSLQTAAFSNLSSLFDSQWHKIMIGVERSSATLFVDCNRIESLPIKPRGPIDIDGFAVLGKLADNPQVSVPFELQWMLI
HCDPLRPRRETCHELPARITPSQTxxxxxxxxxxxxxxxxxxxxxxxxxxxxxxxxxxxxxxxxxxxxxxxxxxxxxxxx
xxxxxxxxxxxxxxxxxxxxxxxxxxxxxxxxxxxxxxxxxxxxxxxxxxxxxxxxxxxxxxxxxxxxxxxxxxxxxxxx
xxxFHDGDPLCPNACxxxxxxxxxxxxxxxxxxxxxxxxxxxxxxxxxxxxxxxxxxxxxxxxxxxxxxxxxxxxxxxxx
xxxxxxxxxxxxxxxxxxxxxxxxxxxxxxxxxxxxxxxxxxxxxxxxxxxxxxxxxxxxxxxxxxxxxxxxxxxxxxxx
xxxxxxxxxxxxxxxxxxxxxxxxxxxxxxxxxxxxxxxxxxxxxxxxxxxxxxxxxxxxxxxxxxxxxxxxxxxxxxxx
xxxxxxxxxxxxxxxxxxxxxxxxxxxxxxxxxxxxxxxxxxxxxxxxxxxxxxxxxxxxxxxxxxxxxxxxxxxxxxxx
xxxxxxxxxxxxxxxxxxxxxxxxxxxxxxxxxxRAPTDQHIKQVCMRVIQEHFAEMAASLKRxxxxxxxxxxxxxxxxx
xxxxxxxxxxxxxxxxxxxxxxxxxxxxxxxxxxxxxxxxxxxxxxxxxxxxxxxxxxxxxxxxxxxxxxxxxxxxxxxx
xxxxxxxxxxxxxxxxxxxFCEPASCTMQLVSEHLTKGLTLERLTAAWLSA
```

D new database search output

Sequences producing High-scoring Segment Pairs:		Score	Probability P(N)	N
CA19_HUMAN	COLLAGEN ALPHA 1(IX) CHAIN PRECURSOR.	1430	2.9e-203	1
CA29_CHICK	COLLAGEN ALPHA 2(IX) CHAIN PRECURSOR (FRAGMENTS).	1002	7.1e-140	1
CA19_CHICK	COLLAGEN ALPHA 1(IX) CHAIN PRECURSOR (FRAGMENTS).	1002	7.1e-140	1
CA1Z_CHICK	COLLAGEN ALPHA 1(XII) CHAIN (FRAGMENT).	136	1.0e-24	2
CA19_RAT	COLLAGEN ALPHA 1(IX) CHAIN (FRAGMENT).	157	7.9e-15	1
CA1Y_HUMAN	COLLAGEN ALPHA 1(XI) CHAIN PRECURSOR.	77	0.0054	1

Caption overleaf

Green, P., Lipman, D.J., Hillier, L., Waterston, R., States, D., & Claverie, J.-M. (1993) *Science* **259**, 1711–1716.

Iris, F., Bougueleret, L., Prieur, S., Caterina, D., Primas, G., Perrot, V., Jurka, J., Rodriguez-Tome, P., Claverie, J.-M., Cohen, D. & Dausset, J. (1993) *Nature Genet.* **3**, 137–145.

Karlin, S. & Altschul, S.F. (1990) *Proc. Natl. Acad. Sci. U.S.A.* **87**, 2264–2268.

Kerlavage, A.R., Adams, M.D., Kelley, J.C., Dubnick, M., Powell, J., Shanmugam, P., Venter, J.C. & Fields, C. (1993) *Proceedings of the 26th Hawaii International Conference on System Sciences*. Los Alamitos, IEEE Computer Society Press, pp. 585–594.

Legouis, R., Hardelin, J-P., Levilliers, J., Claverie, J.-M., Compain, S., Wunderle, V., Millasseau, P., Le Paslier, D., Cohen, D., Caterina, D., Bougueleret, L., Lutfalla, G., Weissenbach, J. & Petit, C. (1991) *Cell* **67**, 423–435.

Martin-Gallardo, A., McCombie, W.R., Gocayne, J.D., FitzGerald, M.G., Wallace, S., Lee, B.M.B., Lamerdin, J., Trapp, S., Kelley, J.M., Liu, L.I., Dubnik, M., Johnston-Dow, L.A., Kerlavage, A.R., de Jong, P., Carrano, A., Fields, C. & Venter, J.C. (1992) *Nature Genet.* **1**, 34–39.

McCombie, W.R., Adams, M.D., Kelley, J.M., FitzGerald, M.G., Utterback, T.R., Khan, M., Dubnick, M., Kerlavage, A.R., Venter, J.C. & Fields, C. (1992) *Nature Genet.* **1**, 124–131.

Oliver, S.G. *et al.* (1992) *Nature* **357**, 38–46.

Pearson, W.R. & Lipman, D.J. (1988) *Proc. Natl. Acad. Sci. U.S.A.* **85**, 2444–2448.

Schmid, R.M., Perkins, N.D., Duckett, C.S., Andrews, P.C. & Nabel, G.J. (1991) *Nature* **352**, 733–736.

Senapathy, P., Shapiro, M.B. & Harris, N.L. (1990) *Methods Enzymol.* **183**, 252–278.

Sulston, J., Du, Z., Thomas, K., Wilson, R., Hillier, L., Staden, R., Halloran, N., Green, P., Thierry-Mieg, J., Qiu, L., Dear, S., Coulson, A., Craxton, M., Durbin, R., Berks, M., Metzstein, M., Hawkins, T., Ainscough, R. & Waterston, R. (1992) *Nature* **356**, 37–41.

Waterston, R., Martin, C., Craxton, M., Huynh, C., Coulson, A., Hillier, L., Durbin, R., Green, P., Shownkeen, R., Halloran, N., Metzstein, M., Hawkins, T., Wilson, R., Berks, M., Du, Z., Thomas, K., Thierry-Mieg, J. & Sulston, J. (1992) *Nature Genet.* **1**, 114–123.

CHAPTER THIRTY-SEVEN

Finding Frameshift Errors in Anonymous DNA

J.W. SHAVLIK
Computer Sciences Department, University of Wisconsin, Madison, WI, USA

37.1 INTRODUCTION

In any project to map and sequence a genome, the interpretation of the final sequence is an important task. However, errors and ambiguities in the sequence greatly complicate such analysis. This chapter describes computational methods for locating sequencing errors, particularly frameshift errors (the insertion or deletion of a nucleotide, which can cause a coding region to 'shift' into an improper frame). Frameshift errors are among the most common, and hardest to resolve, errors encountered in large-scale sequencing. Owing to the triplet nature of the genetic code, frameshift errors can be disastrous if one wishes to align newly sequenced DNA with known protein sequences.

In this chapter, we discuss several approaches to the computational detection of sequencing errors. Several of these approaches are being investigated in conjunction with a project to determine the complete DNA sequence of *Escherichia coli* (Daniels *et al.*, 1992). There are two closely related goals of this work. First, we wish to correct sequencing errors by noting inconsistencies with other biological data, which is the theme of this chapter. Second, we wish to locate and identify those regions of the sequence that encode proteins – both known and heretofore unknown, thereby characterizing 'anonymous' DNA (i.e. DNA sequenced in a laboratory and whose function is unknown). While we will focus on techniques that address the first of these goals, the achievement of the two goals is closely related, as will be seen.

Two basic approaches are discussed in the remainder of this chapter. One involves performing robust similarity matches with known protein sequences – frameshift errors are found by noting when matches span more than one reading frame. The second approach involves analyzing known genes using methods of statistics and machine learning. The results of such analyses can be used to scan anonymous DNA and detect abrupt changes in the likelihood that a particular frame codes for protein; adjacent and complementary discontinuities in different frames suggest frameshift errors.

37.2 THE FIND-IT ALGORITHM

We have developed a similarity-search algorithm that is robust in the presence of frameshift errors. This algorithm builds on the BLAST similarity-search program (Altschul *et al.*, 1990). BLAST efficiently produces homologous protein–protein matches, but

Table 37.1 Algorithm for matching known proteins to new DNA sequences in the presence of frameshift errors.

Given a sequence of DNA:
(1) Translate the DNA in all six reading frames
(2) Apply BLAST (Altschul *et al.*, 1990) to produce partial matches (with gaps) to the *nrdb* database of the US National Library of Medicine. This *non-redundant database* merges the Protein Information Resource (PIR), SwissProt, and translated GenBank databases
(3) Collect all of the matches to each protein encountered
(4) By piecing together matches, look for consistent *coverings* of each protein (see text for details)
(5) Score the combined protein matches, sort, and report the best matches

since these matches are at the protein level, they do not extend across frameshift errors. The method described below coherently combines partial matches (to a given protein) in different reading frames, thereby overcoming missing and extra nucleotides in sequenced DNA.

Our FIND-IT algorithm is described in Table 37.1. Given a sequence of DNA, the algorithm first translates it in all six reading frames. We assume that stop codons simply represent a 21st amino acid, and we give this amino acid a very high negative score (−25) in the PAM partial-match matrix, thereby strongly discouraging alignments that span stop codons. Owing to computer memory-size limitations, we chop up these very long putative peptides at approximately every 1000 amino acids; we perform these cuts at the nearest stop codon.

Next, the FIND-IT algorithm uses a variant of BLAST that allows gaps in alignments to match each 'megapeptide' against a database of known proteins. We convert to peptides, rather than directly matching DNA to GenBank entries, because partial matches to amino acids are possible; we use the PAM 120 matrix (Dayhoff, 1978) to determine the match score between two amino acids. Also, due to the genetic code's degeneracy and the different codon usage patterns of various species, vastly different DNA sequences may lead to quite similar protein sequences.

A useful feature of BLAST is that a peptide need not match in its entirety; rather, it reports matching *sub*sequences. For example, a match may terminate within a peptide due to a frameshift error; in this case the remainder of the protein will match another megapeptide (examples of this follow).

A match returned by BLAST maps a portion of the DNA sequence to a segment of a protein. Figure 37.1 schematically shows three matches; the numbers following the matches indicate their frame. Note that matches A and C provide a consistent 'covering' of the protein, while match B is inconsistent with the other two. Also note that by combining matches A and C, an extra nucleotide in the DNA can be identified (the one marked with an X). Below, we define what it means to be a *consistent* collection of matches to a given protein (which we call a 'covering').

Assume match i maps the DNA segment $[A,B]$ to the protein segment $[P,Q]$, while match j maps the DNA segment $[C,D]$ to the protein segment $[R,S]$. Also assume $C > A$. If matches i and j both belong to a consistent covering, then the following constraints hold:

(1) If $C>A$ then $R>P$. That is, the left-to-right order on the DNA is the same as that on the protein sequence. Also, they must be on the same DNA strand (i.e. forward or reverse).
(2) If matches i and j intersect on one sequence (DNA or protein), they also intersect in the corresponding locations on the other. (Since there may be extra

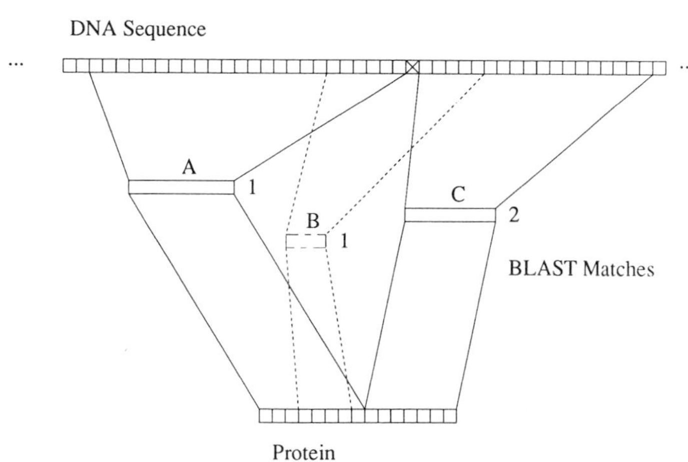

Figure 37.1 Combining BLAST matches to overcome frameshift errors.

or missing nucleotides in the DNA sequence, 'small' discrepancies are ignored.)

(3) This distance between DNA locations *B* and *C* is approximately three times the distance between protein locations *Q* and *R*. That is, the amount of DNA between the end of one match and the beginning of the next roughly corresponds to the number of amino acids between the two matches. (This constraint can be relaxed if this algorithm is applied to eukaryotic DNA. Hence, the approach elegantly extends to the recognition of exons among intervening sequences.)

In our current experience, FIND-IT generally looks for consistent coverings from a set of 10–100 matches; the resulting coverings contain from one to 15 matches. There are of the order of 2^{100} possible combinations when given 100 matches. Although our algorithm is combinatorial, we have devised several heuristics to greatly prune the number of possibilities considered; the time spent looking for consistent coverings is much less than that which BLAST spends producing the matches. The number of matches that BLAST produces grows rapidly as one reduces the threshold for an acceptable score. We are able to manage the combinatorics of combining matches when the minimal acceptable BLAST score is 55 (we use 5 as the gap-open penalty and 3 as the gap-extend cost).

A sketch of the algorithm for finding consistent coverings follows. We first sort all of BLAST's matches to a given protein, according to where the match starts on the protein. We then step through these sorted matches, keeping a list of all potential coverings. For each match, we consider making new potential coverings by adding it to each of the existing possible coverings, keeping those new coverings that satisfy the constraints described above. We also add the singleton covering that starts with this match to the list of all potential coverings, as this might be the first element in an acceptable covering. (Note that we do not *replace* existing coverings when we add a match; rather, we add the new composite covering that results from the addition of one match to an existing covering. Since a later match in the sorted list might fit better, we need to keep the unextended existing covering for later consideration.)

As one can see, this procedure can lead to a doubling of the list of potential coverings at each step. To keep the size of this candidate list manageable, we also require that an acceptable match cover some minimal fraction of the entire protein (currently we require 10% coverage). At each step of the covering algorithm, we prune the list of candidate coverings by discarding those that, even if the rest of the protein was completely matched, would not cover the minimal fraction of the entire protein. We also prune potential coverings that are proper subsets of other potential coverings; for example, if one potential covering contains the matches {I, J, K} and another contains the matches {I, K}, we discard the latter. The fact that we sorted the list of BLAST matches allows us to safely make these decisions about coverings to discard. Finally, we keep singleton matches that score above some threshold even if they do not cover the minimal fraction of the entire protein (unless these singleton matches are part of some successful composite covering).

We first tested our algorithm on phage λ DNA sequence; all of the λ proteins in PIR were found, as well as several matches to non-λ proteins. (Some, but not all, of these homologies are noted in the PIR annotation.) Also, we have introduced insertions and deletions in λ; the algorithm still located the proteins. Since then we have applied FIND-IT to about 50 *E. coli* DNA segments. Generally, FIND-IT covers over half of these DNA segments with strong composite matches (often to *E. coli* genes already in GenBank), leading to the detection and correction of several frameshift errors every 1000 bases. Later in this chapter we present a simple experiment that assesses the ability of FIND-IT to locate proteins in the presence of error-ridden DNA.

37.2.1 Using multiple matches to detect sequencing errors

This section presents an actual composite match produced by the FIND-IT algorithm. Figure 37.2 contains a three-piece match to a rat protein. This match illustrates how sequencing errors can be detected. FIND-IT categorizes matches according to the percentage of exact matches (this percentage is only measured over the portion of the protein covered) – the match in Fig. 37.2 is classified as a moderate one. The algorithm also reports the archive that contains the matching protein. Each BLAST match in the composite is reported. The bottom line in a match is the translated DNA, while the top is the protein sequence. Numbers in the middle of the lines represent amino acid positions in the protein and nucleotide positions in the entire DNA sequence, respectively. The middle line presents the *alignment*; letters represent identical matches, + signs represent positive-scoring partial matches, full stops(.) represent matches that score zero, and blanks represent negative-scoring matches. Dashes (-) represent gaps introduced by BLAST during its matching process. Finally, braces (||) indicate the 'seed' match in the BLAST algorithm. The final two alignments are truncated in this figure for purposes of clarity.

Note that in addition to characterizing a portion of a sequence, a covering suggests frameshift errors. In

```
Protein name:
*3-Hydroxyisobutyrate dehydrogenase precursor - Rat (fragment) #EC-number 1.1.1.31

BLAST hits combined: 3
Category:            similar-gene
Composite Score:     287
Exact Matches:       68/196 = 34.83%
Archive:             PIR3
Strand:              reverse

[1] BLAST score:  73  || Protein            44      69   RSMASKTP|VGFI|GLGNMGNPMAKNLI
                      || Alignment                       R MA+---|++FI|GLG.MG+PMA ..I
    Exacts:      54%  || DNA (frame 4)   19890   19822   RVMAA---|IAFI|GLGQMGSPMATQFI

[2] BLAST score: 159  || Protein            66     186   KNLIKHGYPLILYDVFPDVCKEFKEAGE ...
                      || Alignment                       +NL+.+G .L ++DV ++. + . + G.
    Exacts:      33%  || DNA (frame 5)   19832   19470   RNLLQQGHQLRVFDVNAEAVRHLVDKGA ...

[3] BLAST score:  60  || Protein           289     341   TTLMA|KDLG|LAQDSATSTKTPILLGS ...
                      || Alignment                       +.L  |KDLG|+A D A.    .P+ LG+
    Exacts:      30%  || DNA (frame 6)   19177   19019   SILPH|KDLG|IALDVANQLHVPMPLGA ...
```

Figure 37.2 An alignment to a non-*E. coli* gene.

Fig. 37.2, there were two frame transitions; careful inspection of the boundaries between successive matches leads to the prediction of missing or extra nucleotides. These hypotheses can then be checked by reviewing the original sequencing gels. Also, when FIND-IT reports matches to *E. coli* sequences already in GenBank, other possible sequencing errors and ambiguities can be detected by noting the discrepancies between the protein sequence and the translated DNA.

37.2.2 An experiment: evaluating FIND-IT's noise sensitivity

This section contains an experimental evaluation of our gene-finding method. The experimental method is as follows: a known gene (of length 999) was extracted from the GenBank database (the gene for *replication protein O* in bacteriophage λ) and various amounts of noise added to it in each of 25 experimental runs. Following this, we applied FIND-IT and counted the times it found the initial gene. We investigated three simple noise models and one composite model:

(1) *Replacement* – with probability p, a given nucleotide is replaced with another one.
(2) *Deletion* – with probability p, a given nucleotide is deleted.
(3) *Insertion* – with probability p, a nucleotide is inserted after a given nucleotide.
(4) *Combination* – with probability p, one of the above three changes occurs at a given nucleotide. All three possibilities are equally likely.

These noise models are somewhat simplistic – owing to the nature of the sequencing process, insertions and deletions are most likely to occur within *runs* of the same nucleotide (e.g. . . . AAAAA . . .). Nevertheless, these models are sufficient for our present purposes.

Figure 37.3 contains the results of this experiment. The experiment indicates that, under all four noise models, FIND-IT is unaffected until the noise rate exceeds 3%. Since the estimated error rate in sequencing is less than 1%, the FIND-IT approach should be robustly finding genes.

37.2.3 FIND-IT summary

The FIND-IT algorithm provides a mechanism for finding protein matches to anonymous DNA, even when the DNA contains frameshift and other sequencing errors. Basically, FIND-IT 'sews together', across reading frames, protein–protein matches found by BLAST. This algorithm is being used regularly in the *E. coli* sequencing project. Since we began the work on FIND-IT in the autumn of 1990, we have become aware of a similar approach independently developed by Posfai & Roberts (1992).

We are studying the computational complexity of the 'covering' problem defined above and devising more efficient algorithms for it. Meidanis (1992) has developed an $O(N \log N)$ algorithm for finding the best covering of a given protein. His algorithm provides a tool for rapidly finding consistent coverings, thereby allowing us to reduce the threshold for

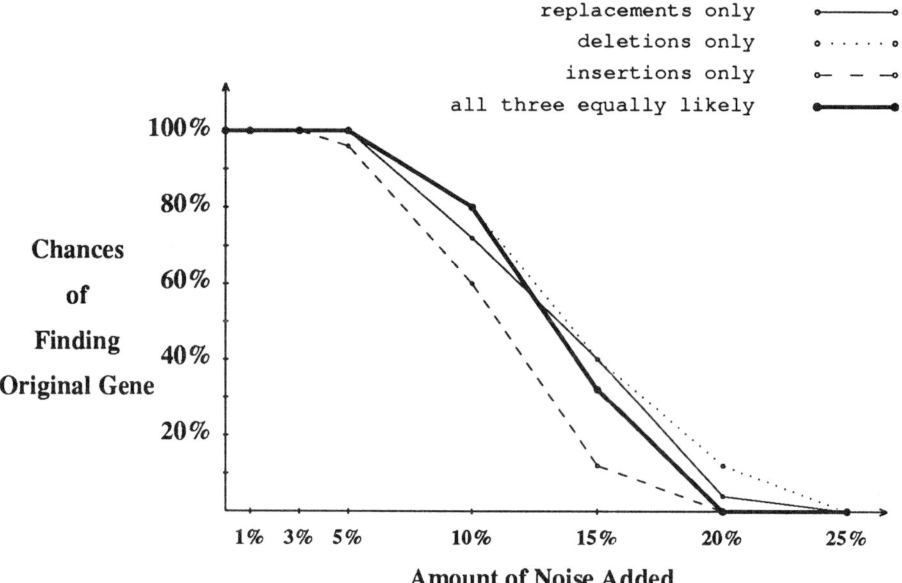

Figure 37.3 FIND-IT's chances of finding a gene as a function of sequence noise (results averaged over 25 experimental runs).

minimal BLAST matches and to eliminate some of the other restrictions discussed above.

We are also developing an alignment program that directly matches DNA to proteins, thereby allowing frameshift errors to be found in the same manner that gaps are found in protein–protein (or DNA–DNA) matches. Specifically, we have defined a scoring function that has penalties for DNA insertions and deletions as well as the standard gap penalties, this scoring function is optimized using dynamic programming (Needleman & Wunsch, 1970). Unlike BLAST, in this program we compute the complete alignment matrix. Hence, to align a 10 kb DNA segment against the nonredundant database (nrdb) of NLM, which contains about 60 000 proteins and 20 million residues, we calculate over 10^{10} entries. We currently have parallel implementations that run on the Wisconsin Condor system (Litzkow *et al.*, 1988), which uses otherwise idle workstations, and the Thinking Machines Corporation 64-node CM-5 parallel computer. It takes less than one day of 'real time' to align a 10 kb DNA segment against nrdb using our parallel algorithms. Similar DNA–protein alignment methods can be found in Knecht & Gonnet (1992) and States & Botstein (1991).

When no similarity matches to known proteins are possible, we need alternative methods for recognizing genes and locating putative frameshift errors. The next section discusses approaches that complement the FIND-IT method.

37.3 FINDING FRAMESHIFTS USING STATISTICS AND MACHINE LEARNING

An alternative way to locate frameshift errors is to utilize some computer program that predicts the likelihood that a given reading frame is a coding region. One can plot this information for each reading frame; a sharp decline in one plot coupled with a sharp increase in another plot suggests a frameshift error. A number of methods, based on either statistics or machine learning, produce predictions about coding regions and, hence, can be used to help locate frameshift errors in regions of DNA that do not match any existing proteins. In the remainder of this section we briefly review how these systems are trained, using known genes, to make their predictions on new DNA. The next section contains a sample plot (see Fig. 37.5) of coding-region predictions for each reading frame in a segment of DNA.

The computer-based approaches to coding-region location operate by generating their predictions based upon a relatively small, fixed-length 'window' of DNA. By sliding this window along a DNA sequence, the algorithms produce continuous signals that show the prediction for each reading frame along the length of the sequence.

The approaches of Fickett (1983), Staden (1990), and Gribskov *et al.* (1984) are based on amino acid

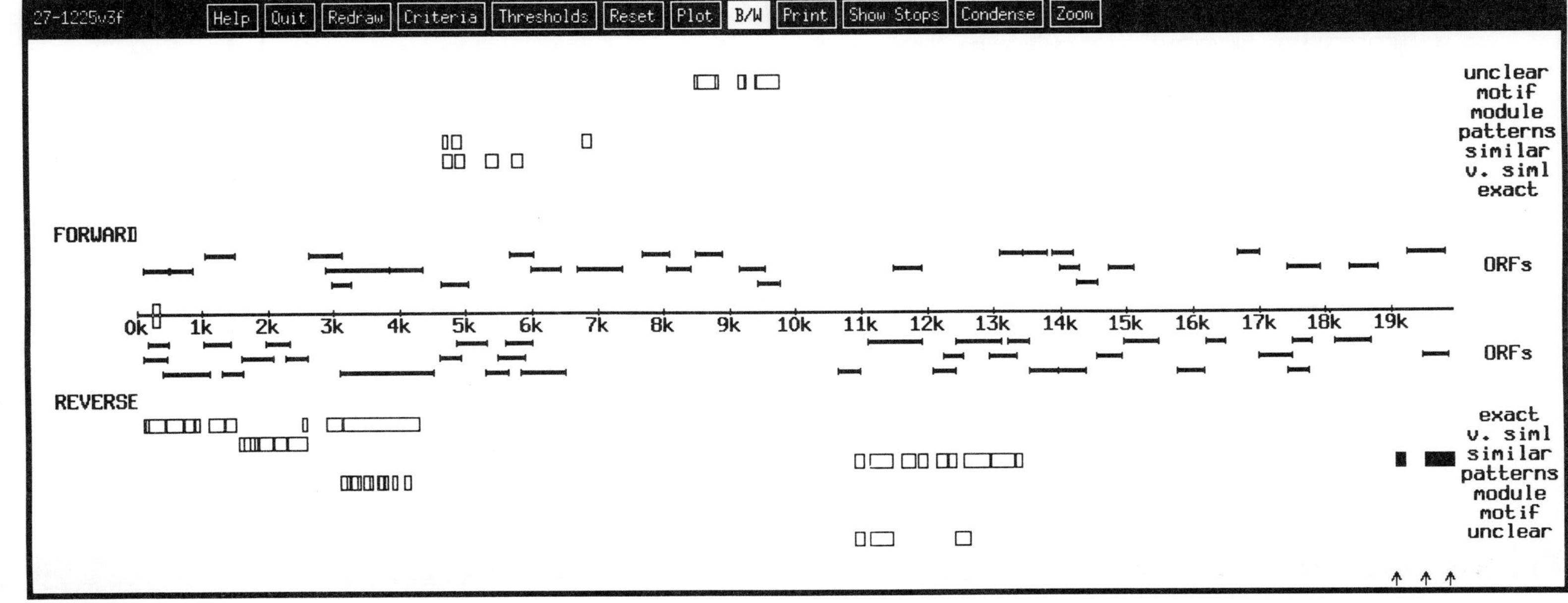

Figure 37.4 The graphical interface to FIND-IT.

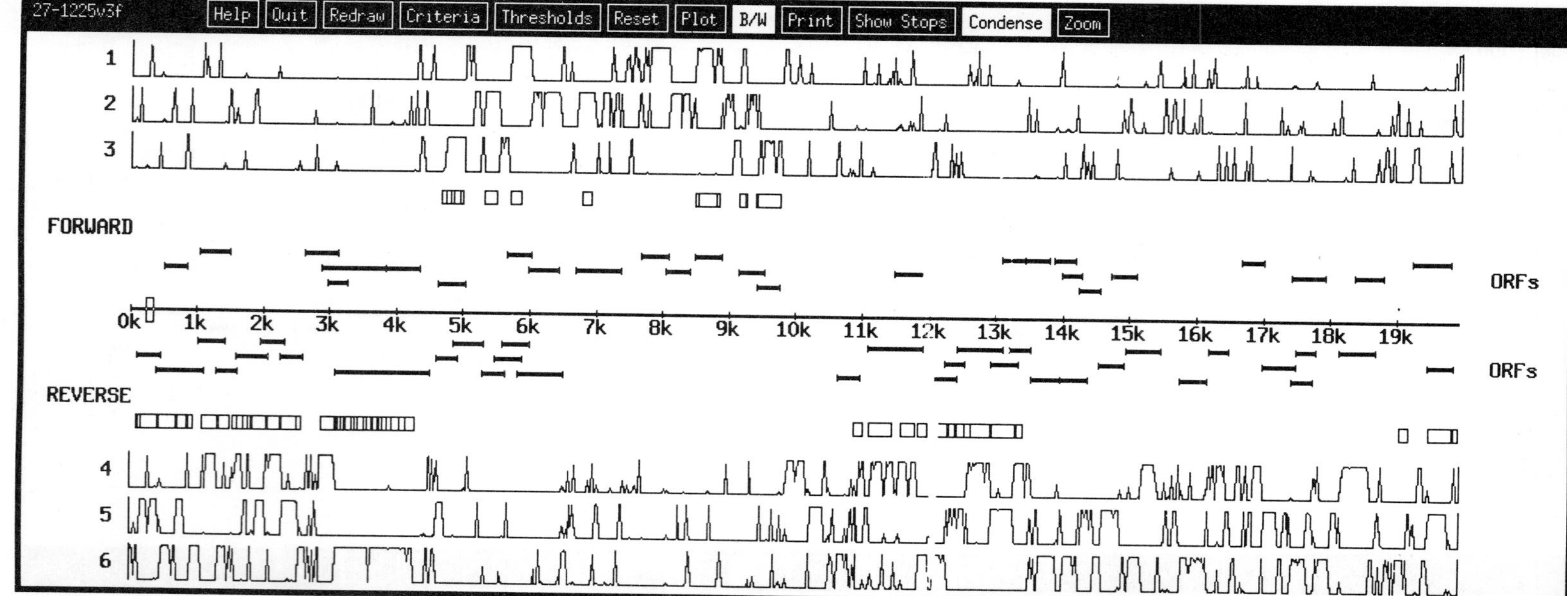

Figure 37.5 Plotting the neural network's predictions of coding regions in all six reading frames.

frequencies and codon usage statistics. To express a given amino acid, an organism usually has several codons among which to choose. The organism rarely chooses among these alternatives uniformly, but rather exhibits a preference, based both on the gene's level of expression and on the organism itself. Roughly speaking, one estimates the probability that a window is (in-frame) over a coding region by multiplying the probabilities that each codon in that window appears in a coding region. These probabilities are estimated from the known genes.

The statistical approaches generally assume that the codons in a window are independent of one another. Along with Lapedes *et al.* (1989) and Uberbacher & Mural (1991), we (Noordewier *et al.*, 1991; Craven & Shavtik, 1993) have investigated using machine learning to learn how to recognize coding regions. These techniques are capable of extracting higher order statistics and do a slightly better job of predicting coding regions in new segments of DNA (Farber *et al.*, 1992; Craven & Shavtik, 1993) than do the statistical approaches that assume independence.

We have trained neural networks, using known *E. coli* genes, to predict whether a window of *E. coli* DNA is an in-frame coding region (Craven & Shavtik, 1993). (Farber *et al.* (1992) also trained networks to predict for each frame.) Generally, people train neural networks to predict whether a window is coding or not, but this does not directly aid the detection of frameshift errors. Instead, we need to know whether the peptide that would result from translating the current window's DNA is likely to be in an *E. coli* gene. The next section contains sample plots of the output of our trained network.

37.4 A GRAPHICAL INTERFACE TO FIND-IT

We have developed a graphical interface, based on X-windows, that allows us to inspect the output of FIND-IT. In order to further illustrate FIND-IT's operation, in this section we discuss some aspects of this interface.

Figure 37.4 contains a sample 'screen dump'. The line across the middle represents a DNA segment, and the open-reading frames (ORFs) on each strand are shown. The composite matches found by FIND-IT are categorized, along the vertical dimension, according to their strength (measured in terms of the percentage of exact amino acid matches) and length (motifs and modules are short but strong matches). The row labeled 'patterns' does not involve FIND-IT; it presents the matches found by the program of Henikoff & Henikoff (1991).

The user can select composite matches with the mouse; the matches become solid colored, arrows appear that indicate the beginning of the individual BLAST matches, and another window containing the alignments opens. Selecting the solid-colored rectangles in Fig. 37.4 leads to the presentation of the alignment information that appeared in Fig. 37.2. (The user can also ask to see the PIR, SwissProt, or GenBank documentation – as appropriate – for the matched protein.) This interface runs on workstations with color screens, and in this case color is used to indicate the reading frame of each individual BLAST match, as well as the ORFs. This greatly simplifies the visual recognition of frameshifts.

The user can also request that the system plot the predictions of the neural networks described in the previous section. Figure 37.5 shows what happens to Fig. 37.4 when this option is chosen. First, all of FIND-IT's matches are condensed to fall in one row. Then the predictions about each frame being coded are plotted, in color when possible. Notice that the plots in Fig. 37.5, for reading frames 4–6 between 17 and 19 kb suggest (a) the presence of a gene not matched by FIND-IT and (b) that this region of DNA contains several frameshift errors.

37.5 CONCLUSION

In this chapter we discussed computational methods for locating errors, especially frameshifts, in anonymous DNA. We are successfully applying our FIND-IT method in support of Blattner's Human Genome project at Wisconsin. FIND-IT builds on the BLAST protein–protein alignment program, extending BLAST so that matches can be found that span several reading frames. Identifying such matches helps locate frameshift errors. We also briefly discussed methods, based on statistics and machine learning, that one can use to find frameshift errors in DNA that does not align with any known protein. Although we have largely developed and tested our approaches on prokaryotic DNA, with small modifications the basic techniques should be applicable to locating errors in eukaryotic DNA.

ACKNOWLEDGMENTS

Discussions with Fred Blattner, Donna Daniels, Guy Plunkett, Mark Craven, Debby Joseph, Joao Meidanis, Prasoon Tiwari, and Eric Bach greatly impacted the material presented in this chapter. Special thanks go to Fred Blattner, director of the *E. coli* sequencing project, for providing the DNA sequences and for pointing out the problem of matching DNA containing frameshifts to protein databases. Kevin Cherkauer, Mark Craven, Charlie Squires, Andy Whitsitt and Derek Zahn helped out at various times with programming the graphical interface to the composite matches found by FIND-IT.

Currently we are working with Edward Uberbacher of Oak Ridge National Laboratory and his colleagues on making our trained neural network (see Section 37.3 and Fig. 37.5) accessible via electronic mail. More specifically, we are creating GRAIL-E, a version of GRAIL for locating coding regions in *E. coli* DNA. This system should be publicly accessible sometime in 1994. Send E-mail to shavlik@cs.wisc.edu for further information.

This work is partially supported by the Department of Energy (grant No. DE-FG02-91ER61129), the National Science Foundation (grant No. IRI-9002413), and the Office of Naval Research (grant No. N00014-90-J-1941 (to the author)) and the National Institutes of Health (grant No. HG00301, to F. Blattner).

REFERENCES

Altschul, S.F., Gish, W., Miller, W., Myers, E.W. & Lipman, D.J. (1990) *J. Mol. Biol.* **215**, 403–410.

Craven, M.W. & Shavlik, J.W. (1993) In *Proceedings of the Twenty-Sixth Hawaii International Conference on Systems Science (Biocomputing Track)* pp. 773–782. IEEE Computer Society Press, Los Alamitos.

Daniels, D.L., Plunkett, G., Burland, V.D. & Blattner, F.R. (1992) *Science* **257**, 771–778.

Dayhoff, M.O. (1978) *Atlas of Protein Sequence and Structure*. National Biomedical Research Foundation, Washington, DC.

Farber, R., Lapedes, A. & Sirotkin, K. (1992) *J. Mol. Biol.* **226**, 471–479.

Fickett, J.W. (1983) *Nucleic Acids Res.* **10**, 5303–5318.

Gribskov, M., Devereux, J. & Burgess, R.R. (1984) *Nucleic Acids Res.* **12**, 539–549.

Henikoff, S. & Henikoff, J.G. (1991) *Nucleic Acids Res.* **19**, 6565–6572.

Knecht, L. & Gonnet, G.H. (1992) *Alignment of Nucleotide with Peptide Sequences* (*Technical Report 184*). Departement Informatick, ETH, Zurich.

Lapedes, A., Barnes, C., Burks, C., Farber, R. & Sirotkin, K. (1989) In *Computers and DNA, SFI Studies in the Sciences of Complexity VII*. Addison-Wesley, New York.

Litzkow, M., Livny, M. & Mutka, M.W. (1988) In *Eighth International Conference on Distributed Computing Systems* pp. 104–111, Washington, IEEE Computer Society Press.

Meidanis, J. (1992) Algorithms for problems in computational genetics. Ph.D. thesis, Department of Computer Science, University of Wisconsin, Madison, WI.

Needleman, S.B. & Wunsch, C.D. (1970) *J. Mol. Biol.* **48**, 443–453.

Noordewier, M.O., Towell, G.G. & Shavlik, J.W. (1991) In *Advances in Neural Information Processing Systems*, R. Lippmann, J. Moody and D. Touretzky (eds). Morgan-Kaufmann, Los Altos, CA, pp. 530–536.

Posfai, J. & Roberts, R.J. (1992)*Proc. Natl. Acad. Sci. U.S.A.* **89**, 4698–4702.

Staden, R. (1990) *Methods Enzymol.* **183**, 163–180.

States, D.J. & Botstein, D. (1991) *Proc. Natl. Acad. Sci. U.S.A.* **88**, 5518–5522.

Uberbacher, E.C. & Mural, R.J. (1991) *Proc. Natl. Acad. Sci. U.S.A.* **88**, 11261–11265.

CHAPTER THIRTY-EIGHT

Software Tools For Primer Site Prediction and Sequence Similarity Searches

B. RAPPAPORT,[1] J. GATEWOOD,[2] C. FIELDS,[3] & N. DOGGETT,[2]

[1] Department of Electrical and Computer Engineering, New Mexico State University, Las Cruces, NM 88003, USA

[2] Life Sciences Division and Center for Human Genome Studies, Los Alamos National Laboratory, Los Alamos, NM 87545, USA

[3] The Institute for Genomic Research, Gaithersburg, MD 20878, USA

38.1 INTRODUCTION

Sequence-tagged sites (STSs) (Green *et al.*, 1991) and expressed sequence tags (ESTs) (Adams *et al.*, 1991) are powerful sequence-based markers for diverse genome mapping applications. Crucial to the development of an STS or the chromosomal assignment of an EST is the selection of oligonucleotide primers for use in the polymerase chain reaction (PCR). Another important component in the analysis of an STS or EST sequence is the comparison of the candidate sequence against GenBank and EMBL DNA sequence databases. Here we describe computer programs that: (1) automate the selection of PCR primers in a DNA sequence; (2) parse repetitive DNA elements from a test sequence; and (3) facilitate searches against GenBank and EMBL databases using either FASTA (Pearson & Lipman, 1988) or BLAST (Altschul *et al.*, 1990) alignment algorithms. An X-window interface tool provides convenient window access to the programs which can run on either a single sequence or on an entire directory of sequence files (batch-mode). The programs are designed to operate on UNIX-based computer workstations, and are available through anonymous ftp at telomere.lanl.gov.

38.2 PROGRAM DESCRIPTIONS

38.2.1 ASTS

ASTS (automated STS) is a computer program that automates the selection of PCR primers in a DNA sequence for the purpose of amplifying a unique site in the genome. This program performs the following functions.

(1) Sequences and their complements are compared using FASTA and/or BLAST with consensus sequences of known human repetitive elements (Jurka *et al.*, 1992), vector sequences, and all sequences in the primate section of GenBank. Regions of similarity are identified using a sliding window with widths and similarity score cutoffs for the repeats, vector and primate databases. All of this information yields a summed similarity histogram for the sequence. Regions of similarity are indicated above the corresponding bases in the histogram using an 'r' for repeats, a 'v' for vector, and '*' for primate sequences (see Fig. 38.1).

(2) The regions of each sequence that are not similar to either consensus repeats, vector, or primate sequences are analyzed for length, separation,

```
Histogram for 22b1d.seq

Parameters Used:
GenBank window width = 25
GenBank percent match = 90
Repeat window width = 20
Repeat percent match = 70
Vector window width  = 25
Vector percent match = 90
Matches allowed in primer = 2
Min primer length = 22
Max primer length = 25
Min primer C+G content = 45
Max primer C+G content = 55
Min primer separation = 60
Max primer similarity = 40
Max temperature difference = 2
C/G at 3' primer ends = Y
Output reject primers = N


    ********** ********** ***
    ********** ********** *** *
    ********** ********** *** *
    ********** ********** *** *
    ********** ********** *** **
    vvvvvvvvvv vvvvvvvvvv vvv **
  0 ccccgggtac cgagctcgaa ttctgccaag aataagggta agactggaag cagattcgtc

                                                                rrrrr
                                                                rrrrr
                                                             r  rrrrr
                        r rrrrr rrrr rrr r  rr               r  rrrrr
 60 cctagagcct ccaggtagaa actcagttca gccaacacct tgatttcacc ttctgacccc

    r rrrrrrrr rrr                            r rrrrrrrrrr r r rrr
    r rrrrrrrr rrr                            r rrrrrrrrrr r r rrr
    r rrrrrrrr rrr                          r r rrrrrrrrrr r r rrr r
    r rrrrrrrr rrr   r                      r r rrrrrrrrrr r r rrr r
120 taagcagaga acctagccat gctatactgg acctctgagc cacaaaacta tgagctacta

                          r rr rrrrr rrrrrrr  r rr
                        r r rr rrrrr rrrrrrr  r rr r
                        r r rr rrrrr rrrrrrr  r rr r
                        r r rr rrrrr rrrrrrr  r rr r r
180 aacgagtgtt ggtttaaggt gctatatttg tggtcacttg ttacacagcc atagaagagt




240 aatacatcta tcaaacaggg ctctagtcac agntaattac tcagggcctt tggttaaccc




300 caaaaacaat ccaagcttcc tcttattccc tacttttctt ggncanancc actttttggg




360 ggtnggttgg natttacccg gngacccngg aaaggtcanc aa
```

Figure 38.1 Histogram output file for a sample sequence. The parameters for window width, fractional match, etc., are listed at the beginning of the file. An *, r, or v above the sequence indicates a window match to primate GenBank, repeat database, or vector database starting at that nucleotide and continuing for the specified window width with at least the specified fractional match for each database. Each level on the histogram containing an *, r, or v indicates a separate match to the databases. Primers which meet the criteria for length, C+G content, separation, similarity, melting temperature, and C/G at 3′-ends are selected from regions of the histogram with up to three bases matching the database cutoffs (bases containing an *, r, or v).

G+C content, melting temperature, ability to form hairpins, and complementarity to each other.
(3) Pairs of oligomers that meet these criteria and that have similar melting temperatures are identified as potential PCR primers pairs. The parameters that control the sensitivities of the database searches, significance of matches, product and primer length, composition, and melting temperatures are set by the user. Equations for determining melting temperatures have been derived from Los Alamos data on melting temperatures for oligonucleotide standards and are based on nearest neighbor nucleotide frequencies within each primer (Dogget N.A. & Ratliff R.A. unpublished).

38.2.1.1 Program usage

The program is started by typing 'asts'. Program options that are available are listed below. These options are activated by typing a space and the option following asts. For multiple options a space must occur between each option. Some options are described below.

-x — *X-Window interface option* – operates the X-Windows version of ASTS.

-e '*xxx*' — *Extension option* – causes ASTS to recognize '*xxx*' as the extension for sequence files that are analyzed. The default format for sequence filenames is *name*.seq.

-d '*filename*' — *Defaults option* – causes ASTS to use the default parameters file '*filename*' for the default parameters instead of the standard defaults file 'asts_defaults'.

-q — *Quiet option* – prevents the output of results to the screen (results are always saved in files). This is valid only for the nonwindows interface version of ASTS.

For example, by typing:

asts -e "txt" -d "my_defaults" -q

ASTS will operate on sequence files with the extension. txt, using the default parameters in the file my_defaults, and the results will not be printed to the screen.

38.2.1.2 Program parameters

After ASTS has been started the program will prompt the user for the following entries. The asts_defaults (or another defaults file selected by the user) file will display default entries or values for each prompt. The user simply enters the return key to use the default entry, or types in a new entry or value.

Input source type – directory or single filename for sequence files. *User enters "d" or "f" for directory or file.*

Input source name – directory name or single filename. *User enters directory or filename using full path names.*

Output source directory – output directory name where the output is printed.
User enters directory name with the full path where the output is to be printed.

GenBank window width – the sliding window width for searching the GenBank primate database versus candidate sequences.
User accepts the default value of 25 nucleotides or enters a new value.

GenBank fraction match – the percent similarity match within the GenBank window width for the region to be marked as a matched region in the summed similarity histogram.
User accepts the default value of 90% or enters a new value.

Repeat window width – the sliding window width for searching the repeat consensus database versus candidate sequences.
User accepts the default value of 20 nucleotides or enters a new value.

Repeat fraction match – the percent similarity match within the repeat consensus window width for region to be marked as match region.
User accepts the default value of 75% or enters a new value.

Vector window width – the sliding window width for searching the vector database versus candidate sequences.
User accepts the default value of 20 nucleotides or enters a new value.

Vector fraction match – the percent similarity match within the vector window width for region to be marked as match region.
User accepts the default value of 90% or enters a new value.

Minimum primer length – the minimum length required for a primer.
User accepts the default value of 20 nucleotides or enters a new value.

Maximum primer length – the maximum length that will be considered for a primer.
User accepts the default value of 25 nucleotides or enters a new value.

Minimum primer CG content – the minimum CG content in each primer pair.
User accepts the default value of 45% or enters a new value.

Maximum primer CG content – the maximum CG content in each primer pair.
User accepts the default value of 55% or enters a new value.

Minimum primer separation – the minimum separation between primer pairs (from first 5′ primer end to second primer 3′ end).
User accepts the default value of 60 nucleotides or enters a new value.

Maximum primer melting temperature difference – the maximum difference in melting temperature between the two primers.
User accepts the default value of 2°C or enters a new value.

Maximum primer matches – the maximum number of sequence similarity hits from the summed similarity histogram that are allowed in a primer.
User accepts the default value of three nucleotide matches or enters a new value.

CG at ends – C or G wanted at 3′ ends of primers.
User accepts the default answer of yes or enters no.

Output rejected primer pairs – primer pairs that were rejected due to GC content, melting temperature, and CG at ends, can also be outputted.
User accepts the default answer of no or enters yes.

38.2.1.3 X-Window interface menu

(1) Change parameter button – clicking the change parameters button pops up the primer parameters menu and database file menu for the user to edit.
(2) Choose primers button – clicking this button executes the primer selection.

38.2.1.4 Outputted files

ASTS outputs five different files for each sequence analyzed with the extensions listed below. In addition, if a directory of sequences is analyzed, a no primer file is made which contains the sequence files for which no primers were chosen. In the outputted files listed below, *xxx* indicates the name of the sequence that was analyzed.

xxx.dna Raw DNA sequence stripped of any header or numbers.
xxx.hist Summed similarity histogram.
xxx.pgb FASTA output comparing candidate sequence with primate GenBank.
xxx.oth FASTA output comparing candidate sequence with vector or other database.
xxx.rep FASTA output comparing candidate sequence with repeat database.
xxx.prm List of primer pairs chosen by ASTS, their locations on the sequence, and their melting temperatures.

38.2.2 Removal tool program

The removal tool program is a parsing program that masks elements of a sequence after comparing the sequence to a user-selected database. Those portions of a sequence that match the database based on a particular window width and percent match are substituted nucleotide for nucleotide with the letter N. In this manner, the spacing of the remaining sequence is unchanged. The program is most useful for removing highly repeated elements (such as Alu or line repeats) from a sequence prior to searching GenBank for significant single copy sequence homologies. The user selects the sliding window width and percent fractional match that is used.

38.2.2.1 X-Windows interface menu and parameters

Input source type – directory or single filename of sequence files.
User clicks on directory or file buttons.

Input source name – directory name or single filename.
User enters directory or filename using full path names.

Output source directory – output directory name where the output is printed.
User enters directory name with the full path where the output is to printed.

Window width – the sliding window width for searching the database versus candidate sequences.
User enters a window width value.

Fraction match – the percentage similarity match within the window width for region to be marked as match region.
User enters a percentage match value.

Remove-it button.
This button executes the parsing program.

38.2.3 Similarity search tool

The similarity search tool allows the user to perform FASTA and/or BLAST comparisons versus a number of databases including GenBank, Consensus Repeat, and other user-defined databases. The similarity search tool allows the user to compare a directory of sequences or a single sequence versus the databases using a convenient windows interface. The similarity search tool takes as input DNA sequence(s) from a file or directory, produces a standard and complemented sequence file and runs either FASTA or/and BLAST. The standard sequence is stored in the output directory with an extension '.dna' and the complement is stored with an extension 'c.dna'. The similarity output files are stored in the output directory with GenBank extensions.

38.2.3.1 X-Windows interface menu and parameters

Input source type – directory or single filename for sequence files.

User clicks on directory or file buttons.

Input source name – directory name or single filename.
User enters directory or filename using full path names.

Output source directory – output directory name where the output is printed.
User enters directory name with the full path where the output is to be printed.

FASTA directory – directory containing the FASTA executables.
User enters directory and path containing the FASTA executables.

Similarity type – FASTA or/and BLAST buttons:
User clicks on FASTA and/or BLAST buttons.

Database sources – databases buttons on which to run similarity comparisons.
User clicks on the various GenBank or user databases buttons.

Change-parameters button.
This button allows the user to change the input parameters (path and file names) for GenBank, Repeat, Other, FASTA and BLAST directories.

Make-similarity button.
This button starts the similarity process.

38.3 SUMMARY AND CONCLUSIONS

The computer programs and tools that have been described here were developed to aid genomics efforts at Los Alamos National Laboratory. Specifically, these programs were designed to automate primer picking and to assist in DNA sequence comparison analysis for large-scale STS and EST production efforts. ASTS automates the selection of PCR primers in candidate STS or EST sequences. The Removal Tool Program parses repetitive or user-defined elements from a sequence to mask their contribution prior to performing a similarity search. The Similarity Tool provides an X-window interface and convenient batch mode operation of FASTA and BLAST search algorithms for rapid comparisons to GenBank or other databases. Copies of these programs are available through anonymous ftp at telomere.lanl.gov.

ACKNOWLEDGMENTS

The authors wish to thank Mark Wilder for administration of the Local Sun Network, and helpful discussions. The authors are also grateful to Jim Jett, Ed Hildebrand, and Bob Moyzis for helpful discussions. This work was supported by the US Department of Energy (grant No. 005063 (F137)) under contract W-7405-ENG-36.

REFERENCES

Adams, M.D., Kelly, J.M., Gocayne, J.D., Dubnick, M., Polymeropoulos, M.H., Xiao, H., Merril, C.R., Wu, A., Olde, B., Moreno, R., Kerlavage, A.R., McCombie, W.R. & Venter, J.C. (1991). *Science* **252**, 1651–1656.

Altschul, S.F., Gish, W., Miller, W., Myers, E.W. & Lipman, D.J. (1990). *J. Mol. Biol.* **215**, 403–410.

Green, E.D., Mohr, R.M., Idol, J.R., Jones, M., Buckingham, J.M., Deaven, L.L., Moyzis, R.K. & Olson, M.V. (1991). *Genomics* **11**, 548–564.

Jurka, J., Walichiewicz, J. & Milosavljevic, A. (1992). *J. Mol. Evol.* **35**, 286–291.

Pearson, W.R. & Lipman, D.J. (1988). *Proc. Natl. Acad. Sci. U.S.A.* **85**, 2444–2448.

CHAPTER THIRTY-NINE

Approaches to Identification and Analysis of Interspersed Repetitive DNA Sequences

J. JURKA
Linus Pauling Institute of Science and Medicine, 440 Page Mill Road, Palo Alto, CA 94306, USA

39.1 INTRODUCTION

Eukaryotic genomes contain vast amounts of repetitive DNA in the form of either large arrays of tandem repeats localized in centromeric and telomeric regions of chromosomes (Brutlag, 1980; Schulman & Bloom, 1991; Blackburn, 1991), or interspersed with the remaining chromosomal DNA. The interspersed repetitive DNA is composed of families of short and long complex sequences referred to as 'SINEs' or 'LINEs' (Singer, 1982). It also includes scattered clusters of tandem repeats of 10 or more base pairs, called 'minisatellites' (Armour & Jeffreys, 1992), or 'VNTRs' (Nakamura *et al.*, 1987), and of microsatellites with tandemly repeated units shorter than in minisatellites (Weber & May, 1989). SINEs and LINEs are increasingly being viewed as families of pseudogenes derived from a limited number of founder genes, with unknown function, called 'master' or 'source' genes (recently reviewed by Deininger *et al.*, 1992). These genes are thought to be expressed in germline cells where most likely their RNA is reverse transcribed to cDNA. Many SINE and LINE families are composed of distinct subfamilies which can be traced back to slightly different founder genes, best represented by the subfamily consensus sequences. Some of the subfamilies may have originated from a single evolving founder gene in outbursts of retropositions millions of years apart (Jurka, 1989; Deininger *et al.*, 1992).

To date, approximately 50 families of interspersed repetitive elements have been identified in the human genome alone. Their sizes range from hundreds to hundreds of thousands of elements per family (Hwu *et al.*, 1986; Jurka, 1990; Kaplan *et al.*, 1991). The total number of elements in all-identified human repetitive families exceeds 1 million, i.e. one repeat every 2–3 kb of DNA. This estimate does not include simple interspersed repeats mentioned above.

Owing to the common occurrence of interspersed repeats in newly sequenced DNA, their identification and elementary analyses will become routine in the near future. At the simplest level, the identification can be used for elimination of the repetitive DNA from the newly obtained sequences prior to database searches. Identification of subfamilies and other standard analyses can be helpful for accurate sequence assembly or probe design. It can also be of importance for specific biological studies like identification of hotspots for recombination, imposition of the time-scale on genetic rearrangements, etc.

39.2 IDENTIFICATION OF REPETITIVE DNA

The basic procedure for identification of repetitive elements is to compare the DNA sequence under investigation with a reference collection of repeats as indicated by the simplified procedure shown in Fig. 39.1. The first stage (fast search) is used for identification of the most obvious similarities and the approximate location in the input DNA. The next step is the alignment of the sequence regions identified by the fast search to determine the biological significance of their similarity and the exact positions of the potential repetitive element within the analyzed sequence. If the presence of a repeat in the input sequence is confirmed, it is replaced by an equally long string of random characters. This 'censoring' procedure is particularly important if several repetitive sequences are clustered side by side or even inserted in one another. In such cases the alignment of the censored input sequence against the reference sequences must be repeated, as indicated by the interrupted arrow, to identify all members of the cluster. The details of this process are outlined below.

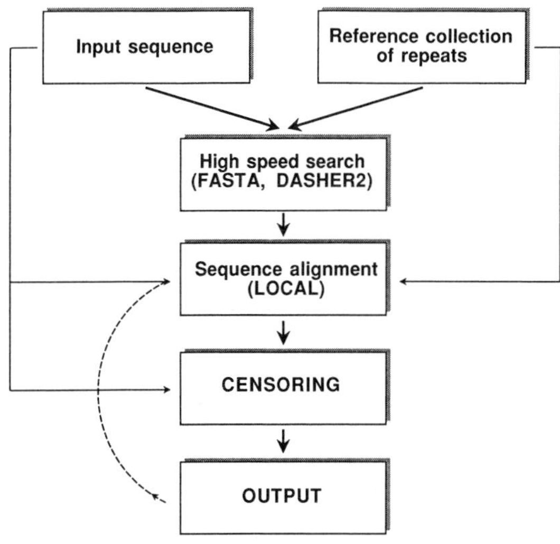

Figure 39.1 A simplified scheme for the identification of repetitive DNA.

39.2.1 Fast searches

Given the growing volume of the sequence data and the limited capacity of the available computers, it is prudent to use fast data searches (Wilbur & Lipman, 1983; Pearson & Lipman, 1988; Altschul *et al.*, 1990) to select preliminary DNA segments similar to the reference sequences. The best-known software based on these algorithms are FASTA and BLAST. In our studies we most extensively used DASHER2 (Faulkner, unpublished) based on the algorithm of Wilbur & Lipman (1983).

The programs used for fast searches usually provide us with a ranked list of scores truncated at some threshold values. Typically, the ranked list identifies reasonably well sequences similar by about 65% or more to the reference repeats. However, as the sequence diversity grows, many biologically significant similarities are increasingly being listed among the lower scores, even below the most conservatively estimated cutoff thresholds. For example, a genetically unstable intron within the BCR gene (Chen *et al.*, 1989) contains a number of repetitive elements, including an unreported fragment of the L1 element at positions 1156–1372 (GenBank locus: HUMBCR22I). This fragment could not be identified by running the BCR intron sequence against the database of prototypic repeats. Some remedies for such situations are: continue the alignment for lower scores; use consensus sequences as reference repeats; or go directly to the alignment procedures without fast searches if more diverse repeats are sought.

39.2.2 Alignment

The alignment is closely associated with determination of the sequence boundaries. So far, this can be best accomplished using the LOCAL algorithm (Smith & Waterman, 1981). LOCAL appears to be effective for a large spectrum of repetitive elements. We also implemented an alignment program based on the minimum length encoding approach (Allison & Yee, 1990; A. Milosavljevic & J. Jurka, unpublished). A version of this software can be run via electronic mail (Jurka *et al.*, 1992). This program can be used for independent evaluation of the alignments obtained with the LOCAL program.

The following critical points deserve additional comments: evaluation of the biological significance of the similarity determined by the sequences alignment, and identification of the reliable boundaries of the aligned repeats.

39.2.2.1 Similarity and homology

There are no absolute rules to determine true homology (i.e. common ancestry) between repeats based on their sequence similarity. Purely statistical criteria cannot be directly applied since there is no intrinsic correspondence between statistical and biological significance (Karlin & Brendel, 1992). However, there is a set of rules based both on statistical analysis and biological knowledge which can quite effectively be applied for practical purposes.

If the stretches of similarity are short, it is usually worthwhile to analyze the data in more detail whenever the number of uninterrupted matches between two sequences is greater than expected by chance. The practical formula for the longest common word between two DNA sequences N_1 and N_2 nucleotides long is: log (N_1N_2)/log 4 (see Karlin *et al.*, 1989).

The most powerful indicator of the homology between two weakly similar repetitive elements is the relative ratio of transition- and transversion-type mismatches determined from the LOCAL alignment. Transitions are defined as A↔G and T↔C base changes in the DNA; all the remaining base substitutions are called 'transversions'. If the base changes in the evolving repeats were to occur at random, one would expect the ratio of transversions to transitions to be about 2:1. This is the expected ratio between two unrelated sequences. However, if the sequences are related, this ratio is often well below 2, due to the fact that among naturally occurring mutations transitions tend to predominate over transversions. This phenomenon is particularly pronounced in sequences containing large number CpG dinucleotides. For practical purposes, the repeats with the LOCAL similarity score above 30 and the ratio of transversions to transitions below 1.5 can be considered homologous. The homology can be reliably determined for even lower LOCAL similarity scores provided that the ratio of transversions to transitions goes down as well. Obviously, the ratio must be based on a reasonable statistical number of substitutions.

An additional test for verification of the identified repeat may be its assignment to the proper subfamily if the relevant information is available.

39.2.2.2 Determination of the sequence boundaries

To determine exact boundaries of a retroposed repeat, one usually studies so-called 'flanking repeats', i.e. short oligonucleotides flanking the inserted sequence at both sides. This approach works relatively well for exact flanking repeats at least 10–12 nt long (depending on the size of the flanked element). The shorter the flanking sequence, the more likely it is to occur by chance. As a result, the exact positions of many retroposed elements remain uncertain or erroneous throughout the literature. There is no good algorithmic remedy to this situation. The most promising approach is comparative analysis of the homologous sequence data from different organisms. For example, Alu repetitive elements are primate-specific (with the likely exception of some old Alu monomers). By comparison of the DNA regions flanking Alu repeats with homologous regions from other mammals, one can determine the exact position of the insertion sites with considerable accuracy. This approach will be more difficult to apply for older repeats and more distant species.

The alignment procedures almost always work on DNA segments of limited length, which may cause artificial truncations of the aligned sequences, especially when repetitive sequences are long or clustered next to one another. A separate analysis is necessary to combine the artificial fragments into a complete sequence.

39.2.3 Censoring

The identified repetitive DNA is replaced by a random string of characters to avoid realignment of the same sequences if the alignment intervals overlap. This censoring can be done automatically based on the sequence alignment, however, one should use conservative homology estimates to avoid possible censoring of DNA which is not homologous to repetitive elements. Subsequent alignments of the precensored sequences with the reference collections can be done using less stringent criteria. They usually reveal more diverged repeats or complex recombinants and sequence fragments. Once the sequence is censored, it gives a more compact list of similarity scores when run against GenBank/EMBL databases. The censored sequence can facilitate identification of unknown repetitive families, as described before (Jurka, 1990).

39.3 REFERENCE COLLECTIONS

GenBank/EMBL databases contain hundreds of repetitive sequences inserted within genes, and often unannotated. As a result, any newly sequenced DNA containing even a single repeat matches hundreds of GenBank entries and the resulting output is very difficult to analyze. A remedy to this situation is a concise organization of the repetitive sequences in a form of well-annotated, electronically available files which can be used in computer-assisted analyses of repeats. The first electronic collection of annotated sequence examples for 53 human repetitive families was compiled recently (Jurka *et al.*, 1992). The repetitive elements are represented by somewhat arbitrarily chosen sequence examples, with the exception of Alu and alphoid DNA represented by known consensus sequences.

39.3.1 Founder genes and consensus sequences

Many, if not all, repetitive sequences are derived from a limited number of founder genes, as shown schematically in Fig. 39.2. Once derived, these

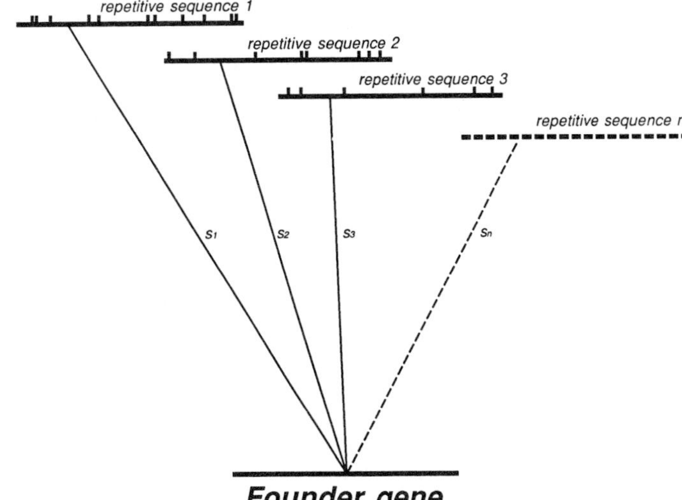

Figure 39.2 Origin of repetitive families according to the founder gene model. Vertical bars indicate mutations relative to the founder gene. $S_1, \ldots, S_n$ denote similarities between the founder gene and the mutated copies. For further explanation see text.

sequences continue to accumulate mutations independently from one another. However, after multiple alignment, sequences with the original nonmutated bases usually form a majority in each position of the aligned set. These majority bases are used to produce consensus sequences which are the best-known reconstructions of founder genes for repetitive families (Jurka & Milosavljevic, 1991). The mutual similarity (S) between two repetitive sequences which have evolved independently from the same founder gene is expressed by the following formula: $S = S1S2 + (1 - S1)(1 - S2)/3$, where S1 and S2 represent similarities between repetitive sequences 1 and 2 and their source gene as indicated in Fig. 39.2. The formula was derived under the assumption that all four bases mutate independently and with equal probabilities $p(A,C,G,T) = 0.25$. If S1 and S2 > 0.25, then $S < S1$ and $S < S2$. This means that the similarity between two repeats is lower than the similarity of either one to the source gene. As indicated above, the best known reconstruction of the source gene is a family consensus sequence. Therefore, the accurate consensus sequence is of greater value for computer searches and probe design than any randomly chosen repetitive sequence. The usage of consensus sequences is particularly important when searching for repeats with low mutual similarity (<65%).

To construct an adequate consensus sequence, the following needs to be taken into consideration. Once repetitive elements have been derived from their founder gene and incorporated back to the genome, they undergo random mutations at neutral or nearly neutral rates in most sequence positions with the exception of CpG dinucleotides (i.e. whenever the 5'-C is followed by G). CpG dinucleotides mutate around 10 times faster than non-CpG doublets (Labuda & Striker, 1989) and mostly via transition-type mutations. CpG dinucleotides sometimes converge to CpA or TpG in the consensus sequences. If the TpG dinucleotides are inherited from the founder gene, the majority of the offspring repeats share TpG in the homologous positions and the consensus sequence is unequivocal. The same applies to CpA. However, if the TpG and CpA dinucleotides are derived via mutations from the CpG dinucleotides in the founder gene, they are present in almost 1:1 ratio in the homologous positions of the offspring repeats. As a result, the so-called 'majority consensus' may accidentally converge to either TpG or CpA, pending statistical variations in the sample. Such positions should be corrected to CpG to adequately reflect the founder gene.

If there is no dominant base in non-CpG positions of the representative set of aligned sequences, or if the second most abundant base is more predominant than in most non-CpG positions of the set, it may indicate that the sequences are composed of two or more subfamilies originating from different founder genes.

39.3.2 'Simple' repetitive sequences

Simple repeats are represented by interspersed, often imperfect, tandem arrays of mono- and oligo-nucleotides (Vogt, 1990) and include mini- and micro-satellites discussed in the introduction. Among the most abundant simple repeats are interspersed poly(A) stretches. Some of them have been introduced into the genome by retroposable elements containing long 'poly(A) tails'. Such tails should be removed from, or at least reduced in, the reference sequences used for computer searches as they can match efficiently any simple A-rich sequence and obscure the identification procedure.

To a limited extent, the identification of simple

sequences can be done using the same approach as the one for SINE and LINE elements outlined above. One can organize a limited collection of potential microsatellites composed of tandemly repeated mono-, di-, tri-, and tetra-nucleotides. This approach becomes increasingly impractical as the number of nucleotides per tandemly repeated unit grows. Not all tandemly repeated *n*-mers are equally abundant in the genome and consideration of all of them is redundant. Unfortunately, there is no reference list which would rank the simple sequences according to their relative abundance in eukaryotes. Until such a list is compiled, the best approach is to eliminate only the microsatellites using the methodology outlined above and compare the rest of the investigated sequence with GenBank/EML databases. Some of the resulting matches will inevitably come from unscreened simple sequences with longer tandem repeats. To distinguish them from more complex output sequences, one can apply a heuristic definition of complexity (e.g. Trifonov, 1990). A more systematic approach to this problem from the algorithmic point of view has been published elsewhere (Milosavljevic & Jurka, 1993).

39.4 FUTURE DIRECTIONS

Identification and analysis of repetitive DNA is in its early infancy. With the accelerating flow of sequence data to electronic banks, more repetitive DNA will have to be detected and analyzed. As the computers become more powerful, the sequence alignment is expected to replace fast searches in the identification of repetitive DNA. Reference collections of repeats for different species will have to be compiled and updated on a regular basis. Identification of subfamilies within repetitive families will be done automatically in ever growing detail. A model tool for identification of subfamily type for individual Alu sequences is already available via electronic mail (Jurka & Milosavljevic, 1991; Jurka *et al.*, 1992). Discovery of yet unknown repetitive families will be done predominantly through the computer-assisted analyses of the growing DNA sequence databases.

ACKNOWLEDGMENTS

This work was supported in part by the US Department of Energy (Human Genome Program Grant No. DE-FG03-91ER61152). I thank Jolanta Walichiewicz and Martha Best for help with preparation of the manuscript.

REFERENCES

Allison, L. & Yee, C.N. (1990) *Bull. Math. Biol.* **52**, 431–453.

Altschul, S.F., Gish, W., Miller, W., Myers, E.W. & Lipman, D. (1990) *J. Mol. Biol.* **215**, 403–410.

Armour, J.A. & Jeffreys, A.J. (1992) *FEBS Lett.* **307**, 113–115.

Blackburn, E.H. (1991) *Nature* **350**, 569–573.

Brutlag, D.L. (1980) *Annu. Rev. Genet.* **14**, 121–144.

Chen, S.J., Chen, Z., d'Auriol, L., Le Coniat, M., Grausz, D. & Berger, R. (1989) *Oncogene* **4**, 195–202.

Day, W.H.E. and Morris, F.R. (1992) *Nucleic Acids Res.* **20**, 1093–1099.

Deininger, P.L., Batzer, M.A., Hutchison III, C.A. & Edgell, M.H. (1992) *Trends Genet.* **8**, 307–311.

Hwu, H.R., Roberts, J.W., Davidson, E.H. & Britten, R. (1986) *Proc. Natl. Acad. Sci. U.S.A.* **83**, 3875–3879.

Jurka, J. (1989) *J. Mol. Evol.* **29**, 496–503.

Jurka, J. (1990) *Nucleic Acids Res.* **18**, 137–141.

Jurka, J. & Milosavljevic, A. (1991) *J. Mol. Evol.* **32**, 105–121.

Jurka, J., Walichiewicz, J. & Milosavljevic, A. (1992) *J. Mol. Evol.* **35**, 286–291.

Kaplan, D.J., Jurka, J., Solus, J.F. & Duncan, C.H. (1991) *Nucleic Acids Res.* **19**, 4731–4738.

Karlin, S. & Brendel, V. (1992) *Science* **257**, 39–49.

Karlin, S., Ost, F. & Blaisdell, B.E. (1989) In *Mathematical Methods for DNA Sequences* M. Waterman (ed.). CRC Press, Boca Raton, FL, pp. 133–157.

Labuda, D. & Striker, G. (1989) *Nucleic Acids Res.* **17**, 2477–2491.

Milosavljevic, A. & Jurka, J. (1993) *CABIOS* **9**, 407–411.

Nakamura, Y., Leppert, M., O'Connell, P., Wolff, R., Holm, T., Culver, M., Martin, C., Fujimoto, E., Hoff, M., Kumlin, E. & White, R. (1987) *Science* **235**, 1616–1622.

Pearson, W. & Lipman, D.J. (1988) *Proc. Natl. Acad. Sci. U.S.A.* **85**, 2444–2448.

Schulman, I. & Bloom, K.S. (1991) *Annu. Rev. Cell Biol.* **7**, 311–336.

Singer, M.F. (1982) *Cell* **28**, 433–434.

Smith, T.F. & Waterman, M.S. (1981) *J. Mol. Biol.* **147**, 195–197.

Trifonov, E.N. (1990) In *Structure and Methods. Vol. 1: Human Genome Initiative & DNA Recombination* H.R. Sarma & M.H. Sarma (eds), Guilderland, Adenine Press, pp. 69–77.

Vogt, P. (1990) *Human Genet.* **84**, 301–336.

Weber, J.L. & May, P.E. (1989) *Am. J. Human Genet.* **44**, 388–396.

Wilbur, W.J. & Lipman, D.J. (1983) *Proc. Natl. Acad. Sci. U.S.A.* **80**, 726–730.

CHAPTER FORTY

Computational Tools for DNA Sequence Analysis

O. WHITE[1] & T. DUNNING[2]

[1] The Institute for Genomic Research, 932 Clopper Road, Gaithersburg, MD 20878, USA

[2] Computing Research Laboratory, Box 30001/3CRL, New Mexico State University, Las Cruces, NM 88003–0001, USA

40.1 INTRODUCTION

A variety of programming tools for the computational analysis of sequences from the DNA sequence databases are described here. All tools were written in the programming languages C and Lisp, and are portable to multiple computers. These tools are ideally suited to the UNIX (UNIX is a computer operating system and registered trademark of AT&T) operating system, but could be used under a number of operating systems. Two categories of sequence analysis are facilitated by these tools: (1) file extraction and parsing of GenBank sequence entries, and (2) compositional analysis of DNA sequences. We chose to adopt this tool oriented approach to allow functionality to be composed, to simplify programming and facilitate exploratory data analysis.

40.2 EXTRACTION AND PARSING OF GENBANK SEQUENCE ENTRIES

The National Center of Biotechnology Information (NCBI) GenBank DNA sequence database is a collection of published DNA sequences in a flat file format. Each file contains annotation fields that describe the publication source, the organism, a description of the gene, and other information related to the DNA sequence. The fields in the GenBank file are used to select the entries that will be used for computational experiments. Selection of particular entries involves choosing appropriate keywords from these fields, and selecting the entries in the GenBank flat file that contain these keywords.

Using the program **munge**, an entire GenBank entry can be converted to a single line of text. This conversion of the GenBank file makes the file more compatible with text-searching programs such as the UNIX pattern-matching commands **grep** or **egrep**. These pattern-matching programs allow great flexibility in selecting entries based on any information in the entries. Most current GenBank extraction methods, such as the NCBI RETRIEVE E-mail server, do not have pattern matching capabilities that use regular expressions. The use of **munge** also allows the user to direct the selected entries of a GenBank file into various UNIX commands without creating intermediate (and often large) disk files. After conversion of the GenBank file (GenBank files are available from ncbi.nlm.nih.gov by anonymous **ftp**) using the program **munge**, entries from GenBank can be selected using the UNIX utility **egrep**. GenBank entries selected by

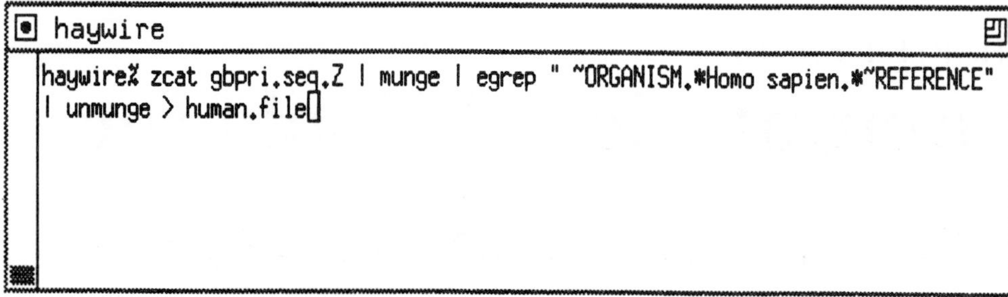

Figure 40.1 A UNIX session using **munge**, **egrep** and **unmunge**. This is a line the user types in at the UNIX system prompt to retrieve all GenBank entries that contain the expression Homo sapien in the ORGANISM field. The GenBank file gbpri.seq is a relatively large file (43 Mbyte) and is stored on disk as a compressed file named gbpri.seq.Z (14 Mbyte). The UNIX command **zcat** uncompresses the GenBank file. The output of **zcat** is directed ('piped') to the next command **munge**, by the use of the character '|'. This redirection enables the user to use multiple commands on the UNIX system prompt. **munge** converts the GenBank data stream so each GenBank entry is contained on a single line. The output of **munge** is piped to the UNIX pattern-matching program **egrep**. The text in quotation marks, '~ORGANISM. *Homo sapien.*~REFERENCE', appearing after the **egrep** command, is the expression that **egrep** uses to search the data stream. The portion '~ORGANISM . . . ~REFERENCE' specifies that the search will be confined to the ORGANISM field of the GenBank entry. The '.*Homo sapien.*' inside of the '~ORGANISM . . . ~REFERENCE' expression specifies that any string containing the character-string 'Homo sapien' will be identified by **egrep**. If **egrep** identifies any lines containing expression '~ORGANISM.*Homo sapien.*~REFERENCE', it passes that line onto the data stream. The data stream is directed into the command **unmunge** to convert the GenBank entry back into its original form. The character '>' takes the data stream from **unmunge**, and writes the data stream to a file on the UNIX disk. Thus, the final destination of the zcat |**munge**|**egrep**|**unmunge** data stream is a file named human.file.

egrep are then converted back to their original form using the tool **unmunge**. Figure 40.1 illustrates a UNIX session using **munge**, **unmunge** and the pattern-matching command **egrep**.

The FEATURES field in each GenBank entry contains the coordinates of all the known biologically relevant sequence components for each entry. These coordinates can be used to extract the DNA strings that correspond to particular sequence elements for analysis. The program **entries** converts these elements into a simpler form which is easily manipulated by Lisp programs to perform a variety of tasks (a public-domain version of Lisp is available from lisp-rt2.slisp.cmu.edu by anonymous **ftp**). This form has a very simple syntax which can be read directly using the input/output procedures which are part of all standard Lisp implementations. Lisp is particularly suited for parsing because of the great flexibility in data structures and the simpler memory allocation and deallocation that Lisp allows. The output resulting from the Lisp-based parsing program **extract_exons** is illustrated in Fig. 40.2. An additional, less flexible tool, **parse_exons**, written in C, will also extract certain forms of annotated introns and exons.

40.3 COMPOSITIONAL ANALYSIS

Software tools have also been written for the compositional analysis of DNA sequences. These programs facilitate three types of DNA analysis: (1) single nucleotide information analysis of aligned sequences; (2) the examination of short DNA subsequences of length *k*, or '*k*-tuples'; and (3) miscellaneous statistical investigations. All tools were written in the programming language C. In addition to producing somewhat faster code, C is preferable to Lisp for numerical applications. All the tools that use DNA strings as input use the format of the type shown in Fig. 40.2.

40.4 SINGLE NUCLEOTIDE INFORMATION ANALYSIS

Information is a measurement of deviation away from random noise (Shannon, 1949). Sequence consensus around the intron splice site, for example, is essentially a measurement of information. Information is a standard method for characterizing binding sites (Schneider *et al.*, 1986) and is related to the DNA-protein binding constant (Berg & von Hippel, 1987). Using the extraction and parsing tools described above, the splice sites of introns can be collected for analysis of consensus and information content. The tool **bf** (base frequency) takes strings of DNA and counts the occurrences of each nucleotide at every position along the set of aligned sequences. Counts can be normalized or not as specified by the user. When normalized, the counts become frequencies. The tool **bf** also notifies the user if illegal characters

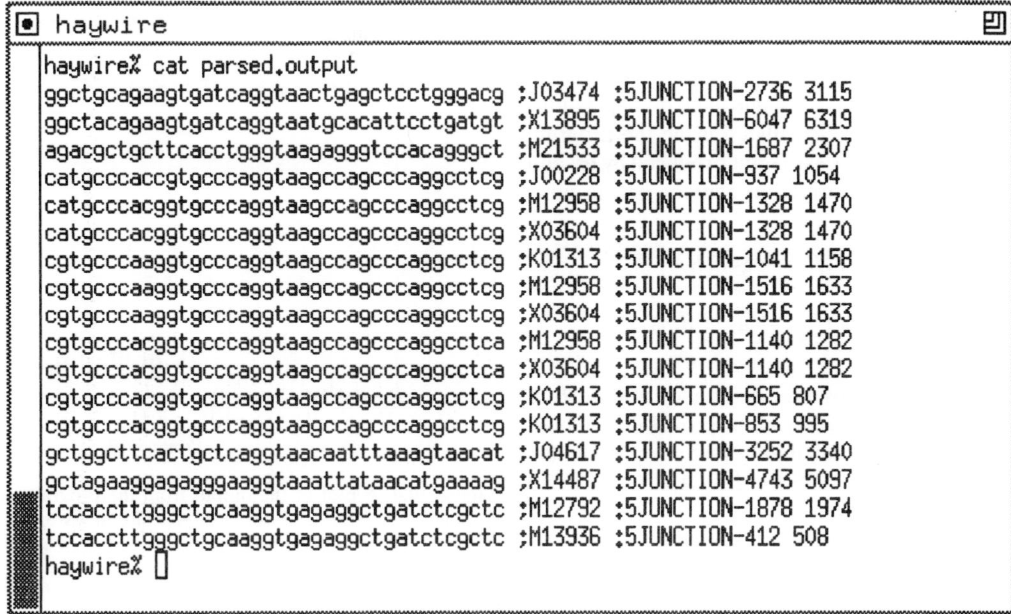

```
haywire
haywire% cat parsed.output
ggctgcagaagtgatcaggtaactgagctcctgggacg ;J03474 ;5JUNCTION-2736 3115
ggctacagaagtgatcaggtaatgcacattcctgatgt ;X13895 ;5JUNCTION-6047 6319
agacgctgcttcacctgggtaagagggtccacagggct ;M21533 ;5JUNCTION-1687 2307
catgcccaccgtgcccaggtaagccagcccaggcctcg ;J00228 ;5JUNCTION-937 1054
catgcccacggtgcccaggtaagccagcccaggcctcg ;M12958 ;5JUNCTION-1328 1470
catgcccacggtgcccaggtaagccagcccaggcctcg ;X03604 ;5JUNCTION-1328 1470
cgtgcccaaggtgcccaggtaagccagcccaggcctcg ;K01313 ;5JUNCTION-1041 1158
cgtgcccaaggtgcccaggtaagccagcccaggcctcg ;M12958 ;5JUNCTION-1516 1633
cgtgcccaaggtgcccaggtaagccagcccaggcctcg ;X03604 ;5JUNCTION-1516 1633
cgtgcccacggtgcccaggtaagccagcccaggcctca ;M12958 ;5JUNCTION-1140 1282
cgtgcccacggtgcccaggtaagccagcccaggcctca ;X03604 ;5JUNCTION-1140 1282
cgtgcccacggtgcccaggtaagccagcccaggcctcg ;K01313 ;5JUNCTION-665 807
cgtgcccacggtgcccaggtaagccagcccaggcctcg ;K01313 ;5JUNCTION-853 995
gctggcttcactgctcaggtaacaatttaaagtaacat ;J04617 ;5JUNCTION-3252 3340
gctagaaggagagggaaggtaaattataacatgaaaag ;X14487 ;5JUNCTION-4743 5097
tccaccttgggctgcaaggtgagaggctgatctcgctc ;M12792 ;5JUNCTION-1878 1974
tccaccttgggctgcaaggtgagaggctgatctcgctc ;M13936 ;5JUNCTION-412 508
haywire%
```

Figure 40.2 An example file of a Lisp-parsed GenBank file. Text appearing after each semicolon refers to the GenBank ACCESSION number, the intron splice site type, and the GenBank DNA coordinates of each sequence. Note these sequences are aligned, that is, at the same position into each line, nucleotide 18, an intron 5′ splice site begins. Position 18 contains the nucleotide G, while position 19 contains the nucleotide T. Alignment of sequences is useful for information analysis and for programs such as **bf**.

```
haywire
haywire% cat parsed.output | bf | nl
     1  183 143 161 225
     2  215 145 160 192
     3  183 122 242 165
     4  191 170 136 215
     5  219 168 126 199
     6  192 156 164 200
     7  168 194 129 221
     8  182 177 121 232
     9  175 152 169 216
    10  209 147 135 221
    11  169 194 151 198
    12  184 174 159 195
    13  224 136  99 253
    14  194 155 153 210
    15  254 223 129 106
    16  430  70  62 150
    17   57  21 572  62
    18    0   0 712   0
    19    0   0   0 712
    20  490  36  77 109
    21  396 105  24 187
    22  162  63 359 128
    23  153  98  68 393
    24  267 118  77 250
    25  241 121  59 291
haywire%
```

Figure 40.3 Output from the program **bf**. The file parsed.output is similar to the aligned sequences in Fig. 40.2. (In this case parsed.output contains 712 aligned sequences). This file is sent into the data stream using the UNIX command **cat**. The data stream is piped into the UNIX command **nl**, that numbers the lines contained in a file. The first column reflects the line numbering from **nl**. The second, third, fourth and fifth columns are the nucleotide counts for A, C, G, and T, respectively. Note that at line 18 all the nucleotides from the aligned sequences were G and at line 19 all the nucleotides were T. These correspond to the positions 18 and 19 of the sequences in Fig. 40.2.

are found in the DNA sequences that were generated from GenBank. The output of **bf** is illustrated in Fig. 40.3.

Information contained at single nucleotide positions of aligned sequences is calculated using the program **calc_I. calc_I** reads the base count produced by **bf**, and then generates the information content $I(n)$ in units of bits/position:

$$I(n) = \sum_{B\varepsilon\,\{A,C,G,U\}} F(B,n)\log_2 F(B,n) - \sum_{B\varepsilon\,\{A,C,G,U\}} P(B,n)\log_2 (PB,n)$$

where $F(B,n)$ is the observed frequency of base B in position n, and $P(B,n)$ is the prior probability of base B in position n. Prior probability is set to equiprobable values (i.e. $P(B,n) = 0.25$ for $B\varepsilon\ \{A,C,G,U\}$ in all positions) by convention.

These measurements of DNA sequences are samples from a much larger source of DNA (i.e. the organism's genome). This type of sampling is by no means an exhaustive or random representation of an organism's genetic content. This means that special care must be taken in drawing conclusions from these data. In particular, the calculation of error bounds is mandatory.

A new method has been created to generate error bars on the value $I(n)$. The previous method (Schneider *et al.*, 1986) of determining the possible range of information assumed that the probability of any nucleotide occurring at a particular position is the same as the probability of that nucleotide occurring at any other position. This assumption does not reflect the nucleotide content changes observed across intron and exon boundaries. Across splice junctions the frequency of guanosine varies from a frequency of 100% to 10% (lines 18 and 20 the file parsed.output in Fig. 40.3). Ironically, information analysis inherently assumes the probabilities of nucleotide occurrences in a particular position changes; that is why these measurements are taken. On the other hand, earlier calculations of expected variation assumed that the frequency of a nucleotide at any position was equal.

To calculate the upper and lower limits of possible variation of information without making this assumption, our program uses the formula proposed by Schneider *et al.* (1986) and the single nucleotide frequencies at each position to calculate each error bar (White *et al.*, 1992). This method can be prohibitively time consuming with large data sets. (One error bar could take several days of computation on Sun SPARC2 workstation.) However, this computationally expensive algorithm involves the calculation and summation of many very nearly identical numbers. The program **calc_H** uses an algorithm that calculates and weighs the sum of many fewer numbers. Knowing that the probability of all possible events is always

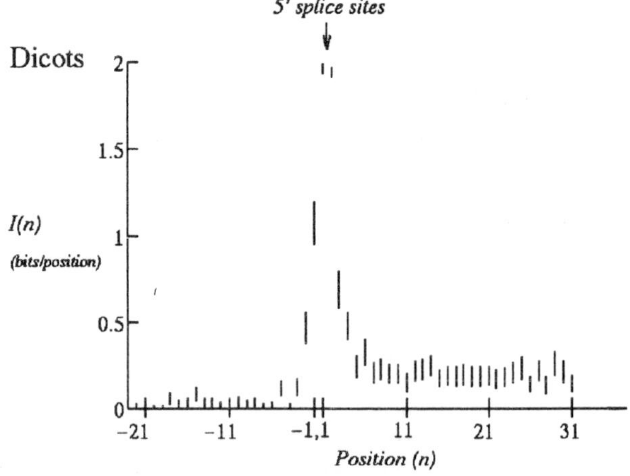

Figure 40.4 A plot of the output from the program **calc_H** of single nucleotide information measurements around the 5′ splice site of the plant group dicots (White *et al.*, 1992). The x axis reflects single nucleotide positions along aligned sequences using typical numbering convention. (At the 5′ splice site nucleotides in the exon (moving upstream from the splice junction) are numbered −1, −2, −3 etc. while nucleotides moving downstream into the intronic region are numbered +1, +2, +3, etc.) Because all 5′ splice sites contain a G at position +1 information ($I(n)$) is maximal at this position. The measure of information is lower at positions away from the splice site because the nonrandom distribution of nucleotides at these positions increases. An example of the possible range in statistical variation contained in information measurements is reflected in the size of error bars at each position. Error bars are of ±2 S.D. (within 95% confidence limits).

one, a parallel calculation is made which checked that the sum of the probabilities of each possible combination was equal to 1. This second sum was used to follow the accuracy of the overall computation in order to obtain a sufficiently precise answer in a practical and short period of time. An example of the possible range in statistical variation contained in information measurements is presented in Fig. 40.4. This program could be improved by using adaptive step methods such as are used by differential equation solvers.

40.5 ANALYSIS OF *k*-TUPLES

The investigation of frequencies of small oligonucleotides has proven useful in the understanding of sequence composition of DNA. A *k*-tuple is an oligonucleotide of length *k*. A hexanucleotide, for example, is a 6-tuple. Historically, the term '*k*-tuple' was first used in mathematical analyses that did not involve DNA, and is thus a more general term than *k*-mer or oligonucleotide. To the extent that the analyses used by our programs apply to more general phenomena, the use of k-tuple is more appropriate. A suite of tools was created to count distinct k-tuples, and perform statistical analysis on k-tuple distributions. We have only used $k \leqslant 6$, but these tools are constructed to handle larger values of k.

The program **counter** takes input of the format shown in Fig. 40.2 and counts the k-tuples found in those sequences. The user specifies which size k is to be measured, and whether **counter** is to measure frequencies or counts. It is important to remember that the counts collected from **counter** are of overlapping k-tuples. **counter** stops reading strings of length l at l-k, thus a caveat to experiments using this tool is that the measurements may be influenced by end-effects. The output of **counter** is in a standardized format to facilitate processing by other programs.

All frequencies of oligonucleotides are influenced by the underlying distribution of subsequences of which they are composed. For example, plant introns are rich in the nucleotides adenosine (A) and thymine (T). Measurements of dinucleotides containing A and T in plant introns would be elevated over the frequencies of G+C-containing dinucleotides. Thus, k-tuple frequencies reflect two separate components: (1) a portion that is the result of random occurrences of one nucleotide after another; and (2) another portion that is due to a dependency of one nucleotide on another. This concept can also be extended to oligomers larger than dinucleotides. Larger frequencies of the 5-tuple GACCG could be the result of elevations in the nucleotide G and the trinucleotide ACC. In the case of dinucleotides, it is easy to model how the underlying distributions of nucleotides may be influencing dinucleotide frequencies. The expected frequencies for dinucleotides are going to be the product of the frequencies of each single nucleotide. A 'normalized frequency' (F' (BB')) for dinucleotide frequencies can be obtained using the formula:

$$F'\ (BB') = \frac{F(BB')}{F(B) \times F(B')}$$

where $F(BB')$ is the observed dinucleotide frequency and $F(B)$ and $F(B')$ are the observed single nucleotide frequencies for two single nucleotides. Deviations of $F(B')$ from 1 in the above formula reflect differences in the observed frequency for a particular dinucleotide from the expected frequency of that dinucleotide, based on its single nucleotide composition. Because changes in the observed frequency of dinucleotides from the expected frequency reflect a dependency of one nucleotide on another, these deviations may be of some biological relevance. Normalized frequencies have been shown to vary in the two evolutionarily distinct plant groups, monocots and dicots (White *et al.*, 1992). A similar analysis when k>2 is not as easy as in the dinucleotide case. In the case of GACCG, increases in its frequency may be the result of elevated G and ACC content, or increases in GA and CCG. In the absence of more sophisticated models, a simple method for determining the expected counts for k-tuples where k>2 has been adopted (this is referred to as an $l-1$ Markov model; Markov models, as they apply to biology, are discussed in Stukle *et al.*, (1990); the use of l instead of k is unfortunate in the present context, but has no real significance). To calculate the k-tuple frequency $F(B_1... B_k)_{exp}$ that might be expected based on underlying smaller k-tuples the following equation can be used:

$$F'(B_1...B_k)_{\rm exp} = \frac{C(B_1...B_{(k-1)}) \times C(B_2...B_k)}{C(B_2...B_{(k-1)}) \times {\rm Total}_{k\text{-tuples}}}$$

Using the example for measuring the expected 5-tuple frequency of ACACG, $C(B_1...B_{(k-1)})$ corresponds to the counts for the tetranucleotide ACAC, $C(B_2...B_k)$ would be the counts for CACG, and $C(B_2...B_{(k-1)})$ would be the counts for the trinucleotide CAC.

Again, deviations of observed counts from expected counts for the same k-tuple are considered to be of possible biological importance. Such a deviation would cause F' to tend to have a value other than one. In the cell, we know that the chromosome is recognized by proteins, such as trans-activating factors or restriction enzymes, which interact with relatively short stretches of DNA. Biologically relevant k-tuples may play a role in DNA–protein recognition interactions or they may contribute to a larger structural pattern in DNA.

Software has been devised that performs this analysis. **corrector** takes three sets of k-tuple counts as input ($k, k-1, k-2$) and calculates Markov-corrected counts for a given k. Another measure of departures of observed counts from expected counts is computed by:

$$\text{std}(B_1...B_k) = \frac{C(B_1...B_k)_{\rm exp} - C(B_1...B_k)_{\rm obs}}{\sqrt{C(B_1...B_k)_{\rm exp}}}$$

This 'normalized deviation' (*std* $(B_1...B_k)$) is useful in that it computes abundance changes in observed versus expected counts more effectively than normalized frequencies. The normalized frequencies of two different k-tuples give the same number if the ratio observed : expected counts is 100:200 or 1:2. However, the expected variation of a count is proportional to the square root of the count. Thus, the expected variation of two is far larger (relatively speaking) than the expected variation of 200. The formula for normalized deviation appropriately reflects the variation between observed/expected counts. The program tool **std** generates normalized deviation values by taking two files, one of observed counts and the other of Markov-corrected counts, and computing the above formula.

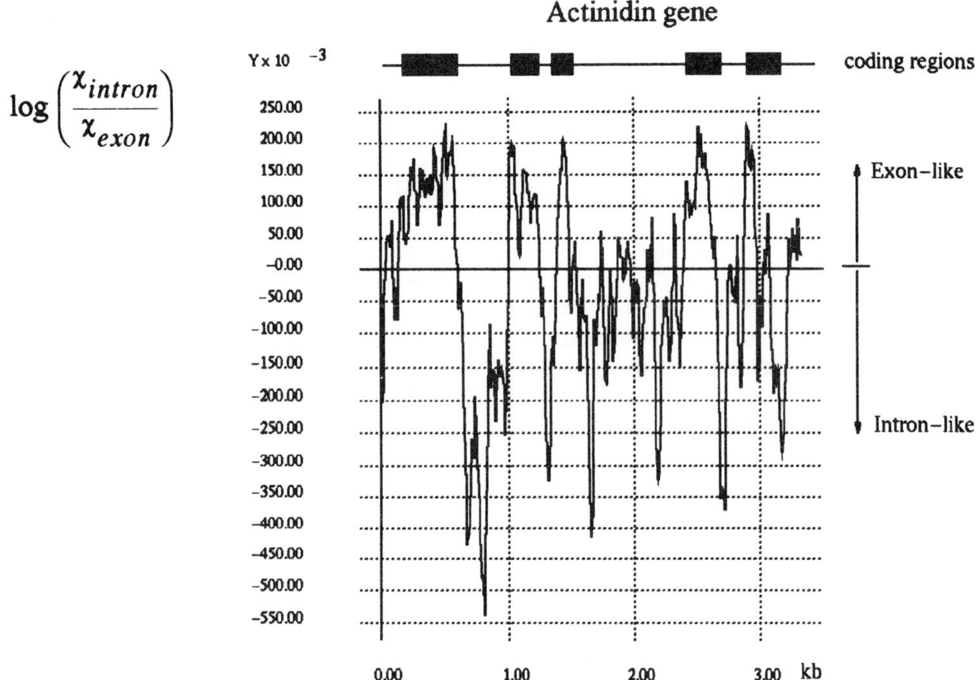

Figure 40.5 A graphical representation of the output from **window_cmp**. Counts of 6-tuples were taken separately from dicot introns and exons. These distributions of 6-tuples were used in χ^2 -tests against DNA sequences of 50-nt windows from the actinidin gene of the dicot, *Actinidia deliciosa.* Measurements were taken at 10-nt intervals using the formula: log $(\chi_{intron}/\chi_{exon})$, where χ_{intron} and χ_{exon} are the results from χ^2 tests between actinidin 6-tuples versus the intron- and exon-specific 6-tuples, respectively. Comparisons of 6-tuple distributions that are more exon-like result in positive values, while comparisons of 6-tuple distributions that are more intron-like result in negative values. Regions that are exon-like correspond to the coding regions of actinidin.

Additional operations can be done on *k*-tuple files using the tool **math**. Two files are supplied to **math** and the user can specify the **math** operations addition, subtraction, multiplication, division and $\log_2 (X/Y)$. Such operations are often useful in the numerical manipulation of *k*-tuple collections.

40.6 STATISTICAL PROGRAMS

Because *k*-tuple distributions are different between the data sets of exons and introns, these distributions could serve in the identification of sequence level determinants of mRNA recognition mechanisms. Figure 40.5 shows a graphical representation of the results from **window_cmp** where *k*-tuples from exons and introns are compared to *k*-tuples from a plant gene sequence using a χ^2 test. Comparisons that show the sample DNA sequence to be more like exon-specific *k*-tuples rather than intron-specific *k*-tuples implicate the test region as coding.

Numerical experiments can be conducted on computer-fabricated DNA sequences to generate normalized deviations. The manufactured DNA will be a collection of one random *k*-tuple after another. These data give the range of values that result from sequences that have structure no greater than *k*. The program **gen** generates manufactured DNA using the output from **counter**. The DNA sequences are synthesized so they have the *k*-tuple frequency and length as the data sets from which the counts were derived. The 5% and 95% confidence limits from distributions of normalized deviations have been determined from sequences created from **gen.** Normalized deviation values of actual data that occur outside of the 95% and 5% confidence limits were considered statistically significant.

Table 40.1 summarizes the features of the programs described in this chapter. Four string handling utilities (**random_clip**, **random_line**, **size_select** and **window_select**) are included in the table. All software is available through anonymous **ftp** at the site **ftp.tigr.org.** Contact **owhite@tigr.org** for additional information.

Table 40.1 A summary of programs.

Program name	*Function*	*Options*
munge	Converts each entry of a GenBank file into a single line for regular expression searching	None
unmunge	Converts munged entry into a typical GenBank file	None
entries	Parses each entry of a GenBank file into a Lisp-readable format	-w – writes an error file to disk
extract_exons	A Lisp based program extracts various gene portions from the output of **entries**	User specifies whether to extract intron, exon, or splice site junctions using the CDS, CDS join, or mRNA join features of the GenBank entry
parse_exons	Parses each entry of a GenBank file and retrieves the exons and introns from CDS join	-e – extract exons -i – extract introns -d – debug, show all parsed information -w – writes an error file to disk
bf	Measures single nucleotide counts from aligned sets of sequences from the output of **extract_exons**	-f – measure base frequency
calc_H	Calculates single position information content from the output of **bf**	None
calc_I	Calculates standard deviation of single position information content from the output of **bf**	None
counter	Counts *k*-tuples from the output of **extract_exons**	-k *n* – measure *k*-tuple size *n* -f – measure frequency
convert_fasta	Converts a fasta file into a form readable by **counter**	None
corrector	Calculates Markov-corrected frequencies for sets of *k*-tuples from the output of **counter**	-k *n* – measure *k*-tuple of size *n* -c – calculate corrected counts -f fname – filename, where files of the format fname.n.counts are stored on disk ($n = k, k-1$ and $k-2$)
std	Calculates the normalized deviation for a set of *k*-tuples using the output from **counter** and **corrector**	-f fname – filename of the Markov-corrected *k*-tuples -F fname – filename of the observed counts
math	Perform simple math functions two files (X and Y) from **counter**	-f fname – filename of file X -F fname – filename of file Y -a – addition -s – subtraction -m – multiplication -d – division (X/Y) -l – $\log_2$ (X/Y)
gen	Generates synthetic DNA strings based using the output of **counter**	-k *n* – use counts of *k*-tuples of size *n* -n *l* – create a string of length *l*
window_cmp	Perform a sliding window scan of a DNA string, and run χ^2 test on window and the counts of two input files. Counts are of the format supplied by **counter**	-k *n* – use counts of *k*-tuples of size *n* -f fname – filename of input count file 1 -F fname – filename of input count file 2 -w *l* – use sliding window of length *l* -o *l* – increment the window forward using an overlap of length *l*
random_clip	Randomly clip a portion of a DNA string	-c *n* – size of chunk to be clipped from string

Table 40.1 Continued

Program name	*Function*	*Options*
random_line	Randomly selects DNA strings from a file of DNA strings	-f fname – filename, where a file containing DNA strings of the format fname.all is stored on disk. **random_line** then writes two new files fname.test and fname.use to disk -d n – divisor, percentage (1–100) of DNA strings that are sent to fname.test
size_select	Selects DNA strings of specified lengths from a file of DNA strings	-l n – select strings longer than length n -s n – select strings shorter than length n
window_select	Select windows from a DNA string of specified distance from either end of string	-R d:n – starting in from a distance d, select from right end of string a window of length n -L d:n – starting in from a distance d, select from left end of string a window of length n

REFERENCES

Berg, O., & von Hippel, P. (1987) *J. Mol. Biol.* **193**, 723–750.

Schneider, T.D., Stormo, G.D., Gold, L. & Ehrenfeucht, A. (1986) *J. Mol. Biol.* **188**, 415–431.

Shannon, C.E. (1949) *The Mathematical Theory of Communication.* University of Illinois Press, Urbana, IL.

Stukle, E.E., Claudio, E., Uhlrich, G. & Nielsen, P.J. (1990) *Nucleic Acids Res.* **18**, 6641–6647.

White, O., Soderlund, C., Shanmugam, P. & Fields, C. (1992) *Plant Mol. Biol.* **19**, 1057–1064.

CHAPTER FORTY-ONE

A Practical Guide to the GRAIL E-mail Server

E.C. UBERBACHER,[1] X. GUAN,[1] & R.J. MURAL[2]

[1] Engineering Physics and Mathematics and [2] Biology Division, Oak Ridge National Library, Oak Ridge, TN 37831–6364, USA

41.1 INTRODUCTION

The methods used for obtaining genome information are undergoing a revolution. One cornerstone of this revolution is an increased reliance on sequence-based computational methods to provide insight into the structure, function, and organization of the genome. This trend will undoubtedly continue as both sequencing and informatics technologies improve. The infrastructure of computational genome analysis has also changed dramatically to a distributed environment where various databases and analytical resources may be called into play to deal with a given analysis problem. In this type of distributed computing environment, it is a significant challenge for even the most skillful genome scientist to gain the expertise necessary to interact with the multitude of DNA and protein, human and model organism, mapping, sequencing, and structure databases and specialized analysis resources.

The introduction of GRAIL in mid-1991 represented a new kind of resource in the genome community based of E-mail access rather than the traditional direct login and command line approach. While E-mail access may not be appropriate for mapping databases requiring complex query capabilities, for more singular types of analyses, E-mail represents a convenient lowest common denominator, and one especially useful for noncomputer specialists. The modularity 2 of E-mail analysis makes it possible to incorporate such systems into software environments which access multiple resources. In addition, E-mail systems have the advantage of requiring minimal resources for implementation and support. Since the advent of GRAIL, other useful E-mail systems (see list in a later section) have been constructed for a variety of purposes. Despite the simplicity of the GRAIL interface, a certain level of skill is required to access the system. The following is a guide which can facilitate using GRAIL in a laboratory environment.

41.2 THE PURPOSE AND USE OF E-MAIL GRAIL

The GRAIL E-mail server is designed to provide an initial automatic localization and characterization of the protein-coding regions of genes from DNA sequence data. Many laboratories sequence in search of particular genes, perhaps mapped to a region, which are of biochemical, genetic, or medical interest. The identification of one or more coding regions by

GRAIL provides a starting point for subsequent computational and experimental investigation, such as isolation and sequencing of a cDNA for the gene or identification and functional analysis of the gene product. Some large-scale sequencing laboratories, sequencing both genomic DNA and cDNAs, use GRAIL as a means of automated sequence annotation placing putative coding characterization (and protein sequence matches) in a database for further analysis. The need for systems capable of accurate automated annotation will increase as high-speed sequencing methods become available, and this use of the GRAIL server is consistent with the longer term goals of the GRAIL project.

The GRAIL E-mail server recognizes coding regions in DNA sequence through a technology generally referred to as 'pattern recognition'. Basically, the system recognizes characteristics of sequence found to occur in protein coding regions. The details of this process have been discussed elsewhere (Uberbacher & Mural, 1991; Mural *et al.*, 1992; Uberbacher *et al.*, 1993). When the user sends a message containing unannotated sequence to the system, pattern recognition methods are applied to each region of sequence and the results for each small sequence neighborhood are combined to provide an overall indicator of coding potential at the given position. One limitation of the method currently employed for pattern recognition is that the system will not find noncoding exons or the parts of exons which contain noncoding sequence, or genes which code for something other than proteins (e.g. RNAs). However, homology or rule-based methods have been developed for identifying RNA genes.

The GRAIL E-mail system is designed to be used by experimental biologists with minimal computer skill. The system provides essential information in a simple format, without unnecessary clutter and without the need for complex interpretation. It has a purposeful conservative bias so that noise is low (although not nonexistent), thus reducing false information which can be costly in laboratory time.

41.3 HOW TO USE GRAIL

41.3.1 Registering to use GRAIL

The GRAIL E-mail server is a cost-free service provided to genome researchers in universities, research hospitals, national laboratories and private companies on a world-wide basis. GRAIL can be accessed and used by sending E-mail to the GRAIL@ornl.gov internet address. Any message sent to this address by unregistered users will receive a response asking the users to send their names and addresses to register. Registration is for book-keeping purposes and so messages and updates can be sent to users. After registering, the GRAIL help file and user ID numbers are returned and users can then submit sequences to GRAIL for analysis.

Registration as a GRAIL user requires only a simple E-mail message sent to the address GRAIL@ornl.gov which is formatted as follows:

Register
Rocket J. Squirrel
Chairman, Moose Genome Project
Frostbite Falls University
Frostbite Falls, Minnesota
(615)–555–1235

The *register* keyword (not case sensitive) signals the system to record a new user. The name and address information is used for communication with the user by mail or phone. The precise form and number of lines in the name/address section is not critical. Upon receiving this message, the system will return a user identification (ID) and help information by E-mail to the user. The help file can also be obtained by sending a message to GRAIL@ornl.gov containing the word '*help*' in the first line. This does not, however, register the user.

41.3.2 Getting help from the GRAIL staff

Additional help may be obtained from the GRAIL staff through E-mail at the address GRAILMAIL@ornl.gov. This mail address is for personal communication, not sequence analysis, and is answered daily. Users sometimes have questions about the significance of GRAIL results or need help understanding the output. The staff welcomes feedback about the system's performance and has found such feedback an important component for improving the system.

41.3.3 How to send sequence to GRAIL

Each sequence submitted for analysis can be a maximum of 100 kb in length and a single E-mail message may contain one or several sequences. The basic procedure to send sequence to GRAIL is a very simple one. First, the sequence must exist in a computer file in ASCII format (standard readable text like DOS text on a PC). This type of sequence file can be created by sequencing equipment and software or extracted from a database using utility programs or commercial software such as the GCG package (Genetics Computer Group, 1991). Sequence files can also be created by hand using a standard text editor or word processor. As described below, the sequence should ideally be in a 'Staden' format, without positional numbers, unusual sequence characters, or extra spaces.

As a second step in preparing a sequence to send, certain information must be placed before the beginning of the sequence. This should be done with the local text editor or word processing software capable of saving a file in ASCII format. Knowledge of a few basic features of the editor, like how to open and save a file and how to type in new lines, is required for this. The precise format of this information to be added is described in Section 41.3.4. Third, some very basic familiarity with the local E-mail system is necessary to E-mail a sequence file. Since mail systems vary from place to place, the local systems' manager should be consulted about this.

41.3.4 Message format

Messages submitted to GRAIL have three basic component parts: (1) a line with user ID and requested analysis options; (2) a header or title line for the sequence to be analyzed; and (3) a line-by-line sequence text in DNA or RNA bases. If more than one sequence is to be analyzed in a single E-mail message, steps (2) and (3) can be repeated multiple times – it is not necessary to repeat line (1). An example of a message containing two sequences to be analyzed might look like this:

```
sequences 2 g1001 -e
> squirrel1.seq
CCCATGATCTGGCTAAGCTGCATCTGC
GTGCGTATGCTGAATTGCGCGTTGT
GGCGTATGTCATGCTTCATCCTGTGTG
CGTAGAACAAAGCGTCGACAGAGTCA
ACNNNNGTGCGTCAT.....
.
.
.
>moose1.pr     second sequence for analysis
AAAAGCTGANNCGTGCATGTGCGCGG
GTGCCCACAAATGTCGT...
.
.
.
```

41.3.4.1 User ID line

When a user registers to use the GRAIL system, as described previously, a user ID is provided starting with a letter followed by a four digit number. This ID number must accompany all requests for analysis. The ID line should contain the keyword 'sequences' (not case sensitive), the number of sequences to be analyzed, and the user ID. This line can also contain several optional 'switches' which specify specific types of analysis. For example, the '-e' switch, above, specifies that a comparison should be made between all 'excellent' putative coding regions found and the SwissProt protein sequence database. Switches also are being incorporated to specify model organism version of GRAIL (rather than the human version). A more detailed description of available switches is provided in a later section. The user ID identifies the user for the GRAIL log, and specifies that following lines contain sequence header and sequence to be analyzed with the options specified. It should be noted that a brief record of each transaction is kept by the system, but the sequence itself is purged from the system immediately after analysis to protect proprietary sequence information.

41.3.4.2 Header line

The header line starts with the special character '>'. This serves to delineate sequences from one another and tells the system it is reading a header line which is a free form comment not to be analyzed as sequence. Any text may be entered within the line. The first part of the line, usually the sequence name, is associated with the return analysis.

41.3.4.3 Sequence format

The sequence is simple text which is not case sensitive. A suitable format is simple 'Staden' format, which contains just sequence – no position numbers, no spaces, etc. Both RNA and DNA alphabets are recognized, and nonstandard letters like 'N', and special characters like '.', '–', and '*' are fixed, although not in a particularly elegant manner. Numbers, such as sequence position numbers, should be eliminated from the mail message. The text lines do not need to be any particular length and can vary in length. Blanks, blank lines, and other 'white spaces' are compressed out (eliminated) during analysis with no harmful effect.

The user can check the range of sequence length in the return analysis to ensure that the sequence was read correctly. If more than one sequence is to be analyzed (as specified next to the keyword 'sequences' in line (1)), subsequent header lines and sequence texts can follow directly below one another. Sequences analyzed by GRAIL must be at least 100 bases in length for a coding/non coding decision to be made.

The sequence sent to the system is considered the 'forward' strand. GRAIL automatically generates the complimentary strand, called the 'reverse' strand, and analyzes it also.

41.3.4.4 Common mistakes made in formatting a GRAIL message

There are several common errors in formatting a file which are responsible for most failures of the system to analyze a message. These include not including user ID, not including '>' symbol at the header line and not including the keyword 'sequences' followed by the proper number of sequences on first line. If a message from a registered user fails to invoke analysis, the system responds with a help message to help the user fix the problem and let the user know that his initial message reached GRAIL. Some E-mail systems have the capability of adding a special trailer to E-mail messages, such as a holiday greeting or quote of the day. Do not do this.

41.4 USING AND INTERPRETING GRAIL OUTPUT

The current GRAIL E-mail service can return three types of output. The first of these is a position-by-position estimate of coding potential in the sequence, calculated at intervals of 10 bases. Figure 41.1(a) shows the coding potential for 100 base windows centered at the given position of the input sequence as determined by the neural network component of the system, the coding recognition module (CRM) (Uberbacher & Mural, 1991). The system examines both strands of the input sequence and returns the location of potential coding regions on the forward strand (that is the sequence as submitted) in the left column of the output and the reverse compliment in the right-hand column. Each strand is numbered from its 5′-end. The arrangement of output is such that the two scores on the same line correspond to the same location in the sequence, one on the forward and one on the reverse strand. The positions of windows with scores below 0.01 are not reported in the output. Often, this saves considerable space in the return message without loss of useful information. This portion of the output also includes the most probable reading frame for windows with scores greater than 0.5 and the region of the sequence over which this reading frame is open.

A number of aspects of pattern which correspond to coding regions are somewhat symmetrical (have similar properties on the sense and nonsense strand). As a result, the neural net system sometimes recognizes the coding character from either strand direction and the scores (left and right column) reflect this. Statistical information is used to resolve this, as described below.

The second part of the output contains a summary table of information regarding putative exons. Figure 41.1(b) is an example of the table generated by the system rule base from the CRM output. The rule base uses information from the CRM and the probable reading frame and open-reading frame extent to combine regions of protein coding potential into estimates of coding exons, make a determination of the strand which encodes each region, and assign a quality score to each predicted coding exon. The quality score 'excellent' corresponds to an actual coding region greater than 90% of the time, 'good' is correct about 65% of the time and 'marginal' is correct about 15% of the time. Both the preferred strand and reading frame determinations are correct about 95% of the time. The probable strand prediction can be further evaluated by examining the strand probability function in the table; confidence in a strand prediction with a score significantly above 0.5 (the switch point for strand determination) is stronger than for a strand prediction near to 0.5. One should be particularly cognizant of situations where several impressive coding regions are found on one strand, but one is predicted on the 'other' strand and with strand probability not far from 50%. It certainly makes sense from an experimental point of view to design experiments around the best exons first and not spend significant effort chasing 'marginal' exons.

The third and optional part of GRAIL output contains the results of sequence comparison. By including a switch (*-s*, *-e*, or *-g*) on the first line of the message submitted to GRAIL, one invokes a search of the translations of potential coding regions from the strand predicted and the frame indicated by the program against the SwissProt protein sequence database (currently release 23, though this is updated periodically). The '*-e*' switch searches only regions which score 'excellent' while '*-g*' searches only 'excellent' or 'good' regions. The '*-s*' switch searches all putative exons (regardless of score) against the database. These searches are done using an implementation of the Smith-Waterman algorithm (Smith & Waterman, 1981) on an Intel iPSC/860 parallel computer. The ten highest scoring matches are returned in the GRAIL output, as shown in Fig. 41.1(c). The score is equivalent to the number of identities over the region matched. Scores are sensitive to length and need to be evaluated with this in mind.

A common question asked about the GRAIL analysis is whether it is sensitive to sequencing errors. Compared to most other algorithms, GRAIL is relatively insensitive. Errors in noncoding regions have virtually no effect although, in the extreme, can increase the noise level. In coding regions, miscalled bases also have virtually no effect. Insertions and deletions of bases in coding regions have a somewhat larger effect, although the system still provides useful coding recognition at a 1% insertion/deletion error rate. In general, we recommend that high-quality sequence data be used.

a

```
.................................................................
3231   0.146    -        - - -      ||  3221   0.061    -          - - -
3241   0.560    3    3036 -  3348   ||  3211   0.129    -          - - -
3251   0.937    3    3036 -  3348   ||  3201   0.030    -          - - -
3261   0.938    3    3036 -  3348   ||  3191   0.046    -          - - -
3271   0.938    3    3036 -  3348   ||  3181   0.143    -          - - -
3281   0.938    3    3036 -  3348   ||  3171   0.549    2     2981 -  3278
3291   0.938    3    3036 -  3348   ||  3161   0.232    -          - - -
3301   0.938    3    3036 -  3348   ||  3151   0.007    -          - - -
3311   0.442    -        - - -      ||  3141   0.003    -          - - -
.................................................................
.................................................................
3541   0.087    -        - - -      ||  2911   0.000    -          - - -
3551   0.105    -        - - -      ||  2901   0.000    -          - - -
.................................................................
.................................................................
3711   0.000    -        - - -      ||  2741   0.064    -          - - -
.................................................................
```

b

```
Potential exons are listed in the following

     pos        strand   strand_prob   frame     quality          orf

1661 -  1791     f          0.84         2      excellent   1655 -  1901
2031 -  2221     f          0.92         2      excellent   1958 -  2222
2341 -  2541     f          0.98         2      excellent   2231 -  2645
3231 -  3311     f          0.94         3      excellent   3036 -  3348
2661 -  2721     r          1.00         3      good        2580 -  2724
2201 -  2241     r          1.00         1      marginal    2197 -  2422
 751 -   780     r          0.65         3      marginal     702 -   780
 161 -   211     r          0.97         1      good           1 -   427
```

c

```
Strand: f, Position:   3231 -   3311, Translation limits:   3204 -   3341

The translated protein is:
LLIPSLLSQGVEDAFYTLVREIRQHKLRKLNPPDESGPGCMSCKC
    1  score:  36.0, pos:   199-   238, >P01115 RASH_MSVHA HARVEY MURINE SARCOMA
    2  score:  36.0, pos:   147-   186, >P23175 RASH_MSVNS MURINE SARCOMA VIRUS
    3  score:  36.0, pos:   147-   186, >P01113 RASH_MSV MURINE SARCOMA VIRUS.<
    4  score:  36.0, pos:   147-   186, >P01112 RASH_HUMAN HOMO SAPIENS (HUMAN).
    5  score:  35.0, pos:   206-   245, >P01114 RASH_RRASV RASHEED RAT SARCOMA V
    6  score:  34.7, pos:   147-   186, >P20171 RASH_RAT RATTUS NORVEGICUS (RAT)
    7  score:  34.7, pos:   147-   186, >P08642 RASH_CHICK GALLUS GALLUS (CHICKE
    8  score:  20.0, pos:   147-   186, >P08556 RASN_MOUSE MUS MUSCULUS (MOUSE).
    9  score:  18.7, pos:   147-   186, >P12825 RASN_CAVPO CAVIA PORCELLUS (GUIN
   10  score:  18.7, pos:   147-   186, >P01111 RASN_HUMAN HOMO SAPIENS (HUMAN).
```

Figure 41.1 Example showing the three parts of typical GRAIL output. (a) The coding probability output for both the forward and reverse strand for a region of the sequence containing a coding exon. Notice that the exon is recognized in both the forward and reverse directions but to a greater extent on the forward strand (the sense strand). The reasons for this partial recognition on the reverse strand have to do with a symmetry of certain pattern aspects in coding regions. (b) The exon summary table, listing the estimated positions and score, strand probability, and corresponding orf for each putative coding region. (c) A protein database search against the translation of the exon in part (a). Note the score is expressed in terms of identity equivalents in the translated sequence shown and derived from the preferred reading frame and strand in (b).

41.5 UPCOMING FEATURES IN GRAIL

41.5.1 Construction of gene models

As an outgrowth of the GRAIL project, technology has been developed to identify gene regions in a sequence and construct relatively accurate models for genes (Uberbacher *et al.*, 1993), including translation start sites and donor and acceptor splice junctions through translation termination. The 'assembly' of a gene model from its recognized component parts provides a largely correct coding message and protein sequence translation and is scheduled to be implemented in mid-1993. Access to this optional capability will be available through a switch on line (1), with specific instructions specified in the help file at that time.

41.5.2 Model organism versions of E-mail GRAIL

Several model organism versions of the E-mail server system are under construction including *Escherichia coli*, dicotyledonous plants, *Caenorhabditis elegans*, and *S. pombe*. These and other special versions will be accessed through a switch on line (1), with the exact details being specified at the appropriate time in the help file.

41.5.3 Stand-alone GRAIL

This version of GRAIL is being constructed to run on workstations as an alternative to the E-mail service. It will include all of the features described here and a graphical user interface which plots output scores, etc., in an X-windows environment.

ACKNOWLEDGMENTS

This research was supported by the Office of Health and Environmental Research, US Department of Energy under contract No. DE-AC05–84OR21400 with Martin Marietta Energy Systems, Inc.

REFERENCES

Genetics Computer Group (1991) *Program Manual for GCG Package, Version 7.*

Mural, R.J., Einstein, J.R., Guan, X., Mann, R.C. & Uberbacher, E.C. (1992) *Trends Biotechnol.* **10**, 66–69.

Smith, T.F. & Waterman, M.S. (1981) *J. Mol. Biol.* **147**, 195–197.

Uberbacher, E.C. & Mural, R.J. (1991) *Proc. Natl. Acad. Sci. U.S.A.* **88**, 11261–11265.

Uberbacher, E.C., Einstein, J.R., Guan, X. & Mural, R.J. (1993) In *Proceedings of the Second International Conference on Bioinformatics, Supercomputing, and Complex Genome Analysis*, St Petersburg Beach, FL, 4–7 June 1992. In press.

CHAPTER FORTY-TWO

Searching for Homologies to Protein Blocks by Electronic Mail

S. HENIKOFF, J.G. HENIKOFF, S. AGUS & J.C. WALLACE
Basic Sciences Division, Fred Hutchinson Cancer Research Center, Seattle, WA 98104, USA

42.1 INTRODUCTION

The BLOCKS electronic mail searcher has been introduced to aid the detection and verification of protein sequence homology. The searcher compares a protein or DNA sequence to the current database of protein blocks. Blocks consist of short multiply aligned ungapped segments corresponding to the most highly conserved regions of proteins. A database of blocks has been constructed by successive application of the automated PROTOMAT system (Henikoff & Henikoff, 1991) to individual entries in the PROSITE catalog of protein groups (Bairoch, 1992) keyed to the SWISS-PROT protein sequence databank (Bairoch & Boeckmann, 1992). BLOCKS version 5.0 consists of 2106 blocks derived from 559 different groups represented in PROSITE vers 9.00 keyed to SWISS-PROT 22. BLOCKS is updated following each significant update of PROSITE, which occurs every 6 months.

The rationale behind searching a database of blocks is that information from multiply aligned sequences is present in a concentrated form, reducing background and increasing sensitivity to distant relationships. This information is represented in a position-specific scoring table or 'profile' (Gribskov *et al.*, 1987), in which each column of the alignment is converted to a column of a table representing the frequency of occurrence of each of the 20 amino acids. The best alignment between a sequence and each entry in the BLOCKS database is noted. If a particular block scores highly, it is possible that the sequence is related to the group of sequences the block represents. Typically, a group of proteins has more than one region in common represented as a series of blocks separated by unaligned regions. If a second block for a protein group also scores highly in the search, the evidence that the sequence is related to the group is strengthened, and is further strengthened if a third block also scores it highly, and so on.

42.2 A SYSTEM FOR CONSTRUCTING A DATABASE OF BLOCKS

The PROTOMAT system is outlined in Fig. 42.1. It consists of modules that can be executed singly or in combination, either manually using user-specified parameters or automatically using parameters determined by the system. The BLOCKS database results from successive execution of PROTOMAT (Henikoff & Henikoff, 1991) on all groups in the PROSITE catalog (Bairoch, 1992).

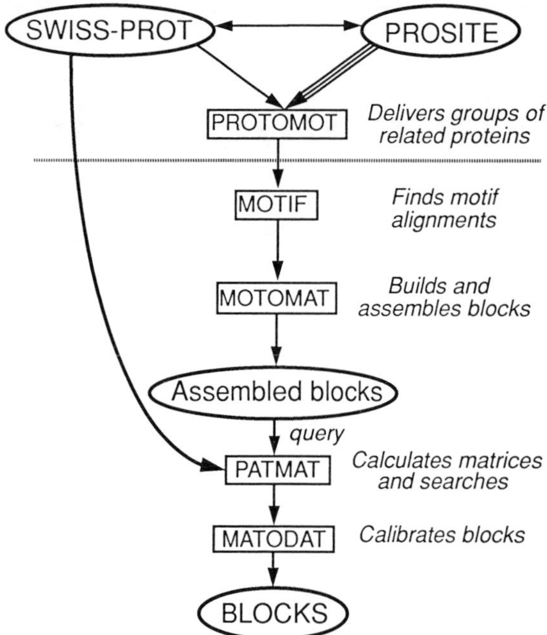

Figure 42.1 Flow chart showing components of the PROTOMAT system. Functions above the line are specific to PROSITE.

To make blocks, PROTOMAT requires a group of two or more related proteins, such as the groups documented in PROSITE. The sequence extraction module, PROTOMOT, reads a PROSITE.DAT file and finds the individual full-length sequences in the SWISS-PROT databank. Using these sequences, PROTOMOT executes a modified version of the MOTIF program of Smith, (1990) for finding motif alignments in an automatic mode where all parameters are determined by the program. MOTIF searches for any set of three amino acids separated by two fixed distances occurring in a large fraction of the protein sequences in the group, where all 20 amino acids and all possible distances up to 17 are considered. Each resulting motif forms the basis for an alignment, against which each of the remaining sequences is aligned. These block alignments are then refined by the MOTOMAT program which merges motifs if they are aligned identically in all of the sequences and extends alignments out in both directions until similarity falls off. If a block exceeds 60 columns in width, it is split into contiguous blocks.

MOTOMAT then uses graph theory techniques to find a best set of blocks for a group, assembling a 'best path' of blocks. A directed graph is constructed with the blocks as its nodes. An arc extends from node b1 to node b2 if block b1 precedes block b2 and does not overlap it in a large fraction of sequences. Different arcs in the graph may include different subsets of sequences. The graph is unrooted, and the restriction that the fraction of sequences in a path be more than half the total number of sequences guarantees an acyclic graph. A topological sort of the graph is performed so that the blocks towards the N-terminus come before blocks towards the C-terminus. A standard recursive depth-first search is used to enumerate all paths through the sorted graph. Because the arcs represent only pair-wise relationships between blocks, evaluation of a path is terminated if, at some point, the path fails to include enough sequences.

MOTOMAT writes out each block to a separate file in a format that resembles a PROSITE entry. These blocks contain only the sequences included in the best path, and the minimum and maximum distances from the preceding block among those sequences is included in the block file. For groups with known repeats, a sequence is included in the best path regardless of the order in which the blocks occur in it, as long as the blocks do not overlap.

Since blocks range in width from 3 to 60 amino acids and include from 2 to over 200 sequences, searching results obtained with them cannot be directly compared unless the blocks are calibrated. This is done by providing two standard scores for comparison, thereby dividing search scores into three regions. The lower calibration score is a value below which search scores are not likely to be interesting and the upper calibration score is a value above which they are. To determine these values, each block is used as a query in a search against the complete SWISS-PROT database using the matrix searching module from the PATMAT program. Every possible sequence segment the width of the block is aligned with a profile derived from the block and is scored. The rank-ordered search results are analyzed by the MATODAT module to separate the scores of sequences that were used to construct the block (considered true positives) from the scores of other sequences (considered true negatives). The 99.5th percentile score of the true negative sequences is used as the lower calibration score to allow for errors and omissions in the protein group used to construct the block. The number of true positive scores might be small and their distribution skewed, so their median is used as the upper calibration score. The ratio of upper to lower calibration scores multiplied by 1000 is referred to as the 'strength' of the block. Strength is a quantitative measure of the ability of a block to discriminate between true positives and true negatives. If blocks are too strong, they will discriminate against distant relatives, whereas if they are too weak, they will fail to exclude chance alignments.

To build a database of blocks, we use the groups of related proteins documented in the PROSITE catalog after removal of redundant groups. A best path of blocks is constructed for each PROSITE entry. MATODAT calibrates the blocks and concatenates them into the BLOCKS database. This database is searched using PATMAT with a sequence as a query and conversion of each block to a profile (Gribskov *et al.*, 1987; Henikoff *et al.*, 1990). Each raw score is normalized by dividing it by the lower calibration score, which is stored in the block, and multiplying by 1000. This normalization of the scores means that in a search of the BLOCKS database, a score below 1000 is below the lower calibration score.

42.3 BLOCKS SEARCHER

PATMAT is used to search the database of blocks (Wallace & Henikoff, 1992). To start, PATMAT aligns the first position of the sequence with the first position of the first block, and a score for that amino acid is obtained from the profile column corresponding to that position. Scores are summed over the width of the alignment, and then the block is aligned with the next position. This procedure is carried out exhaustively for all positions of the sequence for all blocks in the database. For single block hits, a search score below 1000 can generally be ignored, while a score above the block's strength is evidence that the query sequence is related to the sequences represented in the block. Scores in the middle region are suggestive, but usually require corroborating evidence, such as can be provided by a good score for other blocks from the same best path with a reasonable spacing in between. These statements apply to most blocks; however, a few blocks are especially 'weak' (strength <1100) and results should be interpreted more cautiously, with confidence increasing in proportion to block strength. Multiple block hits are detected and analyzed by BLOCKSORT (S. Henikoff & J. Henikoff, *Genomics*, in press), a module that provides the output returned by the E-mail server.

The PATMAT searching tool accepts protein or DNA sequences in FASTA, Genepro, GenBank, EMBL, SWISS-PROT, GCG or PIR formats. DNA sequences are translated in all six frames for searching. When FASTA, Genepro or GCG format is used, PATMAT decides that the sequence is protein if an alphabet character other than G, A, T, C or N (upper or lower case) is found. This will cause misinterpretation if IUBPAC ambiguities are present in DNA sequence. For protein, the 20 amino acid characters are allowed, with the other 6 alphabet characters interpreted as mismatches. Nonalphabet characters including numbers and symbols are ignored. Figure 42.2(A) shows an example of a protein query in FASTA format (in which the first line begins with '>' and is not recognized as sequence). The subject heading is left blank. In the body of the message, only a single sequence is included. The typical search requires only a few minutes if queries are not long DNA sequences. Neither queries sent to the server nor any results sent out are saved. The blocks searcher queries the current version of BLOCKS, noting the 250 highest scoring blocks aligning with a protein query.

42.4 INTERPRETING RESULTS OF A SEARCH USING A PROTEIN QUERY

A score above 1000 is expected for 0.5% of the blocks in the search using a protein query of average size. Since there are about 2000 blocks, a protein of average size is expected to score about 10 blocks above 1000 by chance alone. The top five hits are reported in rank order. If second or third blocks for one of these top five hits are also detected among the 250 highest scoring hits, these are reported as well. For the query example in Fig. 42.2(A) part of the output that is returned is shown in Fig. 42.2(B).

The best hit in the entire database is for BL00059A, the first of three blocks for the family represented in PROSITE entry PS00059, the zinc-containing alcohol dehydrogenases. Note that the two other blocks for this family, BL00059B (ranking 4th with a score of 1064) and BL00059C (ranking 19th with a score of 976) also were among the top 250 hits. All three blocks are 'strong'. The three blocks align with the query sequence in the same order as the sequences that are represented in the blocks, that is, A→B→C. This is most easily seen in the block map. This map also shows that the distance between the three blocks representing this group is not dissimilar from the distance between the segments of the query that align with these blocks. Therefore, the query might be a member of this family. Further evidence that this is the case comes from examination of the alignment of each query segment detected with the closest single member of the group. For BL00059A, which aligns with an N-terminal segment of the query, the best single sequence is an N-terminal segment of ADHX_HORSE. Different members of this family align best with the B and C blocks; nevertheless, the position of the alignment within each sequence is consistent with the position within the query. For example, the distance between A and B varies from 10 to 14 in known members of this family and is 9 for the query.

Intuitively, it seems unlikely that three high scoring blocks would align with correct distances in between by chance alone. But how unlikely? First, assume that the alignment with the A block occurred by chance

A

```
To: blocks@howard.fhcrc.org
Subject:
>YCZ2_YEAST   Hypothetical 40.1 KD protein in HMR 3' region
MKAVVIEDGKAVVKEGVPIPELEEGFVLIKTLAVAGNPTDWAHIDYKVGPQGSILGCDAA
GQIVKLGPAVDPKDFSIGDYIYGFIHGSSVRFPSNGAFAEYSAISTVVAYKSPNELKFLG
EDVLPAGPVRSLEGAATIPVSLTTAGLVLTYNLGLNLKWEPSTPQRNGPILLWGGATAVG
QSLIQLANKLNGFTKIIVVASRKHEKLLKEYGADQLFDYHDIDVVEQIKHKYNNISYLVD
CVANQNTLQQVYKCAADKQDATVVELTNLTEENVKKENRRQNVTIDRTRLYSIGGHEVPF
GGITFPADPEARRAATEFVKFINPKISDGQIHHIPARVYKNGLYDVPRILEDIKIGKNSG
EKLVAVLN
```

B

```
Query=>YCZ2_YEAST  HYPOTHETICAL 40.1 KD PROTEIN IN HMR 3'REGION.,
 Size=368 Amino Acids
Database=blocks.dat, Blocks Searched=2106

1.----------------------------------------------------------------------------
Block      Rank Frame Score Strength       Location Description
BL00059A    1     1   1306  2043           2-    42 Zinc-containing alcohol dehyd
BL00059B    4     1   1064  2100          52-    90 Zinc-containing alcohol dehyd
BL00059C   19     1    976  2229         153-   211 Zinc-containing alcohol dehyd

P= 8e-07 for BL00059B BL00059C in support of BL00059A
                         |-----   97 residues----|
   BL00059 AAAAAAAAAAA:::.BBBBBBBBBB::::::::::::::.......CCCCCCCCCCCCCCC
>YCZ2_YEAS AAAAAAAAAAA::BBBBBBBBBB::::::::::::::CCCCCCCCCCCCCCC

BL00059A     <->A (1,35):1
ADHX_HORSE 9      AAVAWEAGKPVSIEEVEVAPPKAHEVRIKIIATAVCHTDAY
                   ||  | || |  | |         | ||  | |   ||
>YCZ2_YEAS 2      KAVVIEDGKaVVkEgVPIPELEEGfVLIKtLAVAgnpTDwa

BL00059B    A<->B (10,14):9
ADH3_ASPNI 62     PLIGGHEGAGVVVAKGELVKDEDFKIGDRAGIKWLNGSC
                    | |   ||  |  |  |   ||  |||         |
>YCZ2_YEAS 52     GsILGcdAAGqIVKLGPAVdpkDFsIGDyIyGFIhGsSv

BL00059C    B<->C (52,80):62
ADH1_KLULA 158    GVTVYKALKSANLKAGDWVAISGAAGGLGSLAVQYAKAMGYRVLGIDAGEEKAKLFKDL
                       |   |     |      ||     ||     |  |       |     || |
>YCZ2_YEAS 153    LglNLKwepSTpqRnGpiLLwgGAtAVGqSLIQLAnKlnGfTkIIVvASrKhEKLlKEY

2.----------------------------------------------------------------------------
Block      Rank Frame Score Strength       Location Description
BL00086     2     1   1092  2000         162-   193 Cytochrome P450 cysteine heme

                        |-----   13 residues----|
   BL00086 xxxxxxxxxxxxxxxxxxxxxxxxxxxxxxxxxxxxxxxxxxxxxxxxxxxxxxxxxxxxxxxxxxxxxx
>YCZ2_YEAS xxxxxxxxxxxxxxxxxxxxxxxxxxxxxxxxxxxxxxxxxxxxxxxxxxxxxxxxxxxxxxxxxxxxxx

BL00086x     <->x (319,476):161
CP71_PERAE 433    LIPFGAGRRGCPGIAFGISSVEISLANLLYWF
                    |   |     | |       | ||| |  |
>YCZ2_YEAS 162    STPqrNGPiLlwGgAtAVgQSLIqLANkLngF
```

Figure 42.2 Protein query sequence in FASTA format (A) and blocks searcher output (B), showing the first two hits.

(although its position near the N-terminus and the quite high score of 1306 argues against this assumption). We then can ask, what is the probability that the B and C blocks are also chance alignments? Since these were detected independently of the A block, finding a low probability of chance alignment for these two blocks can confirm our intuition that there is a relationship. These probabilities are based on the rank of each block hit, the sizes of the query sequence and the database, and the observed distances between blocks and the number of blocks in a group (for further details see Henikoff, 1992). In this case, the probability that the B and C blocks also have aligned with the query by chance is estimated to be about 1 in a million ('P=8e−07 for BL00059B BL00059C in support of BL00059A'). Combined with the fact that the A block was the best hit in the database, we conclude that the query is a member of the zinc-containing alcohol dehydrogenase family. This is a distant member of a large family, apparently one not easily detected using other approaches. The query is an open reading frame from yeast chromosome III not reported to be a member of any family either in the original study (Oliver *et al.*, 1992) or in a subsequent more intensive analysis of ORFs (open reading frames) from this chromosome (Bork *et al.*, 1992).

The second hit, BL00086, is typical of a chance alignment. A score of 1092 is unremarkable in a search using an average sized query. Also, the alignment with the closest sequence in the block, with 10 identities over a stretch of 32 amino acids, seems unremarkable given lack of any biological information on the yeast chromosome III ORF. Even though the top five block hits in a search are reported, one should be increasingly cautious about single block alignments as ranks decrease. Note that no P values are reported for single block hits. Note also that the P value for blocks in support becomes less meaningful as one goes down the list; P values above about 1/100 in support of a block that does not score at the top should be regarded with caution.

42.5 INTERPRETING RESULTS OF A SEARCH USING A DNA QUERY

For a DNA query the top-scoring 250 blocks aligning with each frame of the sequence are saved. Results are analyzed similarly to those for a protein query, except that here it is assumed that multiple block alignments with a single sequence might be detected in different frames because of frameshift errors in the sequence. In reporting the location of an alignment, each frame is translated as if it were a single protein, so that a 4 kb sequence consists of three predicted proteins of about 1300 amino acids beginning at the 5′-end (+ frames) and three others beginning at the 3′-end of the sequence (− frames).

Figure 42.3 shows the results from a search using the 4 kb sequence of the *Pseudomonas putida dgd* region obtained from GenBank. Note that three of the six highest ranking blocks belong to the short-chain alcohol dehydrogenase family and are correctly spaced. For two of the block alignments, striking regions of identity with the closest segment in the block are seen, further confirming that this sequence includes a member of this family. The fact that for all three blocks the scores are close to the respective strengths indicates that this predicted protein is about as distant from other members of the family as is the typical member. But these blocks are in different frames on the reverse strand, indicating multiple frameshift errors in the query sequence. The second highest ranking block, BL00600A, is supported by two other high-ranking and correctly spaced blocks in the +3 frame. This is a known member of the class III aminotransferase family (the DgdA protein) upstream of and convergent to the dehydrogenase homology. Although the D block was not detected among the top 250 hits, a P value of about 8/100 000 in support of a very high-scoring block confirms the assignment of DgdA to the family. The next highest scoring hit is an alignment with the single block, BL00044, representing the LysR family of bacterial regulatory proteins. Since there can be no blocks in support, one must evaluate this alignment based on whether the score is reasonably high (1461 is more than 200 higher than the next best hit) and whether this makes sense biologically. In fact, this is the N-terminal portion of the DgdR protein, which regulates the *dgdA* gene. Combined with the fact that like known LysR family members, DgdR is just upstream and oppositely oriented from a gene it regulates, and the observation that this region of LysR family members is always N-terminal, we can conclude that the DgdR protein is a member of the family, though very distant from even its closest relative. So the search has detected members of three families oriented: 5′<–LysR––aminotransferase –> <–dehydrogenase– 3′.

42.6 GETTING THE BLOCKS AND PROSITE INFORMATION FOR A GROUP

Following up a potentially interesting hit is often aided by examining the full set of blocks for a group. Furthermore, since each group in BLOCKS corresponds to a group in PROSITE, the excellent annotations from PROSITE give further information and provide useful references. This information can be obtained

```
1.-----------------------------------------------------------------------
Block    Rank Frame Score Strength      Location Description
BL00061A   6    -1   1289  1336       147-    159 Short-chain alcohol dehydroge
BL00061B   3    -3   1542  1359       211-    221 Short-chain alcohol dehydroge
BL00061C   1    -1   1997  2053       265-    316 Short-chain alcohol dehydroge

P=6.7e-08 for BL00061B BL00061A in support of BL00061C
                         |----- 220 residues----|
   BL00061 A:::::.....................................B:::......CCCCCC
P.cepacia                         :::::::::::::::::A:::::B:::::CCCCCC

BL00061A    <->A (2,36):146       BL00061B   A<->B (52,372):51
DHES_HUMAN 3     TVVLITGCSSGIG    PHBB_ZOORA 74    GPIDVLVNNAG
                   ||||||||| |                     |||||||||||
P.cepacia  147   KTVLITGCSSGFG    P.cepacia  211   GPIDVLVNNAG

BL00061C   B<->C (37,81):43
2BHD_STREX 132   GSIVNISSAAGLMGLALTSSYGASKWGVRGLSKLAAVELGTDRIRVNSVHPG
                 | |||  |   |  | |   | |||  |       ||||     |   | ||
P.cepacia  265   GVIVNVTSSVtLKvLPLVgAYrASKAAVNAFTESMAVELePFGVRAHLVLPG

2.-----------------------------------------------------------------------
Block    Rank Frame Score Strength      Location Description
BL00600A   2     3   1680  2346       498-    532 Aminotransferases class-III p
BL00600B   5     3   1352  1656       668-    681 Aminotransferases class-III p
BL00600C 110     3   1032  1237       707-    717 Aminotransferases class-III p

P= 7.9e-05 for BL00600B BL00600C in support of BL00600A
                         |----- 145 residues----|
   BL00600  AAAAAA::::::::::::::::::::::........BB::::.....CC::::....DD
P.cepacia  <AAAAAA::::::::::::::::::::::BB::::CC

BL00600A    <->A (31,84):497
ARGD_ECOLI 33    GSRIWDQQGKEYVDFAGGIAVTALGHCHPALVNAL
                 ||   |  |    ||  |     ||||||  |
P.cepacia  498   GSfVYDaDGRAiLDFTSGemSAVLGHCHPEIVsVi

BL00600B   A<->B (119,169):135    BL00600C   B<->C (25,54):25
GATA_ASPNI 286   VAAIIVEPIQSEGG   GATA_ASPNI 325   DEVQTGVGATG
                  || | ||| | ||                    || ||||| ||
P.cepacia  668   lAAFIaEPIlSsGG   P.cepacia  707   DEaQTGVGRTG

3.-----------------------------------------------------------------------
Block    Rank Frame Score Strength      Location Description
BL00044    4    -2   1461  2175       897-    954 Bacterial regulatory proteins

                         |-----  24 residues----|
   BL00044  xxxxxxxxxxxxxxxxxxxxxxxxxxxxxxxxxxxxxxxxxxxxxxxxxxxxxxxxxxxx
P.cepacia  <xxxxxxxxxxxxxxxxxxxxxxxxxxxxxxxxxxxxxxxxxxxxxxxxxxxxxxxxxxxx

BL00044x    <->x (0,31):896
ICIA_ECOLI 4     PDYRTLQALDAVIRERGFERAAQKLCITQSAVSQRIKQLENMFGQPLLVRTVPPRPTE
                      |     || |     ||     |||| ||  | ||    ||||    | |
P.cepacia  897   sLeiDLlrsfVVIaEvRalSAAARVGRTQSALSQQmKRLEDivDQPLLpAHRPRRgaD
```

Figure 42.3 Blocks searcher output using *Pseudomonas putida dgdA* DNA sequence from GenBank as query, showing the first three hits.

```
ID is from PROSITE, AC is derived from prosite.dat PS#, DE is abstracted from
prosite.dat DE, BL is PROTOMAT information. For each segment, the SWISS-PROT
ID is followed by the position of the first residue in the segment. Segments
are clustered if >=80% of aligned residues match between any pair of segments.
==============================================================================
ID   HTH_LYSR_FAMILY; BLOCK
AC   BL00044; distance from previous block=(0,31)
DE   Bacterial regulatory proteins, lysR family proteins.
BL   ALR motif; width=58; 99.5%=839; strength=2175
AMPR_RHOCA  (    6)  LPLNALRVFEVAMRQGSFTKAAIHLRVTQAAVSHQVARLEDLLGTALFLRTSQGLIPT

CATM_ACICA  (    1)  MELRHLRYFVTVVEEQSISKAAEKLCIAQPPLSRQIQKLEEELGIQLFERGFRPAKVT

CATR_PSEPU  (    1)  MELRHLRYFKVLAETLNFTRAAEILHIAQPPLSRQISQLEDQLGTLLVVRERPLRLTE

GLTC_BACSU  (    7)  WSLRQLRYFMEVAEREHVSEAADHLHVAQSAISRQIANLEEELNVTLFEREGRNIKLT

ICIA_ECOLI  (    4)  PDYRTLQALDAVIRERGFERAAQKLCITQSAVSQRIKQLENMFGQPLLVRTVPPRPTE

ILVY_ECOLI  (    1)  MDLRDLKTFLHLAESRHFGRSARAMHVSPSTLSRQIQRLEEDLGQPLFVRDNRTVTLT

IRGB_VIBCH  (    2)  QDLSAVKAFHALCQHKSLTAAAKALEQPKSTLSRRLAQLEEDLGQSLLMRQGNRLTLT

LEUO_ECOLI  (   22)  VDLNLLTVFDAVMQEQNITRAAHVLGMSQPAVSNAVARLKVMFNDELFVRYGRGIQPT

LYSR_ECOLI  (    4)  VNLRHIEIFHAVMTAGSLTEAAHLLHTSQPTVSRELARFEKVIGLKLFERVRGRLHPT

MLER_LACLA  (    3)  LNLRDLEYFYQLSKLRSFTNVAKHFRVSQPTISYAIKRLETYYDCDLFYKDSSHQVVD
NAHR_PSEPU  (    6)  LDLNLLVVFNQLLVDRRVSITAENLGLTQPAVSNALKRLRTSLQDPLFVRTHQGMEPT

NHAR_ECOLI  (    6)  INYNHLYYFWHVYKEGSVVGAAEALYLTPQTITGQIRALEDALQAKLFKRKGTWSRTQ

NODD_AZOCA  (    6)  LDLNLLVALNALLSEHSVTSAAKSINLSQPAMSAAVQRLRIYFNDDLFTINGRERVFT

OXYR_ECOLI  (    1)  MNIRDLEYLVALAEHRHFRRAADSCHVSQPTLSGQIRKLEDELGVMLLERTSRKVLFT

RBCR_CHRVI  (    3)  VSLRQLRVFEAVARHNSYTRAAEELHLSQPAVSMQVRQLEDEIGLSLFERLGKQVVLT

SYRM_RHIME  (   32)  IDLNLLVDLEALLQYRHITQAAQHVGRSQPAMSRALSRLRGMLKDDLLVAGSRGLVLT

TDCA_ECOLI  (    7)  PKTQHLVVFQEVIRSGSIGSAAKELGLTQPAVSKIINDIEDYFGVELVVRKNTGVTLT

TRPI_PSEAE  (    6)  PSLNALRAFEAAARLHSISLAAEELHVTHGAVSRQVRLLEEDLGVALFGRDGRGVKLT

VRPR_SALDU  (    4)  LINKKLKIFITLMETGSFSIATSVLYITRTPLSRVISGLERELKQRLFIRKNGTLIPT

YFEB_ECOLI  (    7)  TDLKLLRYFLAVAEELHFGRAAARLNMSQPPLSIHIKELENQLGTQLFIRHSRSVVLT

AMPR_CITFR  (    5)  IPLNSLRAFEAAARHLSFTRAAIELNVTHSAISQHVKSLEQQLNCQLFVRGSRGLMLT
AMPR_ENTCL  (    5)  LPLNSLRAFEAAARHLSFTHAAIELNVTHSAISQHVKTLEQHLNCQLFVRVSRGLMLT
CYSB_ECOLI  (    2)  KLQQLRYIVEVVNHNLNVSSTAEGLYTSQPGISKQVRMLEDELGIQIFSRSGKHLTQV
CYSB_SALTY  (    2)  KLQQLRYIVEVVNHNLNVSSTAEGLYTSQPGISKQVRMLEDELGIQIFARSGKHLTQV
METR_ECOLI  (    2)  IEVKHLKTLQALRNCGSLAAAATLHQTQSALSHQFSDLEQRLGFRLFVRKSQPLRFT
METR_SALTY  (    2)  IEIKHLKTLQALRNSGSLAAAAVLHQTQSALSHQFSDLEQRLGFRLFVRKSQPLRFT
NOD1_BRAJA  (    6)  LDLNLLVALDALMTERKLTAAARSINLSQPAMSAAITRLRTYFRDELFTMNGRELVPT
NOD1_RHILP  (    6)  LDLNLLVALDALMTERNLTAAARSINLSQPAMSAAVGRLRTYFNDDLFTMVGRELVPT
NOD2_BRAJA  (    6)  LDLNLLVALDALMTKRSVTAAARSINLSQPAMSAAIARLRTYFGDDLFTMRGRELIPT
NOD2_RHILP  (    6)  LDLNLLVALDALTTERNLTAAARSINLSQPAMSAAIGRLRDYFRDELFTMNGRELRLT
NOD3_RHIME  (    6)  LDLNLLVALDALMTKRSVTAAARSINLSQPAMSSAIARLRSYFQDELFRMQGRELITT
NODD_BRASP  (    6)  LDLNLLVALDALMTERNLTAAARKINLSQPAMSAAIARLRSYFRDELFTMRGRELVLT
NODD_RHILE  (    6)  LDLNLLVALDRLMTERKLTAAARAINLSQPAMSAAISRWRDYFRDDLFIIQRRELNPT
```

Figure 42.4 Blocks server response to 'get BL00044' in the subject line or the body of the message. The corresponding PROSITE.DAT and PROSITE.DOC entries (not shown) are also returned.

for a single group by sending a blank message to blocks@howard.fhcrc.org with the subject heading: 'get BL?????'. For example, to obtain the three blocks representing the zinc-containing alcohol dehydrogenases, the message should be:

To: blocks@howard.fhcrc.org
Subject: GET BL00059

Figure 42.4 shows the single block returned when the message 'get BL00044' is sent. The corresponding entries from PROSITE.DAT and PROSITE.DOC are sent together with the requested blocks.

42.7 OBTAINING HELP

A help file is returned when the single word HELP appears on the subject line of a blank E-mail message or in the body of a message sent to the Internet address: BLOCKS@HOWARD.FHCRC.ORG. The complete BLOCKS database and PROSITE catalog can be obtained from the repository of the National Center for Biological Information via ftp ('ftp ncbi.nlm.nih.gov' log in as 'anonymous', provide an E-mail address as password, then 'cd repository/blocks' or 'cd repository/prosite'). PROTOMAT software and documentation for DOS and UNIX machines are also available from the repository. Ftp instructions are found in the Announce file in repository/blocks.

42.8 SUMMARY

Searches for homology to protein families can be carried out by electronic mail using the BLOCKS E-mail searcher. Blocks are ungapped multiply aligned segments that represent the most highly conserved regions of proteins. An automated system for finding and assembling a best set of blocks for a family has been applied to 559 groups in the current PROSITE catalog, leading to a database of blocks. This database can be rapidly searched for detection or verification of distant protein homologies. The E-mail server accepts a protein or DNA sequence in most popular formats, returning potentially interesting alignments of the sequence with entries in the current database. As multiple block hits can be especially informative, these are mapped and evaluated in the output. Searches may be carried out using raw DNA sequence, allowing for independent detection of multiple blocks representing a family regardless of reading frame.

REFERENCES

Bairoch, A. (1992) *Nucleic Acids Res.* **20**, 2013–2018.

Bairoch, A. & Boeckmann, B. (1992) *Nucleic Acids Res.* **20**, 2019–2022.

Bork, P., Ouzounis, C., Sander, C., Scharf, M., Schneider, R. & Sonnhammer, E. (1992) *Nature* **358**, 287.

Gribskov, M., McLachlan, A.D. & Eisenberg, D. (1987) *Proc. Natl. Acad. Sci. U.S.A.* **84**, 4355–4358.

Henikoff, S. (1992) *New Biol.* **4**, 382–388.

Henikoff, S. & Henikoff, J.G. (1991) *Nucleic Acids Res.* **19**, 6565–6572.

Henikoff, S., Wallace, J.C. & Brown, J.P. (1990) *Methods Enzymol.* **183**, 111–132.

Oliver, S.G., van der Aart, Q.J.M., Agostoni-Carbone, M.L., Aigle, M., Alberghina, L., Alexandraki, D., Antoine, G., Anwar, R. *et al.*, (1992) *Nature* **357**, 38–46.

Smith, H.O., Annau, T.M. & Chandrasegaran, S. (1990) *Proc. Natl. Acad. Sci. U.S.A.* **87**, 826–830.

Wallace, J.C. & Henikoff, S. (1992) *CABIOS* **8**, 249–254.

CHAPTER FORTY-THREE

Integrating Computational and Experimental Methods for Gene Discovery

C. FIELDS

The Institute for Genomic Research, 932 Clopper Road, Gaithersburg, MD 20878, USA

43.1 INTRODUCTION

The primary goal of most large-scale sequencing projects is the discovery of new genes in previously uncharacterized or only partially characterized DNA. In most cases, high sequencing throughput is essential to the viability of such projects, and few if any resources are available for the detailed characterization of newly identified genes. Efficient methods for finding, and learning as much as possible about, new genes contained in anonymous DNA sequences are needed if gene discovery is to keep pace with sequencing. Maintaining high throughput requires that gene discovery be largely automated; user-intensive sequence analysis quickly becomes either a time or resource bottleneck as the production rate for finished sequence increases beyond a few kilobases (kb) per day. Automated sequencers have rendered obsolete the traditional analysis toolkits that require an expert user to apply multiple discrete methods and synthesize the results by hand.

The analysis of truly anonymous sequences, the gene content of which is completely unknown, is still relatively rare; the human cosmid sequencing project of McCombie *et al.* (1992) is a good example. In the *Caenorhabditis elegans* genome project (Sulston *et al.*, 1992), and in many human cosmid projects (e.g. Martin-Gallardo *et al.*, 1992; Iris *et al.*, 1993) the gene content of the region being sequenced is incompletely known, with some genes mapped to the region but otherwise uncharacterized, and other genes characterized partially, e.g. by cDNA sequencing. The following questions can be asked about any region of such a sequence:

(1) Is the region transcribed?
(2) Does the region encode protein?
(3) If the region is a transcription unit, what is its structure?
(4) If the region is protein coding, what is the product?
(5) Does the region contain regulatory elements?
(6) Is the region recognizably intergenic?

Anonymous expressed sequence tag (EST) projects face many of the same questions; given that cDNA libraries may contain unspliced messages and sometimes genomic DNA contaminants, they arguably face all of them.

A range of software tools has been developed to address the questions listed above, several of which are reviewed in this volume. This chapter considers the use of sequence analysis tools in a high-throughput setting, and their integration with experimental methods. The focus throughout is on learning as much as possible about a new sequence with minimal effort.

43.2 THE RELEVANCE OF EST PROJECTS TO GENOMIC SEQUENCE ANALYSIS

The most straightforward way to identify new genes is by sequence similarity to a gene already in the database (Seely *et al.*, 1990). A new gene embedded in a region of genomic sequence can be readily identified by similarity searching if it or a close relative is represented in a searchable database by a cDNA sequence. The complete exon–intron structure of the coding region of the putative human *fos*B proto-oncogene, for example, was identified in a 68.5 kb contig from human chromosome 19q13.3 by a 91% sequence identity to the previously sequenced mouse *fosB* gene (Martin-Gallardo *et al.*, 1992). The use of a local-alignment algorithm such as BLAST (Altschul *et al.*, 1990) is advantageous for this type of comparison, since each exon will be matched separately with correct boundaries. Given tools like BLASTX that automatically search six-frame translations of new DNA sequences against amino acid sequence databases (Gish & States, 1993), new genes can be identified as long as a gene with similar translation is present in the database. A novel protein–phosphatase gene was identified in the 19q13.3 contig, for example, by 49% amino acid similarity between its putative protein product and a rat protein-phosphatase gene. The nucleotide sequence similarity in such cases may be weak to undetectable by conventional similarity algorithms.

The presence of large numbers of EST sequences from many different organisms in the public sequence databases will enormously increase the probability that any gene in an anonymous genomic fragment can be identified by a simple similarity search. A number of genes have been identified by the *Caenorhabditis elegans* genome sequencing project (Sulston *et al.*, 1992) based on similarity to or identity with ESTs. Human ESTs (Adams *et al.*, 1991; see also Chapter 10 in this volume) have identified the 3′ end of the human *fosB* gene discussed above, and have allowed almost complete determination of the structure of a third gene in the 68.5 kb 19q13.3 contig, which was known (as 'B') only by a PCR-amplifiable predicted exon at the time of publication (Martin-Gallardo *et al.*, 1992). Human ESTs have also confirmed the presence of predicted genes in both *C. elegans* and yeast (Oliver *et al.*, 1992) by similarity of their putative protein products. While the number of genes that can be identified by either DNA or amino acid sequence similarities to sequences obtained from other phyla may be limited (Green *et al.* 1993), EST projects are now underway for members of several animal and plant phyla. Each new EST project brings the situation envisioned by Seely *et al.* (1990), in which any new gene could be identified by a similarity search, closer to reality.

43.3 PREDICTING GENES IN ANONYMOUS GENOMIC SEQUENCES

The advent of cosmid-scale sequencing led to a burst of development of automated or semiautomated tools for predicting the locations and structures of protein-coding genes in large anonymous sequences. The first of these tools, **gm** (Fields & Soderlund, 1990; Soderlund *et al.*, 1992) uses a combination of local methods such as splice-site detection by consensus matrices (Stormo, 1988) and global methods such as codon usage, and in **gm** Version 2 hexamer content (Bougueleret *et al.*, 1988), to identify coding exons and introns. These are then assembled into complete, frame-consistent gene models using either exact or greedy (best first with no backtracking; see Chapter 34 in this volume) assembly algorithms. Users of **gm** can specify consensus matrices and compositional measures appropriate to any organism, as well as all of the relevant consensus match and other acceptance criteria for models. Partial information about a gene, such as partial cDNA sequence, can also be supplied as input. **gm** produces all models that are consistent with the inputs and criteria used, of which the user can select subpopulations for further analysis.

A number of variations on the theme of combining local and global methods to assemble gene models have now been developed. Hutchinson & Hayden (1992) developed SORFIND, a partial assembler that locates individual open-reading frames flanked by consensus splice sites. The GeneFinder program developed for the *C. elegans* genome project combines partial assembly with automated BLAST searching (P. Green & L. Hillier, unpublished; briefly described in Sulston *et al.*, 1992). The GeneID system of Guigo *et al.* (1992) and the GAP system of Uberbacher *et al.* (1993) are both complete gene assemblers that employ neural networks for both local and global feature recognition and use rule-based assembly algorithms. The GeneParser system of Snyder & Stormo (1993) uses dynamic programming for assembly to guarantee optimality of the models generated. Searles (1992) has developed a very flexible grammar-based system which can be programmed to recognize a variety of complex structures, including both protein-coding and tRNA genes. Finally, Fichant & Burks (1991) have developed an assembly system specifically for identifying tRNA genes, which was used to good effect to find a new human tRNA gene in the 19q13.3 contig described above (Martin-Gallardo *et al.*, 1992).

All gene structure prediction systems employ consensus representations, whether encoded in consensus sequences, matrices, or neural networks, of features such as splice sites, and all employ compositional tests that are based on the compositional properties of known genes from one or more target organisms. They are, therefore, all reasonably accurate when predicting the locations and structures of genes with sequence characteristics similar to those of known genes, and all are less accurate when predicting the locations and structures of genes with unusual site sequences or compositional properties. Experience and the published reports suggest that the likelihood of finding at least one coding exon of any gene using functional site and compositional criteria appropriate for the source organism using any of these systems is quite high: probably on the order of 90%. The probability of correctly predicting the exact structure of a moderately complex (10–20 exon) gene contained in a cosmid-sized or larger sequence is substantially lower. Improving the accuracy of complete gene structure prediction is likely to require a substantial improvement in our understanding of the range of variation of both the sequence requirements of functional sites and the compositional properties of exons, introns, and intergenic DNA.

Given the difficulty of complete gene structure prediction, many groups employ some form of compositional analysis to detect isolated coding regions as the initial step in sequence analysis. Popular compositional tests include codon preference (Staden, 1984), third-base autocorrelation (Fickett, 1982), and hexamer composition (Bougueleret *et al.*, 1988). The GRAIL system of Uberbacher & Mural (1991) combines these and several other methods, using a neural network to weight the outputs of each. GRAIL is available as an E-mail server, and has become the method of choice for quickly analyzing new human sequences to identify coding regions that cannot be detected by similarity searches. GRAIL is currently being extended to analyze sequences from several other organisms (see Chapter 41 in this volume).

43.4 CONFIRMING PREDICTIONS EXPERIMENTALLY

Predicted genes that do not resemble any previously known genes require experimental confirmation. The polymerase chain reaction (PCR) enormously simplifies this process. High-resolution compositional methods such as GRAIL, and all gene structure prediction methods, yield predicted exon sequences from which PCR primers can be designed. Amplification of a fragment of appropriate size from a cDNA library, using primers selected from one or more predicted exons, confirms expression of the predicted gene in the tissue from which the library was made. PCR products amplified from cDNA using primers from different exons can be sequenced to yield the cDNA sequence between the primers, thus confirming the locations of splice sites, and identifying any intervening exons that may have been missed by the predictions. Interexon PCR from several cDNA libraries was used successfully to confirm the expression and structure of three predicted genes in a 58 kb contig from human chromosome 4p16.3 (McCombie *et al.*, 1992), and three predicted genes from contigs on 19q13.3 (Martin-Gallardo *et al.*, 1992). Rapid Amplification of cDNA Ends (RACE–PCR) provides a convenient method for confirming 5′ and 3′ exon ends (Frohman, 1990).

The ease with which multiple PCR amplifications can be performed, and the resulting fragments sequenced, obviates much of the problem of false-positive predictions. All possible intra- and inter-exon PCR amplifications for 13 predicted exons can be run in a single 96-well plate; the inter-exon predictions for 14 exons can be run in such a plate. This is roughly the number of exons that GRAIL predicts for each strand of the chromosome 4 and 19 contigs discussed above. Testing interexon amplifications only for well-separated exons further reduces the number of amplifications that are needed.

43.5 INTEGRATING SEQUENCE ANALYSIS WITH SEQUENCE PRODUCTION

Genomic and EST sequencing projects have thus far approached sequence analysis with quite different strategies. Genomic sequencing projects have standardly accumulated the raw sequence data, assembled it, and filled most gaps prior to significant analysis. Analysis methods are then applied to the final contiguous sequences. Finalizing the sequence prior to analysis is the traditional strategy of small-scale, single-gene sequencing. EST projects, in which each raw sequence is an independent, effectively final result, have adopted a highly automated analysis strategy in which multiple analysis methods are applied to every sequence as soon as it is produced, with the results written directly into a database for later retrieval and comparison (Adams *et al.*, 1991; Kerlavage *et al.*, 1993). Similarity searching against databases of known genes, known and predicted proteins, and interspersed repeats, together with coding-region predictions, low-entropy repeat identification, and compositional analysis are all performed as part of a single, largely automated process. This strategy effectively sorts the sequences into classes –

known genes, new members of known gene families, unknown protein-coding sequences, unknown non-coding sequences, and interspersed repeats – for later comparative analysis and either experimental or computational follow up. Some genomic STS sequencing efforts have adopted a similar approach to sequence analysis (see Chapter 38 in this volume).

The use of automated analysis to sort raw sequences by class prior to assembly and finishing has much to recommend it, especially in the context of shotgun sequencing projects. Sorting the raw sequences would identify regions likely to pose assembly problems, such as those containing overlapping interspersed repeats, and regions of special interest in which accuracy, and hence redundancy, should be increased. The identification of multiple fragments of a single gene by similarity searching can impose a relative order on those fragments prior to assembly. Finally, PCR can be used as outlined above to amplify cDNA fragments spanning predicted exons, which can be entered into the production line for sequencing together with further genomic fragments. Predicted genes can then be confirmed as they are assembled, not afterwards.

Some assemblers already identify interspersed repeats, and remove them prior to the initial assembly (see Chapter 33 in this volume). Linking an assembler that included some database management capabilities to an automated analysis stream such as that described by Kerlavage *et al.* (1993) would be relatively straightforward. As genome sequencing projects move to higher throughput, the incorporation of automated pre-assembly analysis can be expected to render the shotgun phase of sequencing both more efficient and more informative.

43.6 SUMMARY

The rapid progress of EST projects will result, very likely within the next one to two years, in a situation in which essentially all genes in new genomic sequences will be identifiable by simple database searches. The combination of coding region and gene structure prediction with PCR allows the efficient confirmation of the complete exon–intron structures of new genes. By performing much of the computational analysis before assembly, the structures of most genes contained in a sequence can be determined simultaneously with finishing the sequence.

REFERENCES

Adams, M., Kelley, J.M., Gocayne, J.D., Dubnick, M., Polymeropoulos, M.H., Hong Xiao, Merril, C.R., Wu, A., Olde, B., Moreno, R.F., Kerlavage, A.R., McCombie, W.R. & Venter, J.C. (1991) *Science* **252**, 1651–1656.

Altschul, S., Gish, W., Miller, W., Myers, E. & Lipman, D. (1990) *J. Mol. Biol.* **215**, 403–410.

Bougueleret, L., Tekaia, F., Sauvaget, I. & Claverie, J.-M. (1988) *Nucleic Acids Res.* **16**, 1729–1738.

Fichant, G. & Burks, C. (1991) *J. Mol. Biol.* **220**, 659–671.

Fickett, J. (1982) *Nucleic Acids Res.* **10**, 5303–5318.

Fields, C. & Soderlund, C. (1990) *Comput. Appl. Biosci.* **6**, 263–270.

Frohman, M. (1990) In: *PCR Protocols*, M. Innis, D. Gelfland, J. Sninsky & T. White (eds). Academic Press, San Diego, pp. 28–38.

Gish, W. & States, D.J. (1993) *Nature Genet.* **3**, 266–272.

Green, P., Lipman, D., Hillier, L., Waterston, R., States, D. & Claverie, J.-M. (1993) *Science* **259**, 1711–1716.

Guigo, R., Knudsen, S., Drake, N. & Smith, T. (1992) *J. Mol. Biol.* **226**, 141–157.

Hutchinson, G. & Hayden, M. (1992) *Nucl. Acids Res. 20*, 3453–3462.

Iris, F., Bougueleret, L., Prieur, S., Caterina, D., Primas, G., Perrot, V., Jurka, J., Rodriguez-Tome, P., Claverie, J.M., Dausset, J. & Cohen, D. (1993) *Nature Genet.* **3**, 137–145.

Kerlavage, A., Adams, M., Kelley, J., Dubnick, M., Powell, J., Shanmugam, P., Venter, J.C. & Fields, C. (1993) In *Proceedings of the 26th Hawaii International Conference on System Sciences* (T.N. Mudge, V. Milutinovic & L. Hunter, eds.) IEEE Computer Society Press, Los Alamitos, CA, pp. 585–594.

Martin-Gallardo, A., McCombie, W.R., Gocayne, J., FitzGerald, M., Wallace, S., Lee, B.M., Lamerdin, J., Trapp, S., Kelley, J., Liu, L.-I., Dubnick, M., Dow, L., Kerlavage, A., De Jong, P., Carrano, A., Fields, C. & Venter, J.C. (1992) *Nature Genet.* **1**, 34–39.

McCombie, W.R., Martin-Gallardo, A., Gocayne, J., FitzGerald, M., Dubnick, M., Kelley, J., Castilla, L., Liu, L.-I., Wallace, S., Trapp, S., Tagle, D., Whaley, L., Cheng, S., Gusella, J., Frischauf, A.M., Poustka, A., Lehrach, H., Collins, F., Kerlavage, A., Fields, C. & Venter, J.C. (1992) *Nature Genet.* **1**, 348–353.

Oliver, S.G. *et al.* (1992) *Nature* **357**, 38–46.

Searles, D. (1992) In: *Artificial Intelligence and Molecular Biology*, L. Hunter (ed.). AAAI Press, Cambridge, MA, pp. 47–120.

Seely, Jr, O., Feng, D.-F., Smith, D.W., Sulzbach, D. & Doolittle, R.F. (1990) *Genomics* **8**, 71–82.

Snyder, E. & Stormo, G. (1993) *Nucleic Acids Res.* **21**, 607–613.

Soderlund, C., Shanmugam, P., White, O. & Fields, C. (1992) In: *Proceedings of the 25th Hawaii International Conference on System Sciences*, V. Milutinovic & B. Shriver (eds). IEEE Computer Society Press, Los Alamitos, CA, pp. 653–662.

Staden, R. (1984) *Nucleic Acids Res* **12**, 505–519.

Stormo, G.D. (1988) *Annu. Rev. Biophys. Chem.* **17**, 241–263.

Sulston, J., Du, Z., Thomas, K., Wilson, R., Hillier, L., Staden, R., Halloran, N., Green, P., Thierry-Mieg, J., Qiu, L., Dear, S., Coulson, A., Craxton, M., Durbin,

R., Berks, M., Metzstein, M., Hawkins, T., Ainscough, R. & Waterston, R. (1992) *Nature* **356**, 37–41.

Uberbacher, E. & Mural, M. (1991) *Proc. Natl. Acad. Sci. U.S.A.* **88**, 11 261–11 265.

Uberbacher, E., Einstein, J., Guan, X. & Mural, R. (1993) In: *Proceedings of the Second International Conference on Bioinformatics, Supercomputing, and Complex Genome Analysis*, H. Lim, J. Fickett, C. Cantor & R. Robbins (eds). World Scientific, Singapore pp. 465–477.

PART III

Informatics

C. DATA MANAGEMENT AND DATA BASES

CHAPTER FORTY-FOUR

Design Issues in Developing Laboratory Information Management Systems

S. LEWIS
Lawrence Berkeley Laboratory, Berkeley, CA 94720, USA.

44.1 INTRODUCTION

A laboratory information management system (LIMS) provides assistance in tracking, manipulating, and analyzing laboratory data. It is essentially a sophisticated record keeping system built to answer the question: 'Who did what, when, with which materials?' (Fields, 1992). The aim is to simplify laboratory data management and to reduce the amount of technician time required to maintain accurate and logically consistent laboratory data. This core capability is the focus of the following discussion of the issues involved in developing a LIMS for large scale sequencing laboratories. The following topics will be covered.

(1) Does the project require a LIMS—*or* How big is big?
(2) Some fundamental software engineering principles.
(3) Automation.
(4) Putting the pieces together (system architecture).
(5) The computing environment of the LIMS (network, hardware, OS, etc.).
(6) How a computer represents biology (database options and considerations).
(7) Making the information available (development of a user interface).
(8) Capturing laboratory information (data acquisition).
(9) Making use of the information (providing data to other applications).

44.2 DOES THE PROJECT REQUIRE A LIMS – *OR* HOW BIG IS BIG?

The sole compelling reason to invest in the development of a LIMS is the size of the project and projects are definitely becoming larger. Most sequencing projects are moving to become multimegabase-per-year efforts. DOE intends to successfully sequence 225 Mb per year by 1995, and NIH has similar goals. Successful sequencing is defined in terms of (1) rate of production, (2) low cost (less then $0.50 per base), and (3) accuracy. Using current methods it is impossible to produce, record, verify, and retrieve the data with sufficient speed and accuracy. As the scope of sequencing projects increases corresponding gains must be made in sequencing productivity and accuracy.

The greatest opportunities for increasing productivity lie in those tasks having the greatest influence on production rate. A typical example from our own laboratory shows that sequencing a 3 kb fragment requires 8 working hours for molecular manipulation and 16 hours for data entry, assembly and sequence analysis. Clearly, the bottleneck in production is not the biology but the data handling and analysis. Any percentage savings in this area will have a large impact on productivity. The means of achieving this gain are automating and computerizing the information handling.

The critical factors for accuracy are the rate of error generation and the rate at which it is possible to correct errors. A realistic estimate of error rates using manual record keeping is 5%. In tracking 1000 clones this would mean that, on average, 50 would contain some type of error (lost, mislabeled, incorrectly annotated, etc.). The rate at which someone can correct these mistakes depends on a person's innate ability to keep track of things and how much data they must examine to locate the problem. Most people can keep in mind no more than 7 ± 2 pieces of information (Miller, 1956). Given this limitation, it follows that the more data there are the more difficult and time-consuming it becomes to locate a mistake by searching laboratory notebooks.

The above example does not include errors in the sequence itself introduced during data entry, or by mistaken basecalling. At least 75% of sequencing errors are due to human handling of the data (M. Palazollo & G. Rubin, personal communication). Most presently available methods of semiautomated fragment data management and assembly are file-based systems. This organization of the data precludes efficiently handling projects of more than cosmid size – about 40–80 kb or less. Moreover, these methods require considerable user intervention and time to maintain.

The objectives of a LIMS therefore are: (1) to eliminate the data handling and analysis bottleneck, (2) to reduce the introduction of data errors through automated data entry, and (3) to assist in the search for corrections by organizing and maintaining a database.

44.3 SOME FUNDAMENTAL SOFTWARE ENGINEERING PRINCIPLES

Building a LIMS is a software engineering development task requiring new and novel tools, strategies and technologies. There are currently no off-the-shelf systems that directly address the unique requirements of a genomic laboratory information system. Because custom software is an expensive proposition, a leader of such development efforts must ensure that the effort makes productive use of available resources. The following are the key issues of software project management (Brooks, 1975; Boehm, 1985, 1987).

(1) *Keep it simple.* A LIMS development project will only succeed if the scope is realistic given the constraints of time and budget. Avoid 'gold-plating': building software that consumes extra effort but only marginally increases its utility.
(2) *Eliminate rework via rapid prototyping.* Rapid prototyping tightens the communications loop between developers and users and thus reduces the risk of reworking the software (Lowell, 1992). The basis of growth is a continuous interchange between the biologist (describing desired functions) and the developer (who responds by providing a program that is an interpretation of what the biologist requested).
(3) *Hire the best people.* The largest variation in programmer productivity is due to the talents of the individuals. The best individuals outperform the worst by more than 25 to 1.
(4) *Hire biologically knowledgeable people.* An additional productivity multiplier of more than 2 may apply to those people who are familiar with the applications area. Using people who know the field avoids rework due to belated comprehension of the issues.
(5) *Invest in the development environment.* The cost of office space and workstations is more than recaptured by improvements in productivity.
(6) *Make use of commercially and publicly available software tools.* The best tools are those that provide for the most flexible and efficient implementation and ease of long-term maintenance.

The successful 'STS pipeline' constructed in Eric Lander's laboratory at the Whitehead Institute has all the basic elements of a LIMS: A database containing the STS sequences and associated primers; a network to the computers running the DNA sequencers; automatic data acquisition of the sequences; data exchange with an analysis program for selecting primers, and finally a user interface used to confirm selected primers. It provides an essential service in the laboratory and is a good example of the benefits derived from attention to software engineering issues.

44.4 AUTOMATION

The general problem of computer assisted process control is not new. What varies is how well developed it is within any particular domain. The most well established areas of large-scale computing systems are in banking and airline reservation systems. These industries are now entirely dependent upon automation, as anyone who has tried to use them when the system is 'down' knows. It has taken decades to achieve the efficiency and reliability that we take for granted.

It is only within the last two decades that automation has begun to fully establish itself in other industries. Each process being automated must have a system tailored to meet its unique requirements. Vendors in the automation industry supply or exploit toolkits (based upon the general requirements) that facilitate construction of each custom application. The pharmaceutical industry provides the closest analog to sequencing laboratories. New automation systems are being built and used for drug evaluation and quality control because of the same incentives for increased efficiency and accuracy. Enzyme-linked immunosorbent assays (ELISAs) are one such process that is very common throughout the clinical testing area for which automation and LIMS exist. As another example, one company has a system on-line that specifies, controls, and records experiments in enzyme kinetics (D. Balaban, personal communication).

The benefits of this experience do not come from reusing particular pieces of software, but from supplying the appropriate approaches to use in designing a laboratory system. This is because the hard part of building software is the design of the conceptual construct, not the labor of representing it and testing the accuracy of representation (Brooks, 1987). Two useful techniques (Stankovic & Ramamritham, 1988) are:

(1) *design methodologies* such as data flow or state transition diagrams are useful for describing the flow of laboratory operations; and
(2) *concurrent processing architectures* that are used in real-time systems control (acting as task dispatchers, and instrument monitors) are useful because of the inherent parallel nature of the work in a laboratory.

44.5 PUTTING THE PIECES TOGETHER

The most fundamental issue in designing a LIMS is determining what pieces of software are needed (modules) and how they interact. A software module, first, provides a very specific, circumscribed function and, second, it has a clearly defined external interface. For example, each category of instrument in the laboratory requires at least two software modules, one to control the instrument (usually provided by the vendor) and a second to retrieve the output from that instrument. A third module may be required to translate the output and store it in the database. Automating control of the instrument requires yet another module to relay instructions to the instrument. Integration of the modules is dependent upon careful definition of the module's interface specifications. Figure 44.1 is an extremely simplified example of LIMS architecture.

44.6 THE COMPUTING ENVIRONMENT OF THE LIMS

The power and available support for computers using UNIX operating systems indicate their choice as hardware platforms. Ethernet provides a robust and readily available network service. Both UNIX and the network require a degree of administration to configure and manage. Introductory surveys are readily available (Thomas & Yates, 1982; Stevens, 1990).

The importance of network access to a functioning LIMS must be emphasized. Quite apart from the personal convenience network access provides, a LIMS is dependent upon it to transfer data locally and externally.

(1) *Data acquisition* – transferring sequences directly from the sequencers to the LIMS.
(2) *Data exchange* – to programs that compare new sequences with known sequences to find out if anything already exists they resemble.

Another potential consideration for a LIMS is distribution of the data. Collaborators working in distant laboratories may need access to the information. There are a number of potential mechanisms for distributing the data, with greater and lesser degrees of complexity. The list over provides some potential combinations, followed by a summary of the tradeoffs involved:

database server	user interface	data	usage
local	local		remote login
local	remote		client-server
remote	remote		snapshot
		remote	text dump

Figure 44.1 Access methods for different types of databases or data sources.

(1) Remote logins require that guest accounts be set up on on a local server. The drawbacks are the added administration tasks, potential security risks and lost CPU cycles.
(2) Client–server models have the most benefits by providing the most up-to-date data and transparent use. The drawbacks are the time and expertise required for configuration.
(3) Snapshots are easy to implement using existing file transfer protocol (ftp) services; however, the data and programs become stale between transfers, and versions of each must be synchronized.
(4) Text dumps of the data are the lowest common denominator because compatibility between the user's system and the database are not requirements. However, the consequence is the loss of all the database and user interface features. One such service is Gopher developed at the University of Minnesota (releases available via anonymous ftp from boombox.micro.umn.edu in/pub/gopher. Questions may be sent to gopher@boombox.micro.umn.edu.).

44.7 HOW DOES A COMPUTER REPRESENT BIOLOGY?

44.7.1 Options in DBMS

In the past, scientific applications rarely used database management systems (DBMS). Typically, scientists dealt directly with files structured specifically for their applications. More recently, scientists have recognized that the complexity and the amount of data for scientific applications require more robust data management facilities for keeping track of the data being generated. Any database management system seeks to provide the following capabilities.

(1) *Data independence.* The application data structure provided by the database is logically 'independent' of how the database physically stores it. Ideally, applications can reference the data using constructs equivalent to the users conceptual viewpoint.
(2) *Data connectivity.* The database maintains the relationships among the data, the multiple interdependencies that exist.
(3) *Data integrity.* This is an issue for any database coordinating multiple users, redundant data and dependent values. Mechanisms must be in place to ensure that data values are valid, secure and protected from corruption.
(4) *Data sharing.* The aim is to support simultaneous use of information by many users and minimize storage.
(5) *Query flexibility.* Users must be able to browse through the data using any logical qualification.
(6) *Performance.* Application requirements demand that data be located and retrieved with efficiency.

The choice of a DBMS reflects the relative importance the LIMS places on each of the above considerations. There are currently major efforts that use: commercial relational DBMS; object modeling tools built to shield relational DBMS (the Object–Protocol Model for LIMS, V. Markowitz *et al.*, draft document, Lawrence Berkeley Laboratory, Mailstop 50B–3238 Berkeley, CA 94720 VMMarkowitz@lbl.gov); commercial object-oriented DBMS; and domain-specific databases (i.e. ACeDB (R. Durbin and J. Thierry-Mieg (MRC Laboratory for Molecular Biology, Cambridge, UK and CNRS–CRBM Montpellier, France) available via anonymous ftp from ncbi.nlm.nih.gov.)). These different approaches vary greatly in their emphasis.

A *relational DBMS* organizes all data into tables. New tables can be formed by joining elements of existing tables. This is a uniform, clear, and simple organizing principle. However, particularly for scientific data, there is a mismatch between the tables of the relational system and the application specific constructs required (e.g. a consensus DNA sequence). This mismatch is the main cause for the difficulty of developing and maintaining scientific databases using relational systems. The solution has been to transfer the problem of reconstructing the biological object to the application or user end. The application deals with the reconstruction on an ad hoc basis, based on implicit or direct knowledge of the physical organization of the tables. Thus, the use of relational DBMS violates the objective of data independence. It makes development, maintenance and modification tedious, error prone and time consuming. However, relational DBMS have some very attractive features, primarily in the area of data integrity and query language. These commercial systems come with built-in methods for dealing with multiple users, access security, data validation and data integrity mechanisms. For projects that must support a large user group such features are essential.

One alternative approach is the development of *object modeling tools* to insulate the application from the underlying relational tables by using intermediary object structures, that is, extending current data management technology to accommodate new application data structures and operators. The object–protocol model (OPM) supports the specifications of both object structures and processes such as laboratory protocols (the Object–Protocol Model for LIMS, V. Markowitz *et al.*, draft document, Lawrence Berkeley Laboratory, Mailstop 50B-3238 Berkeley, CA 94720 VMMarkowitz@lbl.gov). Given an input and a set of parameters, a protocol instance (procedure) produces an output. For example, a biologist takes a DNA template (the input), prepares it (the parameters), then runs the DNA out on a sequencer (the protocol instance) which results in a sequence (the output). This technology supports nested specification of protocols, protocol generalization, chaining of protocols and optional protocols. It hides the relational DBMS by completely automating the generation of the corresponding relational tables, integrity constraints, and queries. It best suits those laboratories evaluating and generating new strategies by providing high-level tools for defining, querying and modifying protocols.

The commercially available technology of *object-oriented DBMS* is maturing. The important ideas of this approach are encapsulation, information hiding, polymorphism and inheritance. Encapsulation describes the capability of an OODBMS to compose a new object by aggregating a number of separate items. For example, encapsulation allows us to think of a sequence as a single object that incidentally is composed of a series of individual bases. Information hiding shields the user from the internal implementation of objects. An example of this might be in primer selection: internally the object may either store this information or it may calculate it dynamically. The user only knows that the sequence object can provide primers. Polymorphism expresses the idea that various objects respond to identical requests in different ways. This allows generalizing a fundamental idea, like map position, to be expressed in a variety of ways depending upon the type of object being positioned. Inheritance refers to the ability to create new objects that are extensions of existing objects. The reason these ideas are critical is their inherent efficiency for creating and describing biological objects. An OODBMS clearly satisfies the goal of data independence. It also provides data sharing, data connectivity (by means of encapsulation and inheritance), data integrity and performance. It is weakest in query flexibility because no object-oriented query language analogous to SQL currently exists. This requires hand coding all search and retrieval operations from the database. This task is not necessarily overwhelming if there are a limited number of queries, as happens for most browsing style LIMS. Examples of laboratories using OODBMS are the Whitehead Institute (Nat Goodman), Cold Spring Harbor (Tom Marr) and Baylor College of Medicine (Charles Lawrence).

ACeDB is in use on a number of major projects. In some sense it is a throw-back, for it deals directly with files structured specifically for these applications. The modeling technique lies somewhere between hierarchical style (an early database design in which the data are organized as trees) databases and an object-oriented approach. It excels in providing clean data independence and connectivity because the designers built data structures that can closely fit the biology. Because ACeDB caches (moves the information from the disk into main memory and then keeps it there) the data it undoubtedly has the best performance of all the systems discussed. The query language provided with ACeDB is limited, but in practice this is not an issue because biologists seldom directly use the query facility and when they do use it the queries tend to stay in safe territory. The two biggest problems associated with using ACeDB are data sharing and data integrity. Because the ACeDB user interface is not a separate program from the database, individual users all execute their own private copy of ACeDB, including their own database cache. Security on the database disk files for multiple users with write permission is minimal. Another drawback concerns data integrity and validation. Manual checking and validating the data introduced to ACeDB consume approximately 30% of total maintenance time. There are no facilities for checking allowed values on fields, or backing out from an update. Developers are making efforts to correct these weaknesses and the situation will improve in the future. Figure 44.2 gives a summary of the above information.

For a LIMS, data independence and connectivity are very important technical issues to consider when selecting a DBMS. Developers devote a tremendous amount of effort to modeling the biology; therefore, the closer the fit between the data structures ultimately required and the structures provided by the DBMS the easier this task becomes. The time spent in modeling is substantial and unavoidable because the applications will not work without accurate models. The needs of the applications drive the data models created. In turn, the needs of the laboratory drive the applications.

The remaining consideration in selecting a DBMS is support. The available skills of the development team can out-weigh any particular technical advantages. A commercial relational DBMS is extremely complex to administer and effectively requires an additional staff member to maintain it, quite separate from the database developer. An object-oriented DBMS likewise requires expertise on the perspective of object-oriented programming. One attractive feature of ACeDB is its approachability: A competent C programmer can use it with ease.

44.7.2 How does a computer represent biology? A modeling consideration

The essential kernel of a sequencing LIMS is the DNA being mapped and sequenced. These DNA fragments are the primary object of study, thus it is natural to focus the data structures around these elements. A

	Strengths	Weaknesses
relational DBMS	data integrity data sharing query language	data independence performance
object-oriented DBMS	data independence data connectivity	query language
object protocol model (with relational DBMS)	data independence data integrity query language	performance
ACeDB	data independence data connectivity performance	data integrity query language data sharing

Figure 44.2 Strengths and weaknesses of alternative database management system architecture.

LIMS must also track the experimental operations performed on the DNA samples. This distinction between 'what is done' and 'with which materials' is critical. There is, of course, a mutual dependency between the two, but it is best to avoid the temptation to combine the two aspects.

The systems in use at the Livermore and Los Alamos national laboratories exemplify these data modeling problems. These systems were developed under tight budget constraints and in the context of urgent demands by biologists for quick implementations. Because of these constraints both LANL and LLNL chose to encode their protocols directly into the database schema. The consequence is a loss of adaptability.

Some laboratories are seeking strategic improvements in sequencing procedures (L. Hood & T. Hunkapiller, unpublished). These laboratories require the ability to dynamically introduce new protocol definitions without requiring redesign of the data management system. They also need the ability to edit and modify existing protocols. This situation provides the clearest argument for maintaining the separation between the operations being performed and the biological samples themselves (i.e. to avoid reworking the database every time the protocols change). OPM, mentioned earlier, is a direct response to this protocol modeling issue. Implementing this general case is extremely difficult. Still, flexibility is important even for laboratories using a single approach because strategies will sometimes alter to adopt improvements in technique and the design of the LIMS should be equally adaptable.

44.8 MAKING THE INFORMATION AVAILABLE

'We envision information in order to reason about, communicate, document, and preserve that knowledge' (Tufte, 1983). This statement certainly applies in a sequencing laboratory, whose prime focus is amassing genetic knowledge. An essential aspect of any LIMS is providing an effective interface for biologists to use to make the information available to them.

There are two aspects to all user interfaces: how they appear and how they behave. This is known as the 'look and feel' when comparing window management systems. One involves presentation, that is, effectively conveying information to users. The other describes the response, the predictable behavior of the interface. The issues involved in the design of a user interface are summarized below (Keller & Keller, 1992).

The key elements of an effective data display are the following.

(1) *Clarity by adding detail.* To clarify, all pertinent details must be visible. The information is ambiguous if essential information is elsewhere. For example, it is extremely annoying when reading to continuously flip to another page to see a figure referred to in the text.
(2) *Layering and separation.* Visually, as much can be conveyed in the white space of an image as in the graphics themselves, and the implications of this separation can be either informative or misleading. For instance, on a map of the genome the amount of space between two markers may be proportionate to the distance between them, or may simply be arbitrary if only the ordering is known. Developers must be careful in designing the layout to avoid presentations that lead to false assumptions and illusions.
(3) *Comparability.* The presentation must enable the comparison of differences. The key mechanism for doing this is by presenting, within eyespan, multiple views of the same type, each of which displays a variation of the data. For example, when

resolving misidentified bases in a sequence it helps to have all the individual sequencing runs aligned in the display.

(4) *Color.* Color is useful to label, measure and decorate the display. Color-coded DNA sequence displays are good examples of effective color use. Color can be very distracting, however, and may not be distinguishable to those who are color-blind. Therefore, displays should use it quite sparingly. Bright, strong colors are only appropriate when applied to extremes in the data.

(5) *Spatial narratives.* Each axis on the plane of the monitor is useful to represent alternate types of information. An example is base sequences laid out horizontally and individual sequence runs stacked up beneath them vertically.

The important characteristics of user interface behavior are the following:

(1) *Visibility.* An interface must provide visual cues if a person using the interface is to know that a particular graphic can provoke an action. Active text or graphics are useless if they are indistinguishable from their surroundings and thus camouflaged.

(2) *Obvious functional relationships.* Expected behavior must match actual behavior by following the conventions associated with appearance (e.g. buttons are pushed and handles are pulled).

(3) *Feedback to actions.* Responses should be immediate and apparent. Nothing is more frustrating than clicking repeatedly on an item without any noticeable effect, only to be hit later with a snowball of activity. Furthermore the system should be fail-safe, in that the interface should not punish mistakes in usage by anything more onerous than a beep. It should provide as much help as possible to guide users through this exchange, with innocuous consequences for mistakes.

(4) *Incorporation of the user's basic assumptions.* Sharing a common perspective with users makes the interface predictable. Interfaces should, above all, behave in a way that matches the expectations of users.

The interface is a bridge between users and the database. Its role is to translate and, like human translators, its presence should be as transparent as possible. The generic goals above simply describe how an interface can melt into the background. The concrete realization of this invisibility is dependent upon (1) how well the developer grasps the biologist's viewpoint, (2) how easily the data structures can represent this viewpoint in the database, and (3) what the developer can realistically accomplish given the limitations of the tools available. The first two points simply reiterate issues covered in the database section. First, a shared knowledge of biology makes the critical exchange of information between biologist and developer more efficient. Second, the user interface software must internally maintain data in a form that accurately matches this biology. Therefore, the closer the database model fits the model of the user interface the less work there is to do in reformatting the data. A corollary to this is the mutual dependency of these two alternate models; if one changes then the other must change as well. This lock-step can create problems if the correspondence is manually maintained. There are two alternative solutions to this: either have the database and the user interface share identical models (e.g. both in ACeDB form, or both in Smalltalk, or both in C++), or create a secondary database that describes the models in the primary database and use this description to automatically derive the user interface model (using Metadata to automatically generate user interfaces for genomic databases, M. Zorn (Lawrence Berkeley Laboratory, Mailstop 50B–3238 Berkeley, CA 94720, MDZorn@lbl.gov), 1992, preproposal).

The final element of implementation is the tool used to build the user interface. There are too many possible user interface development tools to discuss all of them. Briefly, some representative examples include: Commercial graphics libraries (e.g. Motif (Open Software Foundation, Inc.), custom graphics libraries (e.g. ACeDB), and Smalltalk (LaLonde & Pugh, 1990). Figure 44.3 provides some examples of their appearance.

All these tools have graphic capabilities: they can draw lines, circles, squares, bit maps, and so forth. Concise summaries of current results require graphical displays. All of these tools require that graphics displays be hand-coded. This graphics code is the core of user interface development, providing there is a solution to the modeling issues.

44.9 CAPTURING LABORATORY INFORMATION

Errors occur when technicians enter data manually, therefore, one goal of a LIMS is to electronically transfer the data from various instruments, particularly the automated sequencing machines.

Apple Macintoshes control the ABI sequencing machines. There is software available for Macintoshes that transparently connects the Mac's file system to a UNIX file system (Apple UNIX File Sharing (AUFS)). From the user's perspective, transferring a sequence data file to the UNIX system is as simple as dragging that file into a particular folder.

	Strengths	Weaknesses
Motif	• Can run on any X terminal • Graphical User Interface Builders are available • Attractive	• Large library of widgets to learn
ACeDB	• Can run on any X terminal • Model is identical to database model • A number of genomic displays already exist	• Limited to ACeDB database
Smalltalk	• Can run a variety of platforms • Model can closely match biological objects • Attractive	• Large number of classes to learn

Figure 44.3 Strengths and weaknesses of alternative graphic user interface development tools.

IBM PCs control the Pharmacia sequencing machines. Their method for collecting sequencing runs requires more direct human interaction. Installing network file server (PC-NFS) software and hardware on the PC enables the transfer of the sequence data files to the UNIX system.

Automating the transfer of the sequence into the LIMS database requires developers to write a program that checks the UNIX directory daily for new sequences and then writes them in the database. One function of this program is collecting the description of the sequencing run from the sequence file header. It is obviously undesirable to require technicians to enter the same information twice, both sociologically and technically. Since the sequencer already has this information it is possible to interpret this header and collect it automatically into the LIMS.

Clones are the other major category of data. Bar codes attached to the microtiter plates containing the clones can track clones in a fully automated laboratory. Until full automation closes the loop, however, manual identification must suffice.

At this time there is no available means to accomplish the reverse; that is, to instruct and control a sequencing machine remotely. Doing this would obviate the need for technicians to re-enter information that is already in the LIMS, instead they would simply verify the information's accuracy.

44.10 MAKING USE OF THE INFORMATION – DATA EXCHANGE WITH OTHER APPLICATIONS

A LIMS can also support other applications that make use of the laboratory data it maintains. A summary of some services a LIMS may provide includes:

(1) It is a tool that can monitor the progress of an entire project. For example, a LIMS can determine the status of a large set of samples that are being simultaneously evaluated by a number of technicians.
(2) It contains mechanisms that record the antecedents of each result in the laboratory. A LIMS provides the ability to trace an individual result's history back through the complete series of procedures carried out to produce the result.
(3) A LIMS can couple fully assembled results (e.g. a consensus sequence) to the set of individual experimental results (individual sequencing runs) used to infer that conclusion. Conceivably, a LIMS would permit a user to select any single element within the assembled result and then directly retrieve all the raw data supporting that result, including any subjective decisions made that affected its determination.
(4) A LIMS supplies a method of physically locating each sample and reagent in the laboratory, for example, by means of bar code identifiers on individual microtiter plates that the LIMS can track and verify.
(5) A LIMS assists in process optimization by enabling biologists to assess the impact of variations in the protocols.
(6) A LIMS provides the ability to diagnose problems by following the trail of recorded procedures to detect faulty instruments or contaminated reagents.
(7) A LIMS supports inventory control, by recording the aggregate usage and rate of use of laboratory reagents. Applications can then use this knowledge to automate materials ordering.
(8) LIMS data assists in statistical analysis of erroneous results to determine what systematic types of data errors are being encountered in the laboratory.

Table 44.1 Databases employed by genome project laboratories.

Name	*Location*	*User interface*	*Database*
ACeDB R. Durbin and J. Thierry-Mieg	Medical Research Council Cambridge, UK (and about 60 other sites)	Custom designed X graphics toolkit	Custom C program (b-trees, UNIX)
mapbase N. Goodman S. Lincoln *et al.*	Whitehead Institute M.I.T. Cambridge, MA	Smalltalk	Object-oriented DBMS (ObjectStore, C++)
Genome Topographer T. Marr *et al.*	Cold Spring Harbor	Smalltalk	Gemstone (Smalltalk)
GRM C. Lawrence, W. Rindone S. Honda *et al.*	Baylor College of Medicine Houston, Texas	Smalltalk	Gemstone (Smalltalk)
ESTDB A.R. Kerlavage *et al.*	The Institute for Genomic Research Gaithersburg, MD	SQL	Relational DBMS (SYBASE)
LIMS V. Markowitz, T. Hunkapiller	University of Wash. Berkeley, CA and Seattle, WA	Motif	OPM (SYBASE)
genetics workbench P. Cartwright	Utah	Custom designed X graphics toolkit	–
'lab notebook' or contig browser E. Branscomb *et al.*	Lawrence Livermore National Lab Livermore, CA	Custom designed X graphics toolkit and SQL	Relational DBMS (SYBASE)

LIMS support an automated mechanism to analyze failure rates and types of failures over the long term for both instruments and particular protocols.

(9) A LIMS can supply accurate estimation and monitoring of laboratory costs. One application of a LIMS is to use the information it contains to calculate what it cost and how long it took to obtain each raw individual result. Biologists could use this information to identify time and cost bottlenecks and make appropriate changes to redirect strategies or to undertake new technology development efforts.

(10) A LIMS supports quality control by associating a level of confidence with any particular result, based upon known error rates.

All of these possible uses depend upon the information the LIMS maintains. Each is a development project in its own right that may involve algorithmic design along with the issues of modeling and user interfaces. In relationship to the LIMS the key issue is: In what format is the information transferred?

The most straight-forward approach is to directly use the syntax and model of the LIMS database. Each application requires two translations: (1) to move the data from LIMS database format into application format, and (2) to move the data from application format back into LIMS database format. One drawback of this approach is that it limits the program's broader applicability. The programs may only execute in conjunction with this particular LIMS. If using the programs for other data sets is an issue, then alternative syntaxes that are more general-purpose are possible. Some examples are ASN.1 (Steedman, 1990) (in use at the National Center for Biotechnology Information), IDL (Snodgrass, 1989) (in use at Los Alamos National Laboratory), or LISP (Whitehead Institute).

44.11 CONCLUSIONS

To be successful, large-scale sequencing projects must: (1) complete their goals in a reasonable time period, and (2) reliably preserve the sequence information they have gathered. LIMS software can expedite research and contribute to a successful conclusion by providing a more accurate and less labor-intensive recording and availability of information. Development of such information systems will not only support current endeavors but will enable future large-scale genome projects to be carried out efficiently and economically.

Some of the larger, successful, or progressive efforts are summarized in Table 44.1.

REFERENCES

Boehm, B.W. (1985) *A Spiral Model of Software Development and Enhancement (TRW Technical Report 21–371–85).*

Boehm, B.W. (1987) *Improving Software Productivity.* IEE.

Brooks, Jr, F.P. (1975) *The Mythical Man-month.* Addison-Wesley, New York.

Brooks, Jr, F.P. (1987) *Computer* **Apr.**

Fields, C. (1992).

Keller, P.R. & Keller, M.M. (1992) *Visual Cues.* IEE Computer Society Press.

LaLonde, W.R. & Pugh, J.R. (1990) *Inside Smalltalk.* Prentice Hall, Englewood Cliffs, NJ.

Lowell, J.A. (1992) *Rapid Evolutionary Development (Software Engineering Practice).* Wiley, New York.

Miller, (1956)

Snodgrass, R. (1989) *Interface Description Language.* Computer Science Press.

Stankovic, J.A. & Ramamritham, K. (1988) *Hard Real Time Systems.* IEE Computer Society Press.

Steedman, D. (1990) *Abstract Syntax Notation One. Tutorial and Reference.* Camelot Press, Trowbridge, Wiltshire.

Stevens, W.R. (1990) *UNIX Network Programming.* Prentice Hall, Englewood Cliffs, NJ.

Thomas, R. & Yates, J. (1982) *A User's Guide to the UNIX System.* McGraw Hill, New York.

Tufte, E.R. (1983) *The Visual Display of Quantitative Information.* Graphics Press.

CHAPTER FORTY-FIVE

A Relational Database Primer for Molecular Biologists

A.J. CUTICCHIA
John Hopkins University School of Medicine, Baltimore, MD 21205, USA.

45.1 INTRODUCTION

In its most general terms, a computerized database can be thought of as an electronic filing system. Just as individuals have stored information in record books for centuries, today increasing numbers of researchers utilize computers as tools for the collection and retrieval of information. Strictly speaking, 'data' refers to the observations that are recorded in a database and 'information' refers to the knowledge derived from the data in a particular context. However, the terms will be used interchangeably here. By utilizing the power of the computer, data retrieval times for the answers to questions, *queries*, have been greatly decreased over those that were obtained through manual record keeping.

The telephone book can be used as a starting point in gaining understanding of some basic principles relating to databases. The telephone book is an excellent tool for its primary purpose, retrieving phone numbers based on names and addresses. It is possible to retrieve the phone numbers and the names of an individual's neighbors by scanning through the entire book and picking out those people who live on that same block, but this could become quite tedious. However, if computerized, this list of neighbors could be generated in about the same amount of time as it takes to perform a retrieval of the telephone number for John C. Smith.

The telephone book shows one view of the data which it contains. A criss-cross directory, a telephone book sorted numerically by telephone numbers, is another view of that same data. By placing the information in a computerized database and providing the appropriate query capabilities, one gains flexibility in the types of questions that can be efficiently answered. A database provides freedom in separating the format in which data are entered from the way data are presented. All the data presented in the telephone book can be printed in a sorted list of two dimensions (*rows* and *columns*) also known as a *table*.

	column 1	*column 2*	*column 3*
row 1	Abraham, John	213 Broken Tree	555–3530
row 2	Anderson, Alvin	101 Windy Lane	555–2847
row 3	Anderson, James	3872 Sherwood Marsh	555–9920
	⋮	⋮	⋮
row *n*	Zwill, XY	2390 Ducks Beak	555–8832

A database in which all the information is stored in a single table is commonly referred to as a 'flat-file database'. Though flat-file databases do possess certain qualities, such as the ability to be sorted by a particular column, they have less flexibility than those databases where data can be broken into multiple tables and contain the relationships between those tables. A database conforming to this structure is called a *relational* database.

One example of the utility of a relational database is the transaction that occurs when an automobile is rented. The rental car company stores information on the cars in its fleet (availability, license numbers, mileage, etc.) and information on its customers (credit card number, reservation date, model requested, etc.). When a rental occurs, the information on the customer is joined to the information on the particular car which has been assigned. Every time a car is rented and returned, the information on that car which has changed (e.g. mileage) is updated. The entire record does not have to be re-entered when another customer is assigned that vehicle since the vehicle information is already stored in the database. Similarly, if a customer frequents a certain rental company on a regular basis, that customer's record can be retrieved and linked to the assigned vehicle each time. There is no need to duplicate information in the database. The two tables of this rental car database examples are as follows:

Table 1	*Table 2*
Customer name	Car make
Customer address	Car model
Credit card information	Mileage
Reservation date	License number
Model requested	Customer assignment
Car assignment	

In making the join between tables 1 and 2, it is important that the relationship be unambiguous. The car assignment field can be used to hold the key to join information from table 1 with table 2 and as the license number is unique to each vehicle, that value can be used for the join. In this manner, table 1 is unambiguously joined with table 2.

As it was shown in the telephone book example, the information in each of these tables can be sorted in a variety of ways. It is important that each table have at least one unique identifier so as to avoid ambiguity within the database which could lead to corruption of the data. In the case of table 2, we have shown that the license number is unambiguous and for this table can serve as a unique identifier, also known as a *primary key*. However, in table 1 there is no unique identifier. There is no certainty that two customers will not have the same name (e.g. 'John Smith'), nor is there certainty that a husband and wife using the same credit card could not rent vehicles at different (or the same) times. In these cases, it is customary that an arbitrary identifier, such as CUSTOMER ID, be assigned. These principles will be examined further in a later example.

45.2 DATABASE HARDWARE AND SOFTWARE

As was stated in the last section, a relational database consists of a set of two or more tables which can be joined unambiguously to one another. There exist database management systems (DBMS) for virtually every computer platform in existence. Software ranges in price from less than $100 for some personal computer packages to well into several thousands of dollars for multiuser database systems developed for use on mainframe computers. The addage 'Don't use a sledge hammer to drive in a nail' applies to database hardware and software decisions. Make certain the tools are fit for the job. Just as a major airline would not handle customer reservations using a $1000 personal computer, a laboratory should not feel compelled to maintain its database of clones using a multimillion dollar supercomputer because of the speed advantage of cutting down search time from 1 s to 10 ms.

Several years ago, the driving force for hardware purchases was the software that was to be run. Today, this is less of a factor. Many major DBMS producers support a variety of platforms. Furthermore, the wide range of products available for most systems insures that there should be a product that meets most, if not all, of the database needs of the laboratory using any readily accessible computer. There are, however, some points to keep in mind when setting up a DBMS.

The first point is that there are different flavors of relational database products. Some are form-based, where the user establishes the tables by filling in forms and produces reports by indicating which fields from which tables should be output. Queries are also performed by presenting the user with a form whereby search parameters for particular fields are entered. An example of this type of database retrieval would be the ability to retrieve all data on an individual stored in the phonebook database by filling in any information known about any of the columns in the phonebook table. With some work, it is possible to produce professional quality database applications using this type of product. They are readily available on both the Macintosh and PC platforms. Other products rely on programming based on the structured query language (*SQL*, pronounced 'sequel'). SQL, originally spelled SEQUEL, was developed by D.D. Chamberlin and other researchers at the IBM Research Laboratory in the early 1970s. After several competing products based on SQL were released, the American National Standards Institute (ANSI) defined, in 1986, the first set of standards (ANSI X3.135) to which SQL products must conform in order to be deemed SQL compliant.

Later, the standards were adopted by the International Standards Organization (ISO). A SQL-based DBMS has the advantage of readily allowing the user to produce any desirable query by formulating simple text commands. The disadvantage of a system such as this rests in the fact that it requires the user to have knowledge of the underlying relationships that constitute the particular data model. This is referred to as the *schema.* Some DBMS have combined the flexibility of an SQL database engine along with the ease of use of form-based queries using *client-server* architecture.

In an SQL client-server system, the database itself is able to be queried using SQL. However, software is also available to allow the person(s) with the responsibility for implementing the database to produce forms (front end) which send the appropriate SQL queries (back end) to process and display the results. Thus, the database engine is a SQL server which processes SQL commands and produces the requested output. The client can be as simple as a text editor that allows the user to enter SQL statements, or a very complex graphically based form package that generates the SQL based on the search parameters entered into the forms. Many personal computer-based packages have been produced to allow PCs to access SQL servers directly.

Another factor to take into account when setting up the database is the required amount of fixed-disk storage. Storage requirements should take into account the expected amount of data as well as the space required for the software. Furthermore, in order to increase the efficiency of retrievals, it may be useful to produce *indexes* for certain tables. These indexes can require significant amounts of storage. Also, for some software packages, the expected amount of storage space is partitioned during the database construction and the expansion of this partition can at times be nontrivial. When choosing a fixed-disk storage device, speed should also be a consideration. As the principle of a relational database rests in the ability to collect data from two or more joined tables, this requires a great deal of disk access. The faster the drive, the quicker the results. The use of a disk cache can aid in efficiency as well. A disk cache is a program that stores information obtained from the fixed-disk in memory in order to increase transfers of data from the fixed-disk to the CPU. Thus, when data are read from this disk, the program reads the next few blocks of information into memory so that if that information needs to be processed it can be accessed directly from memory.

A tape archival device is an important component in any database platform. Archives should be made as frequently as is reasonable for the system. Some databases perform a database archive several times within a single day. More commonly, the rule is to perform an entire database archive once a week and to perform daily incremental archives in between. It is a good practice to store tapes for 2 weeks before they are rewritten. Thus, two full back-ups and 2 weeks of incrementals will be available in the event of a system failure. It is also appropriate to store the tapes off-site from the database in an environment in which the tapes are not readily subject to either fire or theft.

45.3 EXAMPLE DATABASE

With some of the fundamentals of relational databases and the issues of hardware and software requirements presented, an example database relevant to a sequencing laboratory can be examined in depth. This example is extremely simplistic but does demonstrate some of the basic concepts in designing and implementing a small database. The first step in database design is to determine what information is to be stored. Though this statement may seem trivial, time spent in developing a stable data model is well rewarded in minimizing the need to make schema changes to a database already containing data. In this example, information on sequences, clone properties, status of GenBank submissions, references associated with GenBank submissions, and the people involved in the project will be stored.

The SEQUENCE table will be defined to include that information relevant to the actual sequence of the DNA molecule. It will include the sequence itself, the length, the date the sequence was last edited, the person who last edited the sequence, and the clone from which the sequence was determined. A primary key for the sequence table will be defined as the seq_ id or sequence identifier. Therefore, every sequence stored in the database will have a primary key associated with it.

Is it necessary to store information on clones within the SEQUENCE table? The answer is 'No'. One of the goals in setting up the database is to limit tables to those fields that describe one aspect of the entire database. It seems logical that data on the clone (such as supplier and vector) should be put in its own table, the CLONE table. Thus, for now, it is sufficient to place a pointer to this CLONE table, which will be defined shortly, in the SEQUENCE table. This pointer will refer to the primary key for each clone.

The data relevant to the individual who last edited the sequence can also be placed in an additional table. Indeed, it would be possible to place data such as the last name of the individual directly in the SEQUENCE table. By using a pointer to another table, PEOPLE, which will hold information such as full name and address, it will be possible to gain greater flexibility in queries, as will be shown. Therefore, a pointer to the PEOPLE table will be included in the SEQUENCE table. In determining this pointer, some care should be taken in ensuring that it be unique. The pointer could be last name, but as it is possible that two people with the same last name might edit this database over time, the primary key for the PEOPLE table will be used as the pointer. Another advantage in putting information on the individual who last edited the sequence in a different table from the sequence information itself is that it adds efficiency by allowing a central table for PEOPLE to be accessed for any other table which might be joined to information on individuals (such as the supplier of a clone). Thus, if an individual supplies clones and edits sequence there need only be a single record for that individual. If information on that individual changes, it will be necessary to update the data in a single table and have the changes reflected when information on each clone supplier or the last person to have edited a particular sequence is retrieved.

As sequences are collected, it may become useful to submit them to GenBank. Information related to a GenBank submission (accession number, date of submission, whether the data are being kept private, etc.) could be included in the SEQUENCE table. However, it is possible that the sequences from several overlapping clones may be assembled into a single GenBank submission. Therefore, seq_id will be used as a pointer to the GENBANK table.

So far, it has been decided that sequence data will be related to tables containing data on clones, individuals, and the GenBank submission. The CLONE table will contain a primary key, clone_id and pointer to the PEOPLE table for further information on the clone supplier. Basic information on the clone such as 5′ and 3′ restriction sites, the polylinker site, the name of the library with which the clone is associated, and any general comments will also be included in the CLONE table.

The PEOPLE table will include a primary key, pers_id, along with name, address, and telephone information. The GENBANK table will include the gb_id as its primary key and will include a pointer to the SEQUENCE table, seq_id (in this case the same field name as in the SEQUENCE table is used, though any name could be chosen). It will also include a pointer to the PEOPLE table to designate the individual who submitted the sequence to GENBANK (in this case sub_id is the *foreign key* and will point to the pers_id field in the PEOPLE table). The GenBank accession number will also be stored. Reference information could be placed in the GENBANK table; however, in some cases more than one reference may be associated with a particular sequence or more than one sequence may be associated with a particular reference. Therefore, a fifth table, REFERENCE, will be created with ref_id as the primary key and points to the field also named ref_id in the GENBANK table. As fields are designated by both table and field names, it is possible to have fields with the same name in different tables each of which represent different data items.

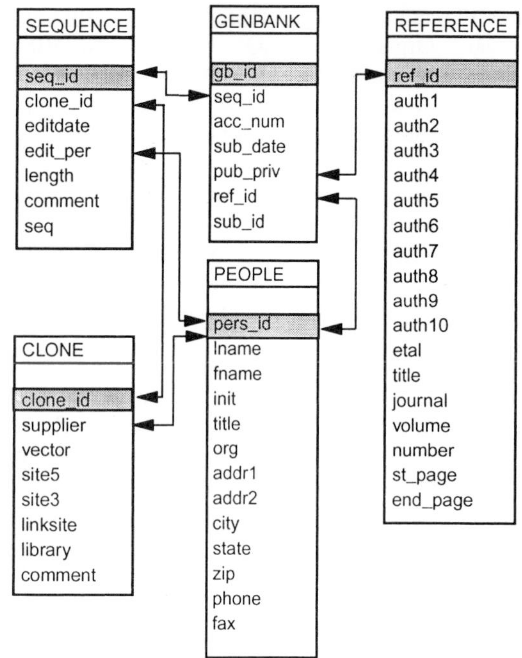

Figure 45.1 Database schema.

The REFERENCE table, in addition to its primary key, will include names for 10 authors and a designator, etal, if more than 10 authors exist for a particular paper. Journal, volume, number and pages are also included in this table. Figure 45.1 shows the tables with the associated fields and their relationships. This type of diagram, which depicts the underlying data model of the database, is commonly referred to as an *entity-relationship (ER) diagram.* The primary key for each table is shaded. Along with an ER diagram, it is useful to have a listing of the tables (also known as *entities*) and fields (also known as *attributes*) associated with each table along with a short description. This listing is referred to as a *data dictionary.* Data dictionaries generally include information on datatypes for each field as well as any rules for allowable values. An abbreviated data dictionary for this example is shown in Fig. 45.2.

Database designers often find it useful to designate the type of relationship between two particular tables. The types are: ONE-TO-ONE, ONE-TO-MANY, MANY-TO-ONE, and MANY-TO-MANY. In the

Table	Field	Description
SEQUENCE	seq_id	sequence identifier
	clone_id	clone identifier
	editdate	date of last sequence edit
	edit_per	person identifier of last editor
	length	length of sequence in bases
	comment	comments on sequence
	sequence	actual sequence
GENBANK	seq_id	sequence identifier
	acc_num	GenBank accession number
	sub_date	date of submission
	pub_priv	whether data is public or private
	ref_id	reference identifier of associated reference
	sub_id	person identifier of submitter
CLONE	clone_id	clone identifier
	supplier	person identifier of supplier of clone
	vector	vector used in library construction
	site5	5' restriction site
	site3	3' restriction site
	linksite	polylinker site
	library	library identifier
	comment	comments on clone
PEOPLE	pers_id	person identifier
	lname	last name
	fname	first name
	init	middle initial
	title	title within organization
	org	organizational affiliation
	addr1	address line 1
	addr2	address line 2
	city	city
	state	state
	zip	zip code
	phone	phone number
	fax	fax number
REFERENCE	ref_id	reference identifier
	auth1	first author
	...	...
	auth10	tenth author
	etal	indicates more than 10 authors
	title	title of article
	journal	journal
	volume	volume
	number	number
	st_page	starting page
	end_page	ending page

Figure 45.2 An abbreviated data dictionary.

case of our example, we can determine the type of relationship between SEQUENCE and GENBANK by asking the following types of questions:

(1) Can more than one sequence in the SEQUENCE table have the same GENBANK accession number? (Yes, it was previously stated that sequences may be concatenated to constitute a single GenBank submission.)
(2) Can more than one accession number in the GENBANK table be associated with a particular sequence? (It will be assumed that no redundant sequences will be submitted, so the answer is 'No'.)

Therefore, the relationship between the SEQUENCE and GENBANK tables is MANY-TO-ONE. By applying similar questions for the other relationships between tables, the following table can be derived.

Table A	*Table B*	*Relationship*
SEQUENCE	GENBANK	MANY-TO-ONE
SEQUENCE	CLONE	ONE-TO-ONE
SEQUENCE	PEOPLE	MANY-TO-ONE
GENBANK	PEOPLE	MANY-TO-ONE
GENBANK	REFERENCE	MANY-TO-MANY
CLONE	PEOPLE	MANY-TO-ONE

Once the database has been designed, it is then possible to create the tables using the database software. In this example, SQL statements are given for generality. There does exist a class of software, computer aided software engineering (CASE), tools that can actually create the database (i.e. draft the create table statements in the appropriate syntax) using information from ER diagrams along with datatype declarations.

In this example, the data types necessary for the database are integers and characters. Though not part of the ANSI standard, a date data type will also be used in this example since that datatype is widely supported in most DBMS. Another datatype, *text*, is commonly used for those fields that can vary in length and are of sizes greater than can be stored as character datatypes (usually 255 characters). In this example, comment fields and the actual nucleotide sequence would require the use of the text datatype. The advantages to using character over text is a utility issue. As a general rule, text datatypes cannot be used as retrieval fields.

To create the SEQUENCE table, the user would execute an SQL statement such as the following:

```
CREATE TABLE Sequence
    (seq_id      integer,
    clone_id     integer,
    editdate     date,
    edit_per     integer,
    length       integer,
    comment      text,
    seq          text);
```

Note that the CREATE TABLE command is followed by the table name and a listing of fields along with their associated datatypes within parentheses. The end of a SQL command is indicated by the use of the semicolon.

As it was stated earlier, indexes can be created to increase performance during retrievals. There are some trade-offs for this increase in retrieval speed. First, although an index is invisible to the database user, it can occupy significant amounts of storage space. Moreover, additions of a record to an indexed table can take longer than additions to nonindexed tables. A general guideline is that, if retrievals occupy a larger proportion of the database use than updates, indexing those tables that are frequently searched will aid the overall efficiency of the system. As an example, if the database contained 10 000 records in the sequence table and the user wished to access the record with seq_id = 7900, the program would normally check each of the records in the database and return those with seq_id = 7900 (since seq_id is unique, it would of course return a single record). However, if the table is indexed on seq_id, the program would jump directly to the record where seq_id = 7900. To create an index for the SEQUENCE table based on the seq_id the command is the following:

```
CREATE UNIQUE INDEX Seqind ON
Sequence (seq_id);
```

The CREATE UNIQUE INDEX command is followed by the name of the index, the word ON, and the field for which the index is based (in parentheses). As stated above, the index is now invisible to the user and will be used when the DBMS processes a query for which the index would increase performance. The use of the UNIQUE in the command ensures that no two records may have the same seq_id. This is an indirect way of forcing seq_id to be a unique primary key. The reason for naming the index is so that it may be dropped in the future if necessary. Dropping an index might be necessary if a large number of records needed to be added at a single time. To drop the index the following command would be typed:

```
DROP INDEX Seqind;
```

The issuance of this command removes the index but does not in any way affect any of the fields in the associated table.

It is also possible to alter a table once it has been created using most database software. Though modification to tables should be avoided if at all possible, it is not uncommon to add a field to a table at a later time. If it was desired to add a field to PEOPLE for a prefix to the name and it was decided that this prefix would have the default value of 'Dr.' the command would take the form of:

```
MODIFY TABLE People ADD
    (prefix    char(5) DEFAULT = "Dr.");
```

In this case, "Dr." would automatically be placed in the prefix whenever a new record for the PEOPLE table is created. The char(5) datatype indicates that data will be five characters or less in length.

Previously, it was shown how a unique index is one indirect way of enforcing the use of unique primary keys. In the creation of the CLONE table, the field clone_id could be directly designated the primary key in the following way:

```
CREATE TABLE Clone
 (clone_id   integer NOT NULL PRIMARY KEY,
  supplier   integer,
  vector     char(25),
  site5      integer,
  site3      integer,
  linksite   integer,
  library    char(25),
  comment    text);
```

After the database has been created, many forms-based packages will allow the user to generate a data entry form and fill out that form as a way to input data into the database. These packages insulate the user from SQL insert statements. As an example, the statement to create a record for Jim Smith in the PEOPLE table would be as follows:

```
INSERT INTO People
VALUES (1001,'Smith','Jim',,'Researcher',
'Jim Smith Laboratory','1003 Willowood',,
'Ellicott City','MD','21042','410–555–6490',
'410–555–9705');
```

Above, it is shown that the data for the fields in the PEOPLE table are placed in the same order as the fields are listed in the table. In the case of init and add2 where no data are to be entered, only a comma indicating the need to pass to the next field is placed in the statement. If a record is to be added with only a few fields entered, it is possible to designate which fields are to be entered in the following manner:

```
INSERT INTO People (pers_id,lname,fname,phone)
  VALUES (1002,'Wilson','Woody','706–555–3465');
```

It is important that the primary key for this table be unique or the update will not occur (e.g. if the field was designated PRIMARY UNIQUE in the creation statement). Most databases can create sequential identifiers for the user automatically by performing a count on the number of records in a table and incrementing that value by one and placing it in the record automatically.

Once the database has been produced and data have been entered (either through SQL, other text commands, or through a form) the data can be queried. Once again, many packages allow the user to generate blank forms and input search parameters for retrieval. Just as in the case of data entry, these forms insulate the user from direct queries to the database relying on knowledge of the schema. However, in some cases it may be necessary to query the database for the answer to a question that cannot be answered using the previously designed forms. A form and its resulting report are nothing more than a previously written query that is lacking search parameters. Sometimes these queries are referred to as *canned queries* to indicate that they have been previously developed by the database programmer during system implementation. Knowledge of SQL and the data model for a particular database will give the user the ability to ask any possible question regarding the data if the database can be directly queried using SQL statements.

In the first example, the code to list the vector, supplier identifier, and library for clone with clone_id = 2100 is as follows:

```
SELECT vector, supplier, library FROM Clone
    WHERE clone_id = 2100;
```

In the previous command, the desired fields are selected from the CLONE table given the constraint of clone_id = 2100. The query could also be written as:

```
Select Clone.vector, Clone.supplier, Clone.library
   FROM Clone
   WHERE Clone.clone_id = 2100;
```

In this case, the fields have been prefixed by the table in which they are found. When performing queries to a single table this redundancy is usually omitted. However, it is necessary to specify *TABLE.field* when queries to two or more tables are performed.

If more information on the supplier, such as name and phone number, is desired, the CLONE and PEOPLE tables could be joined by the following query:

```
SELECT Clone.vector, Clone.supplier, Clone.library,
People.lname, People.fname, People.init
   FROM Clone, People
   WHERE People.pers_id = Clone.supplier
   AND Clone.clone_id = 2100;
```

The query is representative of the construction of a table join. By adding the restriction of WHERE pers_id = supplier, the program returns information from the PEOPLE table that refers to the supplier whose id is associated with the record for **clone_id** = 2100. To show the last names of all suppliers for all clones in the database the following statement would be entered:

```
SELECT Clone.clone_id, People.Iname
    FROM Clone, People
    WHERE Clone.supplier = People.pers_id;
```

The above query is an example of joining tables based on referential integrity.

Thus, by using a simple example, the fundamentals of database modeling, implementation, and query have been presented. Databases are becoming more prevalent in research laboratories each year. There are increasing numbers of database products tailored for the particular data management needs and level of expertise required for programming. By taking the time to develop a database management system for the research laboratory, increased efficiency in daily operations that rely on information retrieval can be achieved. Databases tailored for the non-computer scientist that can run on personal computers are now common. There will, in all probability, always be large databases such as GenBank, PIR, and GDB which serve the needs of the scientific community in general. However, the skill and manpower needed to operate an international database should not be indicative of the efforts necessary to automate the database management needs of an individual research laboratory.

REFERENCES

These selections were chosen for their presentation of generic issues relating to database design, development and query. Once a database product has been purchased, the user should seek out those titles available for their specific software package. Also, many software vendors offer courses specific to their product and tailor classes to the individual needs and level of expertise of the students.

Byers, R.A. (1986) *Everyman's Database Primer Featuring dBase II Plus.* Ashton-Tate Publishing Group, Torrance, CA.

Chorafas, D.N. (1989) *Handbook of Database Management and Distributed Relational Databases.* TAB Books, Blue Ridge Summitt, PA.

Date, C.J. (1986) *An Introduction to Database Systems*, Vol. 1, 4th edn. Addison-Wesley, Reading, MA.

Gruber, M. (1990) *Understanding SQL.* Sybex Inc., San Francisco, CA.

Martin, J. (1981) *An End-User's Guide to Data Base.* Prentice-Hall, Englewood Cliffs, NJ.

Ullman, J.F. (1982) *Principles of Database Systems,* 2nd edn. Computer Science Press, Rockville, MD.

Vetter, M. & Maddison, R.N. (1981) *Database Design Methodology.* Prentice-Hall, Englewood Cliffs, NJ.

CHAPTER FORTY-SIX

ACEDB: A Tool for Biological Information

J.M. CHERRY[1] & S.W. CARTINHOUR[2]

[1] Department of Genetics, Stanford University, Palo Alto, CA, USA.

[2] National Agricultural Library, USDA, Beltsville, MD, USA.

46.1 INTRODUCTION

Traditionally biological data have been managed using a variety of computer software. Much of the data produced by a small research project can easily be stored in word processor and spreadsheet documents as well as several types of database applications. For larger scale projects, particularly those requiring the integration of diverse kinds of data, this solution quickly becomes inadequate. In these cases project coordinators have developed specialized databases to analyze and store information needed for their specific situations, often utilizing one of the commercially available relational database management systems.

The development of a specialized database involves considerable resources and expertise. An important question that might be asked by the director of a *new* project is whether an existing database can be transferred to a different site and quickly reconfigured for new (but similar) purposes, thus saving the time and expense of development. Unfortunately, ease of reuse – in the sense that a biologist without the assistance of specialized computer personnel can attempt the task – has not been a general property of most complex databases. This limitation brings with it two broad consequences. First, many databases are constructed *de novo*, without significant reference to existing databases; essentially, an expensive wheel is reinvented over and over again. Second, and more importantly, biologists yield the direct responsibility for database configuration to computer experts, even though biologists are both the producers and consumers of the data in the database.

Recently a database tool designed specifically for use by biologists has been developed as part of the *Caenorhabditis elegans* genome effort. This database software is known as ACEDB and is the creation of Richard Durbin (MRC–LMB, Cambridge, UK) and Jean Thierry-Mieg (CNRS, Montpellier, France) (Durbin & Thierry-Mieg, 1992). The name ACEDB, A *C. elegans* Data Base, is also the designation of the nematode genome database (which uses the ACEDB software) maintained by Durbin & Thierry-Mieg. ACEDB has many outstanding features which we will review below, but two of them need to be stated briefly at the outset. First, ACEDB is a generalized genome database. It can be used to create new databases without the need for any reprogramming or in fact any sophisticated computer skills. Second, ACEDB permits biologists to describe and organize their data in a manner that closely resembles how they typically think about their information. The result is that ACEDB offers both ease of access for reconfiguration,

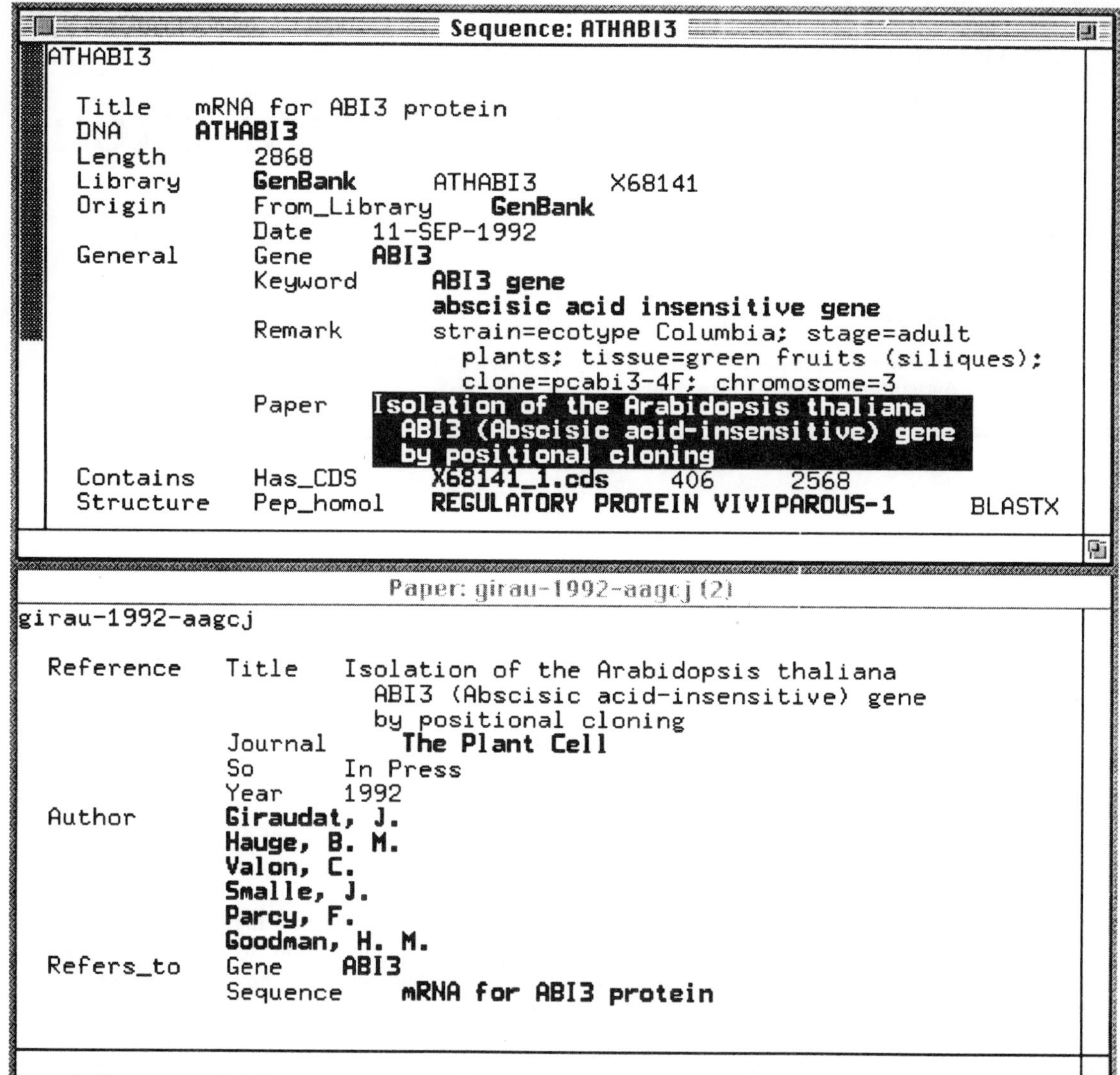

Figure 46.1 Example text displays from the *Arabidopsis* database presented by the ACEDB software. The upper display contains the sequence annotations for a sequence obtained from GenBank. The lower window is the reference information for the sequence. Text in bold typeface indicates a cross-reference to other information.

and ease of reuse for transfer from one project to another. We will return to these points later after introducing the methods used to interact with the ACEDB software.

46.2 ACEDB DISPLAY OF THE INFORMATION

Perhaps one of the most critical features of any database system is the effectiveness of its user interface, i.e. whether or not users can reliably retrieve useful information from it, whether they feel intimidated by it, and indeed whether or not they enjoy using it. Although the ultimate database does not yet exist, ACEDB is becoming widely accepted as providing one of the best examples of an effective and helpful interface. This interface is part of the ACEDB software and requires no programming in order to become active. Below we introduce the major ACEDB display types and some of their properties. This will make clearer the approach ACEDB takes to data display although we must omit many details for the sake of brevity.

ACEDB employs several features that allow the user easy access to the variety of types of information in a genome database. Information is presented using

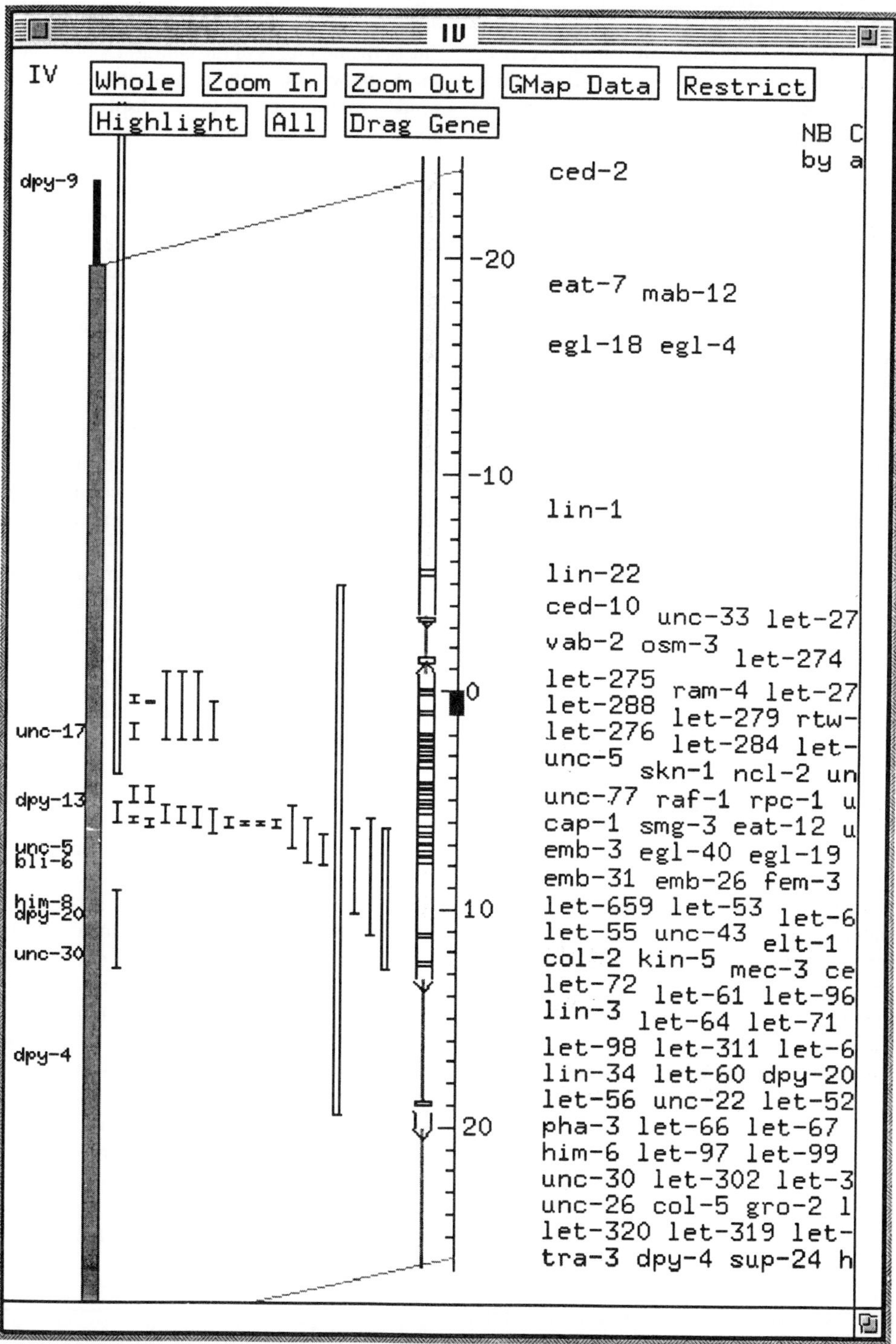

Figure 46.2 Genetic map display from ACEDB presenting a portion of *C. elegans* chromosome IV. On the far left familiar marker genes are listed to provide landmarks. Farther to the right deficiencies and duplications are indicated representing the region involved. Continuing to the right is a cartoon of the chromosome. The wide regions represent areas covered by the overlapping cosmid and YAC physical map. To the right of the genetic distance scale the genetic loci are placed, in this particular view the number of loci is very large and ACEDB automatically spreads out the loci names so that none is occluded. The zoom-in button at the top of this window allows the user to examine a smaller magnified region of the chromosome.

both text and graphic display windows and is navigated through the use of menus and a mouse. Each window used by the software to display information can be used to retrieve related information. ACEDB allows all parts of the database to be cross-referenced with each other thus providing a dense navigable network in which to locate information. Therefore, there is no one path to information in a database constructed with ACEDB.

Since the cross-references can be explored via a mouse click, database *browsing* becomes a simple matter. Figure 46.1 shows two windows containing text displays presented by the ACEDB software. The text items in boldface type indicate a cross-reference to other information contained in the database. The cross-referenced information can be retrieved by clicking with the mouse on the boldface text. In Figure 46.1, a reference in a window containing information about a bibliographic citation was followed to locate and display the appropriate sequence annotation. In this example the user is also one click away from the DNA sequence, peptide homology, and genetic map information.

One of the special features of the ACEDB software is the ability to create a pictorial representation in one of several specialized graphic displays using information contained within the database. Currently available graphic displays include a genetic map, physical map, sequence display, gridded clone display and simulated agarose gel display.

The genetic map display provides a graphical representation of a variety of genetically defined sites such as genetically defined map locations of mutations or molecular genetic markers. The genetic map can be associated with regions covered by mapped deficiencies and duplications, and the contact points between the physical and genetic maps (Fig. 46.2). A locus can be included on more than one genetic map, each genetic map representing a collection of information. The primary results (an estimate of recombination distance) of two-point or three-point recombination experiments can also be presented in a manner that allows each experimental result to be visually compared with the location of the involved markers in any defined genetic map.

The physical map display provides a view of the continuous overlapping set of cosmid and yeast artificial chromosome (YAC) clones referred to as a 'contig'. The associations of these clones with DNA sequences and genetic markers are also displayed (Fig. 46.3). The contig display provides, as do all the graphic displays, a scrolling device which allows easy movement along large contigs. A small summary view is presented in the lower part of the display. The pictorial objects presented in this graphic display can be selected to navigate to other displays. For example, each small box located in the center of the physical map display represents a DNA sequence associated with one of the members of the contig. Double clicking on a small box will cause the appropriate sequence display window to be presented. A text display with information about each of the cosmids or YACs is available from selecting a clone, represented by a line segment. Also, the genetic map display could be reached by selecting the central strip that divides the window in half or one of the loci names just below the central strip.

The sequence display provides a view of the standard sequence features used by the DNA sequence databases such as coding regions, regions of similarity to DNA or peptide sequences, repeat units, promoter and binding sites (Fig. 46.4). Selecting one of the sequence features causes the corresponding region of the sequence to be highlighted with color. Several analysis features are available from the sequence display, including the ability to produce a restriction map, show predicted fingerprint bands and find a specific site in a DNA or protein sequence. ACEDB also produces a codon usage table from all or a subset of sequences from its database and can generate a table of splice-site consensus sequences from its database entries that have the exon–intron boundaries annotated. Beyond those analysis functions the GeneFinder software from Phil Green (Washington University, St Louis) has been incorporated into the ACEDB software and provides a number of predictive analyses useful in the determination of coding regions. The GeneFinder results are displayed graphically in the sequence display window (Fig. 46.5). These include splice-site potentials, codon region potentials, terminator locations, open reading frames and start-site potentials. The GeneFinder functions built into ACEDB will also automatically select possible coding regions and display the resulting exons–intron structure prediction.

ACEDB also provides the ability to trigger the action of an external program and to have this event associated with a specific object. In the *Arabidopsis* genome database this feature is used to present scanned images of autoradiograms. The ACEDB software activates an external graphic image display program to actually display the image. This feature can be used to execute specialized analysis programs or to display information not available to ACEDB.

46.3 DATABASE DESIGN

The process of creating a new database using the ACEDB software involves constructing a 'model' of the information to be represented. Initially the models that are included with one of the released databases based on ACEDB provide a wealth of models for

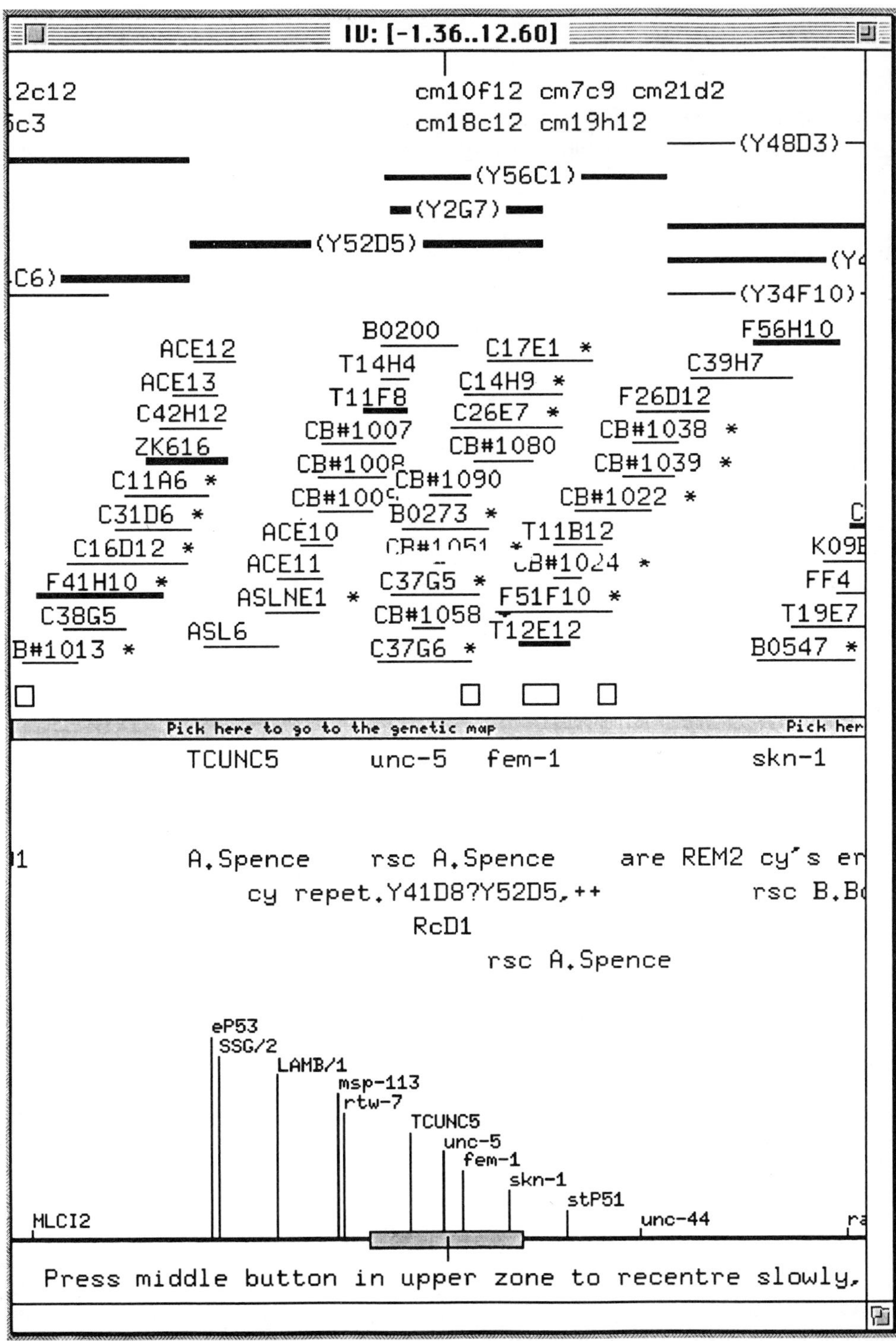

Figure 46.3 Physical map display from ACEDB presenting part of the overlapping cosmid and YAC contig on *C. elegans* chromosome IV. This single window is divided into two regions by the centrally located grey horizontal bar. The physical map display is closely associated with the genetic map. The lower half of the window contains genetically defined loci, comments and remarks, and an overview of the current region on the genetic map. In the upper half of this window the YAC clones are represented with a line segment illustrating the extent of their overlap with other members of the contig. The clones represented by a bold line indicate that these cosmids and YACs are present on one of the gridded clone filter panels included in the database.

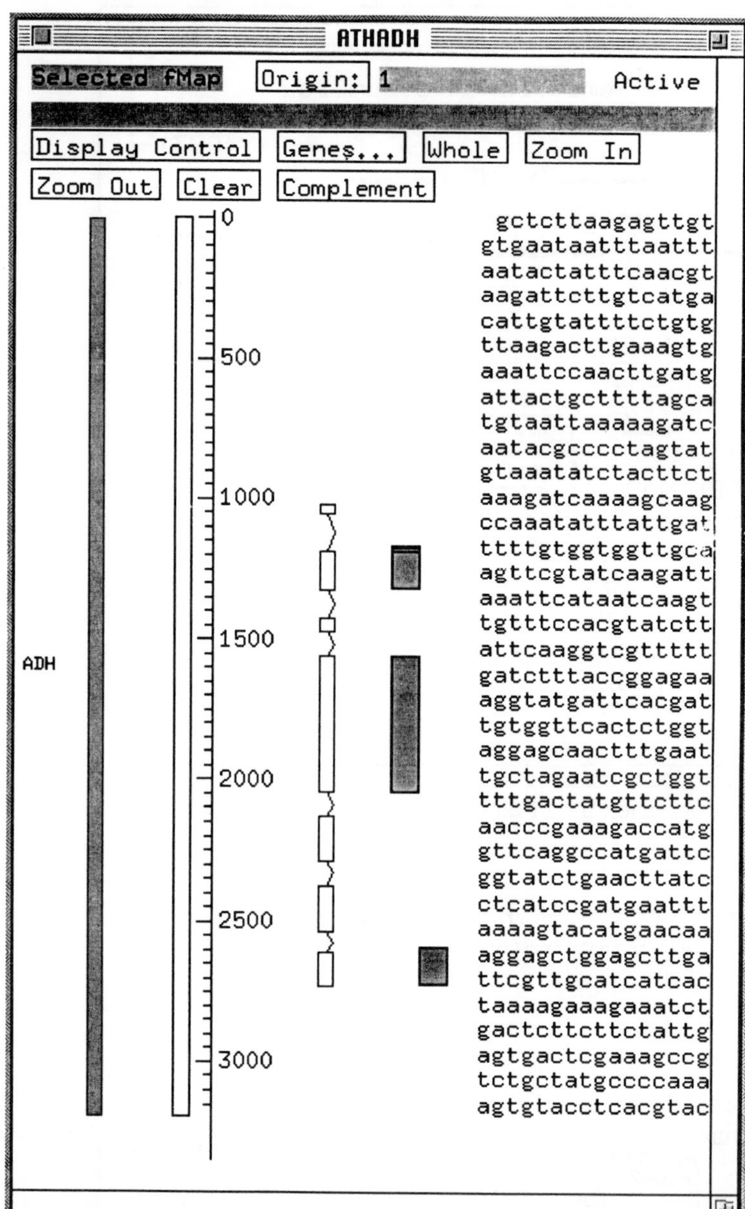

Figure 46.4 Sequence display from the *Arabidopsis* database using the ACEDB software. The exon–intron structure is illustrated in the center of this window. To the right of the coding region cartoon are shaded boxes which represent regions of peptide similarity. These peptide similarity boxes indicate the portion of the sequence that was found similar via a database searching program, such as BLASTX which is used for the *Arabidopsis* sequence. The similarity boxes are located in columns representing the reading frame in which they were found. The DNA sequence and nucleotide number scale are also shown.

immediate use. Later, however, the new database manager will probably elect to modify the models in one or more ways to accommodate specialized data. The following discussion presents what is involved in creating an ACEDB model.

Information in an ACEDB database is stored in classes. A class represents a compartment that defines a collection of information on a common topic. Most ACEDB databases contain a variety of general classes such as 'sequence', 'locus', 'clone', 'chromosome', and 'paper'. These can be modified as required; in addition it is possible to create entirely novel classes tailored for new kinds of data. The basic metaphor is that a compartment (class) can have slots (data entry points) within it that are able to contain data. The task of configuration is to define what those slots contain and how they are connected to other compartments in the database. Although, in principle, it would be possible to configure a database with a single class, in practice this would create a database that would have a dearth of associations and be clumsy to use. It is recommended to use many small classes instead because a modular design allows information to be easily shared through the database.

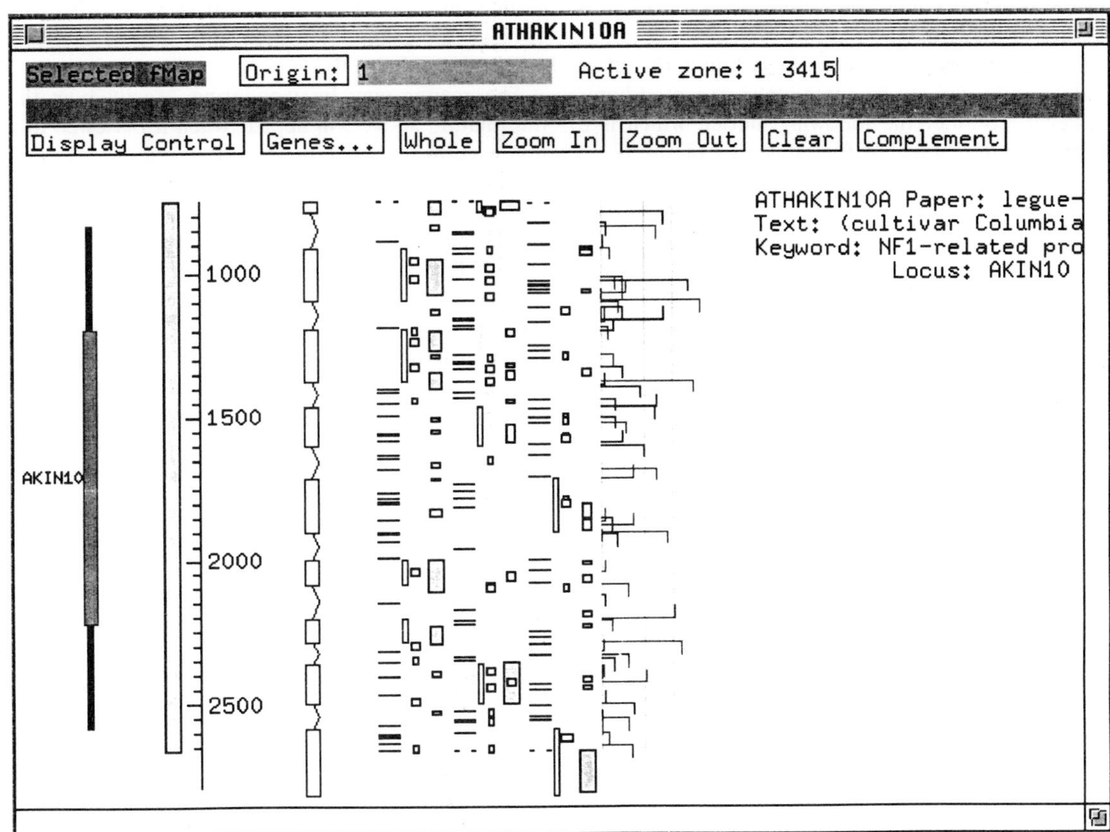

Figure 46.5 GeneFinder results presented in a 'sequence' display. To the right of the nucleotide scale and the exon–intron structure are three sets of columns representing the location of terminator codons, open reading frames, start site potentials and coding sequence potentials calculated by the GeneFinder algorithms incorporated within the ACEDB software. On the right-hand side the splice-site potentials are shown with a longer line segment indicating a higher potential. Donor sites potentials are illustrated by a down-turned flag and acceptor site potentials have an up-turned flag.

The description of a class (known as the model) sets forth the types of information that are appropriate for a given topic. A class can contain unique pieces of information or 'repeating' pieces of information, and either kind can cross-reference information contained within another class. For example, the 'paper' class definition might specify that a particular paper could have a single title, one or more authors, one or more keywords, and have associated with it one or more genes, clones, strains, and so forth. Any of these items could be defined as a cross-reference to another class. As mentioned earlier (Fig. 46.1), it is the cross-references that make it possible to click on one piece of information and move to another related item. The items that cross-reference should be chosen thoughtfully since users will expect them to make sense. It is also important to realize that the description of a class determines, in part, how information in that class will be presented in a text window by ACEDB. This fact may also influence the choices made.

Once a class is defined, data can be entered into it to create data *objects*. To extend the metaphor, a class is just a template which functions to give structure to real data. In ACEDB very little data is required to create an object and, in fact, the only requirement is to supply a unique name. Filling in the other slots is optional although normally an 'empty' object, one with just a name, has little utility. Returning for a moment to the Paper class example, it is sufficient to the ACEDB software to identify a paper object with a name like 'xyz123' without also stating the title or any authors.

When slots are filled with data within a class, the information is placed into a tree of information as defined in the model. The tree structure that describes a class can be built using just a few components and rules. First, each slot is identified by a label which is unique within the class. These labels are called *tags*. Second, the slots themselves come in two types: either they are cross-references to other classes or are simple data items which can be integers, text strings, or decimal numbers. Finally, each tag can be used to label a series of data items or cross-references or a mixture of the two.

ACEDB allows a richness in defining the class models which will not be completely presented here. However, consider the following model, which is a simple example of a class for cataloging plasmids.

?Plasmid	Location	Freezer	Text
		Shelf_box	Int Text Text
	Sequence	?Sequence	
	Reference	?Paper	

In this example, the 'plasmid' class contains the information about the specific freezer location of the plasmid sample and cross-references for two other classes ('sequence' and 'paper'). The anatomy of this class illustrates the points made above. Tags (location, freezer, shelf_box, sequence, and reference) label the slots into which data can be entered. Some tags (for example, freezer) are followed by single data entry points, in this case identified by the term 'text'. Text indicates the kind of data allowed into this slot (123, hello, and 'Hi there!' are all legal 'text' entries). Other tags (for example shelf_box) are followed by multiple slots, in this case a slot for an integer (Int) and two 'text' slots. Finally, some tags (for example 'sequence') are followed by a cross-reference to another class. The '?' in '?Sequence' means that the slot should contain the name of an object in the 'sequence' class.

To create an object in the Plasmid class we must name the object and perhaps populate its slots with data. By far the simplest method for accomplishing this is to create a list of information associated with the plasmid. Here is an example:

Plasmid	pBR322–199a
Reference	shirl–1992–aaaxc
Reference	jones–1990–aabrc
Location	'9th floor freezer'
Sequence	ATCHS

Without going into detail, it is a simple matter to enter data in this form into ACEDB (see Section 46.4). The plasmid entry shows the style required. The entry must start with the object name itself (the plasmid). Thereafter, other data is entered by specifying the slot name (*tag*), then the data itself. A slot in the model may be 'repeating' as is true for the 'reference' slot. In such cases, more than one data item can go into this slot as long as each entry is preceded by the slot name. This is nearly a complete description of the necessary syntax. Notice also what is *not* required: some information (for example, the 'shelf_box') was not supplied. It could be added later, if desired. Additionally, the order in which the tags are listed is not identical to the order of the tags in the 'plasmid model' ('reference' is last in the model). Data order – other than the all-important object name (pBR322–199a in this example), which must appear first – is irrelevant to ACEDB.

The reader may be wondering exactly how the information above will appear when it is viewed in a text window in ACEDB. First, the order of information will follow the order specified in the model. However, *empty slots will not appear*. This means that for the plasmid example the window will not display a 'shelf_box' tag followed by an empty space. Missing data are not shown and the space it would occupy is removed. The result is a much more compact display. It is critical to understand the beneficial consequences this has on model design. Specifically, there is no penalty, in terms of cluttering a window with empty fields, for defining slots that will be rarely filled with data. This is a feature designed with biologists in mind, since most biological data are *sparse* – meaning that, while much can be known about an object, little usually is.

Second, the cross-references in the plasmid window will be boldface as they are in Fig. 46.1. In our example, these will be 'sequence' and 'paper' object names (there will be two distinct paper names, one under the other). However, while we have deliberately chosen cryptic names for these objects, the database is not constrained to use them. When a cross-reference to another class occurs, the name of the cross-referenced object is shown by default. That default can be changed for all occurrences of the cross-referenced class so that the name that is displayed is more informative. The 'paper' and 'sequence' classes can be configured to display the title of the paper or sequence if a title is available as is the case in Fig. 46.1.

Once a class is created it can be referenced by another class. For example, a Sequence object could also contain a cross-reference to the 'plasmid' class. These cross-references are automatically managed by the database. If a 'sequence' is entered and the slot cross-referenced to the 'plasmid' class is filled, the information associated with the 'plasmid' object will be associated with the 'sequence'. All that is required is that the cross-reference be specified in the relevant model.

46.4 INPUTTING INFORMATION

Earlier we mentioned that data can be entered into ACEDB via simple lists in which object slots are specified and filled. This method is extremely efficient and allows large quantities of data to be input in bulk form. Files containing information in this format are referred to as ace files. The format of these files is simply the tag name at the beginning of the line, a space and then the data item(s) or the unique name of the object in another class associated with that tag. The simple text file format can easily be produced for existing sources of information with the use of UNIX utilities or programmable editors such as Emacs. A growing community of ACEDB databases is providing helpful assistance to new users and provides access to specialized utility programs that aid in the conversion of standard data formats.

A major advantage of ACEDB is the ability to add tags to a class without rebuilding the database. In other words, the models file can be edited to add the new tag, the database started, and the new information (which references the new tag) read in. However, for more radical changes to the database model it is best to have ACEDB write out all of its known information into an ace file, edit the models and ace file to match the classes and tags, and then read the information into a newly initialized database.

For sharing the database with others a mechanism to produce and then incorporate updates is provided by ACEDB. The update files are text files in the ace format. These files can be distributed via a variety of methods and allow separate groups to update their copy of the database. Each of these groups could also store their own private data in their copy of the database.

46.5 QUERIES AND THE TABLE EDITOR

One of the strongest features of ACEDB is that it facilitates browsing. Browsing is a useful feature because it takes advantage of an important phenomenon in learning, namely, that even when the information one is trying to remember is not immediately available for recall, it can be recognized when it is seen. ACEDB facilitates this 'I'll know it when I see it' method of retrieval by making it easy to follow cross-references with the mouse. Nonetheless, there are times when more formal queries are useful.

A powerful query language, available using query interface, is provided by the database software. Any piece of information can be located by searching for text or numerical values associated with a specific data item, or even searching for the presence or absence of a tag. The answer to a query is expressed as a list of objects that contain information satisfying the query. It is also possible, once an 'answer' is generated, to use the answer as the input to another query. The next answer may contain a list of objects from a completely different class. For example, by querying the 'author' and 'paper' classes (assuming the cross-references are in place) it is simple to find all the papers ever published by author John Doe (a list of 'paper' objects) and then find all the authors who ever co-published with him (a list of 'author' objects).

A new query interface is available with the 1.9 version of ACEDB that provides easy access to all the information in the database without knowing the names of the tags in the database. The new query interface written by Gary Aochi (Lawrence Berkeley Laboratory) provides menus that aid the user in requesting just the right information and minimizes the need to understand formal query syntax.

Another query feature provided by ACEDB is the ability to search the database for specific data items and generate a table of the results. The table can display the names of objects (even from different classes) as well as the information in slots within objects. For example, all genetically mapping loci contained on chromosome 4 that also have been placed on the physical map could be presented in a table containing columns labelled genetic marker name, genetic map location in cM, name of the clone it is associated with on the physical map and map location of this clone on the physical map. This provides a useful method of presentation for several types of information. Once the specification for a table is created it can be saved and used again later. The interface to the table editor is similar to the query interface where the user makes choices from menus to easily build a complicated multiple query.

46.6 AVAILABILITY OF ACEDB AND CURRENT GENOME DATABASE PROJECTS

Currently the ACEDB software is available for a variety of Unix workstations utilizing the X Windows environment, (including Sun Microsystems SPARC-station, Silicon Graphics IRIS, NeXT, Digital Equipment DECstation, and IBM R6000). The C program source and the application executables are available free of charge via Anonymous FTP through the Internet computer network from the host *ncbi.nlm.nih.gov*.

To date, two species databases have been released that utilize the ACEDB software: the nematode database ACEDB and the *Arabidopsis* database AAtDB (Cherry *et al.*, 1992). Both these databases are available via anonymous FTP from ncbi.nlm.nih.gov or on the NCBI Repository CD-ROM available from the National Center for Biotechnology Information at the National Library of Medicine. For more information about the NCBI and the Repository CD-ROM send electronic mail to info@ncbi.nlm.nih.gov or to the postal address: NCBI Data Repository; NCBI/NLM/NIH; Building 38A, Bethesda, MD 20894 USA.

Several other species or specialized databases are under development using the ACEDB software including soybean (SoyBase), wheat (GrainGenes), pine tree (APtDB), canine (DogBase), mycobacterium (MycDB), and human chromosome 21 and 22. It has also been announced that the *Drosophila* genome database project (FlyBase) will use ACEDB as one of the means to distribute its information.

46.7 EXPERIENCES OF THE AAtDB PROJECT

The *Arabidopsis* genome database, AAtDB, uses exactly the same software as the *C. elegans* genome database, ACEDB. No programming was involved to change from an animal database to a plant genome database. The update mechanism provided by the ACEDB software is utilized by the AAtDB Project to distribute information to over 40 sites worldwide.

It is difficult to summarize the variety of experiences involved in configuring the AAtDB database and populating it with information. At the risk of oversimplification, we offer a few observations. First, we would recommend that the guiding principle of model design be simplicity. Although it may seem possible to specify every possible kind of data in advance, in practice it is not and such effort is wasted. Instead, allow real data to drive the configuration process so that the models reflect what is actually happening in the laboratory. ACEDB imposes little penalty on trying new configurations so there is no need to get everything 'right' the first time.

Second, we strongly recommend that personnel involved in processing large amounts of data learn how to use Emacs, a text editor available for both Unix and VMS computers. The time required to learn Emacs will be repaid manyfold because of the power this editor provides. Several excellent books are available and are appropriate for the new user, including the GNU Emacs manual itself available from the Software Freedom Foundation.

Finally, we urge anyone interested in reconfiguring ACEDB to contact the growing community of ACEDB database groups. The tone in this community is cooperative and friendly. These groups provide support to new and experienced groups in their design of classes and inform each other of new ACEDB features as they appear.

ACKNOWLEDGMENT

We wish to acknowledge the authors of ACEDB, Richard Durbin and Jean Thierry-Mieg, who have created an exceptional tool which is revolutionizing the ability of biologists to store and manipulate their data.

REFERENCES

Cherry, J.M., Cartinhour, S.W. & Goodman, H.M. (1992) *Plant Mol. Biol. Rep.* **10**, 308–309, 409–410.

Durbin, R. & Thierry-Mieg, J. (1992) *A C. elegans Database: I, User's Guide, II, Installation Guide, III, Configuration Guide; Syntactic Definitions for the ACEDB Database Manager.* ACEDB Distribution Kit, available by anonymous ftp from ncbi.nlm.nih.gov (IP 130.14.20.1), crim.crim.fr (IP 192.54.151.100), cele.mrc-lmb.cam.ac.uk (IP 131.11.84.1).

CHAPTER FORTY-SEVEN

The Integration of Curated Biological Databases

R. OVERBEEK & M. PRICE
Mathematics and Computer Science Division, Argonne National Laboratory, Argonne, IL 60439–4801, USA

47.1 BACKGROUND

With the advent of the Genome Project, substantial advances have been made in the availability of genomic data. Not only are databases for retention of DNA sequences growing rapidly, but also specialized databases for peptides, enzymes, structural data, motifs, alignments, metabolic pathways, and two-dimensional protein gels are also now available; and one can expect that the quality and variety of such databases will continue to improve rapidly.

The support role played by biological databases is changing. It can be viewed as progressing through three distinct stages:

(1) In the first stage, biological databases were primarily archives of raw data. In the cases of sequence repositories (such as GenBank, PIR, EMBL, and the Swiss Protein Data Bank), these databases were primarily used to support a very limited set of critical services; in particular, the most widespread use involved locating sequences similar to a specified sequence, and then extracting annotations relating to the similar sequences.
(2) In the second stage, databases were enhanced to include references to other repositories, support systems were established to allow users to enter queries via E-mail, centralized services were built that allowed a user to access data from a number of databases through convenient menus, and client/server architectures started to appear.
(3) In the nascent third stage, we will see a more seamless integration of large numbers of curated databases.

Products relating to the first stage of database support have reached the commercial market. Where the demands of the user community could not be satisfied by standard database technology (largely because of the peculiar requirement of rapid, sensitive similarity searches), both software solutions (Altschul *et al.*, 1990; Pearson, 1990) and high-performance hardware solutions (Intelligenetics, Inc. & MasPar Computer Corp., 1992) have emerged. While this area continues to attract a great deal of serious research and development effort, the critical capabilities are now accessible to a majority of the biological community and have become commonplace in biological laboratories.

Software products relating to the second stage of database support are now robust enough for distribution, and the resulting technology can be expected to enter the commercial marketplace shortly. Initially, these capabilities will be supported at a limited number of well-funded sites, but with the increasing use of networks they will also rapidly become available to the entire community.

The third stage, however, is still within the prototyping phase. It will be several years before systems robust enough for distribution appear. In this paper, we provide insight into why integration is important, what is required for integration, and how efforts are likely to proceed.

47.2 INTEGRATION OF DATABASES

Integration of distinct curated databases involves establishing a single query mechanism that offers access to data represented in the ensemble of integrated databases. First, let us illustrate what we mean; then we will discuss the significance of constructing such functionality.

Consider a researcher wishing to investigate expression of genes. Clearly, much of the work will be done in the wet lab. However, after examining the available literature relating to a specific topic, one might reasonably wish to peruse the available raw data before investing substantial efforts in the lab. The following sorts of questions might occur.

(1) Does some detectable phenomenon occur in the regions immediately upstream of genes induced by phosphate starvation that does not occur upstream of most genes expressed under normal conditions?
(2) Consider long sections of known sequence from distinct organisms. In how many cases do we have long, known homologies between three or more organisms? Which sections in these homologous regions both occur outside expressed genes and appear abnormally conserved?
(3) Consider just divergent genes. First, how easy is it to reliably locate the Pribnow box and -35 signal for these genes? Second, do any pairs of such genes show unusual structural similarity in their first 20 amino acids? (Here, we would settle for an initial attempt to detect such similarity by checking for sections that show unusual similarity to the same initial section of a single three-dimensional profile.)

One can certainly answer such questions using existing facilities. Yet, a level of difficulty is introduced when data from one source must be evaluated and then used to formulate queries against another source of data. For example, to address the first question, one must:

(1) go to a database like ECO2DBASE (produced by the *Escherichia coli* Gene–Protein Database Project) to determine the set of genes induced by phosphate starvation (as well as a set that is not induced by phosphate starvation);
(2) extract the corresponding entries from GenBank to gain access to the sequence data;
(3) extract the appropriate substrings of sequence for the set of genes that have sequenced upstream regions; and
(4) use a set of tools for searching sequence to locate some phenomenon.

None of these steps is either conceptually difficult or overwhelmingly complex; yet the process is characterized by enough impedence to make the overall task time consuming and error prone.

In the context of an integrated database, a single query would be used to extract the upstream regions of genes induced by phosphate starvation, a second query would be used to extract upstream regions of genes not induced by phosphate starvation, and then an interactive investigation using sequence analysis tools would be used to search for some distinguishing phenomenon. The process of extracting the relevant data should take just minutes; and, in fact, it did take us less than 15 min to extract the relevant sequence data from a prototype system.

Similar comments could be made about the second and third questions posed above. In each of these cases, if the type of question were common, then one could construct a simple program to solve that class of questions. The central point, however, is that there are almost endless questions that will necessarily be posed about biological data in the effort to extract meaning. *Ad hoc* programs to solve specific questions represent a relatively expensive way to address the needs of the community. What is required is a *coherent framework*, an *abstract query facility*, and a *user interface* that is suitable for mass consumption.

47.3 THE CONCEPTUAL FRAMEWORK REQUIRED FOR INTEGRATION

A number of groups have started projects attempting to integrate genomic data. These efforts, often focusing on one limited subclass of biological data, have yielded a number of insights (Baehr *et al.*, 1992; Yoshida *et al.*, 1992). The most common approach is to construct an object-oriented framework using any of a wide variety of available implementation technologies. In some of these implementations, the full capabilities of object-oriented systems have been used. In others, a more constrained notion of object has been employed and features such as inheritance have been dropped.

47.4 REPRESENTATION OF DATA AS RELATED OBJECTS

The prototype system we have developed currently supports the following conceptual model.

(1) The database contains information about *typed objects*. Typed objects have attributes of the form *keyword(value)*.
   (a) *Curated objects* all have unique identifiers. For example, GenBank may be thought of as describing objects of several types, including *sequence_fragment, cds, tRNA, rRNA*; the Swiss Protein Data Bank describes objects of type *peptide*; and the Enzyme Data Bank and BioBank describe abstract objects of type *enzyme*. Each object from any of these databases has an associated unique identifier.
   (b) *Constructed objects* have a more transient existence and include *intervals* of other objects, *vectors, sets, lists*, and so forth.

(2) Objects are connected by *relationships*. Thus, an object of type *cds* might relate to an object of type *peptide*, which in turn might relate to objects of type *enzyme* and *peptide_alignment*. Occasionally, relationships have attached items of data, which we call em arguments. For example, an object of type em cds might relate to several distinct genetic maps, and the location of the gene on each map would be carried as an argument of the relationship.

(3) A mechanism exists for describing positional relationships between objects in the same genome. When one knows exactly where an object occurs relative to another (e.g. where a tRNA occurs on a *sequence_fragment*, a *sequence_fragment* occurs on a *chromosome*, or a *cds* occurs on a *chromosome*), then one asserts that the first object is *precisely bound* to the second object. When position can be isolated only to an interval, then the first object is said to be *imprecisely bound* to the second. This is clearly just a special case of the more general notion of relationship; however, it plays such a central role in many critical uses of the data that we have designated a fixed mechanism for encoding such relationships.

During the past few months, we have cast existing databases within this framework. The resulting database includes data relating to a variety of viruses, microorganisms, and eukaryotes. The classes of typed objects now include the following:

(1) *Genome* objects represent an entire genome (and relate to the chromosomes, plasmids, etc.) for specific organisms.
(2) *Chromosome* objects represent specific chromosomes for specific organisms.
(3) *Sequence_fragment* objects represent a section of DNA sequence that has been captured. These data comes from GenBank.
4) *Enzyme* objects represent an 'abstract enzyme' in they can relate to many distinct peptides and genes (from many organisms). The data associated with this type of object now come from the Enzyme Data Bank created by Amos Bairoch. We hope eventually to include the wealth of information captured in DBEMP (a database developed in Pushchino, Russia).
(5) *Peptide* objects represent specific peptide sequences. Most of these data currently come from the Swiss Protein Data Bank (again, developed by Amos Bairoch) and the Protein Identification Resource (PIR).
(6) *Prosite* objects represent the peptide motifs compiled by Amos Bairoch.
(7) Objects of types *cds, rRNA, tRNA*, and *misc_RNA* represent specific genes (from specific organisms). Much of the data associated with these types of objects comes directly from GenBank.
(8) *Map* objects are used to represent physical or genetic maps (and most of the information one gets from accessing these objects will be through relationships to the objects contained in the map). The system includes maps of several bacterial organisms.
(9) Objects of type *eco2dbase* capture the data provided by Fred Neidhardt's project to develop data relating to expression of *E. coli* genes.
(10) *Rebase_entry* objects describe the sites cut by restriction enzymes. This data comes from the database distributed by Rich Roberts.
(11) *Pdb_entry* objects contain data relating to the coordinates of atoms within a specific peptide for which the crystal structure has been determined. This data has been extracted from the Protein Data Base distributed by Brookhaven National Laboratory.
(12) *Peptide_alignment* objects contain alignments of peptides. Most of the currently available alignments were acquired from the Protein Information Resource.
(13) *Nucleotide_alignment* objects contain alignments of DNA or RNA sequences. These have come from several sources including those distributed

by the Ribosomal Database Project and the Berlin Data Bank.

(14) *Phylogenetic_tree* objects assert phylogenetic relationships between organisms. These objects come from a variety of sources. The largest is distributed by the Ribosomal Database Project and includes approximately 500 microorganisms.

(15) Objects of type *compound, reaction* and *pathway* are used to encode the reactions in metabolic pathways. The compound information is largely from Peter Karp's database, and the metabolic pathway information was assembled by Murali Raju (who started from a set of reactions provided by Ray Ochs). We hope eventually to include the much richer class of data available in DBEMP.

This is only a partial list, and we find that we add new object types frequently, since the body of curated data is expanding so rapidly.

47.5 AN ABSTRACT QUERY FACILITY

The central issue in establishing effective access to this growing body of objects is to establish an extensible framework for formulating queries. If the abstract mechanism is adequate, more usable interfaces can be layered on top of it; if it is not, then efforts to adapt the database to include new types of objects will be costly.

We have elected to build an abstract interface upon the notion of 'evaluating expressions'. If an expression can be evaluated, then its value will be an object. In many cases, an expression could evaluate to any of a set of values; in such cases, our system will allow the user to examine all possible values, one at a time. For example:

```
obj([type=cds,genome='E.coli'])
```

is an expression that evaluates to any protein coding gene in *E. coli* (so, the expression can evaluate to any of a large set of objects).

We have constructed a broad set of operators. Thus, an expression of the form:

```
obj([type=cds,genome='E.coli'])* cds_to_protein
```

computes one gene in *E. coli*, and then crosses a relationship to an object representing the translated polypeptide; it evaluates to an object of type *peptide.* The operator in this case is the binary functor */2, which takes an object as its first argument and a relationship as its second argument. Complex expressions can be used effectively to migrate through the database extracting desired pieces of data. For example:

```
dna_sequence(
    interval(
        require(obj([type=eco2dbase]),
                X in has_attribute(X, 'PSI')
                ) * eco2dbase_to_cds,
                -50, -1))
```

is a complex term that evaluates to the DNA sequence of the 50 bp upstream region of an *E. coli* gene induced by phosphate starvation. What is important in this example is not the precise syntax; rather, it is the fact that a complex search and extraction operation can be expressed concisely using a limited set of well-chosen operators. In English, the corresponding operation might be described as 'Find an ECO2DBASE entry corresponding to a gene induced by phosphate starvation. From this gene, locate the corresponding sequence fragment from GenBank that contains the gene (if there is one). Then, compute the interval corresponding to the upstream region (again, you must check to make sure that the upstream region has been sequenced and you must get the correct strand). Return the sequence fragment corresponding to this interval'.

We have implemented numerous operators for manipulating objects of the types described earlier. Often a single operator can be applied to objects of a variety of types. By carefully selecting appropriate operators, one gradually determines a powerful semantics for querying the database. At this point, we have operators for:

(1) the simple data types (Boolean, numeric, list, and atomic);
(2) manipulation of sequences (match patterns from a fairly rich class, locate common subsequences, locate and align similar subsequences, compute the occurrences of all *k*-mers of a given length);
(3) access to alignments; and
(4) a general set of operators for accessing genomic objects and traversing relationships.

We realize that a richer set of operators will be needed to properly address the numerous issues relating to genetic and physical maps, phylogenetic trees, structural data and the numerous detailed types required to organize and interpret biological data.

47.6 USER INTERFACES

Our initial goal is to produce a database that supports integrated access to a wide class of biological objects. Once this is accomplished, the next goal, naturally, is to reduce the effort required to use the database by including an appropriate user interface. In the past, we have experimented with a limited English interface (Baehr *et al.*, 1991). In our current system, a graphical interface seems more appropriate.

Since we have constructed the underlying abstract semantics in terms of expressions, a query is actually a tree in which each node is an operator that takes several inputs. The most convenient way to construct queries might well involve drag-and-drop icons that represent the operators, connecting the output of one operator to an input for another operator. Queries become a tree of icons with connections between outputs and inputs. The root of the tree represents the highest level operator, and the output from that node is the output for the entire query.

Using a graphical interface of this sort, one can type check arguments as outputs are connected to inputs. For some operators, the user would be required to type in arguments, and these could also be validated. If a query were to produce unexpected results, the user could simply extract subtrees to verify that they are evaluating to reasonable output streams.

47.7 SUMMARY

Sequence data, hybridization data, structural data, and inferred data (such as motifs, phylogenetic trees, and maps) are becoming available at an increasing rate. We are spending substantial resources to derive such data and to make that data available to the biological community. The most commonly required database services relate to similarity searches and straightforward access to archived data. These are now provided in commercially available software and have become accessible to a significant portion of the community. The second stage of requirements relates to access over the network, timely updates of data, and sites supporting sets of the most commonly accessed databases. These services are also progressing rapidly. The third stage of service will focus on integration of numerous curated databases. This stage introduces a variety of substantial problems. Certainly, resolution of incompatible nomenclature and semantic models (topics that have been widely discussed in the literature) represent significant barriers to integration. However, the adoption of an object-oriented technology, along with an implementation strategy based on rapid prototyping, will allow researchers to produce a number of systems during the next 2 to 3 years. It will take time for these systems to develop suitable semantics and user interfaces, but the resulting capabilities will far exceed those currently available.

ACKNOWLEDGMENT

This work was supported by the Office of Scientific Computing U.S. Department of Energy, under contract No. W–31–109–Eng–38.

REFERENCES

Altschul, S.F., Gish, W., Miller, W., Myers, E.W. & Lipman, D.J. (1990) *J. Mol. Biol.* **215**, 403–410.

Baehr, A., Hagstrom, R., Joerg, D. & Overbeek, R. (1991) *Querying Genomic Databases (Argonne Technical Report ANL/MCS–TM–155).* Argonne National Laboratory.

Baehr, A., Dunham, G., Ginsburg, A., Hagstrom, R., Joerg, D., Kazic, T., Matsuda, H., Michaels, G., Overbeek, R., Rudd, K., Smith, C., Taylor, R., Yoshida, K. & Zawada, D. (1992) *An Integrated Database to Support Research on Escherichia coli (Argonne Technical Report ANL–92/1).* Argonne National Laboratory.

Intelligenetics, Inc. & MasPar Computer Corp. (1992) *A Massively Parallel Sequence Similarity Search Program for Molecular Biologists (Advanced Product Information Bulletin* **Jun.**).

Pearson, W.R. (1990) *Methods Enzymol.* **183**, 63–98.

Yoshida, K., Smith, C., Kazic, T., Michaels, G., Taylor, R., Zawada, D., Hagstrom, R. & Overbeek, R. (1992) In *Proceedings of the International Conference on Fifth Generation Computer Systems*, Tokyo, Japan, 1–5 June 1992, pp. 307–320.

Appendix: Sources of Sequence Analysis Software

The software described in Part III is available from the sources listed below. For programs with an E-mail contact, information can be obtained from the author at the E-mail address listed. For programs available by ftp, from a computer on the Internet, use ftp to connect to the Internet address listed. Log in as 'anonymous' and enter your E-mail address as the password. Programs may have to be configured to match the local environment and compiled prior to use.

FINDIT (Shavlik): E-mail to shavlik@cs.wisc.edu
ASTS (Rapport *et al.*): E-mail to asts@telomere.lanl.gov
FASTA (Pearson): anonymous ftp to uvaarpa.virginia.edu
BLAST (Altschul *et al.*): anonymous ftp to ncbi.nlm.nih.gov
BLAST (Altschul *et al.*): E-mail server at blast@ncbi.nlm.nih.gov
PLSEARCH (Smith *et al.*): anonymous ftp to mbcrr.harvard.edu
SORFIND (Hutchinson *et al.*): anonymous ftp to ulam.generes.ca
gm (Soderlund *et al.*): anonymous ftp to ftp.tigr.org
GeneId (Guigo *et al.*): E-mail server at geneid@darwin.bu.edu
GRAIL (Uberbacher *et al.*): E-mail server at grail@ornl.gov
Repetitive sequences (Jurka *et al.*,): E-mail server at pythia@anl.gov
BLOCKS (Henikoff *et al.*): E-mail server at blocks@howard.fhcrc.org
Statistical analysis tools (White and Dunning): anonymous ftp to ftp.tigr.org

Index